Design and Applications of Nanoparticles in Biomedical Imaging

Jeff W.M. Bulte • Michel M.J. Modo
Editors

Design and Applications of Nanoparticles in Biomedical Imaging

Springer

Editors
Jeff W.M. Bulte
Russell H. Morgan Department of Radiology and Radiological Science
The Johns Hopkins University School of Medicine
Baltimore, MD, USA

Michel M.J. Modo
Department of Radiology
University of Pittsburgh
Pittsburgh, PA, USA

ISBN 978-3-319-42167-4 ISBN 978-3-319-42169-8 (eBook)
DOI 10.1007/978-3-319-42169-8

Library of Congress Control Number: 2016948724

Printed on acid-free paper

This Springer imprint is published by Springer Nature
The registered company is Springer International Publishing AG
The registered company address is: Gewerbestrasse 11, 6330 Cham, Switzerland

Preface

Nanoparticles have emerged as an exciting technology platform that can be exploited for diagnostic, therapeutic, and their combined theranostic applications. Their varied chemical compositions can be exploited for use with different imaging modalities, such as magnetic resonance imaging (MRI), magnetic particle imaging (MPI), computer tomography (CT), positron emission tomography (PET) and single-photon emission computed tomography (SPECT), ultrasound (US), as well as optical (fluorescent) imaging. The key determinants for successful use of nanoparticles are sensitivity, specificity, pharmacokinetics, and safety. In the 21 chapters of this book, world-leading experts describe the chemistry, imaging acquisition and processing, as well as the requirements for clinical translation. Applications range from cell tracking in regenerative medicine to the early detection of cancer and inflammation across all major imaging modalities. We hope that this book will provide a fundamental basis for further innovation and synergy of the fields of nanomedicine timely and diagnostic imaging.

Baltimore, MD, USA — Jeff W.M. Bulte
Pittsburgh, PA, USA — Michel M.J. Modo

Contents

Nanoparticles as a Technology Platform for Biomedical Imaging

Jeff W.M. Bulte and Michel M.J. Modo

Imaging and therapeutic delivery is increasingly relying on nanoparticles as a key technology platform. This common technology interface affords new opportunities to combine the diagnosis and treatment into a "unified" one-step theranostic approach. Nanoparticles are now emerging as a main innovation driver in developing novel imaging applications, especially in nanomedicine.

1 Origins of Nanoparticles in Biomedical Imaging

The use of nanoparticles dates back to ancient history with clay minerals providing color in pottery [1], with gold and silver particles being incorporated in opulent ceramics by Mesopotamians [2]. The word "tattoo" comes from the Polynesian word "tatau," where the Pacific Islanders have used nanoparticles for thousands of years to mark one's genealogy, societal hierarchy, and personal achievements. These visual properties of nanoparticles were also documented by Michael Faraday, who was experimenting with different metal particles and their effects on light [3]. These nanoparticles are natural products from organic (e.g., polysaccharides), as well as inorganic elements (iron oxides). These specific applications revealed their unique

J.W.M. Bulte (✉)
Division of MR Research, Russell H. Morgan Department of Radiology and Radiological Science, The Johns Hopkins University School of Medicine,
217 Traylor Bldg, 720 Rutland Avenue, Baltimore, MD 21205, USA
e-mail: jwmbulte@mri.jhu.edu

M.M.J. Modo
Department of Radiology, University of Pittsburgh,
3025 East Carson Street, Pittsburgh, PA 15203, USA
e-mail: modomm@upmc.edu

J.W.M. Bulte, M.M.J. Modo (eds.), *Design and Applications of Nanoparticles in Biomedical Imaging*, DOI 10.1007/978-3-319-42169-8_1

properties that led to detailed investigations to more systematically produce particles exhibiting the desired characteristics.

A major characteristic of particles of <100 nm is their ability to form colloids (i.e., lack of sedimentation when suspended in a liquid phase) [4]. To account for size, nanoparticles were initially termed ultrafine particles (1–100 nm in size) to contrast these with fine (100–2500 nm) and coarse particles (2500–10,000 nm). The term *nano* is derived from the Greek language, meaning "dwarf." It was only added in the 1990s to provide further emphasis on size as being a key characteristic that distinguishes these ultrafine particles from larger ones. This is reflected in the definition of the International Union of Pure and Applied Chemistry (IUPAC), which determines a size between 1 and 100 nm as the key characteristic, although under certain circumstances particles larger than 100 nm can also behave like nanoparticles [5].

Nanoparticles have a long-standing tradition in biomedical imaging. The first contrast agent was used in 1905 for X-ray imaging by Walter Cannon. For this, naturally occurring high-density metal salts, notably bismuth- or barium-based nanoparticles, were mixed with food to noninvasively visualize the mechanics of the digestive tract [6]. The generation of novel radioactive particles was a by-product of the nuclear arms' development that in 1946 was declassified as part of the Atomic Energy Act for the civilian development of radioactive-based therapies and imaging [7]. Imaging of ^{198}Au colloids was subsequently used to investigate its organ distribution, revealing an accumulation in the kidneys, spleen, and liver [8]. This provided the first imaging of the reticuloendothelial system (RES). The first specific nanoparticle preparations for electron microscopy consisted of natural horse spleen ferritin that afforded the specific detection of antigen [9]. With the development of liposomes in the 1960s, a new era was heralded in nanoparticle design in which controlled delivery of pharmaceutical compounds [10, 11] and the incorporation of imaging agents, such as ^{131}I-labeled albumin, became feasible [12].

The 1970s saw a rapid expansion of the use of nanoparticles with their first use for biomedical imaging of myocardial perfusion using single-photon emission tomography (SPECT) [13], as well as adaptation to other imaging modalities, such as near-infrared (NIR) optical imaging [14]. The rapid adaptation of computer tomography (CT) in hospitals and the requirement of contrast material, such as iodine [15], further stimulated nanoparticle research with evidence of their major impact in diagnostic radiology. The emergence of positron emission tomography (PET) was dependent on the generation of new radioligands that did not provide anatomical images per se, but were geared towards molecular targets [16]. One of the first such developments visualized staphylococcal abscesses using ^{99}MTC-technetium liposomes [17]. In contrast to CT and PET, the emergence of magnetic resonance imaging (MRI) was not dependent on contrast materials or tracer agents, as the magnetic relaxation properties of ^{1}H provided the signal for image construction. Nevertheless, in the late 1970s, it was discovered that small metallic particles can influence this relaxation rate [18] and could be used to image the liver and spleen [19], as well as specific antigens [20]. In the mid-1990s, magnetic nanoparticles were approved by the FDA and have seen a plethora of uses in MRI [21], including clinical cell tracking [22].

2 The Emergence of a Synergy Between Therapeutics and Imaging

These developments provided not only the foundation for the rapid development of nanoparticle-based clinical imaging during the 1990s, but also new tools for basic scientists. Indeed, it marked the emergence of an interdisciplinary field, where physicists were driving the advances in image acquisition, biochemists were engineering new imaging agents, and biologists/clinicians were exploiting new frontiers of what could be visualized in living subjects [23]. Easy access routes of administration through ingestion or intravenous delivery sufficed for most imaging requirements, such as the gastrointestinal tract and the RES. However, this afforded a limited penetration into tissue (and cells), where many pathological targets are found, especially in the brain where we have the blood-brain barrier and access through other methods is very limited. The pharmaceutical sciences faced a similar issue in terms of delivery of drugs and hence novel means were sought that could cross the vascular wall and permeate into tissues [24], as well as approaches to block uptake by the RES in order to prolong nanoparticle blood half-life, which is necessary for specific antigen-based targeting applications.

Originally, Paul Ehrlich conceived of this approach as a "magic bullet" that will only affect those cells that are targeted [25]. Liposomes developed in the 1960s suited this nanoscopic vision, but apart of the development of polymeric nanoparticles little progress was seen in the development of nanoparticles for drug delivery until the 1990s, when several developments overcame fundamental challenges [24]. Foremost of all, a significant obstacle for targeting of nanocarriers (i.e., material carrying drug for delivery), such as nanoparticles, was escaping the rapid uptake through the RES. A size of <200 nm facilitated retention in the bloodstream, but was insufficient to provide adequate circulating time for extravasation into target tissues. Non-covalent attachment or amalgamation of polyethylene glycol (PEG), so-called PEGylation [26], and its widespread adaptation were the first major advances in targeting by creating a "stealth" mode for molecules to evade the host's immune system and afforded the prolonged circulation of nanoparticles [27]. Active (e.g., antibodies) and passive (e.g., enhanced permeability retention) targeting approaches provided the second component to ensure that nanoparticles accumulate in a desired location [28]. Nanoparticles provide key characteristics for drug delivery, notably improved bioavailability through aqueous solubility (i.e., forming a colloid), increased blood circulation time, and potential for tissue and cell targeting [29].

These advances in delivering therapeutics using nanoparticles cumulated in the realization that engineered nanocarriers could carry not only therapeutic drugs, but also contrast agents that would afford a localization and potential monitoring of such delivery [30]. This conceptual advance of therapeutics and diagnosis led to the formulation of the portmanteau word theranostic at the start of the new millennium [31, 32]. As with most nanoparticle drug delivery and imaging systems, macrophages constituted the first easy target [33], due to their natural properties to rapidly phagocytose particulate material. Further synergies also became apparent in that certain drugs could be tagged with a radioligand [34, 35] and nanoparticles based on elements, such as gold [36, 37], could provide a core technology platform for creating multicomponent, multimodal, and multifunctional agents [38]. Increasingly complex possibilities are emerging with multiple imaging moieties, stealth and targeting functionalities, as well as multiple timed release of therapeutic drugs [39, 40].

3 An Outlook on Challenges and Future Opportunities

One of the most significant advances has been the rapid development of optical and ultrasound nanoparticles [41]. Especially the introduction of quantum dots, as well as the use of near-infrared probes and highly sensitive detectors, have now enabled imaging of deeply seated tissue structures [42], allowing clinical optical imaging [43]. The availability of calcium-sensitive agents, for instance, allows an in vivo imaging approach that bridges the gap between conventional single-cell electrophysiological recording and macroscopic activity recording, such as functional MRI [44]. Light-sensitive theranostic nanoparticles can also be used to monitor reaching a treatment site, with a specific light wavelength triggering the release of drug in just this area, hence providing a very targeted treatment [45]. These approaches further lend inspiration to the development of probes for other modalities, such as MRI, that currently still dominate the clinical arena. However, optical imaging is currently seeing a more rapid development of nanoparticles than any other biomedical imaging modality. The shift beyond near infrared reduces tissue light scatter and greater organ coverage will eventually dominate biomedical imaging in smaller species to drive a deeper understanding of biology. Still, it remains unclear if optical imaging can indeed deliver on whole-organ imaging in larger species, such as primates and humans. Modalities, such as MRI and SPECT, might hence still remain the dominating nanoparticle-based clinical imaging techniques.

Further challenges to clinical applications are the growing considerations for toxic side effects of nanoparticles, the so-called nanotoxicity [46]. Many of the constituent parts of nanoparticles do not exhibit toxicity in their bulk form, but due to the emergent properties at the nanoscale (e.g., increased cell membrane permeation), cytotoxic effects can become apparent [47]. However, there is also support to indicate that the nanosize by itself is insufficient to determine toxicity and that a more detailed general consideration of particle toxicity is needed [48, 49]. The combination of nanoparticles with biologicals, such as stem cells, further raises concerns as to their potential to induce unwanted side effects that might only become apparent over time [50, 51]. An unanswered question remains if materials should be biodegradable and cleared over time or if biological inertness is more desirable [52]. Indeed, these issues raise concern regarding a premature clinical translation and what framework of evidence is needed to ensure safety [53]. Beyond the regulatory framework, the potential for scale-up and cost-efficient production at an industrial scale will also require further investment and might refine quality control procedures, especially in relation to monitoring potential adverse effects [54].

To conclude, nanoparticles are hence a powerful technology platform that affords the integration of imaging and drug delivery. Their increasing sophistication delivers exciting new opportunities to disentangle complex biological questions at the systems level [55], but also constitutes a major step forward to the concept of a "magic bullet," where a drug can be delivered to a very focused area, and potentially even to specific single cells [56]. With the increasing number and versatility of probes for the various imaging modalities, the future for biomedical imaging promises to be exciting [57]. These multimodal and -functional nanoparticles are also likely to be the catalyst for an eventual unification of diagnostic medicine and imaging based on more specific and sensitive tissue- and fluid-based biomarkers that improve early disease detection and classification.

References

1. Colomban P, Gouadec G. The ideal ceramic-fibre/oxide-matrix composite: how to reconcile antagonist physical and chemical requirements? Ann Chimie Sci Materiaux. 2005;30(6):673–88.
2. Sciau P, Mirguet C, Roucau C, Chabanne D, Schvierer M. Double nanoparticle layer in a 12th century lustreware decoration: accident or technological mastery? J Nano Res. 2009;8:133–9.
3. Faraday M. Experimental relations of gold (and other metals) to light. Philos Trans R Soc Lond. 1857;147:145–81.
4. Buzea C, Pacheco II, Robbie K. Nanomaterials and nanoparticles: sources and toxicity. Biointerphases. 2007;2(4):MR17–71.
5. Aleman J, Chadwick AV, He J, Hess M, Horie K, Jones RG, Kratochvil P, et al. Definitions of terms relating to the structure and processing of sols, gels, networks, and inorganic-organic hybrid materials (IUPAC recommednations 2007). Pure Appl Chem. 2007;79(10):1801–29.
6. Cannon W. The mechanical factors of digestion. New York: Longmans; 1911.
7. Wagner HN. A perosnal history of nuclear medicine. New York: Springer; 2006.
8. Bertrand JJ, Qaine H, Tobias CA. Distribution of gold in the animal body in relation to arthrithis. J Lab Clin Med. 1948;33:1133–8.
9. Rifkind RA, Hsu KC, Morgan C, Seegal BC, Knox AW, Rose HM. Use of ferritin-conjugated antibody to localize antigen by electron microscopy. Nature. 1960;187:1094–5.
10. Papahadjopoulos D, Miller N. Phospholipid model membranes. I. Structural characteristics of hydrated liquid crystals. Biochim Biophys Acta. 1967;135(4):624–38.
11. Papahadjopoulos D, Watkins JC. Phospholipid model membranes. II. Permeability properties of hydrated liquid crystals. Biochim Biophys Acta. 1967;135(4):639–52.
12. Gregoriadis G, Ryman BE. Fate of protein-containing liposomes injected into rats. An approach to the treatment of storage diseases. Eur J Biochem. 1972;24(3):485–91.
13. Ashburn WL, Braunwald E, Simon AL, Peterson KL, Gault JH. Myocardial perfusion imaging with radioactive-labeled particles injected directly into the coronary circulation of patients with coronary artery disease. Circulation. 1971;44(5):851–65.
14. Flower RW, Hochheimer BF. Clinical infrared absorption angiography of the choroid. Am J Ophthalmol. 1972;73(3):458–9.
15. Havron A, Seltzer SE, Davis MA, Shulkin P. Radiopaque liposomes: a promising new contrast material for computed tomography of the spleen. Radiology. 1981;140(2):507–11.
16. Nutt R, 1999 ICP Distinguished Scientist Award. The history of positron emission tomography. Mol Imaging Biol. 2002;4(1):11–26.
17. Morgan JR, Williams KE, Davies RL, Leach K, Thomson M, Williams LA. Localisation of experimental staphylococcal abscesses by 99MTC-technetium-labelled liposomes. J Med Microbiol. 1981;14(2):213–7.

18. Ohgushi M, Nagayama K, Wada A. Dextra-magnetite: a new relaxation reagent and its application to T2 measurements in gel systems. J Magn Reson. 1978;29(3):599–601.
19. Mendonca Dias MH, Lauterbur PC. Ferromagnetic particles as contrast agents for magnetic resonance imaging of liver and spleen. Magn Reson Med. 1986;3(2):328–30.
20. Renshaw PF, Owen CS, Evans AE, Leigh Jr JS. Immunospecific NMR contrast agents. Magn Reson Imaging. 1986;4(4):351–7.
21. Bulte JW, Kraitchman DL. Iron oxide MR contrast agents for molecular and cellular imaging. NMR Biomed. 2004;17(7):484–99.
22. Bulte JW. In vivo MRI cell tracking: clinical studies. Am J Roentgenol. 2009;193(2):314–25.
23. Sun Z, Ng KH, Ramli N. Biomedical imaging research: a fast-emerging area for interdisciplinary collaboration. Biomed Imaging Interv J. 2011;7(3):e21.
24. Hoffman AS. The origins and evolution of "controlled" drug delivery systems. J Control Release. 2008;132(3):153–63.
25. Kreuter J. Nanoparticles—a historical perspective. Int J Pharm. 2007;331(1):1–10.
26. Abuchowski A, van Es T, Palczuk NC, Davis FF. Alteration of immunological properties of bovine serum albumin by covalent attachment of polyethylene glycol. J Biol Chem. 1977;252(11):3578–81.
27. Torchilin VP, Trubetskoy VS. Which polymers can make nanoparticulate drug carriers long circulating. Adv Drug Deliver Rev. 1995;16:141–55.
28. Bi Y, Hao F, Yan G, Teng L, Lee RJ, Xie J. Progress on actively targeted nanoparticles for drug delivery to tumor. Curr Drug Metab. 2016.
29. Caldorera-Moore M, Guimard N, Shi L, Roy K. Designer nanoparticles: incorporating size, shape and triggered release into nanoscale drug carriers. Expert Opin Drug Deliv. 2010;7(4):479–95.
30. Kelkar SS, Reineke TM. Theranostics: combining imaging and therapy. Bioconjug Chem. 2011;22(10):1879–903.
31. Ozdemir V, Williams-Jones B, Glatt SJ, Tsuang MT, Lohr JB, Reist C. Shifting emphasis from pharmacogenomics to theragnostics. Nat Biotechnol. 2006;24(8):942–6.
32. Picard FJ, Bergeron MG. Rapid molecular theranostics in infectious diseases. Drug Discov Today. 2002;7(21):1092–101.
33. McCarthy JR, Jaffer FA, Weissleder R. A macrophage-targeted theranostic nanoparticle for biomedical applications. Small. 2006;2(8-9):983–7.
34. Fani M, Maecke HR, Okarvi SM. Radiolabeled peptides: valuable tools for the detection and treatment of cancer. Theranostics. 2012;2(5):481–501.
35. van der Veldt AA, Smit EF, Lammertsma AA. Positron emission tomography as a method for measuring drug delivery to tumors in vivo: the example of [(11)C]docetaxel. Front Oncol. 2013;3:208.
36. Curry T, Kopelman R, Shilo M, Popovtzer R. Multifunctional theranostic gold nanoparticles for targeted CT imaging and photothermal therapy. Contrast Media Mol Imaging. 2014;9(1):53–61.
37. Spivak MY, Bubnov RV, Yemets IM, Lazarenko LM, Tymoshok NO, Ulberg ZR. Gold nanoparticles—the theranostic challenge for PPPM: nanocardiology application. EPMA J. 2013;4(1):18.
38. Chen PC, Liu X, Hedrick JL, Xie Z, Wang S, Lin QY, et al. Polyelemental nanoparticle libraries. Science. 2016;352(6293):1565–9.
39. Arias JL. Advanced methodologies to formulate nanotheragnostic agents for combined drug delivery and imaging. Expert Opin Drug Deliv. 2011;8(12):1589–608.
40. Fang C, Zhang M. Nanoparticle-based theragnostics: integrating diagnostic and therapeutic potentials in nanomedicine. J Control Release. 2010;146(1):2–5.
41. Samanta A, Medintz IL. Nanoparticles and DNA—a powerful and growing functional combination in bionanotechnology. Nanoscale. 2016;8(17):9037–95.
42. Montalti M, Cantelli A, Battistelli G. Nanodiamonds and silicon quantum dots: ultrastable and biocompatible luminescent nanoprobes for long-term bioimaging. Chem Soc Rev. 2015;44(14):4853–921.
43. Radenkovic D, Kobayashi H, Remsey-Semmelweis E, Seifalian AM. Quantum dot nanoparticle for optimization of breast cancer diagnostics and therapy in a clinical setting. Nanomedicine. 2016;12(6):1581–92.

44. Bai R, Klaus A, Bellay T, Stewart C, Pajevic S, Nevo U, et al. Simultaneous calcium fluorescence imaging and MR of ex vivo organotypic cortical cultures: a new test bed for functional MRI. NMR Biomed. 2015;28(12):1726–38.
45. Yavlovich A, Smith B, Gupta K, Blumenthal R, Puri A. Light-sensitive lipid-based nanoparticles for drug delivery: design principles and future considerations for biological applications. Mol Membr Biol. 2010;27(7):364–81.
46. Azhdarzadeh M, Saei AA, Sharifi S, Hajipour MJ, Alkilany AM, Sharifzadeh M, et al. Nanotoxicology: advances and pitfalls in research methodology. Nanomedicine (Lond). 2015;10(18):2931–52.
47. Oberdorster G, Oberdorster E, Oberdorster J. Nanotoxicology: an emerging discipline evolving from studies of ultrafine particles. Environ Health Perspect. 2005;113(7):823–39.
48. Donaldson K, Poland CA. Nanotoxicity: challenging the myth of nano-specific toxicity. Curr Opin Biotechnol. 2013;24(4):724–34.
49. Maynard AD, Warheit DB, Philbert MA. The new toxicology of sophisticated materials: nanotoxicology and beyond. Toxicol Sci. 2011;120 Suppl 1:S109–29.
50. Modo M, Kolosnjaj-Tabi J, Nicholls F, Ling W, Wilhelm C, Debarge O, et al. Considerations for the clinical use of contrast agents for cellular MRI in regenerative medicine. Contrast Media Mol Imaging. 2013;8(6):439–55.
51. Bulte JW, Kraitchman DL, Mackay AM, Pittenger MF. Chondrogenic differentiation of mesenchymal stem cells is inhibited after magnetic labeling with ferumoxides. Blood. 2004;104(10):3410–2.
52. Moros M, Mitchell SG, Grazu V, de la Fuente JM. The fate of nanocarriers as nanomedicines in vivo: important considerations and biological barriers to overcome. Curr Med Chem. 2013;20(22):2759–78.
53. Yong KT, Law WC, Hu R, Ye L, Liu L, Swihart MT, et al. Nanotoxicity assessment of quantum dots: from cellular to primate studies. Chem Soc Rev. 2013;42(3):1236–50.
54. Paliwal R, Babu RJ, Palakurthi S. Nanomedicine scale-up technologies: feasibilities and challenges. AAPS PharmSciTech. 2014;15(6):1527–34.
55. Alivisatos AP, Andrews AM, Boyden ES, Chun M, Church GM, Deisseroth K, et al. Nanotools for neuroscience and brain activity mapping. ACS Nano. 2013;7(3):1850–66.
56. Pavlov AM, Sapelkin AV, Huang X, P'Ng KM, Bushby AJ, Sukhorukov GB, et al. Neuron cells uptake of polymeric microcapsules and subsequent intracellular release. Macromol Biosci. 2011;11(6):848–54.
57. Weissleder R, Nahrendorf M. Advancing biomedical imaging. Proc Natl Acad Sci U S A. 2015;112(47):14424–8.

Basic Principles of In Vivo Distribution, Toxicity, and Degradation of Prospective Inorganic Nanoparticles for Imaging

Jelena Kolosnjaj-Tabi, Jeanne Volatron, and Florence Gazeau

1 General Introduction

Inorganic nanosized particles (NPs) made of iron oxide (iron oxide nanoparticles (IONPs)), heavy metals, and heavy metal-free nanoparticles, such as quantum dots (QDots), silver, and gold NPs, have been the subject of intense preclinical imaging research, substantially (but not exclusively) for the detection of different types of cancers or, ideally, precancerous lesions [1–4]. While conventional medical imaging is currently relatively insensitive and/or unspecific for the detection of most types of early-stage cancers, research efforts have started developing novel imaging agents and modalities that might allow early diagnosis, classification of specific tumors into different subtypes, and/or concomitant treatment of malignant growths [2]. Some inorganic nanoparticles are currently considered as prospective imaging agents [5], which are particularly attractive for spectral CT, magnetic resonance or magnetic particle imaging, as well as photonic, plasmonic, and fluorescent imaging. On the one hand, they can improve the potential of conventional imaging modalities or offer novel imaging modalities for biomolecular and cellular imaging, and on the other hand allow bigger platforms that could combine imaging and therapeutic functions in a single medicinal device. In addition, on the basis of localized surface plasmon resonance, metallic nanoparticles have been recently proposed as contrast agents for photoacoustic imaging and lately also suggested for in vivo surface-enhanced Raman spectroscopy (SERS) [6]. Although SERS spectroscopy is well known as an in vitro analytical tool, the feasibility of in vivo cancer imaging with SERS probes has been demonstrated. Major advantages of SERS imaging should yield higher sensitivity and higher signal specificity than most current imaging modalities while also offering multiplexing capabilities.

J. Kolosnjaj-Tabi • J. Volatron • F. Gazeau (✉)
Laboratoire Matières et Systèmes Complexes, UMR 7057 CNRS/Université Paris Diderot, 10 rue Alice Domon et Léonie Duquet, Paris, Cedex 13 75205, France
e-mail: florence.gazeau@univ-paris-diderot.fr

J.W.M. Bulte, M.M.J. Modo (eds.), *Design and Applications of Nanoparticles in Biomedical Imaging*, DOI 10.1007/978-3-319-42169-8_2

1.1 Applying Inorganic Nanoparticles for Imaging

Inorganic nanoparticles could be used in whole-body imaging [7], in cellular (molecular) imaging [8] and in quantitative cell analysis [9]. In biomedically functional formulations for in vivo applications, the inorganic nanoparticles should be administered in stable, monodisperse colloidal suspensions, where particles, often stabilized with synthetic (organic) coatings, are dispersed in aqueous solvents, adjusted to physiological pH with appropriate osmolarity. The synthetic coatings generally have two main roles. The first is to stabilize the particle (in suspension and/or to increase the circulation time in the blood), and the second one is to serve as linker for additional targeting or therapeutic molecules or complementary imaging molecules that allow imaging with additional diagnostic modalities. Once the particles are applied, the chemical component of the nanoparticle's core remains inconspicuous and the fate of the particle is governed by other features, which include, but are not limited to, particle's size, shape/architecture, surface charge, hydrophobicity, functionalization/targeting ligands, aggregation state, or exposure protocol (e.g., repeated injections). These characteristics directly affect the nanoparticle's interaction with bodily biomolecules and their behavior in the organism [10].

1.2 Nanoparticles' Fate in a Nutshell

The term pharmacokinetics describes and studies the processes by which a *drug* is absorbed, distributed, metabolized, and eliminated by the body. As the mentioned inorganic nanoparticles for imaging (with the exception of some dextran- and silica-coated IONPs) have yet to become drugs, the term "particokinetics" has been chosen here to refer to their in vivo fate [11].

Particokinetics is specific to the administration route (e.g., intravenous, oral, intraperitoneal, subcutaneous), particles' intrinsic and biologically acquired physicochemical characteristics, quantity and repeatability of the administration, and the species to which the particles are applied [12, 13]. Nevertheless, some general features are generally applicable to inorganic nanoparticles. Soon after the administration to a living organism, the nanoparticles are "tagged" with the host's biomolecules and led to the organs of the mononuclear phagocytic system (mainly the macrophages of the liver and spleen), which process the administered nanoparticles [14]. In addition, a fraction of inorganic nanoparticles might accumulate in the skin and/or adipose tissue macrophages, or be blocked within thin vessels of the lungs and get captured by alveolar macrophages. Nevertheless, the liver remains the main organ of metabolism and detoxification and its role in particle processing is essential [12, 13].

Among the parameters determining the clearance of nanoparticles from the blood, the effect of the coating and the hydrodynamic size of circulating nanoparticles seem to determine the particle's biodistribution kinetics [13]. Generally, particles with a larger hydrodynamic diameter (d_H), approximately around 100 nm, will not only be cleared faster than the smaller ones, but if the d_H exceeds 200 nm will be taken up preferentially by the spleen. In contrast, very small nanoparticles ($d_H < 10$ nm) [15] might avoid the phagocytic pathway and be directly eliminated by the kidney. Nevertheless, as small nanoparticles might also aggregate in vivo, they are not systematically eliminated via the kidney [13].

Other factors that impact the particokinetics include particles' curvature, roughness, charge, hydrophilicity, and cross-linking of the coating molecules. Interestingly, while PEG might seem a perfect coating option, it has been suggested that second run injections of PEGylated nanoparticles do not have the same blood half-life, as anti-PEG antibodies might form after the first injection and bind to subsequently injected nanoparticles, resulting in a faster uptake after recurrent injections [16, 17]. Particles' circulation time might also increase due to an age-related decrease in phagocytic activity. Indeed specific coatings, injection of nanoparticle-labeled cells, or pretreatment of phagocytic cells with certain cytokines [18] might also affect the homing preferences of nanoparticles and thus be exploited in imaging of different diseases [19]. Nevertheless, longer circulation times might be important for the imaging of cancerous tissues or metastases or vascular angiography, while shorter blood half-life might be more appropriate for the imaging of various inflammation-mediated diseases, where uptake by macrophages, other than Kupffer cells, is required. Nanoparticles eventually undergo biotransformation (that is, degradation and metabolization processes), which enables nanoparticles to gradually dissolve and eliminate their components via urine and bile, or recycle the material and reuse it, as it happens for iron. Since biotransformation pathways are element specific, they will be described in detail in Section 1.7. Particokinetics can be assessed by imaging, spectroscopic, or magnetometric techniques. While radiolabeling has also been proposed for the follow-up of nanoparticles, the detachment of radiolabels might compromise qualitative and quantitative measures of particles' biodistribution.

1.3 The Nanoparticle-Protein Associates

The biomolecules, which more or less tightly encompass the nanoparticle, form the "protein corona" [20]. The latter involves different smaller molecules and proteins, such as albumin, immunoglobulins, apolipoproteins, fibrinogen, transferrin, alpha-1-antitrypsin, to name a few but the most abundant. Protein binding to nanoparticles is governed by proteins' stoichiometries and binding affinities, and differs among nanoparticles, bodily fluids, cellular compartments, and cell phenotypes [14]. The proteins' binding and unbinding to nanoparticles determine the interaction of nanoparticle-protein complexes with cellular receptors, impact the opsonization processes of the nanoparticles, determine their distribution, influence their enzymatic attack, and govern their biological effects. The protein shell can also disguise specific ligands (such as transferrin or folate), attached to particles' surface, and thus compromise the imaging or targeting potential of nanoparticles [21]. Conversely, the corona can also promote some specific interactions and help crossing biological barriers such as the blood-brain barrier [22].

Nanoparticle-protein associates can be assessed in vitro by size-exclusion chromatography or dynamic light scattering method, which determines the hydrodynamic volume of nanoparticle-protein associates. While these methods are nonspecific and can be applied to all inorganic nanoparticles, other methods can be applied to specific types of inorganic nanoparticles, and will be presented later on in this chapter.

1.4 Nanoparticle Coating: The Essential But Transient Element in Particle Stabilization and Follow-Up

To characterize and follow the integrity of nanoparticle composites, made of the core, shell, and coating, it is necessary to distinguish each of its components from the endogenous compounds. Different strategies have been developed, such as the use of xenobiotics like gold, cadmium, or lanthanides, which are not naturally present in the organism and can be quantified by elemental analysis, or the labeling of the nanoparticle core/shell and/or of the coating with radioisotopes (^{59}Fe, ^{111}In, ^{14}C, ^{51}Cr, ^{65}Zn, etc.).

Radiolabeling strategies are particularly attractive as they allow quantification, in vivo or ex vivo, and have high sensitivity and high specificity. Nevertheless, they do suffer, among other things, from the time limitation of the experiment duration, due to the finite lifetime of radiotracers. The difficulty of these methods lies in the choice of appropriate label that would allow the follow-up of nanoparticle core, the coating, or both. The labeling of the iron oxide core with ^{59}Fe [23, 24] was historically used to study the biodistribution of MRI contrast agents. Post-synthetic techniques using neutron or proton activation were tested on gold or on iron oxide leading to a co-labeled iron core. For monodisperse particles made by thermal decomposition methods, Freund et al. [25] proposed a post-synthetic technique to label oleic-acid-stabilized IONPs by simple incubation and exchange with a water-free tracer dose of $^{59}FeCl_3$, as hard γ-emitter. The labeling was stable in vitro and therefore could help tracking of the IONP core and iron by-products ex vivo. An exchange-labeling technique with heterologous isotopes has also been used to mark IONPs with ^{51}Cr and QDots with ^{65}Zn [26]. IONPs could be successfully tagged with ^{51}Cr due to the similarity between the Fe(III) and Cr(III)-oxide chemistry. However, as soon as degradation of the IONPs occurred, the different transport paths for Fe^{2+}/Fe^{3+} and the tracer Cr^{3+} become a concern [26]. The intracellular processing of Cr^{3+} is mostly unknown. Although a binding peptide, chromodulin has been identified and could induce renal excretion of Cr^{3+}. Interestingly, the excretion and metabolism of ^{51}Cr from IONPs were very different from those of $^{51}CrCl_3$. The excretion via the kidney and gastrointestinal tract was very slow for ^{51}Cr coming from IONPs in comparison to $^{51}CrCl_3$, and the ^{51}Cr coming from IONPs was eliminated mainly in feces and less so in urine. After 4 weeks, a small part of ^{51}Cr originating from IONPs was transferred from the liver to the spleen, gastrointestinal tract, and carcass of treated animals. When compared to the massive degradation and elimination of ^{59}Fe from liver, it has been suggested that ^{51}Cr may be trapped in the liver to a greater extent, due to the lack of a specific Cr^{3+} exporter [26]. The differences between the metabolism of iron and chromium can be used to study the absorption of intact nanoparticles from the gastrointestinal tract. Comparing ^{59}Fe-labeled and ^{51}Cr-labeled IONPs given by gavage to mice, Bargheer et al. showed 5 % absorption of ^{59}Fe in the intestines and undetectable activity of ^{51}Cr. This excludes that intact particles absorb in the intestine, but rather suggests that iron from IONPs is partially digested in the stomach [26].

With similar methodology, CdSe/CdS/ZnS QDots labeled with ^{65}Zn were compared to $^{65}ZnCl_2$ injected intravenously at a trace dose. Zn has a fast turnover being eliminated from all organs with fast exchange kinetics within 4 weeks. In contrast, ^{65}Zn signal persisted in liver and spleen 4 weeks after injection of QDots, confirming that the particles were not fully degraded in these organs. Otherwise Zn released from QDots was distributed in a similar manner to the control compound, the $^{65}ZnCl_2$ [26].

Radioactive labeling offers unique opportunities to track the fate of the particle coating with respect to the nanoparticle cores. Moreover, it is also mandatory to verify that all the tracers or targeting agents that are assembled on the same nanoplatform for multimodal imaging actually reach the same targets in vivo. Radiolabeled IONPs as PET/MRI or SPECT/MRI imaging agents are common [27], but only recent studies evaluated the degree to which the association between the radioactive label and the metallic core remained intact in vivo.

To evaluate respective distributions of nanoparticles' core and shell, Wang et al. investigated ^{59}Fe-labeled IONPs coated with ^{14}C-labeled oleic acid and ^{111}In-labeled phospholipids (DMPE/DTPA-^{111}In) (Fig. 1) [28]. Using a related strategy, Kreyling et al. looked at ^{198}Au NPs with their amphiphilic polymer shell labeled with ^{111}In (Au-DTPA-^{111}In) [29].

The biodistribution of ^{59}Fe and ^{111}In was broadly similar, with highest uptake in liver and spleen, especially at 24 h post-injection (Fig. 1). The greatest discrepancy was found in the kidneys, with an eightfold higher concentration of ^{111}In compared to ^{59}Fe. ^{111}In was recovered at each time point in the urine suggesting the loss of $^{111}In^{3+}$ ions or other ^{111}In species. The ^{14}C exhibited much lower concentration in liver and spleen, confirming a previous study by Freund et al. showing that oleic acid could detach from the iron oxide core in vivo in the bloodstream [25]. Three control experiments were necessary to assess the distribution of the different components of the shell: injection of the labeled phospholipid (DMPE-DTPA-^{111}In), of the citrated label (^{111}In-citrate), and of the labeled ligand (DTPA-^{111}In). The phospholipid DMPE-DTPA-^{111}In had a similar biodistribution as the ^{111}In-labeled IONPs, but exhibited a lower accumulation in liver, spleen, and bone and a higher accumulation in the blood, heart, and lungs. In contrast, ^{111}In citrate was found in the kidney and DTPA-^{111}In in the urine. These results suggest that ^{111}In is not released from NPs in ^{111}In-citrate or DTPA-^{111}In forms but rather as DMPE-DTPA-^{111}In, which forms micelles with the same biodistribution as the nanoparticles. In agreement with this hypothesis, the $^{59}Fe/^{111}In$ ratio decreases from 2:1 at 10 min to 1.25 at 24 h

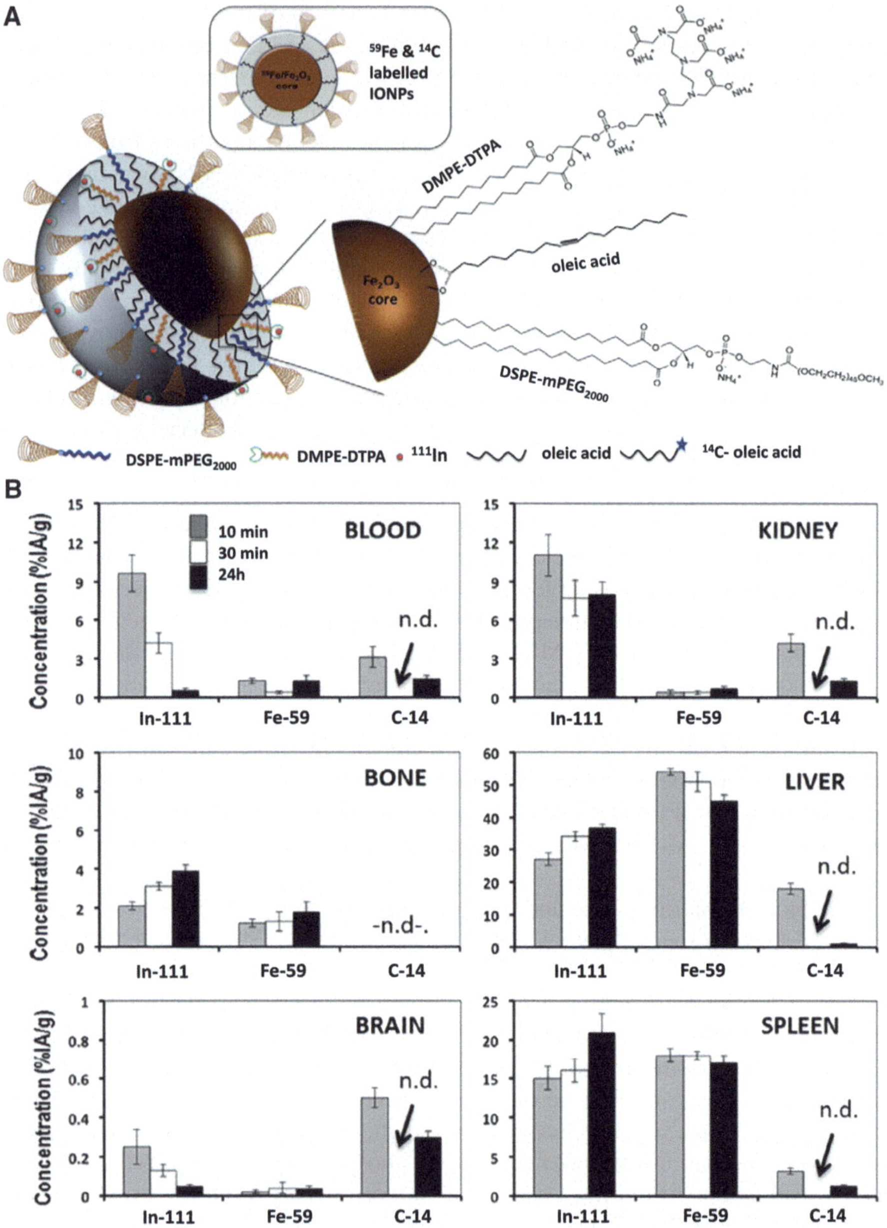

Fig. 1 Radioactive follow-up of nanoparticle core and shell in vivo. (**a**) The nanoparticles consist of ^{59}Fe-labeled IONPs stabilized with ^{14}C-labeled oleic acid and modified with DSPE-mPEG$_{2000}$ and ^{111}In-DMPE-DTPA phospholipid. (**b**) Comparison of the ^{59}Fe, ^{111}In, and ^{14}C tissue concentrations (expressed as percentage of the injected radioactivity per gram) at different time points after intravenous injection of labeled IONPs. Adapted from Wang et al. [28] with permission

in liver, confirming an incomplete and early dissociation of phospholipids from nanoparticles. In the work by Kreyling et al. the ^{198}Au:^{111}In ratio was also analyzed and revealed differences in biodistribution between the two isotopes, suggesting a partial dissociation of the coating from the nanoparticles [29]. The control ^{111}In-DTPA was rapidly excreted in urine, while the ionic form of ^{111}In accumulated in the liver. Most of the ^{111}In and ^{198}Au administered in particulate form are accumulated in the liver. However, the results could not indicate if the ^{111}In detached from Au NPs was in the ionic form or still attached to the polymer [29]. A minor dissociation of the coating occurred in the bloodstream, but the liver could accumulate both intact NPs and separated components [29].

To go further in the analysis, in vitro tests have been performed to mimic the behavior of the NPs and their coating in the liver [29]. Polymer-coated Au NPs and CdSe QDots were accumulated in the lysosomes of Kupffer cells and endothelial cells. The co-localization of QDots and of the polymer shell labeled with DY495 was investigated by fluorescence, showing that the DY495 has been partly displaced from QDots and subsequently exocytosed after several hours [29]. This result suggested an intracellular enzymatic degradation of the polymer shell [30], but could not conclude whether there is a dissociation of the polymer from the nanoparticles or of the DY495 from the polymer or both.

Another relevant parameter to study the fate of the complex formed by nanoparticles and their coating is the presence of the protein corona, which evolves dynamically throughout the journey of the nanoparticle within the body. To understand the influence and the fate of the protein corona in vivo, Bargheer et al. used ^{59}Fe-labeled PEGylated IONPs and ^{125}I-labeled model proteins such as transferrin, covalently bound or adsorbed to the surface of the nanoparticles (Fig. 2) [26]. Importantly, both ^{125}I and ^{59}Fe labels were cleared synchronously from the blood and the biodistribution of ^{125}I transferrin that was bound or adsorbed to IONPs was clearly different from the biodistribution of free ^{125}I transferrin. This result suggests that the protein corona around the nanoparticles was not lost in the bloodstream in the first hour post-administration [26]. However, the ratio $^{125}I/^{59}Fe$ decreased in all organs after 2 h except in stomach and blood, suggesting a substantial recycling and transfer of the transferrin from the liver to the blood and other tissues. Interestingly, there was no significant difference between covalently bound and adsorbed transferrin in this experiment [26]. It must be noted that transferrin was among the adsorbed proteins, which show the highest abundance (>10 %) in coronae of many nanoparticle types [31].

1.5 *Cellular "Conditioning" of Administered Nanoparticles and Its Impact on Imaging Outcomes*

By analogy with the extracellular environment (as for example the circulating blood plasma), which constantly supplies and exchanges proteins surrounding the nanoparticles, the cells continuously modify the conditions of their microenvironments and

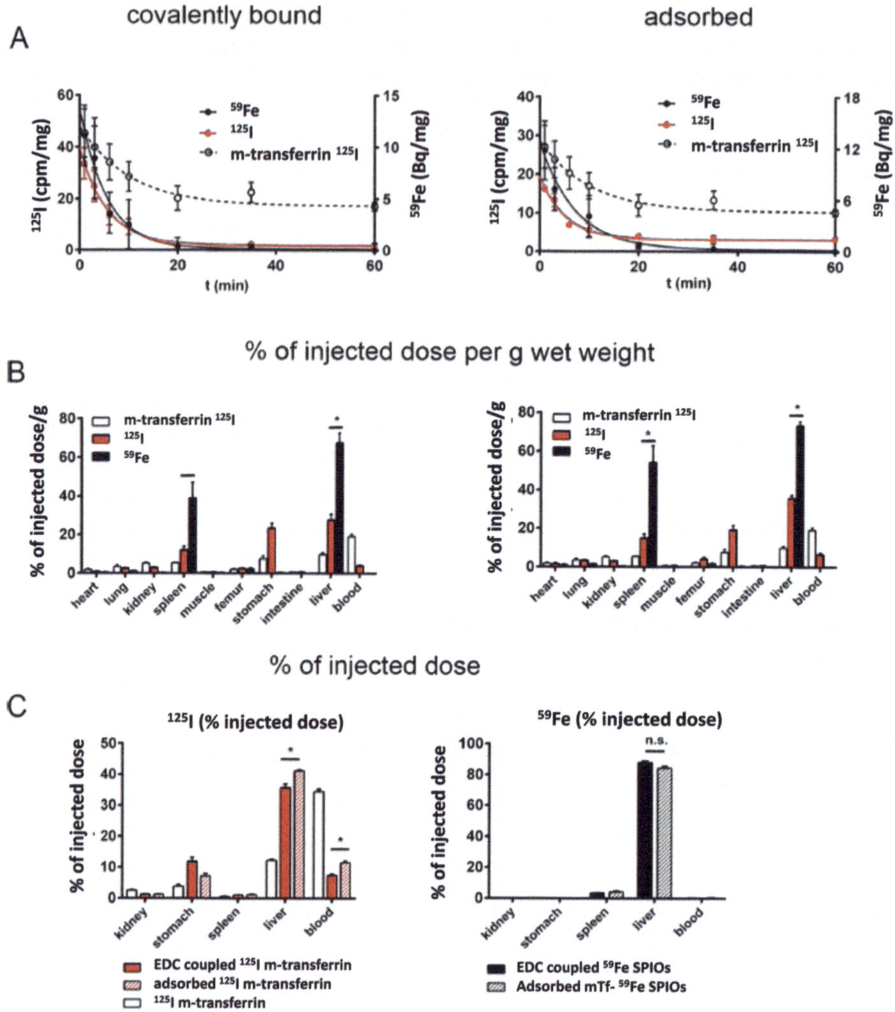

Fig. 2 Radioactive follow-up of nanoparticle core and protein corona in vivo. ^{59}Fe-labeled IONPs with a preformed corona of adsorbed (**a**, **b**, right side) or covalently bound (**a**, **b**, left side) ^{125}I-mouse transferrin, i.v. injected to mice in comparison to free ^{125}I-mouse transferrin. (**a**) Activity of ^{59}Fe and ^{125}I (1–120 min) in blood. (**b**, **c**) Activity of ^{59}Fe and ^{125}I in organs 120 min after injection of IONPs or free transferrin. Adapted from Bargheer et al. (open access) [26]

thus continuously reshape the protein corona of cell-internalized nanoparticles. Wang et al. studied the intracellular processing of fluorescently labeled protein corona at the cellular scale [32]. The adsorbed layer of proteins, resulting from incubation of positively charged polystyrene nanoparticles with full fluorescent serum, was strong enough to be retained on the nanoparticles as it entered the cells and was trafficked into the lysosomes together with nanoparticles (Fig. 3). Within the lysosomes, a progressive decrease of the protein corona fluorescence was observed,

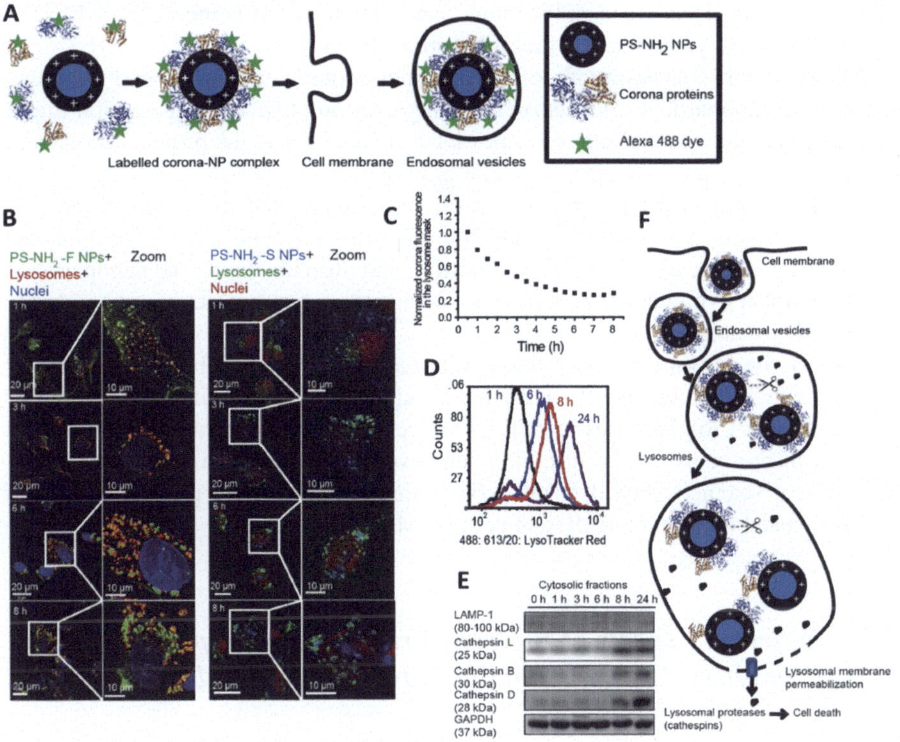

Fig. 3 Intracellular fate of the protein corona. (**a**) Green fluorescent protein corona formed on the blue positively charged amino-modified polystyrene nanoparticles (PS-NH_2 NPs, 50 nm) is retained after cellular uptake. (**b**) Time-resolved localization of PS-NH_2 NPs and their impact on lysosomes in 1321N1 cells, showing increasing accumulation in the lysosomes and subsequent progressive swelling of lysosomes. (**a** PS-NH_2 NPs (*green*), lysosomes (*red*: LAMP-1 antibody), nuclei (*blue*, DAPI staining); (**b**) PS-NH_2 NPs (*blue*), lysosomes (*green*: LAMP-1 antibody), nuclei (*red*: Draq 5 staining)). (**c**) Kinetics of the degradation of the fluorescent-labeled protein corona of PS-NH_2 NPs. (**d**) Flow cytometry distribution of LysoTracker Red staining after exposure to PS-NH_2 NPs, indicating increasing volume of lysosomal volume and gradual formation of a second population with loss of lysosomal integrity. (**e**) Western blot of different cathepsins in cytosolic fraction of treated cells, demonstrating progressive release of cathepsin in the cytosol. (**f**) The green fluorescent protein corona of PS-NH_2 NPs is trafficked to the lysosomes and is degraded, while the lysosomes are swelling and become permeable, allowing cytosolic release of cathepsins. Adapted from Wang et al. [32] with permission

indicating the degradation of the protein corona in the lysosomes over 6–8 h after exposure. Interestingly, the nanoparticles were able to convey a much higher amount of proteins than a simple incubation of cells with proteins would provide [32]. This suggests that enzymatic digestion by proteases at the lysosomal level could be activated by nanoparticle uptake. Importantly, the clearance of protein fluorescence was accompanied by a lysosomal swelling and increased lysosomal permeability, with subsequent damages on cell organelles and initiation of apoptotic signaling pathways (Fig. 3) [32]. Thus the protein corona masks for a while and delays the deleterious effect of positively charged polystyrene nanoparticles until being degraded in the lysosomes by enzymes, such as the cathepsin L [29, 30].

The different concentrations of intracellular proteins, nutrients, and ions also destabilize nanoparticles and set on their aggregation [14]. The aggregation might also be triggered after cellular enzymes have stripped away the protein corona and the particles' organic coating [33]. The aggregation phenomenon is the principal cause that might affect the physical properties, implicated in imaging protocols [34]. Conversely, nanoparticles' interactions with different cellular components might also result in a rearrangement—sorting and individualizing the nanoparticles in the periphery of endo-lysosomic compartments, in order to set on or accelerate particle degradation [35].

Finally, the cells are not only able to internalize the particles, but can also, under specific conditions, release the particles to adjacent or distal cells of the same or different types [36]. Particles' expulsion via cell microvesicles should be taken into account in protocols involving highly sensitive modalities, such as high-field MR scanners or scanners provided with cryogenically cooled MRI probes, which are currently used mainly in preclinical cellular MRI [37].

1.6 Different Aspects of Biotransformation: Therapeutic Properties, Loss of Functionality, and Potential Toxicity

Bio-interactions and bioprocessing of the particle are the *sine qua non* of a particle's life cycle within the organism. Particles' surface (enriched with synthetic or biologically acquired components, such as the protein corona) governs particles' early biodistribution in the short run [35]. Nevertheless, once the biomolecules and parts of the synthetic coating are stripped off, the remnants of the synthetic coatings, particles' rearrangement within cellular compartments, and the nature of the core start governing the long-term persistence within the organism, and the degradation and the elimination processes [29, 35, 38]. The biotransformations of nanoparticles, as well as their potential toxicity, tightly depend on the particles' physical and chemical properties. Yet, some common features characterize the fate of inorganic nanoparticles regardless of their composition, structure, and size. Engineered nanoparticles are mainly captured by macrophages of the liver and spleen and thus tightly packed in their endo-lysosomes, which induce adsorption of biomolecules, particles' aggregation, and gradual erosion of the coating and inorganic core, affecting in turn the physical and chemical properties that are required for imaging or therapeutic protocols [34, 39]. In addition, the degradation and processing might alter the toxicological profile of the administered material [40], and what used to be a small benign object now starts releasing its chemical (molecular), mainly metallic, components (ions), which are more or less rare in the organism and might trigger a cascade of adverse reactions, as the ones induced by the generation of reactive oxygen species (ROS) [41, 42]. These molecules might, under certain conditions, alter the physiological processes of the cell by inducing the lipid peroxidation of the cell membrane, set on mitochondrial and lysosomal damage, or damage the endoplasmic reticulum, which may result in programmed cell death. Nevertheless, antioxidant enzymes (such as superoxide dismutase, glutathione peroxidase, catalases), vitamins (such as α-tocopherol), or hormones (such as melatonin) are some of the antioxidant molecules present in biological systems that are capable of scavenging the ROS [43]. Be that as it may, "*sola dosis facit venenum*" (i.e., the dose makes the poison) [44] and the cooperation between the gradual occurrences of chemical factors (e.g., gradual dissolution of inorganic nanoparticles and progressive reduction of ROS by local antioxidants) and physical (e.g., aggregation and thus delayed degradation by hindering the access of biomolecules to particle surface) and biological phenomena (e.g., up-regulation of genes coding for specific proteins) [45] accompanying particles' degradation over time might counterbalance or moderate the impact of the foreign materials' burden to cells/organisms. Thus, in order to assess the factual toxicity of nanoparticles, their biotransformations have to be monitored under dynamical conditions over relevant periods of time. In addition, the conditions under which some inorganic nanoparticles might have an increased toxicity due to accelerated dissolution, such as UV illumination of QDs [40], should be considered and, if appropriate, avoided.

1.7 Material-Specific Degradation Patterns of Inorganic Nanoparticles In Vivo and Over Time

The specific chemical composition of the nanoparticle's core unconditionally determines nanoparticle's degradation pathways. Organisms possess a large variety of proteins, including enzymes, which are involved in the biotransformation of xenobiotics. As biotransformation allows the elimination or recycling of nanoparticle components, the progressive processing of a crystal to ions might alter the toxicological profile of the administered material.

While different degradation patterns are certainly intuitive, empirical evidence, obtained from gold/iron oxide heterodimers [38] or silver/gold nanohybrids [46], is particularly conclusive. The follow-up of the fate of gold/iron oxide heterodimers [38] evidenced the evolution of the ratio of magnetic iron over gold content and highlighted the distinct distribution and fate of gold and iron oxide within the organs. Over time the ratio was diminishing in the liver, while it remained almost constant in spleen, reflecting the prominent degradation of iron oxide in the liver. Spleen preferentially accumulates gold and is less efficient to degrade both materials. A fraction of heterodimers was able to fragment to individual gold or iron oxide subunits at early time points and the separated subunits had distinct distribution paths. Interestingly, such element-specific biodistribution has also been observed for NIR-absorbing hollow Ag-Au nanoshells, used for photothermal therapy in tumor-bearing mice [46]. The Ag/Au ratio varied considerably between organs, suggesting a size-dependent fragmentation of the heteroparticles at some time point post-injection. Gold was more present in liver and spleen, whereas Ag predominated in tumors for the smallest particles of 43 nm. The thinner shell allowed the 43 nm particles to fragment quicker and release Ag from the core-shell particles. This study again illustrated the different fate of materials and the importance of architecture on the life cycle of particles. Another important aspect is that the fate of NPs used for diagnostic or therapy should be investigated also after stimulation (i.e., laser exposition for photothermal ablation), as the external stimuli may also induce structural transformations of the particles.

2 Iron Oxide Nanoparticles

IONPs are mostly made of magnetite (Fe_3O_4) and/or maghemite (γ-Fe_2O_3) nanocrystals, which may exhibit different shapes (spheres, cubes, rods, rings, flowers, pyramids, octapods, nanoworms, etc.). These IONPs have superparamagnetic properties, and thus might be used in magnetic resonance imaging (MRI) or magnetic particle imaging (MPI). The IONPs might be applied either systemically by direct application (e.g., intravenous injection) of the suspension of nanoparticles or by the administration of heterologous or homologous cells, which were previously labeled in vitro with IONPs [47]. In addition to the imaging functionality, these nanoparticles could potentially be used for cancer treatment, either for magnetic targeting [48] or magnetic hyperthermia [49], or as magnetothermal effectors for deep brain stimulation [50].

2.1 Role, Effect, and Fate of IONP Synthetic (Polymer) and Biological (Corona) Coating

IONPs are commonly coated with organic polymers, such as polyethylene glycol (PEG), oleic acid, dextran, chitosan, phospholipids, and other amphiphilic polymers. The latter increase the stability of IONP suspensions and increase the circulation time in blood after intravenous administration. Administered nanoparticles' surfaces become covered by proteins [10], which are either bound to the synthetic coating or, if the polymer is stripped off, adhere directly to the particles' core.

Magnetic IONPs and protein associates can be specifically probed with a method, which relies on the particles' potential of orienting in the direction of the magnetic field [51]. When the magnetic field is switched off, the particles disorient and their Brownian relaxation time is proportional to their hydrodynamic volume. This relaxation time can be determined by the measurements of the magnetically induced optical birefringence signal. Combined to magnetic sorting and protein electrophoresis, the measurements of the magnetically induced optical birefringence signal [51] confirmed that protein adsorption strongly depends on surface coating and that the composition of the corona differs when IONPs are incubated in a medium supplemented with 10 % of plasma (generally used for cell culture) or when particles are suspended in pure plasma, which resembles in vivo conditions (Fig. 4). Moreover, some proteins, namely albumin and apolipoprotein, have a stabilizing ("individualizing") effect on the nanoparticles, while other proteins, such as fibrinogen, trigger

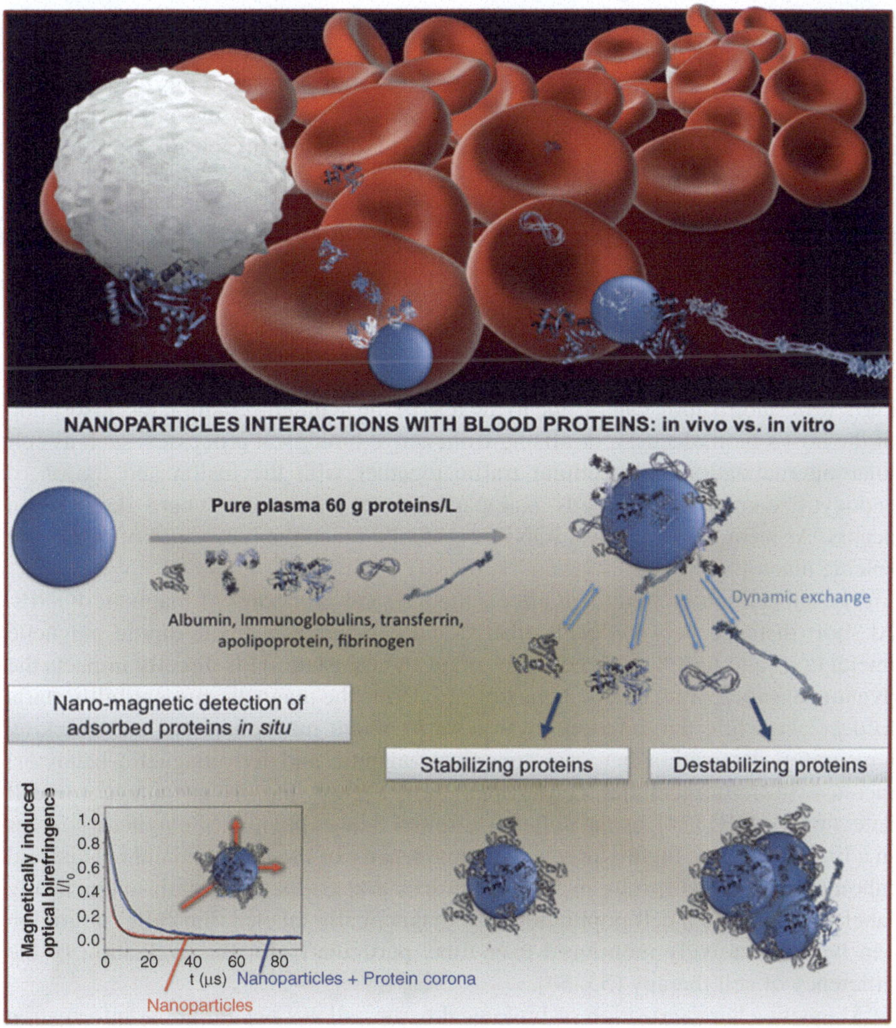

Fig. 4 Nanoparticles' interactions with blood proteins in vivo and in vitro. The adsorption of proteins on magnetic nanoparticles can be determined specifically by measurements of the relaxation of the magnetically induced optical birefringence of magnetic nanoparticles' suspensions [51], which reveals different populations of particles with distinct biological (nature of protein corona) and physical states (aggregation) that interact differently with macrophages

particles' aggregation [51]. Remarkably, the particles with different protein corona not only exhibit different physical states (single particles or aggregates), but can also coexist in plasma and are differently processed by immune cells. Macrophages will neither react nor capture these populations in the same manner or time course. This experiment clearly exemplifies how interactions between biomolecules and nanoparticles can drastically influence the aggregation pattern, distribution, and fate of intravenously injected IONPs [51].

While polymer coatings increase the stability of nanoparticles, it has also been shown that they impact the long-term degradation of IONPs, as PEG-coated IONPs degraded more rapidly in comparison to amphiphilic polymer-coated IONPs, monitored over 1 year [38]. Nevertheless, as we emphasized earlier in this chapter, studies also show that once applied, the nanoparticle core and the polymeric shell dissociate [25, 28]. While the particles remain in the liver and spleen, some components of the coating can be eliminated in urine [28].

2.2 Cell Internalization of IONPs and Its Effects on Nanoparticles' Magnetic Properties

Particle destabilization can be due to the loss of the synthetic coating, the adsorption of the host's biomolecules, or arising from active biological processes, such as cellular internalization. Intracellular traffic together with the fusion and fission of endocytic compartments leads nanoparticles to lysosomes, where degradation occurs. As most nanoparticles, IONPs eventually reach the lysosomes of hepatic or splenic macrophages.

The subcellular confinement affects the physical properties of magnetic IONPs. At short distance between each other, IONPs experience dipole-dipole magnetic interactions and have less freedom to rotate and translate. This directly impacts the dynamical superparamagnetic behavior of IONPs: the magnetic susceptibility starts to decrease while the magnetic hysteresis (remnant magnetization) and blocking temperature (transition between superparamagnetic and ferromagnetic behavior) increase [52]. Intracellular confinement nevertheless has its advantage for cell detection by MRI [34], as the cell that confines a large amount of magnetic sources in a limited volume highly affects the homogeneity of the magnetic field. The local inhomogeneity thus results in MR signal loss that co-localizes with magnetically labeled cell. Once a cell population was magnetically labeled, the cells' migration can be noninvasively monitored over time, particularly for the evaluation of the efficiency of cell therapy [53, 54].

Magnetic characterization of biological tissues allows obtaining the information on local confinement of magnetic nanoparticles within different organs [55]. Furthermore, magnetic measurements are consistent with morphological findings observed by TEM, which show that lysosomes of hepatic/splenic macrophages concentrate a higher number of nanoparticles than the lysosomes of circulating monocytes/macrophages that can be recruited to sites of inflammation [52]. The follow-up of nanoparticles' magnetic properties also provides important information on nanoparticles' biotransformation within organs, which directly impact those nanoparticles' properties that are necessary for therapeutic and diagnostic applications.

2.3 *Degradation and Recycling of IONPs*

In order to characterize the degradation of administered nanoparticles over time in the body, we require specific methods of quantification, which allow detecting an infinitesimal fraction (ppm or ppb) of material per kg of total body weight (bw). Ideally, the method of quantification should permit to distinguish between nanoparticles and their degradation products.

The quantification or the evaluation of the distribution of IONPs and their degradation products in vivo is complicated by the quantity of endogenous iron (about 50 mg/kg b.w.), which is greater than the amount of iron that is administered for MRI (typically 2.5 mg/kg). Consequently, elemental analysis is inappropriate for IONP quantification. On the other hand, methods based on nanomagnetism such as electron paramagnetic resonance (EPR) allow us to specifically and sensitively quantify IONPs [45]. Measurements of magnetization as function of strength of magnetic field or temperature may supplement the data obtained by EPR, as they provide information about particles' size and aggregation state. Besides, MRI provides a mean for monitoring IONPs' distribution over time in the same animal, although quantification remains challenging, since the effect of IONPs on MR relaxivity depends on the local environment and physical state of particles. Multiscale and pluri-method scanning should therefore be used in order to track the distribution and transformation of IONPs.

Different methods have been proposed to monitor the degradation of IONPs with different shape, organization, and surface coatings. Some of them included 7–8 nm spherical IONPs with a hydrophilic glucose-derivative coating; 20 nm IO nanocubes, coated with polyethylene glycol (PEG) (Fig. 5); and 20 nm hetero-structures (dimers) made of gold/iron oxide, coated with an amphiphilic polymer or PEG. Following intravenous administration (2.5 mg/kg b.w.), IONPs mainly reach the liver and, to a lesser extent, the spleen. As expected, the nature of IONP coating determined the initial uptake in the reticuloendothelial organs [45, 49]. The quantity of superparamagnetic iron diminished with time, more rapidly in the liver than in the spleen, as evidenced by MRI and EPR analysis. A potential translocation/elimination of intact particles out of the organs, together with nanoparticle degradation within the organs, account for the superparamagnetic property loss, which results in a decrease of magnetic iron [45].

In complement to macroscopic findings, transmission electron microscopy (TEM), electron diffraction, and associated spectroscopy provide information about the evolution of the subcellular distribution and morphology of IONPs over

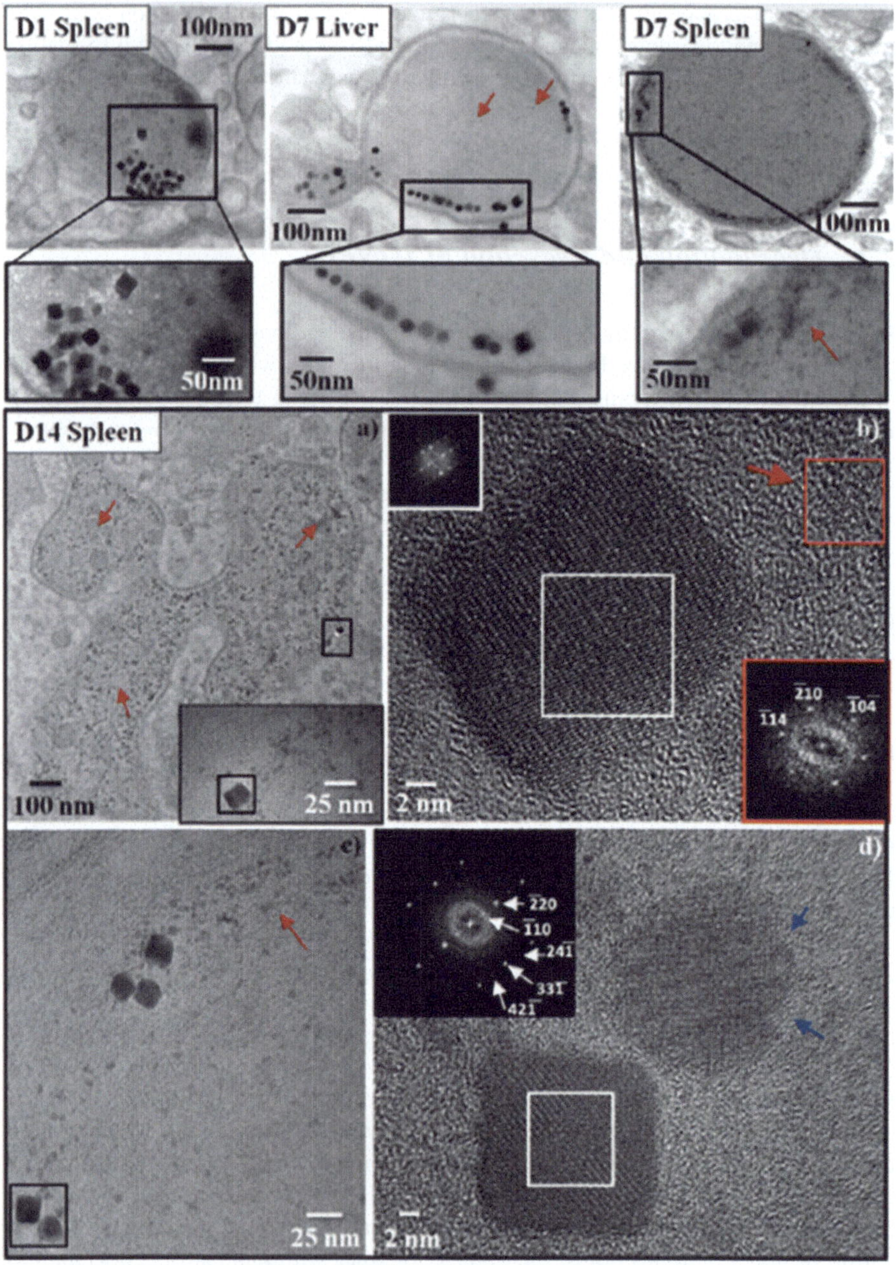

Fig. 5 Intracellular degradation of PEG-coated IONPs in murine liver and spleen after intravenous application, evidencing the transfer of iron into ferritin proteins. The intravesicular distribution of nanoparticles evolves over time: early after injection (day 1), dense assemblies of nanocubes avoid the cell from particle degradation and release of cytotoxic free iron ions. At longer times (days 7 and 14), the nanoparticles are transferred into protein-rich lysosomes in which they are

time after intravenous injection. Interestingly, while nanoparticles are mainly found as clusters within lysosomes of splenic and hepatic macrophages 1 day post-injection, particle sorting was observed in the periphery of lysosomes or in electron-dense areas of the lysosomes at later time points (Fig. 5) [35]. Nanoscale observations also showed that IONPs become increasingly surrounded by monodisperse electron-dense nano-sized (6 nm) objects, which are rich in iron, but poorly crystalline (Fig. 5) [35, 45]. These nanostructures, also known as ferritin, an endogenous iron-storing protein or apoferritin, when not combined with iron, consist of 24 protein subunits that form a hollow shell, where a maximum of 4500 iron atoms can be stored in mineral crystallites of iron oxyhydroxide (mainly ferrihydrite) and other forms of iron oxide. Ferritin binds iron ions and regulates the occurrence of Fenton's reactions, in which ferrous ions react with hydrogen peroxide and form reactive oxygen species that would damage the cells. Iron homeostasis is a finely regulated process where sets of proteins detect, transport, regulate, and store the metal, in order to provide it to the body at its demand (i.e., for hemoglobin synthesis) [56]. The coexistence of ferritins situated proximally to IONPs suggests a mechanism of local iron transfer from degraded IONPs to endogenous iron-storing proteins (Fig. 5). Although the morphological degradation of polydisperse 7-8 nm spheres is difficult to ascertain in vivo, nanocubes and gold/iron oxide heterostructures show evident features of erosion in liver and spleen 7 days after administration.

While the magnetic iron decreases, the nonmagnetic iron (total iron measured by elemental analysis subtracted by superparamagnetic iron measured by EPR) increases, confirming the local transformation of superparamagnetic particles into nonmagnetic iron species. In addition, as the magnetic signature of IONPs fades away due to degradation, the low-temperature magnetic signature for ferrihydrite appears [45]. The local degradation of IONPs and iron transfer in ferritins occurs within the lysosomes on nanoparticles' aggregate margins, where the iron remains bound in crystals until the proteins are synthesized and recruited in the vicinity of nanoparticles. Studies performed more than two decades ago have evidenced that IONP degradation is followed by the recycling of IONP compounds [23, 24]. The labeling of IONP core with ^{59}Fe showed that the iron from NPs incorporated within the hemoglobin of red blood cells in rats [23] and mice [24].

Fig. 5 (continued) more dispersed and susceptible to degradation. *Red arrows* show the coexistence of monodisperse iron-rich ferritin protein of 5.3 ± 0.8 nm diameter with some degraded or resilient nanocubes, suggesting a protein-triggered degradation and local iron transfer (**a**). High-resolution imaging performed 14 days after injection shows that degraded and resilient nanocubes maintain their initial crystalline structure (the spinel inverse and the vacancy-ordered γ-Fe_2O_3 structures along the [103] and [116] zone axes on the images (**b**, **d**), respectively). The *blue arrows* highlight the rough surfaces of the degraded nanostructures. The Fourier transform of the atomic structure of the ferritin core in image (**b**) (spleen day 14 post-injection, *red squares*) shows unambiguously a hematite structure oriented along the [481] zone axis. Adapted with permission from Lartigue et al. [35]

2.4 In Vivo Toxicity of IONPs

IONPs are generally considered as biocompatible, safe, and nontoxic. The median lethal dose (LD_{50}) of intravenously applied citrate-coated ultrasmall (8.6 nm) IONPs in mice was reported to be greater than 949 mg (17 mmol) Fe/kg body weight (b.w.) [57], which is almost 400 times more than what is required for MRI imaging. For comparative purposes note that the LD_{50} of cisplatin, a commonly used anticancer therapeutic agent, is 11 mg/kg b.w. after i.v. injection in mice [58]. Indeed the LD_{50} values of IONPs vary in other species, when other coating agents are used or when IONP cores and/or hydrodynamic sizes differ. As an example, the LD_{50} of (8 ± 3 nm) IONPs after i.v. administration to rats was 36.42 mg/kg b.w. [59].

As the liver (either due to the phagocytic activity of its macrophages, the Kupffer cells, or due to specific ligands that target hepatocytes) remains the main accumulation organ after administration of most types of IONPs, the administration of such nanoparticles, which may induce oxidative stress, might result in an increased level of liver enzymes. Several studies published thus far reported on low or minor transient changes in liver enzyme levels after dextran- and pluronic-coated IONPs (dose 10 mg iron/kg b.w.) [60]. The administration of dextran-coated IONPs to humans did not induce any clinically significant adverse reactions.

3 Quantum Dots

QDots are semiconductor nano-crystallites typically made of lead sulfide, lead selenide, cadmium selenide, cadmium telluride, cadmium sulfide, indium arsenide, or indium phosphide that are emerging as potential fluorescent probes for biomolecular and cellular imaging. Due to their outstanding brightness, high photostability, and broad absorption spectrum coupled with size-dependent narrow emission bands, QDots have been particularly attractive as efficient multiplexing probes for cell labeling and tracking. Their potential biomedical applications mainly concern in vivo imaging for early cancer detection. Unlike conventionally used organic dyes and fluorescent proteins, QDots have very large molar extinction coefficients, on the order of $0.5–5 \times 10^6$/M/cm which is 10–50 times greater than those of organic dyes [2], and could thus allow sensitive detection of analytes at a lower concentration, which is especially important for detection of cancerous lesions at their early stage. In order to efficiently fluoresce, QDot cores should be protected with a shell. The latter can be made of ZnS, silica, or different polymers.

3.1 Cellular Processing and Fate of QDots

The optical properties of QDots are affected by cell internalization: agglomeration into endosomes and lysosomes, as well as intralysosomal acidic pH. These changes in overall fluorescence of the QDots complicate the quantification of intracellular

QDots. For example drastic reductions in QDot photoluminescence quantum yield may result from oxidation or surface defects upon exposure of QDots to low pH. Oxidative dissolution of the QDots and rapid reduction in fluorescence emission have been observed despite the presence of organic shells on the surface of the QDots [40].

From the biological point of view, persistent localization of CdTe-based QDots in lysosomes might result in an enlargement of the lysosomal compartment as an adaptive response to the presence of QDots, and an enhancement of the antioxidant response mediated by intracellular glutathione. Activation of the transcription factor TFEB, a key regulator of lysosome biogenesis, shows that cells are able to sense the overload of lysosomes with nanoparticles and adapt their response to diminish oxidative stress [61].

Pioneering studies [62, 63] used either the fluorescence emission of QDots or the elemental quantification of cadmium to track QDots ex vivo and report on their particokinetics and biodistribution. Long-term experiments demonstrated that the QDots remained fluorescent after at least 4 months in vivo suggesting that they keep their integrity [63]. Fischer et al. observed the sequestration of lysine-coated and BSA-coated CdSe/ZnS in RES cells, but no short-term excretion in urine and feces, nor degradation of the inorganic part of QDots [62]. To assess degradation, they used ultracentrifugation on the digested organs to separate the intact QDots in tissue from isolated ions. Fitzpatrick et al. investigated the 2-year fate CdSe/ZnS Qdots coated with PEG 5000 intravenously injected in mice [64]. The fluorescence emission of QDots was observed in the blood immediately after injection and later in the liver, spleen, lymph, and bone marrow. However, the emission in liver faded after 5 days. Signal fade-out was slower in the bone marrow (3–6 months) and even slower in lymph nodes (up to 2 years). It was not clear whether this fading was due to particle dissolution, darkening, or excretion mechanisms. Surprisingly, two-photon spectral confocal microscopy indicated the presence of QDot clusters in liver, spleen, and lymph nodes 2 years post-injection, but their fluorescence was significantly blue-shifted [64], which differs from previous reports reporting red-shifting attributed to QDot clustering. Interestingly, the blue shift in peak position was not seen at early time points, neither attributed to storage and was due solely to the long residence time in living animals, indicating partial degradation of the semiconductor crystals (Fig. 6). In spite of the ZnS shell and polymer coating, the blue shift could be due to the loss of core materials, a change in the size/shape ratio of the nanocrystal, or the degradation of the ZnS shell. The absence of apparent toxicity of QDots, injected at a concentration that would show lethal toxicity as soluble cadmium ions (8.5 mg/kg of Cd, equivalent to about two lethal doses of Cd^{2+} injected intravenously), and the persistent fluorescence signal indicate that some particles remained almost intact after 2 years without substantial leakage of cadmium, but that they had experienced some crystal or surface modification explaining the blue shift. In the mentioned study, nanoscale examination of the morphology and elemental composition of the persistent particles would be necessary to provide insight into the long-term biotransformation of QDots.

In another study, the chemical fate of aqueous synthesized CdTe/ZnS injected intravenously in mice was investigated over the course of a 28-day period [65].

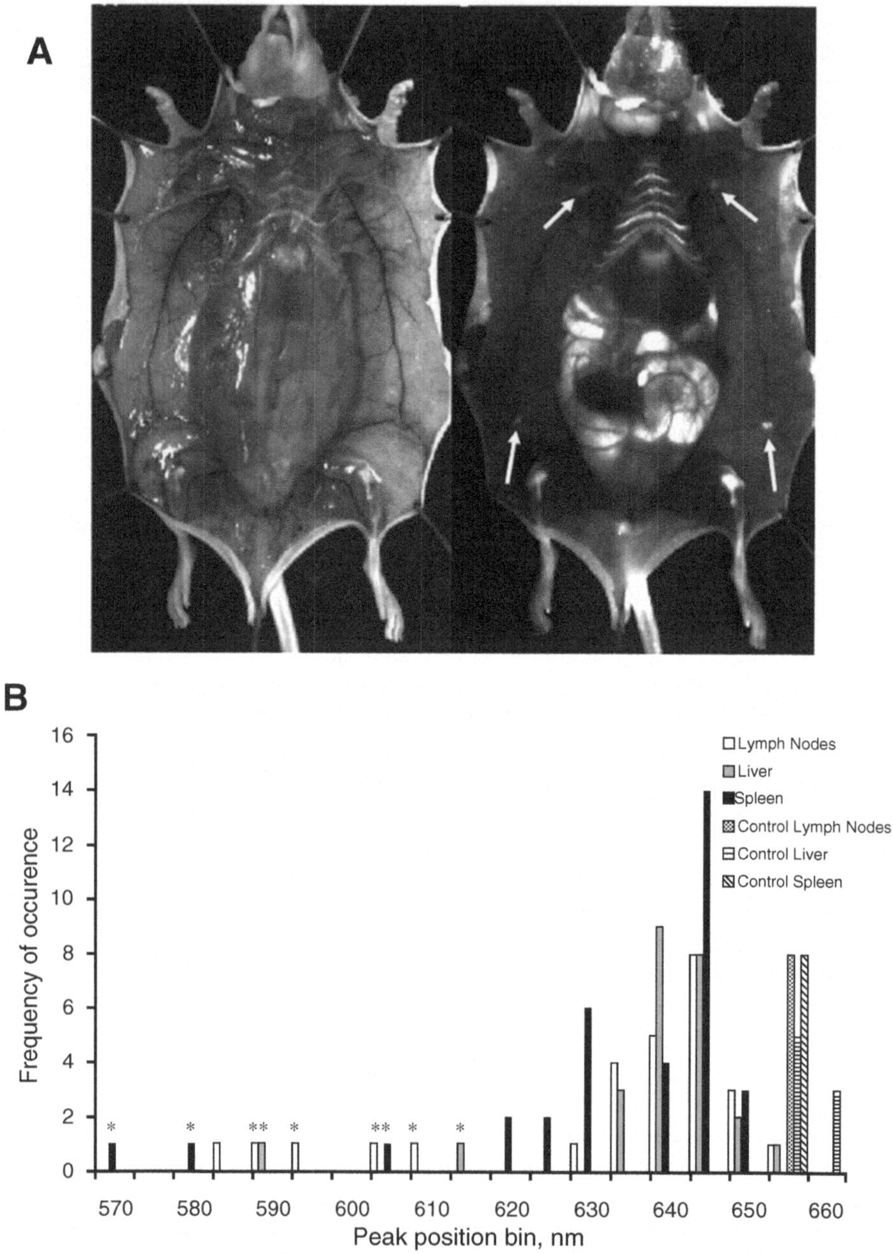

Fig. 6 Two-year fate of CdSe/ZnS Qdots coated with PEG 5000 intravenously injected in mice. (**a**) Mouse visualized 2 years post-injection (visible light on the *left*, 655 nm fluorescence on the *right*), showing signal coming from QDots in the inguinal and axillary lymph nodes, indicated by *white arrows*. Autofluorescence is seen in the intestines and in the cartilage in the kneecaps. (**b**) Comparative histograms of the fluorescence peak positions of lymph nodes, liver, and spleen from freshly injected control mice (24 h post-injection of QDots) and from 2-year mice. The QDot fluorescence in liver, spleen, and lymph nodes 2 years post-injection is significantly blue-shifted, indicating partial degradation of the semiconductor crystals. Adapted from [64] with permission

ICP-MS was used to simultaneously measure the concentrations of cadmium and tellurium in the blood and tissues and the molar ratio of Cd:Te was used as a marker for QDot degradation. Interestingly, Cd and Te displayed different blood kinetics and biodistribution profiles. While the Cd:Te ratio did not vary compared with intact QDs within the first hours in the blood, it decreased gradually over time from the 6-h time point suggesting that the QDots were not chemically stable and partially break down in the blood. Furthermore the biodistribution of Cd and Te was different in organs: Cd accumulated in the liver, kidneys, and spleen, whereas Te was distributed primarily to the kidneys. Cd:Te ratio in various organs differed over time, and more particularly in liver and kidneys [65]. These findings indicated that CdTe/ZnS could degrade differently depending on the organs and that Cd and Te have different fates in vivo. One hypothesis is that metallothionein, a protein inducible by various metallic elements, could form stable complexes with Cd released from QDots and accumulate in the liver and later in kidneys. Indeed the half-life of Cd appeared to be very long in kidneys. This study illustrates how different the fate of NPs can be in vitro and in vivo, as the same QDots were chemically stable in vitro. The distinct distribution of Cd and Te suggests that the modified/degraded QDots (with different Cd:Te ratio as the original QDots), their residues, and the soluble-released ions can have different clearance mechanisms [65]. The mechanisms by which the covalent bonds of QDots dissolved in biological extra- or intracellular environments have not been elucidated and require further investigations. The fates of soluble species in comparison to NP residues remain unknown. This study nevertheless highlights the complex mechanisms of NP processing in the body, leading to time-dependent and organ-dependent oxidation of QDots, dissolution, remediation by endogenous protein, and persistence or clearance [65]. Clearly complementary methods are urgently needed (including electron microscopy) to identify and track the NPs and their by-products long-term and to relate their life cycle to potential toxicological outcomes.

3.2 Skin: The Underemphasized Organ of Nanoparticle's Accumulation

Intravenously administered inorganic nanoparticles generally and expectedly end up within the liver and the spleen. Nevertheless, another organ, which is much more accessible and prone to biopsies, has been recently highlighted. Sykes et al. recently described how ZnS-capped CdSeS QDots and Au NPs could be visually detected and quantified in the skin (Fig. 7) [66]. Administered xenobiotics were then quantified with inductively coupled plasma atomic emission spectroscopy (ICP-AES) measurements, which determined that the dose of administered materials was linearly proportional to the material that accumulated in the skin. In addition, evidence was provided that non-degraded nanoparticles accumulated in dermal macrophages and dendritic cells at lower QDots and Au NP doses (4.4 or 0.07 pmol g/b.w., respectively), and within the pericellular space of the dermis and the subcutaneous tissue at higher doses (80–240 or 0.67–6.67 pmol g/b.w., respectively).

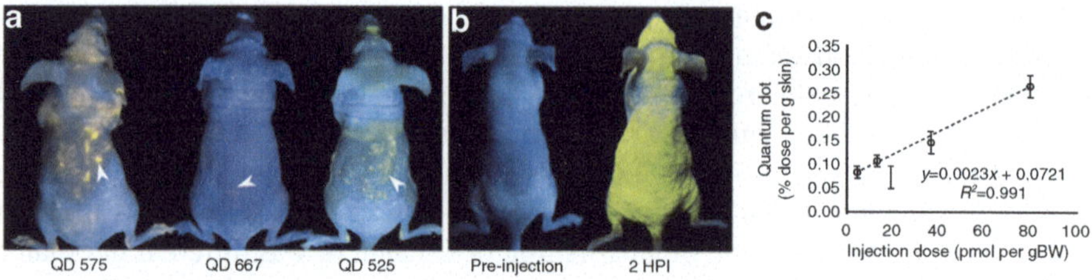

Fig. 7 Accumulation of QDots in mice skin after intravenous injection to mice. (**a**) Mice injected with QDots of different emission wavelengths (575, 667, and 525 nm) simultaneously imaged at 2 HPI under ultraviolet light illumination. *White arrows* point to zones of QDot accumulation. (**b**) An example of a mouse pre- (*left*) and post- (*right*) injection of QDots, where fluorescence covers the entire skin. (**c**) ICP-AES quantification of QDots in the skin is linearly related to injection dose. Error bars denote the standard error of the mean values ($n>3$). Adapted from Sykes et al. with permission [66]

3.3 *In Vivo Toxicity of QDs*

Cadmium-based QDots raise specific toxicity issues given the high toxicity of cadmium ions in humans, as hepatic, renal, neurologic, and genetic toxicities have been reported [67]. The degradation of cadmium-based QDots and the subsequent release of free cadmium ions (Cd^{2+}) may cause their overall toxicity. Cadmium-free QDots are being developed, but their optical properties are not as effective and their toxicity has not been fully addressed [68].

QDots were shown to increase intracellular ROS and induce autophagy. *N*-acetylcysteine, an antioxidant inducer, reduced intracellular ROS and consequently reduced autophagy and enhanced cell death of QDot-treated cells. Independent inhibition of autophagy also reduced cell viability, suggesting that lysosomal biogenesis and induction of autophagy could be a defense mechanism elicited by intracellular accumulation of QDots [69].

As for IONPs and Ag NPs, it remains rather difficult to distinguish the biological effects of intact NPs from that of released ions. Toxicity of QDots correlates with surface oxidation, decrease of size, and disruption of crystal lattice, which also depend on the protective effect of different shell layers and the chemical composition of the QDots [40, 70]. The gradual release and accumulation of Cd^{2+} will likely contribute to toxicity, but in a delayed manner. Indeed, cellular effect would occur only if the concentration of heavy metal ions becomes too high to be processed by cells, which is more probable in nonproliferating cells [71, 72]. Nevertheless, NP-mediated ROS induction could have more rapid effect as reported above. We highlight again that since cultured cells do not metabolize complex chemicals, in vitro tests do not replicate the actual fate of NPs in vivo and studies on animals remain unavoidable.

4 Silver Nanoparticles (Ag NPs)

Silver nanoparticles have been mostly developed for their potent antimicrobial properties. In the past, their use in vivo mainly concerned the treatment of infected wounds or promoting wound healing. Owing to their plasmonic properties, they have been lately suggested as promising platforms for SERS [73] and photoacoustic imaging [74] or as theranostic tools for photothermal or antiviral therapy. The optical properties of Ag NPs make them good candidates for biosensing and imaging in vivo. In addition to the targeted imaging of cancer cells [3], other potential applications of Ag NPs include the identification of pathogens by SERS [75] and the detection of biomarkers in Alzheimer's disease [76]. In addition, as the size and shape of Ag NPs determine their plasmon spectral response, changes in the particle morphology could generate multicolor colloids that could also be used as biological labels [77].

4.1 Therapeutic Properties and In Vivo Behavior of Ag NPs

The antimicrobial and cytotoxic properties of silver nanoparticles have been extensively studied. Nevertheless, the source of these properties still remains unclear due to the complex partitioning of silver species in biological media. The antimicrobial effects of Ag NPs are mostly mediated by Ag^+ ions released by particles. Indeed anaerobic conditions precluding the oxidation of Ag NPs abolish their antibacterial effects [78].

As for IONPs, Ag NPs' fate depends on particle size, shape, and surface chemistry, and their stability can be compromised in biomedical formulations and/or after intracellular processing [79, 80]. While Ag NPs mostly reside in lysosomal compartments, released Ag^+ ions or ultrasmall Ag clusters might translocate out of the lysosomes and eventually transfer from cell to cell, as it happens for IONPs [36].

Silver nanoparticles provide a long-term source for continuous release of Ag^+ ions, delaying the biological outcome until the intracellular Ag^+ concentration becomes high enough and the local environment governs the kinetics of Ag NP degradation. The dissolution rate of Ag NPs was estimated to be 50 times faster in cell lysosomes than in water due to higher ionic strength and acidic pH [81].

Silver nanoparticles might also elicit adverse reactions through interactions of Ag^+ with thiol groups from proteins [82]. Glutathione has a major role in Ag^+ distribution, as Ag^+ interacts with its sulfhydryl groups; thus an organ's ionic silver uptake might be considered proportional to the relative organ glutathione concentration [43]. Empirical evidence obtained from in vitro studies suggests that silver is transformed to silver sulfide species [43]. Thus, a potential scenario has been assumed for Ag NPs fate: (1) Ag NPs dissolve and release soluble silver species, and (2) Ag NPs are directly transformed to silver sulfide particles [83, 84].

Interestingly the nature of the surface coating, as well as the oxygen availability or the NP size or shape, can modulate the NP dissolution and the release of Ag^+. For example, polymers, such as polyvinylpyrrole, offer a better protective shield against degradation than small citrate ligands [85]. Similarly, the dissolution rate of Ag NPs can be reduced by the binding of thiol ligands and the formation of sulfidic coatings on the surface of particles [84].

4.2 *Toxicity of Ag NPs*

The lack of general conclusions regarding the fate and toxicity of Ag NPs mainly comes from the difficulty to assess their cellular or organ uptake (the standard quantification method being ICP-MS), and to distinguish between the effects of Ag NPs and Ag^+ ions released by NPs. This is a general problem for highly reactive and corrosive NPs, as already mentioned for QDots or IONPs.

The differential intracellular distribution of Ag NPs and their ionic residues elicits distinct cell responses, ranging from oxidative stress to binding and reacting with proteins and inducing DNA damages [82]. In addition, high protein concentration in cell culturing media could decrease the in vitro cytotoxicity of nanoparticles and nanosilver's antibacterial activity could be blocked by thiol-containing agents [82]. For example, cell treatment with cysteine, a potent ionic Ag^+ ligand, could also, at least partially, overcome the chromosomal damage in human hepatocarcinoma cells treated with Ag NPs [86].

The acute versus subchronic injury response mostly depends on the dissolution rate of Ag NPs, which should be finely controlled. Overall, the potential of silver NPs for imaging and for therapy resides in a subtle balance between the differently toxic silver species and their partitioning over time as function of their biodistribution. The slow and continued size-dependent degradation of Ag NPs might elicit a delayed toxic effect, as it was the case in a study evaluating Ag NPs' effects after inhalation, where a mild pulmonary fibrosis took place only 3 weeks after administration of Ag NPs [82, 87].

5 Gold Nanoparticles (Au NPs)

Gold salts have been used in the treatment of rheumatoid arthritis [88, 89] for a long time; however recent medical promises involving plasmonic Au NPs include diagnosis with optical imaging and spectroscopy, SERS imaging, spectral CT and photoacoustic imaging, photothermal ablation therapy, and use as "biocompatible" photoactive drug vehicles.

5.1 Role, Effect, and Fate of Au NPs

In contrast to Ag NPs, QDots, and IONPs, Au NPs are generally considered chemically inert and thus not prone to degradation. Although the clearance and excretion profile of Au NPs depend on their size, shape, and surface functionalization, Au NPs display a very low reactivity to their environment, resulting in long-term persistence of intact particles [12, 90]. Persistent NPs, such as Au NPs, thus raise different toxicity concerns from biodegradable NPs. The main concerns are therefore not related to the released ions, but to the accumulation of non-degradable nano-objects that would reside within intracellular compartments and alter basic functions of cell homeostasis.

5.2 Cell Internalization of Au NPs

Although Au^0 is considered relatively inert, corrosion and degradation could occur in the hostile environment of lysosomes. While acidic media has no impact on gold, thiols can bind to the Au surface, and may lead to pulling out of Au atoms via the ligand from the Au surface [91, 92]. Thiols are available in cells, namely in the glutathione. In addition, the oxidation of Au^0 in vivo could be mediated by some amino acids. Thus, in vivo, the Au^0 from NPs could (very slowly) dissolve to soluble Au^+ and Au^{3+} species, especially if the surface ratio is high such as in very small particles or in NPs presenting sharp pikes. Thermodynamic laws might also govern shape transitions of Au NPs, stabilizing different shapes or favoring degradation on preferential faces that are less stable than others. Thus, even for "inert" materials, such as Au, degradation or crystalline reorganization might occur to some extent [38].

5.3 Long-Lasting, But Not "Eternal": The In Vivo Biopersistence of Au NPs

Our group recently investigated the 1-year fate of gold/iron oxide heterodimers injected intravenously in mice. In line with previous studies investigating the fate of gold particles [12], we found coating-dependent accumulation in liver and spleen, with a larger uptake of dimers coated with amphiphilic polymers in comparison to PEG-coated dimers, confirming the impact of the original particle shell [93]. However in comparison to magnetic iron oxide, the gold concentration in liver and spleen decreased only slightly, diminishing by a twofold factor 1 year following the administration. This illustrates the long-term persistence of gold nano-seeds and is in line with previous investigations showing very slow and size-dependent elimination of PEG-coated Au NPs from these organs [94]. This is also consistent with in vivo nanoscale observations proving the long-lasting persistence of the gold core of nanohybrids (NHs) at the same time as iron oxide being degraded (Fig. 8). Hence we clearly observed a material-dependent elimination process, which is related to the differential reactivity of gold and iron oxide to the local environment, and is probably also connected to distinctive biological requirements to degrade/reuse/eliminate the constituents of these two materials.

Nevertheless we clearly observed that gold nanoparticles gradually erode, even if the erosion rate is slow, and gold elimination was correlated with the erosion of the 5 nm gold cores, or their disintegration into smaller particles as observed by TEM in organs (Fig. 8). The kinetics of retention of NHs-derived gold is comparable with a previous report on 4 nm PEG-coated gold particles, which peaked at 7 days post-injection and decreased faster than for larger particles [94]. Moreover, TEM observations of Au NP erosion can account for their long-term elimination, if we assume that very small gold particles or molecular gold can be excreted from the liver through hepatobiliary clearance. It has been shown before that small gold NPs of 10 nm showed the most widespread organ distribution in comparison to larger particles [90], suggesting possible translocation between organs. Semmler-Behnke et al. also reported that the translocation of gold NPs from lungs to secondary target organs was inversely proportional to the NP diameter [95, 96] and that the hepatobiliary clearance from liver at 24 h post-injection strongly increased for 2.8 nm-sized and smaller nanoparticles [97].

5.4 *In Vivo Toxicity of Au NPs*

The acute toxicity of Au NPs is deemed to be very low, and the LD50 of gold clusters (1.9 nm) injected intravenously to mice was as high as 3200 mg of Au/kg b.w. [98].

Nevertheless, there is a growing body of studies suggesting that biopersistent nanomaterials can impact the endo/lysosomal and autophagic pathways [99]. Lysosomal dysfunction could result from nanoparticle biopersistence or inhibition of lysosomal enzymes, such as inhibition of phospholipase resulting in phospholipidosis or inhibition of lysosomal protein degradation resulting in lysosomal overload.

Fig. 8 (continued) day 7 after injection, in some particles, the iron oxide shell has been locally degraded leaving resilient gold core (*white arrows*). Concomitantly, the magnetic properties of the nanohybrids are progressively lost over time in liver and spleen, confirming the dissolution of iron oxide. *Bottom:* Second stage, degradation of gold particles and formation of gold residue assemblies observed on D90 in spleen. (**a**, **b**) Bright-field micrographs, showing intact and degraded heterostructures close to large assemblies of very small gold particles. Some chains of degraded nanohybrids are still apparent. (**c**) STEM-HAADF micrograph of (**b**) showing different contrast of gold residues (*blue* window) and iron-rich ferritin proteins (*red* window). (**d**) STEM-EDX nanoanalysis of the blue window in (**c**) identifies the electron-dense residues as gold particles (**e**) The size distribution of initial gold cores (mean diameter of 5.01 ± 0.86 nm at day 1 in vivo) is shifted to lower sizes (mean diameter of 3.05 ± 0.64 nm) after dissolution of iron oxide at day 90, revealing the gradual erosion of the gold nanocrystals. This study illustrates the different fates of different nanomaterials accumulated together in the same lysosomes and demonstrates non-ambiguously for the first time intralysosomal degradation of iron oxide NPs. Adapted from Kolosnjaj-Tabi et al. [38]

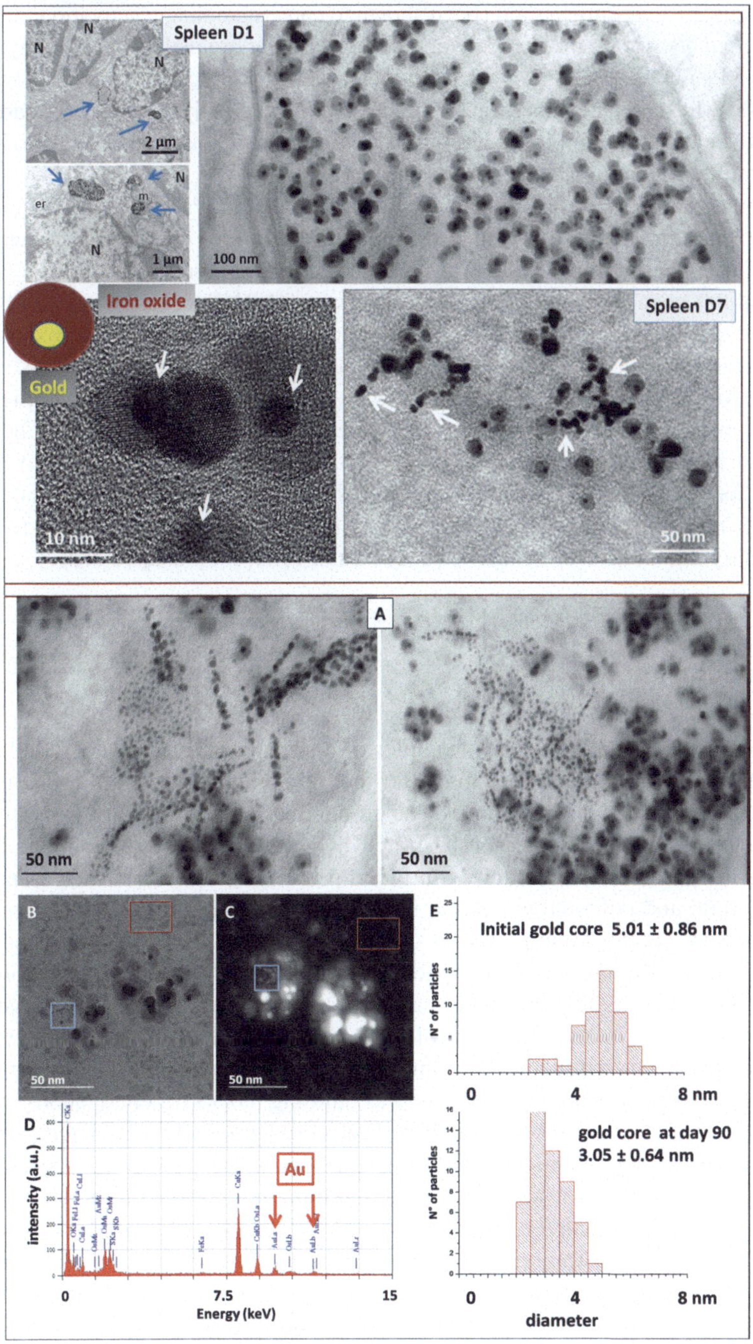

Fig. 8 Example of two-stage local intralysosomal nanodegradation of iron oxide/gold nanohybrids injected intravenously to healthy mice. *Top:* Gold iron/oxide heterostructures are rapidly sequestered in lysosomes of the macrophages in spleen and liver following intravenous injection. The intact heterostructure (gold core and iron oxide shell) is evident at day 1 post-injection. From

Nanoparticles' exposure has also been shown to cause autophagic activation, resulting in increased lysosomal-mediated degradation of cellular organelles. Autophagy induction by NPs may be an attempt to degrade what the cell perceives as foreign or aberrant [100]. In turn, accumulation of biopersistent NPs in lysosomal compartments may induce enzyme inhibition and blockade of autophagy flux, due to NP overload and disruption of autophagosome–lysosome fusion [101].

There are several aspects in the toxicity risks of gold NPs, which are also related to their coating on one part, and shape and size effect on the other part [33]. Cetyltrimethylammonium (CTAB) has been commonly used for synthesis of gold NPs, but has been shown to be the major source of Au NP toxicity [102]. In CTAB-free Au NPs or Au NPs, passivated by polymer shell, the suggested toxicity was highly size dependent. For example, Pan et al. [103] found that ultrasmall 1.2 and 1.4 nm Au NPs elicited apoptosis and necrosis, whereas smaller gold compounds (Tauredon), or larger NPs of 15 nm, were comparatively nontoxic. Chen and coworkers [104] reported in vivo toxicity with spherical Au NPs from 8 to 37 nm, but not for smaller (3 and 5 nm) or larger NPs (50 and 100 nm). The effect of shape (spheres, rod, stars, etc.) has also been reported, but remains rather difficult to assess thoroughly due to diverse synthetic processes of differently shaped NPs.

6 Conclusion

This chapter describes the basic principles of distribution, toxicity, degradation and elimination, or recycling, applicable after the in vivo administration of prospective inorganic nanoparticles for imaging. When nanoparticles enter the organism, they undergo a variety of changes orchestrated by the biological environment, which dictates the particles' interaction with blood proteins, cell internalization, intracellular sorting, and potential expulsion from the cell. After administration, the core-shell (or core-coating) complexes or composite materials in nanohybrids separate and undertake different processing pathways that depend on the chemical composition of the material. The physical characteristics (such as magnetic or plasmonic properties) of nanoparticle cores evolve. In due course a degradation takes place with nanoparticle constituents (the ionic compounds or small clusters of the xenobiotic material) either being recycled (used as building blocks for endogenous components, such as the hemoglobin in red blood cells) and gradually eliminated or merely excreted from the organism after enduring reactions with intracellular biomolecules. While the mechanism of degradation unconditionally depends on the composition of the material, we should be aware of the biological processes that will follow after nanoparticles' administration. On the one hand, a sudden and massive degradation might represent a heavy burden to the organism, which might result in a toxic outcome. On the other hand, a persistent load of poorly reactive nanoparticles might physically hinder and/or alter the physiological processes within the cells, and affect homeostasis. Specific approaches and studies are therefore necessary to strategically design particles in order to harness their biodistribution and persistence, as well as warrant a safe translation into the clinic.

References

1. Wang Y-XJ, Hussain SM, Krestin GP. Superparamagnetic iron oxide contrast agents: physicochemical characteristics and applications in MR imaging. Eur Radiol. 2001;11:2319–31.
2. Gao X, Dave SR. Quantum dots for cancer molecular imaging. Bio-applications of nanoparticles. New York: Springer; 2007. p. 57–73.
3. Tai S-P, Wu Y, Shieh D-B, Chen L-J, Lin K-J, Yu C-H, et al. Molecular imaging of cancer cells using plasmon-resonant-enhanced third-harmonic-generation in silver nanoparticles. Adv Mat. 2007;19:4520–3.
4. Jain PK, Lee KS, El-Sayed IH, El-Sayed MA. Calculated absorption and scattering properties of gold nanoparticles of different size, shape, and composition: applications in biological imaging and biomedicine. J Phys Chem B. 2006;110:7238–48.
5. Huang H-C, Barua S, Sharma G, Dey SK, Rege K. Inorganic nanoparticles for cancer imaging and therapy. J Control Release. 2011;155(3):344–57.
6. Andreou C, Kishore SA, Kircher MF. Surface-enhanced Raman spectroscopy: a new modality for cancer imaging. J Nucl Med. 2015;56:1295–9.
7. Na HB, Song IC, Hyeon T. Inorganic nanoparticles for MRI contrast agents. Adv Mat. 2009;21:2133–48.
8. Bulte JW, Kraitchman DL. Iron oxide MR contrast agents for molecular and cellular imaging. NMR Biomed. 2004;17:484–99.
9. Gao X, Nie S. Molecular profiling of single cells and tissue specimens with quantum dots. Trends Biotechnol. 2003;21:371–3.
10. Monopoli MP, Aberg C, Salvati A, Dawson KA. Biomolecular coronas provide the biological identity of nanosized materials. Nat Nanotechnol. 2012;7:779–86.
11. Teeguarden JG, Hinderliter PM, Orr G, Thrall BD, Pounds JG. Particokinetics in vitro: dosimetry considerations for in vitro nanoparticle toxicity assessments. Toxicol Sci. 2007;95:300–12.
12. Khlebtsov N, Dykman L. Biodistribution and toxicity of engineered gold nanoparticles: a review of in vitro and in vivo studies. Chem Soc Rev. 2011;40:1647–71.
13. Arami H, Khandhar A, Liggitt D, Krishnan KM. In vivo delivery, pharmacokinetics, biodistribution and toxicity of iron oxide nanoparticles. Chem Soc Rev. 2015;44:8576–607.
14. Albanese A, Walkey CD, Olsen JB, Guo H, Emili A, Chan WC. Secreted biomolecules alter the biological identity and cellular interactions of nanoparticles. ACS Nano. 2014;8:5515–26.
15. Alexis F, Pridgen E, Molnar LK, Farokhzad OC. Factors affecting the clearance and biodistribution of polymeric nanoparticles. Mol Pharm. 2008;5:505–15.
16. Ishihara T, Takeda M, Sakamoto H, Kimoto A, Kobayashi C, Takasaki N, et al. Accelerated blood clearance phenomenon upon repeated injection of PEG-modified PLA-nanoparticles. Pharm Res. 2009;26:2270–9.
17. Lila ASA, Kiwada H, Ishida T. The accelerated blood clearance (ABC) phenomenon: clinical challenge and approaches to manage. J Control Release. 2013;172(1):38–47.
18. Jones SW, Roberts RA, Robbins GR, Perry JL, Kai MP, Chen K, et al. Nanoparticle clearance is governed by Th1/Th2 immunity and strain background. J Clin Invest. 2013;123:3061.
19. Beduneau A, Ma Z, Grotepas CB, Kabanov A, Rabinow BE, Gong N, et al. Facilitated monocyte-macrophage uptake and tissue distribution of superparmagnetic iron-oxide nanoparticles. PLoS One. 2009;4:e4343.
20. Lundqvist M, Stigler J, Elia G, Lynch I, Cedervall T, Dawson KA. Nanoparticle size and surface properties determine the protein corona with possible implications for biological impacts. Proc Natl Acad Sci U S A. 2008;105:14265–70.
21. Salvati A, Pitek AS, Monopoli MP, Prapainop K, Bombelli FB, Hristov DR, et al. Transferrin-functionalized nanoparticles lose their targeting capabilities when a biomolecule corona adsorbs on the surface. Nat Nanotechnol. 2013;8:137–43.
22. Zensi A, Begley D, Pontikis C, Legros C, Mihoreanu L, Büchel C, et al. Human serum albumin nanoparticles modified with apolipoprotein AI cross the blood-brain barrier and enter the rodent brain. J Drug Target. 2010;18:842–8.

23. Weissleder RA, Stark D, Engelstad B, Bacon B, Compton C, White D, et al. Superparamagnetic iron oxide: pharmacokinetics and toxicity. Am J Roentgenol. 1989;152:167–73.
24. Pouliquen D, Le Jeune J, Perdrisot R, Ermias A, Jallet P. Iron oxide nanoparticles for use as an MRI contrast agent: pharmacokinetics and metabolism. Magn Reson Imaging. 1991;9(3):275–83.
25. Freund B, Tromsdorf UI, Bruns OT, Heine M, Giemsa A, Bartelt A, et al. A simple and widely applicable method to 59Fe-radiolabel monodisperse superparamagnetic iron oxide nanoparticles for in vivo quantification studies. ACS Nano. 2012;6:7318–25.
26. Bargheer D, Nielsen J, Gébel G, Heine M, Salmen SC, Stauber R, et al. The fate of a designed protein corona on nanoparticles in vitro and in vivo. Beilstein J Nanotechnol. 2015;6:36–46.
27. Thomas R, Park I-K, Jeong Y. Magnetic iron oxide nanoparticles for multimodal imaging and therapy of cancer. Inter J Mol Sci. 2013;14:15910.
28. Wang H, Kumar R, Nagesha D, Duclos RI, Sridhar S, Gatley SJ. Integrity of 111 In-radiolabeled superparamagnetic iron oxide nanoparticles in the mouse. Nucl Med Biol. 2015;42:65–70.
29. Kreyling WG, Abdelmonem AM, Ali Z, Alves F, Geiser M, Haberl N, et al. In vivo integrity of polymer-coated gold nanoparticles. Nat Nanotechnol. 2015;10:619–23.
30. Sée V, Free P, Cesbron Y, Nativo P, Shaheen U, Rigden DJ, et al. Cathepsin L digestion of nanobioconjugates upon endocytosis. ACS Nano. 2009;3:2461–8.
31. Walkey CD, Chan WCW. Understanding and controlling the interaction of nanomaterials with proteins in a physiological environment. Chem Soc Rev. 2012;41:2780–99.
32. Wang F, Yu L, Monopoli MP, Sandin P, Mahon E, Salvati A, et al. The biomolecular corona is retained during nanoparticle uptake and protects the cells from the damage induced by cationic nanoparticles until degraded in the lysosomes. Nanomedicine. 2013;9:1159–68.
33. Soenen SJ, Parak WJ, Rejman J, Manshian B. (Intra)Cellular stability of inorganic nanoparticles: effects on cytotoxicity, particle functionality, and biomedical applications. Chem Rev. 2015;115:2109–35.
34. Lévy M, Wilhelm C, Devaud M, Levitz P, Gazeau F. How cellular processing of superparamagnetic nanoparticles affects their magnetic behavior and NMR relaxivity. Contrast Media Mol Imaging. 2012;7:373–83.
35. Lartigue L, Alloyeau D, Kolosnjaj-Tabi J, Javed Y, Guardia P, Riedinger A, et al. Biodegradation of iron oxide nanocubes: high-resolution in situ monitoring. ACS Nano. 2013;7:3939–52.
36. Silva A, Wilhelm C, Kolosnjaj-Tabi J, Luciani N, Gazeau F. Cellular transfer of magnetic nanoparticles via cell microvesicles: impact on cell tracking by magnetic resonance imaging. Pharm Res. 2012;29:1392–403.
37. Faraj AA, Luciani N, Kolosnjaj-Tabi J, Mattar E, Clement O, Wilhelm C, et al. Real-time high-resolution magnetic resonance tracking of macrophage subpopulations in a murine inflammation model: a pilot study with a commercially available cryogenic probe. Contrast Media Mol Imaging. 2013;8:193–203.
38. Kolosnjaj-Tabi J, Javed Y, Lartigue L, Volatron J, Elgrabli D, Marangon I, et al. The one year fate of iron oxide coated gold nanoparticles in mice. ACS Nano. 2015;9:7925–39.
39. Hardman R. A toxicologic review of quantum dots: toxicity depends on physicochemical and environmental factors. Environ Health Perspect. 2006:165–72.
40. Derfus AM, Chan WC, Bhatia SN. Probing the cytotoxicity of semiconductor quantum dots. Nano Lett. 2004;4:11–8.
41. Halliwell B. Reactive oxygen species in living systems: source, biochemistry, and role in human disease. Am J Med. 1991;91:S14–22.
42. Nel A, Xia T, Mädler L, Li N. Toxic potential of materials at the nanolevel. Science. 2006;311:622–7.
43. Bachler G, von Goetz N, Hungerbühler K. A physiologically based pharmacokinetic model for ionic silver and silver nanoparticles. Int J Nanomed. 2013;8:3365.
44. Lison D, Vietti G, van den Brule S. Paracelsus in nanotoxicology. Particle Fibre Toxicol. 2014;11:35.

45. Levy M, Luciani N, Alloyeau D, Elgrabli D, Deveaux V, Pechoux C, et al. Long term in vivo biotransformation of iron oxide nanoparticles. Biomaterials. 2011;32:3988–99.
46. Goodman AM, Cao Y, Urban C, Neumann O, Ayala-Orozco C, Knight MW, et al. The surprising in vivo instability of near-IR-absorbing hollow Au–Ag nanoshells. ACS Nano. 2014;8:3222–31.
47. Kolosnjaj-Tabi J, Wilhelm C, Clément O, Gazeau F. Cell labeling with magnetic nanoparticles: opportunity for magnetic cell imaging and cell manipulation. J Nanobiotechnol. 2013;11:S7.
48. Béalle G, Di Corato R, Kolosnjaj-Tabi J, Dupuis V, Clément O, Gazeau F, et al. Ultra magnetic liposomes for MR imaging, targeting, and hyperthermia. Langmuir. 2012;28:11834–42.
49. Kolosnjaj-Tabi J, Di Corato R, Lartigue L, Marangon I, Guardia P, Silva AK, et al. Heat-generating iron oxide nanocubes: subtle "destructurators" of the tumoral microenvironment. ACS Nano. 2014;8:4268–83.
50. Chen R, Romero G, Christiansen MG, Mohr A, Anikeeva P. Wireless magnetothermal deep brain stimulation. Science. 2015;347:1477–80.
51. Lartigue L, Wilhelm C, Servais J, Factor C, Dencausse A, Bacri J-C, et al. Nanomagnetic sensing of blood plasma protein interactions with iron oxide nanoparticles: impact on macrophage uptake. ACS Nano. 2012;6:2665–78.
52. Levy M, Wilhelm C, Luciani N, Devaux V, Gendron F, Luciani A, et al. Nanomagnetism reveals the intracellular clustering of nanoparticles in the organism. Nanoscale. 2011;3:4402–10.
53. Bulte JW, Kraitchman DL. Monitoring cell therapy using iron oxide MR contrast agents. Curr Pharm Biotechnol. 2004;5:567–84.
54. Modo M, Kolosnjaj-Tabi J, Nicholls F, Ling W, Wilhelm C, Debarge O, et al. Considerations for the clinical use of contrast agents for cellular MRI in regenerative medicine. Contrast Media Mol Imaging. 2013;8:439–55.
55. Levy M, Gazeau F, Bacri JC, Wilhelm C, Devaud M. Modeling magnetic nanoparticle dipole-dipole interactions inside living cells. Phys Rev B. 2011;84:075480.
56. Beaumont C, Delaby C. Recycling iron in normal and pathological states. Semin Hematol. 2009;46:328–38.
57. Wagner S, SCHNORR J, Pilgrimm H, Hamm B, Taupitz M. Monomer-coated very small superparamagnetic iron oxide particles as contrast medium for magnetic resonance imaging: preclinical in vivo characterization. Investig Radiol. 2002;37:167–77.
58. Lewis RJ, Irving N. Sax's dangerous properties of industrial materials. Van Nostrand Reinhold. 2003 Sax's Dangerous Properties of Industrial Materials, 5 Volume Set, 12th Edition.
59. Volkovova K, Handy RD, Staruchova M, Tulinska J, Kebis A, Pribojova J, et al. Health effects of selected nanoparticles in vivo: liver function and hepatotoxicity following intravenous injection of titanium dioxide and Na-oleate-coated iron oxide nanoparticles in rodents. Nanotoxicology. 2015;9:95–105.
60. Jain TK, Reddy MK, Morales MA, Leslie-Pelecky DL, Labhasetwar V. Biodistribution, clearance, and biocompatibility of iron oxide magnetic nanoparticles in rats. Mol Pharm. 2008;5:316–27.
61. Neibert KD, Maysinger D. Mechanisms of cellular adaptation to quantum dots—the role of glutathione and transcription factor EB. Nanotoxicology. 2012;6(3):249–62.
62. Fischer HC, Liu L, Pang KS, Chan WCW. Pharmacokinetics of nanoscale quantum dots: in vivo distribution, sequestration, and clearance in the rat. Adv Funct Mat. 2006;16:1299–305.
63. Ballou B, Lagerholm BC, Ernst LA, Bruchez MP, Waggoner AS. Noninvasive imaging of quantum dots in mice. Bioconj Chem. 2004;15(1):79–86.
64. Fitzpatrick JAJ, Andreko SK, Ernst LA, Waggoner AS, Ballou B, Bruchez MP. Long-term persistence and spectral blue shifting of quantum dots in vivo. Nano Lett. 2009;9(7):2736–41.
65. Liu N, Mu Y, Chen Y, Sun H, Han S, Wang M, et al. Degradation of aqueous synthesized CdTe/ZnS quantum dots in mice: differential blood kinetics and biodistribution of cadmium and tellurium. Particle Fibre Toxicol. 2013;10:37.

66. Sykes EA, Dai Q, Tsoi KM, Hwang DM, Chan WC. Nanoparticle exposure in animals can be visualized in the skin and analysed via skin biopsy. Nat Comm. 2014;5.
67. Bernhoft RA. Cadmium toxicity and treatment. Sci World J. 2013. doi:10.1155/2013/394652.
68. Soenen SJ, Manshian BB, Aubert T, Himmelreich U, Demeester J, De Smedt SC, et al. Cytotoxicity of cadmium-free quantum dots and their use in cell bioimaging. Chem Res Toxicol. 2014;27:1050–9.
69. Luo YH, Wu SB, Wei YH, Chen YC, Tsai MH, Ho CC, et al. Cadmium-based quantum dot induced autophagy formation for cell survival via oxidative stress. Chem Res Toxicol. 2013;26:662–73.
70. Zhu Z-J, Yeh Y-C, Tang R, Yan B, Tamayo J, Vachet RW, et al. Stability of quantum dots in live cells. Nat Chem. 2011;3:963–8.
71. Cho SJ, Maysinger D, Jain M, Röder B, Hackbarth S, Winnik FM. Long-term exposure to CdTe quantum dots causes functional impairments in live cells. Langmuir. 2007;23:1974–80.
72. Soenen SJ, Demeester J, De Smedt SC, Braeckmans K. The cytotoxic effects of polymer-coated quantum dots and restrictions for live cell applications. Biomaterials. 2012;33:4882–8.
73. Chen Y, Chen G, Feng S, Pan J, Zheng X, Su Y, et al. Label-free serum ribonucleic acid analysis for colorectal cancer detection by surface-enhanced Raman spectroscopy and multivariate analysis. J Biomed Opt. 2012;17:0670031–7.
74. Homan KA, Souza M, Truby R, Luke GP, Green C, Vreeland E, et al. Silver nanoplate contrast agents for in vivo molecular photoacoustic imaging. ACS Nano. 2012;6(1):641–50.
75. Wang Y, Lee K, Irudayaraj J. Silver nanosphere SERS probes for sensitive identification of pathogens. J Phys Chem C. 2010;114:16122–8.
76. Haes AJ, Hall WP, Chang L, Klein WL, Van Duyne RP. A localized surface plasmon resonance biosensor: first steps toward an assay for Alzheimer's disease. Nano Lett. 2004;4:1029–34.
77. Mock J, Barbic M, Smith D, Schultz D, Schultz S. Shape effects in plasmon resonance of individual colloidal silver nanoparticles. J Chem Phys. 2002;116:6755–9.
78. Xiu Z-M, Zhang Q-B, Puppala HL, Colvin VL, Alvarez PJJ. Negligible particle-specific antibacterial activity of silver nanoparticles. Nano Lett. 2012;12:4271–5.
79. Stebounova LV, Guio E, Grassian VH. Silver nanoparticles in simulated biological media: a study of aggregation, sedimentation, and dissolution. J Nanopart Res. 2011;13:233–44.
80. Skebo JE, Grabinski CM, Schrand AM, Schlager JJ, Hussain SM. Assessment of metal nanoparticle agglomeration, uptake, and interaction using high-illuminating system. Int J Toxicol. 2007;26:135–41.
81. Singh RP, Ramarao P. Cellular uptake, intracellular trafficking and cytotoxicity of silver nanoparticles. Toxicol Lett. 2012;213:249–59.
82. Chen X, Schluesener H. Nanosilver: a nanoproduct in medical application. Toxicol Lett. 2008;176:1–12.
83. Liu J, Pennell KG, Hurt RH. Kinetics and mechanisms of nanosilver oxysulfidation. Environ Sci Technol. 2011;45:7345–53.
84. Liu J, Sonshine DA, Shervani S, Hurt RH. Controlled release of biologically active silver from nanosilver surfaces. ACS Nano. 2010;4:6903–13.
85. Tejamaya M, Römer I, Merrifield RC, Lead JR. Stability of citrate, PVP, and PEG coated silver nanoparticles in ecotoxicology media. Environ Sci Technol. 2012;46(13):7011–7.
86. Kawata K, Osawa M, Okabe S. In vitro toxicity of silver nanoparticles at noncytotoxic doses to HepG2 human hepatoma cells. Environ Sci Technol. 2009;43(15):6046–51.
87. Wang X, Ji Z, Chang CH, Zhang H, Wang M, Liao Y-P, et al. Use of coated silver nanoparticles to understand the relationship of particle dissolution and bioavailability to cell and lung toxicological potential. Small. 2014;10:385–98.
88. Thakor A, Jokerst J, Zavaleta C, Massoud T, Gambhir S. Gold nanoparticles: a revival in precious metal administration to patients. Nano Lett. 2011;11:4029–36.
89. De Wall SL, Painter C, Stone JD, Bandaranayake R, Wiley DC, Mitchison TJ, et al. Noble metals strip peptides from class II MHC proteins. Nat Chem Biol. 2006;2:197–201.
90. De Jong WH, Hagens WI, Krystek P, Burger MC, Sips AJAM, Geertsma RE. Particle size-dependent organ distribution of gold nanoparticles after intravenous administration. Biomaterials. 2008;29:1912–9.

91. Paulsson M, Krag C, Frederiksen T, Brandbyge M. Conductance of alkanedithiol single-molecule junctions: a molecular dynamics study. Nano Lett. 2008;9:117–21.
92. Krüger D, Rousseau R, Fuchs H, Marx D. Towards "mechanochemistry": mechanically induced isomerizations of thiolate–gold clusters. Angew Chem Int Ed. 2003;42:2251–3.
93. Lipka J, Semmler-Behnke M, Sperling RA, Wenk A, Takenaka S, Schleh C, et al. Biodistribution of PEG-modified gold nanoparticles following intratracheal instillation and intravenous injection. Biomaterials. 2010;31:6574–81.
94. Cho W-S, Cho M, Jeong J, Choi M, Han BS, Shin H-S, et al. Size-dependent tissue kinetics of PEG-coated gold nanoparticles. Toxicol App Pharmacol. 2010;245:116–23.
95. Semmler-Behnke M, Kreyling WG, Lipka J, Fertsch S, Wenk A, Takenaka S, et al. Biodistribution of 1.4- and 18-nm gold particles in rats. Small. 2008;4:2108–11.
96. Kreyling WG, Hirn S, Möller W, Schleh C, Wenk A, Celik G, et al. Air–blood barrier translocation of tracheally instilled gold nanoparticles inversely depends on particle size. ACS Nano. 2013;8:222–33.
97. Hirn S, Semmler-Behnke M, Schleh C, Wenk A, Lipka J, Schäffler M, et al. Particle size-dependent and surface charge-dependent biodistribution of gold nanoparticles after intravenous administration. Eur J Pharm Biopharm. 2011;77:407–16.
98. Hainfeld J, Slatkin D, Focella T, Smilowitz H. Gold nanoparticles: a new X-ray contrast agent. Br J Radiol. 2006;79:248–53.
99. Stern S, Adiseshaiah P, Crist R. Autophagy and lysosomal dysfunction as emerging mechanisms of nanomaterial toxicity. Particle Fibre Toxicol. 2012;9:20.
100. Song W, Soo Lee S, Savini M, Popp L, Colvin VL, Segatori L. Ceria nanoparticles stabilized by organic surface coatings activate the lysosome-autophagy system and enhance autophagic clearance. ACS Nano. 2014;8:10328–42.
101. Ma X, Wu Y, Jin S, Tian Y, Zhang X, Zhao Y, et al. Gold nanoparticles induce autophagosome accumulation through size-dependent nanoparticle uptake and lysosome impairment. ACS Nano. 2011;5:8629–39.
102. Qiu Y, Liu Y, Wang L, Xu L, Bai R, Ji Y, et al. Surface chemistry and aspect ratio mediated cellular uptake of Au nanorods. Biomaterials. 2010;31:7606–19.
103. Pan Y, Neuss S, Leifert A, Fischler M, Wen F, Simon U, et al. Size-dependent cytotoxicity of gold nanoparticles. Small. 2007;3:1941–9.
104. Chen Y-S, Hung Y-C, Liau I, Huang GS. Assessment of the in vivo toxicity of gold nanoparticles. Nanoscale Res Lett. 2009;4:858–64.

Imaging and Therapeutic Potential of Extracellular Vesicles

Max Piffoux, Florence Gazeau, Claire Wilhelm, and Amanda K.A. Silva

1 Introduction

Extracellular vesicles (EVs) are membrane-delimitated subcellular entities released by cells in a constitutive manner or in response to stress [1–3]. EVs contain membrane proteins and lipids, as well as cytoplasm components in a pattern which depends on the type of stimulation and physiopathology of parental cells [4, 5]. EVs constitute a far-reaching intercellular communication pathway controlling cell signaling, a unique feature that stems from their extraordinary ability to transfer material between neighbor and distal cells. Once released, EVs can be internalized by neighbor cells or circulate to attain distal cells mediating an information transfer that may be either homotypic or heterotypic [6, 7] (Fig. 1).

Formerly considered as cellular dust [8], EVs are now recognized as active players participating in homeostasis and disease [9]. Indeed, EVs act as biological effectors in physiologic processes, such as coagulation [10], immune response [11], pregnancy [12], as well as pathologic ones related to infection [13] and cancer [14, 15]. If EVs play a pivotal role in the regulation of physiological and pathological processes, it is because they are competent to mediate intercellular communication via the transfer of proteins and genetic information to recipient cells.

In addition to the intrinsic properties, EVs may be engineered to display exogenous imaging and therapeutic properties. The possibility of customizing EVs with exogenous imaging tracers and therapeutic drug/nanoparticles has opened up a wide range of exciting perspectives. On the one side, there is an enormous need to decipher the complex fate of EVs and decode their trafficking in the organism. The design of EV displaying imaging tracers will assist in the understanding of their biodistribution and interplay with recipient

M. Piffoux • F. Gazeau • C. Wilhelm (✉) • A.K.A. Silva (✉)
Laboratoire Matière et Systèmes Complexes, UMR 7057, CNRS and Université Paris Diderot, 10 rue Alice Domon et Léonie Duquet, Paris, Cedex 13 75205, France
e-mail: claire.wilhelm@univ-paris-diderot.fr; amanda.silva@univ-paris-diderot.fr

J.W.M. Bulte, M.M.J. Modo (eds.), *Design and Applications of Nanoparticles in Biomedical Imaging*, DOI 10.1007/978-3-319-42169-8_3

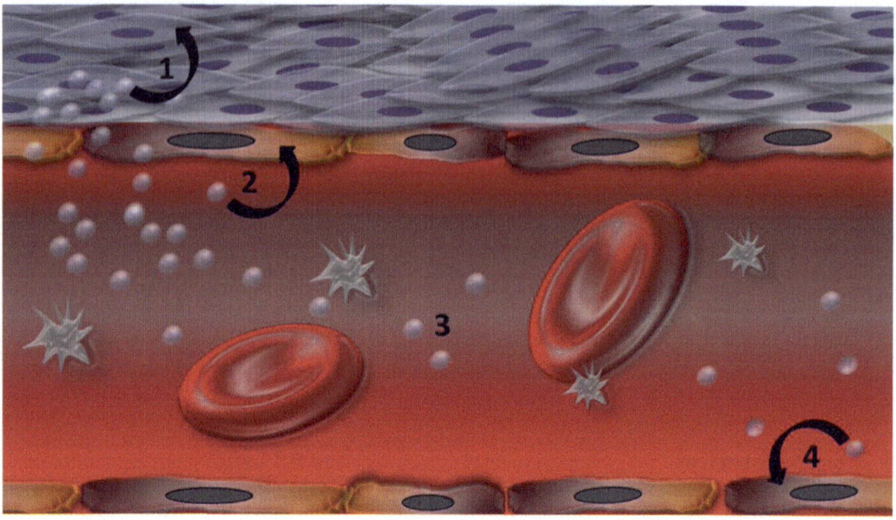

Fig. 1 Extracellular vesicles can interact with neighbor cells delivering their cargo in a homotypic cell transfer (1) or in a heterotypic cell transfer (2). Vesicles may also circulate (3) and be uptaken by distal cells (4)

cells, shedding light on their role in mechanisms related both to homeostasis and disease. On the other side, EV loading with drugs, nucleic acids, or heating nanoparticles represents a unique opportunity to translate EVs into intrinsically biocompatible bio-inspired therapeutic delivery systems. Indeed, EVs feature advantageous attributes of delivery vehicles. First of all, they represent the most physiological carrier choice considering the natural role of EVs as conveyors of biological cargoes and mediators of information transfer. Indeed, EVs are constituted of natural components, whose size and flexibility enable them to travel across biological membranes [16]. Additionally, the ability of EVs to shelter their internal cargo, such as proteins and genomic material, from the harsh extracellular environment space also makes them promising carriers.

Herein, EV engineering for imaging or therapy purposes will be overviewed. First, the biogenesis of EVs is outlined. Production, loading, isolation, and characterization methods are presented and discussed. EV biodistribution studies based on vesicle engineering with imaging agents are commented. The engineering of EVs for designing bio-camouflaged delivery systems is equally highlighted. Our concluding remarks summarize current challenges in the perspective of clinical translation.

2 EV Classification and Biogenesis

The classification of EVs remains a matter of debate. Currently, the proposed classification takes into account their size, content, and mainly their biological origin. Herein, we discuss EV classification into exosomes, microvesicles, and apoptotic bodies.

The term exosomes was coined by Johnstone's group which observed multivesicular bodies (late endosomes) releasing their inner vesicles into the extracellular medium [17, 18]. As the process was in the opposite flow of endocytosis, the term exosomes was used. Exosomes can be considered as an endocytosis end product. The endocytotic pathway is overviewed in Fig. 2. First of all, endocytic vesicles are formed at the plasma membrane and fuse with early endosomes. Their content may then undergo recycling, degradation, or exocytosis. When their content should be recycled, early endosomes regress to the plasma membrane (recycling endosomes). Otherwise, early endosomes follow a maturation step to become late endosomes featuring vesicles (30–100 nm) that bud into the lumen. Such late endosomes or multivesicular bodies (MVBs) may further fuse with lysosomes if the fate of their content is degradation. Alternatively, if their content should be exported, late endosomes fuse with the plasma membrane and excrete their intraluminal vesicles into the extracellular space. Such released vesicles are exosomes [19, 20]. Exosomes originated from different cell types share common groups of proteins: (1) proteins related to antigen binding and

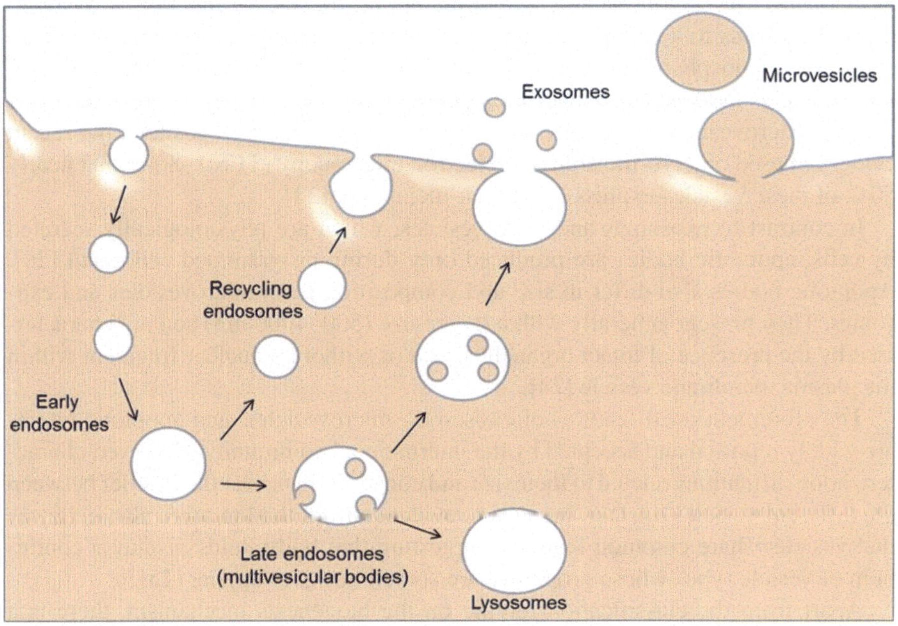

Fig. 2 Exosome and microvesicle formation. Exosomes are a subproduct of the endocytotic pathway originating from endosomes. Endosomes may be classified into early endosomes, late endosomes, and recycling endosomes. When their content is destined for recycling, it is sorted to the plasma membrane via recycling endosomes. When their content is destined for degradation or exocytosis, early endosomes then undergo a series of transformations, namely the formation of 30–100 nm vesicles that bud into the lumen of late endosomes. For this reason, these late endosomes are also known as multivesicular bodies (MVBs). The late endosomes are then destined to fuse with either lysosomes (content degradation) or the plasma membrane (content secretion). The vesicles released into the extracellular space are exosomes. In contrast, microvesicles originate from the direct outward budding and fission of the plasma membrane

presentation such as MHC class I and II proteins; (2) proteins involved in membrane fusion as well as transport such as annexins and rab proteins; (3) proteins involved in cell adhesion such as integrin proteins; (4) proteins from the cytoskeleton such as tubulin and actin; (5) metabolic enzymes such as peroxidases, pyruvate, and lipid kinases; and mainly (6) tetraspanins including CD9, CD63, CD81, and CD82 [21]. Current literature suggests that endosomes are enriched in markers such as CD63 and CD9. However, these biomarkers do not fully define exosomes [20].

Microvesicles display a biogenesis mechanism quite distinct from exosomes. In opposition to the biogenesis of exosomes, microvesicles originate from the direct outward budding and fission of the plasma membrane. Microvesicles are larger in size (50–1000 nm) when compared to exosomes. However, small microvesicles and large exosomes overlap in size [20]. The biogenesis of microvesicles is an outcome of phospholipid redistribution and cytoskeletal protein contraction. In the plasma membrane, there is a natural phospholipid asymmetric distribution that is tightly regulated by aminophospholipid translocases that transfer phospholipids from one leaflet of the plasma membrane to the other [22]. When there is a significant increase of cytosolic Ca^{2+} accompanying cell stimulation, translocase function is dysregulated. This leads to a collapse of the membrane asymmetry culminating in surface exposure of phosphatidylserine (PS). Such a process is followed by the release of microvesicles induced by cytoskeleton degradation via Ca^{2+}-dependent proteolysis [2, 20]. Microvesicles were previously described as annexin V positive, but recent evidence based on cryo-transmission electron microscopy (TEM) shows that nearly 50 % of these vesicles express PS at their membrane [23].

In contrast to exosomes and microvesicles, which are physiologically secreted by cells, apoptotic bodies are produced only during programmed cell death [20]. Apoptotic bodies also differ in size and composition from microvesicles and exosomes. They present generally with a larger size (500–4000 nm) and are characterized by the presence of intact organelles with or without a nuclear fragment within the plasma membrane vesicle [24].

The aforementioned features of exosomes, microvesicles, and apoptotic bodies are widely reported and accepted by the international community. However, characterization difficulties related to their size and constitution render the frontier between the different vesicles unclear. Increasing evidence from the literature also indicates that vesicles share common features suggesting that body fluids display a continuum of vesicle types whose properties are sometimes overlapping [25].

Apart from the classification relying on the biogenesis mechanism, there is a parallel classification based on the source of isolation. Indeed, EVs have been isolated from diverse body fluids, including semen [26], blood [27], urine [28], saliva [29], breast milk [30], amniotic fluid [31], ascites fluid [32], and bile [33], just to name a few. In this way, the terms epididimosomes (vesicles from epididymal fluid), prostasomes (vesicles from seminal fluid) [26], matrix vesicles (vesicles in bone, cartilage, and atherosclerotic plaques) [34], synaptic vesicles (vesicles from neurons) [35], dexosomes (exosomes released from dendritic cells) [36], oncosomes/texosomes (tumor cell-derived exosomes) [37], and outer membrane vesicles (vesicles derived from bacteria) [38] have been used.

3 EV Production

EVs are spontaneously produced during cell culture or released in response to a biological, chemical, or physical trigger.

Exosomes are spontaneously released by cells in culture with complete medium. However, as the complete medium naturally contains exosomes, serum must be previously depleted from its bovine-exosome content by ultracentrifugation. Exosome release is strongly influenced by the producer cells. In a comparative basis, the relative amount of exosomes secreted in the conditioned culture media of five different cells or cell lines was evaluated. Mesenchymal stem cells derived from human embryonic stem cells were found to produce the highest amounts of exosomes, followed by the human embryonic kidney cell line (HEK) [39]. The parent cell type from which exosomes are derived may also induce different effects on the recipient cells. The interested reader may refer to a recent review paper [40].

An important parameter not taken into account in previous studies is the fact that EV production and release in the medium is a dynamic process, which comprises EV recapture by cells. It means that the EV number will increase until it reaches a plateau, depending on many parameters such as the cell production, the cell recapture, and the volume of media. It has been reported that the time until reaching the plateau was in the 8–12-h range for breast cancer cells [41].

Apart from the spontaneous exosome release, cells may be stimulated to induce EV release via serum or oxygen deprivation. For example, serum deprivation enhanced EV release from RPMI 8226, U266, and KM3 cells by 2.5-, 4.3-, and 3.8-fold, respectively, compared to culture in complete medium [42]. Hypoxia is another stress factor triggering EV release. For instance, MCF7, SKBR3, and MDA-MB 231 cell culture at 0.1 % O_2 with complete medium for 24 h resulted in a nearly twofold increase in exosome release compared to the normoxic control, considering nanoparticle tracking analysis (NTA) data [43]. Both serum deprivation and hypoxia are quite straightforward methods for EV production as no further processing is required to eliminate the vesiculation trigger agent.

Cell activation is also known to trigger EV release. As an example, 20-min stimulation of neutrophils with TNF-a (50 ng/ml), IL-8 (50 ng/ml), and leukotriene B4 (LTB4; 10 nM) induced EV release to double compared with resting cells, according to imaging flow cytometry [44]. Interestingly, EVs released upon cell activation display a phenotype different from EV released under serum starvation. EVs from endothelial cells expressed constitutive markers, such as CD31 and CD105, when the triggering stimulus was serum deprivation. In contrast, inducible markers such as CD54 and CD62E were increased for EVs only when endothelial cells were submitted to TNF-α activation stimulus [5].

EV release is also influenced by chemical agents. Cytochalasin B, which inhibits actin polymerization, is able to induce the release of EVs comprising functional cell surface receptors on the membrane and cytosolic proteins in its inner compartment [45, 46]. For instance, cytochalasin B (2 μM) incubation enhanced EV release by freshly isolated neutrophils by a factor of about 2 [44]. Ethanol is another chemical agent able to enhance EV release both in vitro and in vivo. Hepatocarcinoma cells (Huh7.5 cells) presented an increase in the number of released exosomes in response to ethanol incubation in a time- and dose-dependent manner. The enhancement was near tenfold when cells were incubated at 100 mM concentration, according to NTA data [47].

Recent approaches to produce EVs relate to physical and mechanical methods. Machluf et al. proposed the use of a liposome extruder to transform ghost cells or loaded cells into submicron vesicles [48, 49]. Cell ghosts were produced from rat MSCs and human smooth muscle cells hypotonically treated with tris-magnesium buffer to allow cytosol removal. For EV production, ghosts were extruded through 0.4 μm polycarbonate membranes [49]. However, the paper does not present any data about the efficiency and the yield of this method. W Jo et al. reported the effect of high shear-stress to induce EV release. They designed a microfluidic chip featuring 37 μm sized channels, nearly in the cell diameter. Murine embryonic stem cells were pushed by a syringe pump into the channels at high speed (0.04 m/s). The authors indicate that cell deformation by the shear-stress induced EV budding. The authors argue a highly efficient production of vesicles based on protein assay for EV quantification [50]. Jang and colleagues reported the use of sequential filtration to process cells into vesicles. The process comprises the use of micrometric filters (10, 5, and then 1 μm pore size) to create EVs [51]. The use of centrifugal force (2000 rpm) to constrain cells to pass through filters with micro-sized pores (10 and 5 μm pore) was also reported for EV production. The authors indicate that their technique induces a friction force between the cell and the polycarbonate surface of the filter as both lipid heads and the filter membrane are hydrophilic. The tension makes the plasma membrane elongate until rupture is achieved. The lipid bilayer fragments that are planar immediately after membrane rupture spontaneously self-assemble generating EVs of about 100 nm in size. The yield is about 250 higher than spontaneous exosome release in complete medium, according to protein assay data [52]. As it will be further discussed in the next section, the choice of the protein assay to quantify EV release is controversial. In a related approach, the use of nano-blades in a microfluidic system has been reported to slice cell membrane, creating nanovesicles of 100–300 nm [53]. It has to be noted that a similar chip with similar nano-blades is used for rapid intracellular protein extraction by lysis [54], which supports the assumption that these chips are probably allowing the lysis of the cell, thus freeing organelles. The authors report the collection of 1.5×10^{10} vesicles per million cells, corresponding to 20 μg of protein.

Although there are several methods for EV production, the lack of comparison between them and inconclusive EV characterization data makes it difficult to point out which approach is the most suited for EV production.

4 EV Loading

EVs have been successfully loaded with drugs, nanoparticle, radiotracers, micro (mi)RNA, and fluorescent dyes (Fig. 3). The approaches to confer exogenous imaging or therapeutic properties to EVs roughly fall into three categories. One of them relates to load the parent cell so that the EVs released from it inherit their

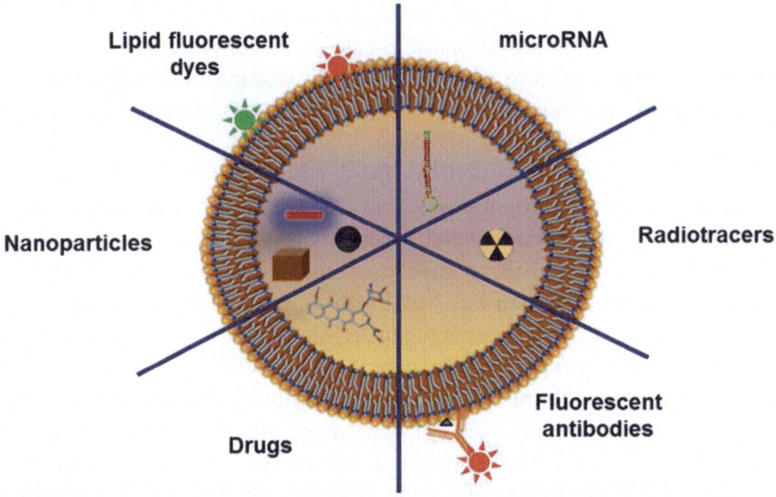

Fig. 3 Engineering extracellular vesicles. Vesicles could be engineered to enclose several molecules, macromolecules, or particles for imaging or therapeutic purposes

cargo. The other strategy concerns the loading of EV at the moment they are produced from physically squeezed parent cells. The last method relates to the direct loading of EV after its production and isolation. All three methods will be discussed in the following subsections.

4.1 Loading the Parent Cell Before EV Release

In a top-down procedure, our group has engineered nanoparticle-loaded EVs from precursor cells previously loaded with the cargo. For this purpose, parent HUVEC cells were first incubated with nanoparticles and allowed to internalize them. Starvation stress was employed to induce the release of vesicles hijacking the cargo from their precursor cell. By this method, EVs could encapsulate a set of nanoparticles regardless of their chemistry or shape, such as iron oxide nanoparticles, iron oxide nanocubes, gold/iron oxide nanodimers, gold nanoparticles, and quantum dots [55]. Different hybrid nanovesicles were designed: magnetic, magnetic-fluorescent, and magnetic-metallic vesicles, either single component or multicomponent.

Dual-drug/nanoparticle EV loading was also feasible by this method. We demonstrated that vesicles from THP-1 cells could be loaded with iron oxide nanoparticles and different therapeutic agents irrespective to their molecular weight, hydrophobic, hydrophilic, and amphiphilic character [56]. Thereby, magnetic vesicles were loaded with a chemotherapeutic drug (doxorubicin), anticoagulant protein (tissue-plasminogen activator (t-PA)), or two photosensitizers (disulfonated tetraphenylchlorin (TPCS2a) [56] or mTHPC [57].

In a related approach, Mao and colleagues loaded 3T3 or HEK293 precursor cells with quantum dots and hydrophobic calcein acetoxymethyl ester as a model drug. Chloroquine was used to induce lysosome swelling liberating quantum dots into the cytoplasm. The release of EVs inheriting the dual cargo from the parent cells was triggered by cytochalasin B [58].

The strategy of spontaneous cell loading for posterior cargo exported by EVs was also tested for miRNA or nucleic acid EV loading. For instance, HEK293T cells were transfected with enhanced green fluorescent protein (EGFP) plasmid using polyethylenimine. EGFP-positive exosomes could be obtained from transfected cells [59]. This approach has been used to incorporate short interfering (si) RNA [60], miRNA [61], and messenger (m)RNA into exosomes. As mRNA transfection results in protein expression by parent cells, both the protein and the mRNA were found to be packaged into exosomes [62]. Although endosomal sorting complexes required for transport (ESCRT) (which mediates exosome budding in multivesicular bodies [63]) have been shown to preferentially address some nucleic acid sequences (up to 57-fold [64]) to EVs, the obtained EV loading yield may be low. For instance, Kanada and colleagues [65] reported the loading of cells with lipofectamine with either miRNA or plasmids coding for this miRNA in the attempt to enable further exosome and microvesicle loading. The authors indicated that they were not able to detect any loading in the exosome fraction. Only a low loading process with the entire plasmid in the microvesicle fraction was detected.

In case the cargo is a peptide or protein, an interesting strategy consists in fusing the protein/peptide of interest with an EV-enriched protein. For instance, there is a near-100-fold enrichment in lamp2b or tetraspanins in EVs compared to the cell itself. Alvarez-Erviti and colleagues [66] used such an approach to engineer a lamp2b protein fused to the neuron-specific RVG peptide. Thereby, they succeeded in decorating exosomal membranes with this peptide coffering neuron targeting. In a similar approach, a protein cargo of interest may be modified to contain a membrane-targeting sequence (acylation domain, myristoylation domain, prenylation domain, or palmitoylation), so that it will localize in the plasma membrane and also in EVs by consequence. Such plasma membrane anchors have been used to target GFP to exosomes [67].

4.2 Loading During EV Production

Another approach is to load vesicles at the moment they are produced. This strategy mainly relies on physical methods of EV release. For instance, EV loading with doxorubicin was performed by extruding parent U937 monocytic cells through a series of polycarbonate membranes with pore sizes of 10, 5, and finally 1 μm in the presence of doxorubicin [51]. Similarly, rat MSCs and smooth muscle cells were extruded through 0.4 μm polycarbonate membranes in the presence of tumor-necrosis-factor-related apoptosis inducing ligand (TRAIL) in order to produce TRAIL-loaded EVs [49]. In a related strategy, concomitant EV loading and production were reported for the nano-blade approach in a microfluidic system mentioned above. After the cell membrane was sliced, plasma membrane fragments self-assembled enveloping exogenous polystyrene latex beads present in the buffer solution to produce bead-loaded EVs, even if there was some unspecific bead adsorption [53].

4.3 Direct Loading or Decoration of EVs

In addition to the nanoparticle cell loading strategy to produce nanoparticle-loaded vesicles, our group also reported the direct vesicle decoration with nanoparticles. For this purpose, endothelial EVs were incubated with anionic iron oxide nanoparticles that were able to bind to the vesicle surface by electrostatic interaction [68]. EV decoration was also reported with fluorescent antibodies. For instance, human platelet-derived EVs could be double-stained by incubation with annexin V- and platelet-specific fluorescent antibodies (anti-CD61, anti-CD63, or anti-CD62P, respectively) [69]. However, it should be noted that the constant of dissociation of these antibodies may enable them to leave their exosome antigen and interact with endogenous targets, misleading exosome biodistribution investigation.

Drugs may also interact directly with the membrane of EVs enabling loading. This is the case of the lipophilic drug curcumin [70, 71] that may be loaded into EVs via hydrophobic interactions. The same applies to lipophilic dyes such as PKH67 and PKH26 or 1,1-dioctadecyl-3,3,3,3-tetramethylindotricarbocyanine iodide (DiR) to confer EVs with a fluorescent label [43, 72]. However, it is important to mention that lipid dye staining should be performed avoiding lipid excess. Otherwise, lipid dyes can form micelles and be co-purified as a contaminant in EV preparations. EVs may equally be loaded with low-molecular-weight molecules such as paclitaxel and doxorubicin via single-step incubation at room temperature. Drug loading data evaluated by HPLC indicated nearly 7 ng of paclitaxel per 1 μg protein and 132 ng doxorubicin for 1 μg protein [73]. EVs may also be directly loaded with radiotracers. Erythrocyte EVs were incubated with sodium ^{51}Cr-chromate at 37 °C. EVs were then washed under ultracentrifugation to remove free chromate [74]. Loading took place as the tracer is able to cross the erythrocyte membrane and bind to hemoglobin. Once in the erythrocyte, the tracer is reduced by glutathione becoming trapped as the reduced form is not able to cross back plasma membrane [75].

Electroporation was proposed as a quite promising method for EV loading according to a highly cited paper [73] that featured successful siRNA into exosomes using electroporation to knock down a therapeutic target in Alzheimer's disease. However, 2 years later, Kooijmans et al. [76] published a study explaining that the obtained results were indeed an artifact. In fact, electroporation created metal ions from the electrodes inducing the formation of siRNA aggregates. These aggregates were then co-purified with exosomes. Therefore, siRNA transfer and efficient RNA silencing were not EV mediated, but induced by siRNA aggregates. Once aggregation was inhibited by adding EDTA ion chelator or by coating the electrodes, siRNA was no more co-purified in the exosome fraction. Haney and colleagues compared different techniques to directly load catalase (a large antioxidant protein) into exosomes: incubation at room temperature, freeze-thawing cycle method, permeabilization with saponin, sonication, and electroporation. Sonication and extrusion, as well as permeabilization with saponin, resulted in a high loading efficiency and the obtained vesicles were in a size range of 100–200 nm [77].

In overall, although there are different loading methods, the lack of comparative studies and thorough characterization render it difficult to point out the most effective one in terms of encapsulation efficiency. Vesicle constitution and integrity after the loading process also remain unclear. It seems that there is still a need to develop scalable methods preserving vesicle constitution and integrity while enabling highly efficient loading.

5 EV Isolation

The complexity of biological fluids renders the isolation of EVs extremely difficult. Ultracentrifugation is the most common method to isolate EVs. Typically, centrifugal accelerations of about 200–1500 × *g* have been applied in order to remove cells and cellular debris. It is necessary to reach 10,000–20,000 × *g* to pellet vesicles larger than 100 nm, while 100,000–200,000 × *g* are required to pellet vesicles smaller than 100 nm [25]. A major hurdle of using ultracentrifugation-based purification methods is the impact of G force on vesicles. The acceleration force may lead EVs to fragment, leak their cargo, or become activated. Besides, centrifugal acceleration of 100,000–200,000 × *g* may induce vesicle fusion and protein sedimentation [78]. All these effects may influence EV properties and purity. In order to avoid cross-contamination between vesicles of different size ranges, it is recommended to perform further purification using a sucrose cushion [16]. Such an additional purification step is expected to eliminate large protein aggregate contaminants, which are sedimented by centrifugation, but do not float on a sucrose gradient [79]. However, this method is time consuming while presenting low yield.

Although ultracentrifugation is the most common purification method, there is currently no consensus about the optimal protocol for the isolation of pure vesicle populations [19]. Other methods may also be considered. New promising techniques are coming to the field, such as the use of ultrafiltration techniques, using 500–1000 kDa filters to retain EVs whereas most proteins are not retained. This technique was used in the first clinical trial using exosomes [80]. Immunoaffinity capture has also been proposed for EV isolation [81, 82]. For this, magnetic beads conjugated with antibodies to bind specifically proteins overrepresented on EVs are used. Tauro and colleagues isolated human colon cancer cell line LIM1863-derived exosomes by means of anti-EpCAM-coated magnetic beads [83]. In this study, immunoaffinity was evaluated to be the best method to capture exosomes, as it was able to isolate a population enriched with exosome markers, and exosome-associated proteins by at least twofold more than ultracentrifugation and density gradient separation. However, this method is expensive and exosomes lacking specific antigens will not be recovered [16].

Immunoaffinity may be conjugated to a microfluidic approach to isolate EVs. Chen et al. reported a microfluidic immunoaffinity method based on their selective binding to anti-CD63-coated surfaces. They demonstrated the feasibility of isolating and extracting exosomal RNA from 100 to 400 μl serum samples within an hour [84]. A cutting-edge technique recently developed couples microfluidics to acoustic purification to isolate EVs. This approach is based on the principle that larger particles move faster as the acoustic force is proportional to their volume. Therefore, larger vesicles and cells move on the side of the channel, while nanometer-sized vesicles are retained in the center flow. This approach is label free and the size cutoff can be controlled electronically in situ, enabling versatile size selection. Although the resolution of this technique allows a 90 % separation yield, it enables EV separation from cells or large debris, but not from protein contaminants [85].

Size-exclusion chromatography was also used to isolate EVs and separate them from contaminating proteins. Böing and colleagues reported that EVs with a diameter larger than 75 nm could be isolated from plasma by single-step size-exclusion chromatography. In contrast to ultracentrifugation, size-exclusion chromatography does not induce vesicle aggregation and there is no risk of protein complex formation induced by acceleration [86]. However, a recent paper described that this method was not as efficient as previously thought, as it leads to low vesicle recovery [87].

Our team described a magnetic sorting method to isolate EVs previously labeled with magnetic nanoparticles. A strong permanent magnet that creates a magnetic field of $B = 650$ mT, and a magnetic field gradient $gradB = 55T\ m^{-1}$ in the volume of the syringe, was used for this purpose. This protocol, dedicated to magnetic EVs, was quite straightforward enabling EV purification in a single step without adding reagents nor antibodies [57].

6 Characterization

Several methods have been used to characterize EVs in terms of size, morphology, and constitution. They will be overviewed herein. However, it is important to mention that none of these techniques singly provide thorough biophysical and biochemical characterization of vesicles and their content [19]. Techniques must be combined systematically in the attempt to perform EV characterization as complete as possible.

Dynamic light scattering (DLS) measurements are user friendly, fast, and rather straightforward. Additionally, DLS instruments also enable the determination of the zeta potential, which is the electric potential difference between the medium and the stationary ion layer bound to EVs [88]. DLS size is determined from fluctuations in scattered light intensity due to the Brownian movement of the particles. The size distribution is calculated by measuring the scattered light fluctuation intensity and applying a mathematical model derived from light scattering and Brownian motion theory [89]. Although DLS gathers many advantages, results may be biased by the presence of large particles in the sample [88, 90] and also by the presence of proteins.

Nanoparticle tracking analysis (NTA) has been increasingly used to provide the size, concentration, and zeta potential measurements of EVs. Particles in a sample are visualized due to the light they scatter when exposed to laser light. The light scattered is then captured by a digital camera. The Brownian motion of each particle is tracked from frame to frame by a dedicated software and the rate of particle movement is calculated through the Stokes-Einstein equation. The technique calculates particle size on a particle-by-particle basis providing important statistical power. EVs from 30 to 1000 nm within a concentration range of 10^8–10^9 can be sized and counted with relatively high sensitivity [91]. NTA circumvents one of the key problems associated to DLS (i.e., polydispersity) as it resolves and accurately measures samples featuring multiple size populations well by discriminating peaks of distinct size. NTA has the additional advantage of enabling vesicles to be analyzed in suspension, avoiding shrinkage and fixation artifacts as it may occur for microscopy analysis. Besides, NTA can detect vesicles in a smaller size range compared to conventional flow cytometry (~300 nm detection limit) [92]. Fluorescent mode detection is also an important asset. Analysis in fluorescence mode provides specific results for labeled EVs. This feature enables users to detect, analyze, and count only a specific population to which the fluorescent marker is bound [91].

Resistive pulse sensing (RPS qNano) operates by detecting transient changes in the ionic current generated by the transport of particles through a nanopore into a nonconductive membrane which separates two fluid cells. As the relative change in current is proportional to the volume of the particle crossing the pore, RPS can accurately determine the diameter of EVs [89]. qNano provides quantitative analysis in samples whose EV size spans from 70 nm to 1 μm at concentrations from 10^5 to10^{12} ml [91]. However, pore clogging and pore stability represent some RPS concerns [93].

The size and morphology of EVs have been widely investigated by transmission electron microscopy (TEM) and atomic force microscopy. Atomic force microscopy enables the investigation of EV size and morphology to be performed directly in suspension [94]. However, as the analysis requires surface adhesion, changes from spherical to hemispherical or flat structure may mislead size and morphology interpretation [95]. TEM uses electrons to create an image. Considering that the wavelength of electrons is more than three orders of magnitude shorter than the wavelength of visible light, the resolution of TEM is much higher than that optical microscopy and it can be lower than 1 nm. However, fixation and dehydration are major pre-analytical steps that may affect the size and morphology of EVs [96]. For instance, the cup-shaped morphology described for exosomes is now recognized as an artifact from pre-analytical steps [79]. Despite these artifacts, TEM is a technique of choice for analyzing EVs engineered to encapsulate nanoparticles. Depending on its constitution, nanoparticles are clearly identified in TEM micrographs as electron-dense spots, acting as tracers. This is the case of quantum dots, iron oxide, and gold nanoparticles.

In order to estimate magnetic nanoparticle loading into EVs and other nanocontainers, our group designed a miniaturized straightforward experimental approach [97]. The setup is on a glass slide/coverslip chamber to which a microtip was integrated as a magnetic attractor. The magnetophoretic velocity of EVs moving towards the magnetic tip was observed with an optical microscope connected to a CCD camera and a computer. The quantitative analysis of magnetophoresis enabled to infer both the magnetophoretic velocity and the magnetic content of the nanocontainers. Additionally, the nano-magnetophoresis experiment under fluorescence microscopy provided information on the constitution of the systems, attesting the co-encapsulation of nanoparticles with a fluorescent drug in the core or the encapsulation of magnetic nanoparticles within the membrane of EVs [56, 97].

Flow cytometry represents a high-throughput multi-parametric method for the analysis of EVs in terms of size, concentration, constitution, and fluorescent drug loading. Flow cytometers detect scattered light and fluorescence that are measured by detectors facing forward and perpendicular to the laser. Measurements are performed in a hydrodynamically focused fluid stream at a rate of hundreds or thousands of events per second. Flow cytometry is the most widely used method to detect vesicles in clinical samples and efforts towards standardization of EV measurements have been reported to palliate discrepancies in results when operating on different flow cytometer devices. For instance, Lacroix et al. on behalf of the International Society of Thrombosis and Haemostatic have proposed a submicrometer bead (Megamix beads; BioCytex, Marseille, France) gating strategy in order to define EV region in a reproducible and standardized set [98]. In addition to result variability when using different flow cytometer devices, even higher variability is obtained when comparing different characterization methods. In a comparative study, vesicle concentration determination differed markedly depending on the used technique (RPS, NTA, or flow cytometry). The authors indicate that such divergence was imparted by the variability on the minimum detectable vesicle sizes of each technique. They estimated that the minimum detectable vesicle sizes were 70–100 nm for RPS, 70–90 nm for NTA, and 270–600 nm for conventional flow cytometry [93]. Brisson's team has recently compared EV detection by conventional flow cytometry and electron microscopy. More precisely, data from cryo-transmission electron microscopy combined with receptor-specific gold labeling was confronted to flow cytometry data, in which the detection of EVs was triggered on the forward scatter parameter, taken into consideration EVs labeled with either annexin 5-Fluo or anti-CD235a-Cy5. Results showed that only 1 % of phosphatidylserine-exposing EVs observed by electron microscopy could be detected by conventional flow cytometry [23]. Interestingly, this group showed that flow cytometry analysis with annexin 5-fluorescence triggering enhanced EV detection by a factor 55 when compared to forward scatter triggering [99]. Also using the fluorescence triggering approach, our group pioneered the detection and imaging of EVs by multispectral imaging flow cytometry. Annexin 5-fluorescence detection was combined to m-tetrahydroxyphenylchlorin (mTHPC) fluorescence detection to confirm that this drug was encapsulated into EVs [57].

In addition to the above-mentioned techniques, emerging "omic" approaches (proteomic, lipidomic, metabolomic, and microarray profiling) brought along unprecedented advances in molecular profiling in the attempt to decipher the roles of EVs [91]. Metabolomics has opened new opportunities to provide a global view of metabolic mechanisms related to EVs as it enables the qualitative and quantitative measurement of thousands of small molecules (<2000 Da). Metabolomic profiling can be performed using different high-throughput and high-sensitivity analytical techniques, including gas chromatography/mass spectrometry and liquid chromatography/mass spectrometry [100]. The same techniques have enabled the comprehensive characterization and quantification of the repertoire of lipid and protein species present in EVs via proteomics and lipidomics [101].

Despite these advances, it is important to highlight that the depth of molecular profiling provided by "omic" approaches is tightly related to the purity of EVs. The same applies to other techniques described herein and this is also the case for protein dosage (mini Bradford, microBCA), which is one of the most widely used methods for EV quantification [102]. Exosomes are classically purified from cell culture medium after a few days of culture. These culture media are enriched in fetal bovine serum at nearly 10 %. In order to avoid bovine exosome contamination, serum is previously depleted from its bovine-exosome content by ultracentrifugation. Even in these conditions, serum is highly enriched in proteins, and cell culture exosome purification leads to purified exosomes contaminated by proteins. Therefore, protein dosage may lead to an overestimation in EV quantification. Indeed, most of the characterization techniques are not able to distinguish EVs, and more particularly exosomes, from protein aggregates. In order to account for this difficulty, Weber and colleagues have reported an EV purity index based on the ratio of nanovesicle counts to protein concentration (number of particles by NTA)/micrograms of proteins by Bradford assay). This ratio is well correlated with the exosome purification steps as the purer the preparation, the higher the ratio. It is important to mention that physical production methods may enable EV production at high yield, but vesicle purity may be a main concern. For instance, authors report the use of nano-blades for obtaining 1.5×10^{10} vesicles per million cells, corresponding to 20 μg of protein [53]. In this case, we can compute a purity ratio of 7.5×10^{8}, corresponding to an impure preparation. Therefore, such purity ratio is expected to assist in EV quality control and may be quite useful when comparing or standardizing vesicle purity [103].

7 Engineering Vesicles for Imaging: Biodistribution Investigation

EV engineering with image tracers is expected to provide important data about the trafficking of EVs in the organism. This represents a unique opportunity to gain insights into the biodistribution of EVs in vivo in the attempt to decipher their complex interplay with different recipient cells both in homeostasis and disease. Willekens and colleagues investigated the fate of ^{51}Cr-labeled erythrocyte vesicles once injected in rats [74]. It was observed that 80 % of the radioactivity was cleared from the circulation within 5 min, with simultaneous uptake by the liver at 55 %. Liver uptake could be inhibited by preinjection of polyinosinic acid and phosphatidylserine, which are scavenger-receptor ligands. These results suggested that erythrocyte vesicles are eliminated by Kupffer cells from the liver via an interaction mediated by the scavenger-receptor.

Our group was the first to propose MR imaging to investigate the biodistribution and clearance kinetics of EVs in mice [104]. Endothelial cells were first labeled with iron oxide nanoparticles and then stressed by serum deprivation to release magnetically labeled EVs. The labeled EVs were injected by IV to mice at a concentration similar to the naturally circulating EV concentration. The biodistribution of endothelial cell-derived EVs was different from that of the iron oxide injected alone. EVs were mostly trapped in the spleen within 5 min after injection and to a lesser concentration in liver in contrast to free iron oxide. The clearance of magnetic label from these organs was also faster when loaded within EVs. Fluorescently labeled EVs were used to assess the relative EV uptake in different immune cell types in spleen, liver, blood, and lung, by flow cytometry. Macrophages and monocytes were the cells that internalized the highest amount of endothelial cell-EVs (Fig. 4).

In a quite complete study, Wiklander and colleagues [59] investigated the influence of the route of administration and cell type of origin on the biodistribution of fluorescently labeled EVs. Biodistribution data is summarized in Table 1. Biodistribution kinetics was also assessed. Slight biodistribution variation was observed as a function of the time from 5 min post-injection to 48 h. The most noticeable changes were observed for the lungs (20 % of signal detection to 7 % from 5 to 30 min), as well as for pancreas and

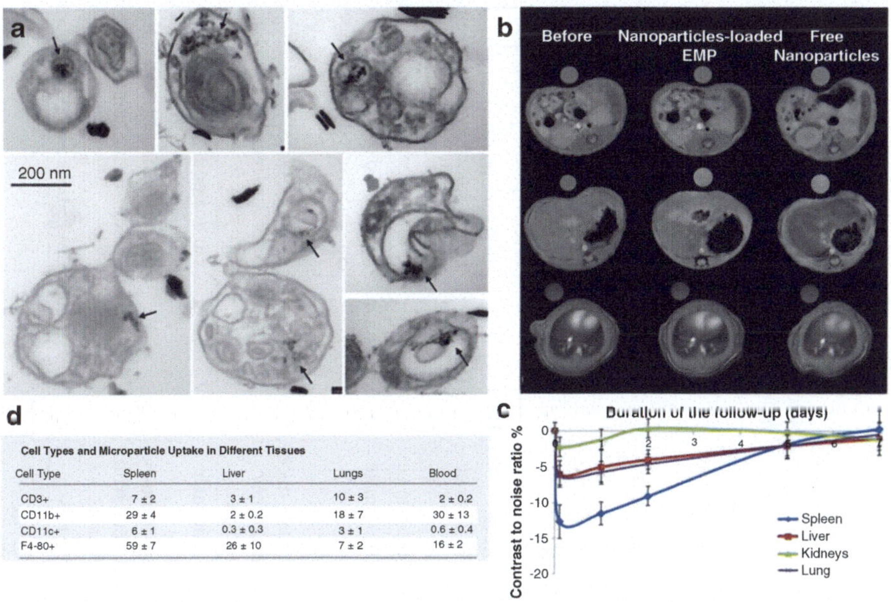

Cell Type	Spleen	Liver	Lungs	Blood
CD3+	7 ± 2	3 ± 1	10 ± 3	2 ± 0.2
CD11b+	29 ± 4	2 ± 0.2	18 ± 7	30 ± 13
CD11c+	6 ± 1	0.3 ± 0.3	3 ± 1	0.6 ± 0.4
F4-80+	59 ± 7	26 ± 10	7 ± 2	16 ± 2

Fig. 4 *In vivo* monitoring of EV distribution. (**a**) Magnetically labeled EVs were derived from endothelial cells after endocytosis of iron oxide nanoparticles and starvation. The nanoparticles within EVs are indicated by arrows on TEM micrographs. (**b**) MRI images of spleen, liver, and lung before and 5 min after IV injection of magnetically labeled endothelial EVs (EMP) or free iron oxide nanoparticles. (**c**) MRI contrast-to-noise ratio in the different organs after injection of EVs. (**d**) Percentage of fluorescent cells in organs 5 min after injection of endothelial EVs labeled with CFSE fluorescent dye (adapted for Rautou et al., Radiology, 2012 [104]

Table 1 Biodistribution of EVs

Cell type source	Animal model	Administration route	Biodistribution	Imaging method	Reference
HEK293T (human embryonic kidney)	NMRI or C57BL/6 mice	IV	24-h time point: Liver>gastrointestinal tract>Spleen>lungs>pancreas	Fluorescence imaging	[59]
HEK293T (human embryonic kidney)	NMRI or C57BL/6 mice	SC	24-h time point: Gastrointestinal tract>liver>pancreas>lungs>spleen	Fluorescence imaging	[59]
HEK293T (human embryonic kidney)	NMRI or C57BL/6 mice	IP	24-h time point: Gastrointestinal tract=liver>pancreas>spleen>lungs	Fluorescence imaging	[59]
C2C12 (mouse myoblast)	NMRI or C57BL/6 mice	IV	24-h time point: Liver>spleen>gastrointestinal tract>lungs>pancreas	Fluorescence imaging	[59]
B16-F10 (mouse skin melanoma)	NMRI or C57BL/6 mice	IV	24-h time point: Liver>spleen>gastrointestinal tract>lungs>pancreas	Fluorescence imaging	[59]
Primary mouse dendritic cells	NMRI or C57BL/6 mice	IV	24-h time point: Liver>spleen>gastro-intestinal tract=lungs>pancreas	Fluorescence imaging	[59]
Mouse endothelial cells	C57BL/6 mice	IV	5 min to 7 days: Spleen>liver>lungs>kidneys	Magnetic resonance imaging and fluorescence imaging	[105]
Bone marrow-derived MSC	CD1 nude mice (acute kidney injury model)	IV	24-h time point: Liver>spleen>injured kidney>lung>normal kidney	Fluorescence imaging	[106]
HEK293 (human embryonic kidney)	Athymic nude mice subcutaneously implanted with Gli36 cells	IV	2-h timepoint: Spleen>liver>lung>kidneys>others Retention after PBS washing: kidney>liver>lung>heart>brain>muscle>others	Multimodal imaging	[107]
Rat erythrocyte	Wistar rats	IV	Liver>bone>blood>skin>muscle>spleen>kidney>lung	TEP	[74]

gastrointestinal tract (2 % of signal detection to 20 % in 48 h). The dose injected was also an important investigated parameter. Relative liver accumulation decreased when EV administration dose increased, probably due to mononuclear phagocyte system saturation resulting in more effective liver bypass at higher doses [59].

The biodistribution of MSC-derived EVs was investigated in a murine acute kidney injury model. At a 24-h time point, EVs were found to locate mainly in the liver, the spleen, and the injured kidney, being present only at a low concentration in the lung. Experiments performed with healthy mice indicated very low EV presence in the kidney, suggesting that kidney targeting was enhanced in acute kidney injury [105].

In another study, the biodistribution of HEK293 EVs was investigated by tracking a multimodal reporter combining both fluorescence and bioluminescence in nude mice bearing heterotopic Gli36 tumor. Most of the signal was located in the liver, spleen, lung, and kidney. In order to investigate signal retention, EV-injected animals were perfused with PBS transcardially before organs/muscle were collected at different time points. It was found that the perfused spleen showed minimal EV signal, suggesting that EVs were not efficiently taken up despite the high amount of EVs present in blood passing through this highly vascularized organ. After perfusion, EV signal was highly present in liver correlating with a described renal and hepatic elimination, with a peak fluorescent signal in the urine at 60 min post-injection [106].

8 Engineering Vesicles for Therapy and Theranosis

Most of the drug delivery systems currently used in clinical practice are chemically synthesized using lipids with liposomes representing the vector of choice [107, 108]. In spite of the remarkable advances in the design of synthetic drug carriers, there is an increasing recognition that natural carriers feature highly desired attributes of drug delivery vehicles [109]. EVs are one of the most imminent examples. Indeed, EVs represent the most physiological carrier choice considering their natural role as conveyors of biological cargoes and mediators of information transfer between cells. The endogenous origin of EVs is a remarkable advantage compared to synthetic nanocarriers, conferring them natural blood stability and natural targeting properties [110, 111]. In addition to their natural intrinsic properties, EVs have the plasticity to be engineered to display exogenous imaging and therapeutic properties. However, a main challenge in designing EVs for therapeutic purposes is whether the production may be scalable or reproducible. Indeed, vesicle yield will directly impact the final production cost. In this respect, the choice of parental cells is critical [112]. For example, MSCs were found to produce large amounts of exosomes compared to other cell lines, suggesting that they may be promising candidates considering a production in a clinically applicable scale [39]. MSC EVs may also feature natural targeting properties inherited from their parent cells that may confer favorable homing to tumor sites. Another important point is that the parent cell should produce EVs devoid of immunological stimulation to prevent inflammatory responses, while also being stable in the circulation to enable cargo delivery. In this respect, MSC and immature dendritic cells have favorable properties justifying them to be considered as eligible parent cells [113].

EVs from dendritic cells or other cell types have demonstrated encouraging therapeutic potential. For instance, doxorubicin-loaded exosomes from dendritic cells enabled drug delivery and inhibited tumor growth in mice. In this study, mouse immature dendritic cells were used to produce exosomes loaded with doxorubicin featuring Lamp2b fused to targeting RGD peptide. Targeting efficiency was demonstrated and a remarkable cytotoxic effect was observed in vivo in an orthotopic mammary tumor model [114]. In a related study, doxorubicin-loaded exosomes from Raw264.7 cells were produced and tested in an ectopic mouse colon adenocarcinoma model. Interestingly, in vivo experiments showed that these exosomes reduced tumor growth without the adverse effects observed for the free drug. Furthermore, compared with commercial doxorubicin-loaded liposomes (Doxil), exosomes presented a more efficient tumor growth reduction [51]. Curcumin also represents an important example of a small therapeutic molecule encapsulated in EVs with successful therapeutic results in vivo. Curcumin-loaded exosomes succeeded in delivering their drug cargo to microglia cells in a lipopolysaccharide-induced brain inflammation model in mice. Following intranasal route administration, exosomes located in the brain within 1 h via a transport along the olfactory pathway likely involving extracellular flow along perineuronal and/or perivascular channels for direct delivery to the brain parenchyma. Results showed that mice treated with curcumin-loaded exosomes were protected from brain inflammation [70]. The potential of exosomes to deliver curcumin to colon cancer is now under clinical investigation (clinical trial no. NCT01294072). In this clinical trial, plant-derived exosomes loaded with curcumin will be administered orally to colorectal cancer patients. The choice of vegetal exosomes was oriented by the need of EV production in sufficient amount for therapeutic application in the clinical setting [112].

In addition to small molecules, macromolecules could also be tested for EV-mediated therapy. For instance, catalase-loaded exosomes were investigated for the management of Parkinson's disease in a murine model in order to inactivate reactive oxygen species that play a detrimental pathological role. Exosomes could be readily internalized by neuronal cells in vitro. Considering in vivo experiments, a considerable amount of exosomes was detected in the mouse brain following intranasal administration. Catalase-loaded exosomes provided significant neuroprotective effects increasing neuronal survival in a mouse model of Parkinson's disease. Interestingly, catalase alone did not decrease inflammation in mice, indicating that exosomes protected catalase against protease degradation [77].

Many studies focused on the transfer of interfering RNAs mediated by EVs. For instance, exosomes loaded with tumor-suppressor let-7a miRNA could be successfully taken up by tumor cells in vivo and were found to inhibit tumor development in a human breast cancer model [61]. Katakowski and colleagues engineered MSCs with a miR-146b, an anti-glioma miRNA cargo. Following an intra-tumoral injection, exosomes were found to reduce glioma xenograft growth in an orthotopic rat model of brain tumor [115]. In an elegant study, Mizrak and colleagues reported the design of EVs carrying the suicide mRNA/protein for cytosine deaminase fused with uracil phosphoribosyltransferase, which is a potent prodrug activator. The approach was combined with a systemic treatment with the prodrug (5-fluorocytosine). In this strategy, when cancer cells take up EVs and the prodrug, cytosine deaminase converts 5-fluorocytosine to 5-fluorouracil (5-FU), which is further converted by the fused uracil phosphoribosyltransferase to the highly toxic 5-fluoro-deoxyuridine monophosphate (5-FdUMP). Tumor growth was completely inhibited in most of the treated mice bearing an orthotopic sciatic nerve tumor model [62].

Our group reported the first success in the design of theranostic EVs. We designed different hybrid EVs from endothelial cells: magnetic, magnetic-fluorescent, and magnetic-metallic vesicles, either single component or multicomponent, combining the advantageous properties of each integrant particle. These hybrid vesicles were able to generate heat when submitted to an alternating magnetic field and could be monitored by fluorescence imaging or MRI [55]. Although in vivo experiments were not carried out for these nanoparticle-loaded vesicles, the obtained data demonstrated their potential for image-guided magnetic hyperthermia.

In addition to theranostic EVs containing only nanoparticles, we also designed drug/nanoparticle-loaded EVs from THP1 macrophages and we investigated their targeting potential in vitro. Using photosensitizer-loaded magnetic vesicles, it was shown that EV uptake by cancer cells could be spatially controlled under magnetic field. Magnetic targeting could be attested by investigating the spatial distribution of different components of these EVs: membrane, drug cargo, and iron oxide nanoparticles. A magnetic targeting effect was demonstrated via anisotropic drug uptake and also by subsequent anisotropic cell death. Targeting could be confirmed at the macroscopic level by MRI, while further proof at the subcellular scale was provided by electron microscopy [56]. In addition to the in vitro investigation, the theranostic potential of these photosensitizer-loaded magnetic EVs was tested in vivo in a murine tumor model. Vesicles could be tracked in vivo by dual-mode imaging, combining optical imaging and MRI. The engineered EVs were found to induce an efficient photodynamic action, as evidenced by tumor growth curves and histological analysis (Fig. 5) [57].

9 Conclusions

EVs are multifaceted subcellular entities whose engineering holds great promises for EV imaging and EV-mediated therapy. Despite the extensive progress in the field, there is a long road until clinical translation. Main challenges include the development of robust and reproducible methods for EV production and isolation in a scale-up perspective. Other important issues to be further addressed are the choice of EV parent cell and the loading method in order to enable cargo encapsulation at high yield. Although much information is now available on EV biodistribution and EV-mediated therapy, researchers must be aware of potential biases in the imaging or therapeutic effects of the cargo alone (devoid of EV membrane). Therefore, special attention should be focused to thoroughly characterize the engineered EV to investigate systematically

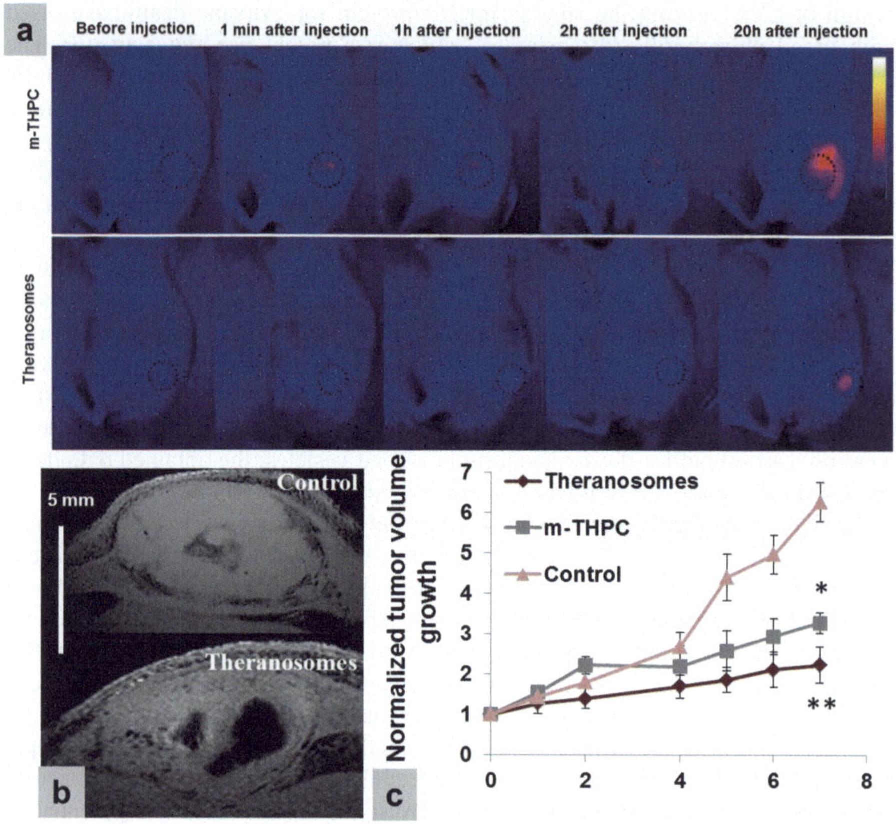

Fig. 5 Theranosome dual-mode imaging in vitro and in vivo and theranosome-mediated therapeutic effect in vivo. (**a**) False-color images of fluorescence emission from nude mice bearing TC-1 tumor before and after theranosome or m-THPC intratumoral injection, as a function of time. (**b**) MRI scans of TC-1 tumors 48 h after PBS intra-tumoral injection (control group) or 48 h after theranosome intra-tumoral injection. (**c**) Tumor growth curves (tumor volume is normalized to day 0) for mice intra-tumorally injected with theranosomes (25 μM m-THPC concentration) or m-THPC (25 μM) both followed by light exposure ($\lambda = 630$ nm − 30 J/cm^2: 77 mW/cm^2 for 390 s) 20 h later, compared with untreated control. All error bars reflect the SEM ($n = 5$). * and ** indicate $P < 0.05$ and $P < 0.01$, respectively, versus untreated control (no injection of therapeutic agent, no light exposure) at day 7. Adapted with permission from Magnetic and Photoresponsive Theranosomes: Silva A. K. et al. Translating Cell-Released Vesicles into Smart Nanovectors for Cancer Therapy. ACS Nano, 7 (6), pp 4954–4966, 2013. Copyright 2015 from American Chemical Society

endogenous and exogenous properties while ensuring an accurate quality control. Until now, studies provide the foundation of EV engineering for imaging or therapy purposes. Our ability to customize EVs to deliver nanoparticles, proteins, nucleic acid, and drugs is rapidly growing. Future investigations may afford an ultimate proof to ensure the safety and efficacy of EVs loaded with exogenous cargo, including rigorous pharmacokinetic and toxicological studies. Future advances are expected to overcome current challenges in the field and progress engineered EVs to the clinical practice.

References

1. Ratajczak J, Wysoczynski M, Hayek F, Janowska-Wieczorek A, Ratajczak M. Membrane-derived microvesicles: important and underappreciated mediators of cell-to-cell communication. Leukemia. 2006;20(9):1487–95.
2. Hugel B, Martínez MC, Kunzelmann C, Freyssinet J-M. Membrane microparticles: two sides of the coin. Physiology. 2005;20(1):22–7.
3. Camussi G, Deregibus MC, Bruno S, Cantaluppi V, Biancone L. Exosomes/microvesicles as a mechanism of cell-to-cell communication. Kidney Int. 2010;78(9):838–48.
4. Mause SF, Weber C. Microparticles: protagonists of a novel communication network for intercellular information exchange. Circ Res. 2010;107(9):1047–57.
5. Jimenez JJ, Jy W, Mauro LM, Soderland C, Horstman LL, Ahn YS. Endothelial cells release phenotypically and quantitatively distinct microparticles in activation and apoptosis. Thromb Res. 2003;109(4):175–80.
6. Boelens MC, Wu TJ, Nabet BY, Xu B, Qiu Y, Yoon T, Azzam DJ, Twyman-Saint Victor C, Wiemann BZ, Ishwaran H. Exosome transfer from stromal to breast cancer cells regulates therapy resistance pathways. Cell. 2014;159(3):499–513.
7. Camacho L, Guerrero P, Marchetti D. MicroRNA and protein profiling of brain metastasis competent cell-derived exosomes. PLoS One. 2013;8(9):e73790.
8. Wolf P. The nature and significance of platelet products in human plasma. Br J Haematol. 1967;13(3):269–88.
9. Krause M, Samoylenko A, Vainio SJ. Exosomes as renal inductive signals in health and disease, and their application as diagnostic markers and therapeutic agents. Front Cell Dev Biol. 2015;3:65.
10. Owens AP, Mackman N. Microparticles in hemostasis and thrombosis. Circ Res. 2011; 108(10):1284–97.
11. Cloutier N, Tan S, Boudreau LH, Cramb C, Subbaiah R, Lahey L, Albert A, Shnayder R, Gobezie R, Nigrovic PA. The exposure of autoantigens by microparticles underlies the formation of potent inflammatory components: the microparticle-associated immune complexes. EMBO Mol Med. 2013;5(2):235–49.
12. Holder BS, Tower CL, Jones CJ, Aplin JD, Abrahams VM. Heightened pro-inflammatory effect of preeclamptic placental microvesicles on peripheral blood immune cells in humans. Biol Reprod. 2012;86(4):103.
13. Prakash PS, Caldwell CC, Lentsch AB, Pritts TA, Robinson BR. Human microparticles generated during sepsis in patients with critical illness are neutrophil-derived and modulate the immune response. J Trauma Acute Care Surg. 2012;73(2):401–7.
14. Al-Nedawi K, Meehan B, Micallef J, Lhotak V, May L, Guha A, Rak J. Intercellular transfer of the oncogenic receptor EGFRvIII by microvesicles derived from tumour cells. Nat Cell Biol. 2008;10(5):619–24.
15. Valenti R, Huber V, Iero M, Filipazzi P, Parmiani G, Rivoltini L. Tumor-released microvesicles as vehicles of immunosuppression. Cancer Res. 2007;67(7):2912–5.
16. Kharaziha P, Ceder S, Li Q, Panaretakis T. Tumor cell-derived exosomes: a message in a bottle. Biochim Biophys Acta. 2012;1826(1):103–11.
17. Johnstone RM, Adam M, Hammond JR, Orr L, Turbide C. Vesicle formation during reticulocyte maturation. Association of plasma membrane activities with released vesicles (exosomes). J Biol Chem. 1987;262(19):9412–20.
18. Pan B-T, Johnstone RM. Fate of the transferrin receptor during maturation of sheep reticulocytes in vitro: selective externalization of the receptor. Cell. 1983;33(3):967–78.
19. Lee Y, El Andaloussi S, Wood MJA. Exosomes and microvesicles: extracellular vesicles for genetic information transfer and gene therapy. Hum Mol Genet. 2012;21(R1):R125–34.
20. Akers J, Gonda D, Kim R, Carter B, Chen C. Biogenesis of extracellular vesicles (EV): exosomes, microvesicles, retrovirus-like vesicles, and apoptotic bodies. J Neurooncol. 2013; 113(1):1–11.

21. Thery C, Zitvogel L, Amigorena S. Exosomes: composition, biogenesis and function. Nat Rev Immunol. 2002;2(8):569–79.
22. Leventis PA, Grinstein S. The distribution and function of phosphatidylserine in cellular membranes. Annu Rev Biophys. 2010;39:407–27.
23. Arraud N, Linares R, Tan S, Gounou C, Pasquet JM, Mornet S, Brisson AR. Extracellular vesicles from blood plasma: determination of their morphology, size, phenotype and concentration. J Thromb Haemost. 2014;12(5):614–27.
24. Elmore S. Apoptosis: a review of programmed cell death. Toxicol Pathol. 2007;35(4):495–516.
25. van der Pol E, Böing AN, Harrison P, Sturk A, Nieuwland R. Classification, functions, and clinical relevance of extracellular vesicles. Pharmacol Rev. 2012;64(3):676–705.
26. Utleg AG, Yi EC, Xie T, Shannon P, White JT, Goodlett DR, Hood L, Lin B. Proteomic analysis of human prostasomes. Prostate. 2003;56(2):150–61.
27. Caby M-P, Lankar D, Vincendeau-Scherrer C, Raposo G, Bonnerot C. Exosomal-like vesicles are present in human blood plasma. Int Immunol. 2005;17(7):879–87.
28. Pisitkun T, Shen R-F, Knepper MA. Identification and proteomic profiling of exosomes in human urine. PNAS. 2004;101(36):13368–73.
29. Ogawa Y, Miura Y, Harazono A, Kanai-Azuma M, Akimoto Y, Kawakami H, Yamaguchi T, Toda T, Endo T, Tsubuki M. Proteomic analysis of two types of exosomes in human whole saliva. Biol Pharm Bull. 2011;34(1):13–23.
30. Admyre C, Johansson SM, Qazi KR, Filén J-J, Lahesmaa R, Norman M, Neve EPA, Scheynius A, Gabrielsson S. Exosomes with immune modulatory features are present in human breast milk. J Immunol. 2007;179(3):1969–78.
31. Asea A, Jean-Pierre C, Kaur P, Rao P, Linhares IM, Skupski D, Witkin SS. Heat shock protein-containing exosomes in mid-trimester amniotic fluids. J Reprod Immunol. 2008;79(1): 12–7.
32. Andre F, Schartz NEC, Movassagh M, Flament C, Pautier P, Morice P, Pomel C, Lhomme C, Escudier B, Le Chevalier T, Tursz T, Amigorena S, Raposo G, Angevin E, Zitvogel L. Malignant effusions and immunogenic tumour-derived exosomes. Lancet. 2002;360(9329): 295–305.
33. Masyuk AI, Huang BQ, Ward CJ, Gradilone SA, Banales JM, Masyuk TV, Radtke B, Splinter PL, LaRusso NF. Biliary exosomes influence cholangiocyte regulatory mechanisms and proliferation through interaction with primary cilia. Am J Physiol Gastrointest Liver Physiol. 2010;299(4):G990–9.
34. Tanimura A, McGregor DH, Anderson HC. Matrix vesicles in atherosclerotic calcification. Exp Biol Med. 1983;172(2):173–7.
35. Fischer von Mollard G, Mignery GA, Baumert M, Perin MS, Hanson TJ, Burger PM, Jahn R, Südhof TC. rab3 is a small GTP-binding protein exclusively localized to synaptic vesicles. PNAS. 1990;87(5):1988–92.
36. Le Pecq J-B. Dexosomes as a therapeutic cancer vaccine: from bench to bedside. Blood Cells Mol Dis. 2005;35(2):129–35.
37. Chaput N, Schartz N, Andre F, Zitvogel L. Exosomes for immunotherapy of cancer. New trends in cancer for the 21st century. New York: Springer; 2003. p. 215–21.
38. Shen Y, Torchia MLG, Lawson GW, Karp CL, Ashwell JD, Mazmanian SK. Outer membrane vesicles of a human commensal mediate immune regulation and disease protection. Cell Host Microbe. 2012;12(4):509–20.
39. Yeo RWY, Lai RC, Zhang B, Tan SS, Yin Y, Teh BJ, Lim SK. Mesenchymal stem cell: an efficient mass producer of exosomes for drug delivery. Adv Drug Deliv Rev. 2013;65(3): 336–41.
40. Lai RC, Chen TS, Lim SK. Mesenchymal stem cell exosome: a novel stem cell-based therapy for cardiovascular disease. Regen Med. 2011;6(4):481–92.
41. Riches A, Campbell E, Borger E, Powis S. Regulation of exosome release from mammary epithelial and breast cancer cells—a new regulatory pathway. Eur J Cancer. 2014;50(5):1025–34.
42. Sun L, Wang H-X, Zhu X-J, Wu P-H, Chen W-Q, Zou P, Li Q-B, Chen Z-C. Serum deprivation elevates the levels of microvesicles with different size distributions and selectively

enriched proteins in human myeloma cells in vitro. Acta Pharmacol Sin. 2014;35(3): 381–93.
43. King HW, Michael MZ, Gleadle JM. Hypoxic enhancement of exosome release by breast cancer cells. BMC Cancer. 2012;12(1):421.
44. Headland SE, Jones HR, D'Sa AS, Perretti M, Norling LV. Cutting-edge analysis of extracellular microparticles using imagestreamx imaging flow cytometry. Sci Rep. 2014;4:5237.
45. Pick H, Schmid EL, Tairi A-P, Ilegems E, Hovius R, Vogel H. Investigating Cellular Signaling Reactions in Single Attoliter Vesicles J Am Chem Soc. 2005;127(9):2908–12.
46. Mao Z, Cartier R, Hohl A, Farinacci M, Dorhoi A, Nguyen T-L, Mulvaney P, Ralston J, Kaufmann SHE, Möhwald H, Wang D. Cells as Factories for Humanized Encapsulation Nano Lett. 2011;11(5):2152–6.
47. Momen-Heravi F, Bala S, Kodys K, Szabo G. Exosomes derived from alcohol-treated hepatocytes horizontally transfer liver specific miRNA-122 and sensitize monocytes to LPS. Sci Rep. 2015:5.
48. Machluf M, Bronshtein T. Liposomal compositions and uses of same patent. 2010; US20120164214 A1.
49. Toledano Furman NE, Lupu-Haber Y, Bronshtein T, Kaneti L, Letko N, Weinstein E, Baruch L, Machluf M. Reconstructed stem cell nanoghosts: a natural tumor targeting platform. Nano Lett. 2013;13(7):3248–55.
50. Jo W, Jeong D, Kim J, Cho S, Jang SC, Han C, Kang JY, Gho YS, Park J. Microfluidic fabrication of cell-derived nanovesicles as endogenous RNA carriers. Lab Chip. 2014;14(7):1261–9.
51. Jang SC, Kim OY, Yoon CM, Choi D-S, Roh T-Y, Park J, Nilsson J, Lötvall J, Kim Y-K, Gho YS. Bioinspired Exosome-Mimetic Nanovesicles for Targeted Delivery of Chemotherapeutics to Malignant Tumors ACS Nano. 2013;7(9):7698–710.
52. Jo W, Kim J, Yoon J, Jeong D, Cho S, Jeong H, Yoon Y, Kim S, Gho Y, Park J. Large-scale generation of cell-derived nanovesicles. Nanoscale. 2014;6(20):12056–64.
53. Yoon J, Jo W, Jeong D, Kim J, Jeong H, Park J. Generation of nanovesicles with sliced cellular membrane fragments for exogenous material delivery. Biomaterials. 2015;59:12–20.
54. Yun S-S, Yoon SY, Song M-K, Im S-H, Kim S, Lee J-H, Yang S. Handheld mechanical cell lysis chip with ultra-sharp silicon nano-blade arrays for rapid intracellular protein extraction. Lab Chip. 2010;10(11):1442–6.
55. Silva AKA, Di Corato R, Pellegrino T, Chat S, Pugliese G, Luciani N, Gazeau F, Wilhelm C. Cell-derived vesicles as a bioplatform for the encapsulation of theranostic nanomaterials. Nanoscale. 2013;5(23):11374–84.
56. Silva AK, Luciani N, Gazeau F, Aubertin K, Bonneau S, Chauvierre C, Letourneur D, Wilhelm C. Combining magnetic nanoparticles with cell derived microvesicles for drug loading and targeting. Nanomed Nanotech Biol Med. 2015;11(3):645–55.
57. Silva AK, Kolosnjaj-Tabi J, Bonneau S, Marangon I, Boggetto N, Aubertin K, Clément O, Bureau MF, Luciani N, Gazeau F. Magnetic and photoresponsive theranosomes: translating cell-released vesicles into smart nanovectors for cancer therapy. ACS Nano. 2013;7(6): 4954–66.
58. Mao Z, Cartier R, Hohl A, Farinacci M, Dorhoi A, Nguyen T-L, Mulvaney P, Ralston J, Kaufmann SH, Möhwald H. Cells as factories for humanized encapsulation. Nano Lett. 2011;11(5):2152–6.
59. Wiklander OPB, Nordin JZ, O'Loughlin A, Gustafsson Y, Corso G, Mäger I, Vader P, Lee Y, Sork H, Seow Y, Heldring N, Alvarez-Erviti L, Smith CIE, Le Blanc K, Macchiarini P, Jungebluth P, Wood MJA, Andaloussi SEL. Extracellular vesicle in vivo biodistribution is determined by cell source, route of administration and targeting. J Extracell Vesicles. 2015:4. 10.3402/jev.v4.26316.
60. Wahlgren J, Karlson TDL, Brisslert M, Vaziri Sani F, Telemo E, Sunnerhagen P, Valadi H. Plasma exosomes can deliver exogenous short interfering RNA to monocytes and lymphocytes Nucleic Acids Res. 2012:1–12.

61. Ohno S-I, Takanashi M, Sudo K, Ueda S, Ishikawa A, Matsuyama N, Fujita K, Mizutani T, Ohgi T, Ochiya T. Systemically injected exosomes targeted to EGFR deliver antitumor microRNA to breast cancer cells. Mol Ther. 2013;21(1):185–91.
62. Mizrak A, Bolukbasi MF, Ozdener GB, Brenner GJ, Madlener S, Erkan EP, Strobel T, Breakefield XO, Saydam O. Genetically Engineered Microvesicles Carrying Suicide mRNA/Protein Inhibit Schwannoma Tumor Growth Mol Ther. 2013;21(1):101–8.
63. Raposo G, Stoorvogel W. Extracellular vesicles: exosomes, microvesicles, and friends. J Cell Biol. 2013;200(4):373–83.
64. van Balkom BW, Eisele AS, Pegtel DM, Bervoets S, Verhaar MC. Quantitative and qualitative analysis of small RNAs in human endothelial cells and exosomes provides insights into localized RNA processing, degradation and sorting. J Extracell Vesicles. 2015;4:26760.
65. Kanada M, Bachmann MH, Hardy JW, Frimannson DO, Bronsart L, Wang A, Sylvester MD, Schmidt TL, Kaspar RL, Butte MJ. Differential fates of biomolecules delivered to target cells via extracellular vesicles. PNAS. 2015;112(12):E1433–42.
66. Alvarez-Erviti L, Seow Y, Yin H, Betts C, Lakhal S, Wood MJ. Delivery of siRNA to the mouse brain by systemic injection of targeted exosomes. Nat Biotech. 2011;29(4):341–5.
67. Shen B, Wu N, Yang J-M, Gould SJ. Protein Targeting to Exosomes/Microvesicles by Plasma Membrane Anchors J Biol Chem. 2011;286(16):14383–95.
68. Vats N, Wilhelm C, Rautou P-E, Poirier-Quinot M, Péchoux C, Devue C, Boulanger CM, Gazeau F. Magnetic tagging of cell-derived microparticles: new prospects for imaging and manipulation of these mediators of biological information. Nanomedicine. 2010;5(5): 727–38.
69. Rank A, Nieuwland R, Crispin A, Grützner S, Iberer M, Toth B, Pihusch R. Clearance of platelet microparticles in vivo. Platelets. 2011;22(2):111–6.
70. Zhuang X, Xiang X, Grizzle W, Sun D, Zhang S, Axtell RC, Ju S, Mu J, Zhang L, Steinman L, Miller D, Zhang H-G. Treatment of brain inflammatory diseases by delivering exosome encapsulated anti-inflammatory drugs from the nasal region to the brain. Mol Ther. 2011;19(10):1769–79.
71. Sun D, Zhuang X, Xiang X, Liu Y, Zhang S, Liu C, Barnes S, Grizzle W, Miller D, Zhang H-G. A novel nanoparticle drug delivery system: the anti-inflammatory activity of curcumin is enhanced when encapsulated in exosomes. Mol Ther. 2010;18(9):1606–14.
72. Peinado H, Alečković M, Lavotshkin S, Matei I, Costa-Silva B, Moreno-Bueno G, Hergueta-Redondo M, Williams C, García-Santos G, Nitadori-Hoshino A, Hoffman C, Badal K, Garcia BA, Callahan MK, Yuan J, Martins VR, Skog J, Kaplan RN, Brady MS, Wolchok JD, Chapman PB, Kang Y, Bromberg J, Lyden D. Melanoma exosomes educate bone marrow progenitor cells toward a pro-metastatic phenotype through MET. Nat Med. 2012;18(6):883–91.
73. Yang T, Martin P, Fogarty B, Brown A, Schurman K, Phipps R, Yin V, Lockman P, Bai S. Exosome Delivered Anticancer Drugs Across the Blood-Brain Barrier for Brain Cancer Therapy in Danio Rerio Pharm Res. 2015;32(6):2003–14.
74. Willekens FL, Werre JM, Kruijt JK, Roerdinkholder-Stoelwinder B, Groenen-Döpp YA, van den Bos AG, Bosman GJ, van Berkel TJ. Liver Kupffer cells rapidly remove red blood cell–derived vesicles from the circulation by scavenger receptors. Blood. 2005;105(5):2141–5.
75. Aaseth J, Alexander J, Norseth T. Uptake of 51CrChromate by human erythrocytes–a role of glutathione. Acta Pharmacol Toxicol. 1982;50(4):310–5.
76. Kooijmans SA, Stremersch S, Braeckmans K, de Smedt SC, Hendrix A, Wood MJ, Schiffelers RM, Raemdonck K, Vader P. Electroporation-induced siRNA precipitation obscures the efficiency of siRNA loading into extracellular vesicles. J Control Release. 2013;172(1):229–38.
77. Haney MJ, Klyachko NL, Zhao Y, Gupta R, Plotnikova EG, He Z, Patel T, Piroyan A, Sokolsky M, Kabanov AV, Batrakova EV. Exosomes as drug delivery vehicles for Parkinson's disease therapy. J Control Release. 2015;207:18–30.
78. György B, Módos K, Pállinger É, Pálóczi K, Pásztói M, Misják P, Deli MA, Sipos Á, Szalai A, Voszka I, Polgár A, Tóth K, Csete M, Nagy G, Gay S, Falus A, Kittel Á, Buzás EI. Detection and isolation of cell-derived microparticles are compromised by protein complexes resulting from shared biophysical parameters. Blood. 2011;117(4):e39–48.

79. Théry C, Amigorena S, Raposo G, Clayton A. Isolation and characterization of exosomes from cell culture supernatants and biological fluids. Curr Protoc Cell Biol. 2006:3.22.1–3.9.
80. Lamparski HG, Metha-Damani A, Yao J-Y, Patel S, Hsu D-H, Ruegg C, Le Pecq J-B. Production and characterization of clinical grade exosomes derived from dendritic cells. J Immunol Methods. 2002;270(2):211–26.
81. Mathivanan S, Lim JW, Tauro BJ, Ji H, Moritz RL, Simpson RJ. Proteomics analysis of A33 immunoaffinity-purified exosomes released from the human colon tumor cell line LIM1215 reveals a tissue-specific protein signature. Mol Cell Proteomics. 2010;9(2):197–208.
82. Tauro BJ, Greening DW, Mathias RA, Mathivanan S, Ji H, Simpson RJ. Two distinct populations of exosomes are released from LIM1863 colon carcinoma cell-derived organoids. Mol Cell Proteomics. 2013;12(3):587–98.
83. Tauro BJ, Greening DW, Mathias RA, Ji H, Mathivanan S, Scott AM, Simpson RJ. Comparison of ultracentrifugation, density gradient separation, and immunoaffinity capture methods for isolating human colon cancer cell line LIM1863-derived exosomes. Methods. 2012;56(2): 293–304.
84. Chen C, Skog J, Hsu C-H, Lessard RT, Balaj L, Wurdinger T, Carter BS, Breakefield XO, Toner M, Irimia D. Microfluidic isolation and transcriptome analysis of serum microvesicles. Lab Chip. 2010;10(4):505–11.
85. Lee K, Shao H, Weissleder R, Lee H. Acoustic Purification of Extracellular Microvesicles ACS Nano. 2015;9(3):2321–7.
86. Böing AN, van der Pol E, Grootemaat AE, Coumans FAW, Sturk A, Nieuwland R. Single-step isolation of extracellular vesicles by size-exclusion chromatography J Extracell Vesicles. 2014:3. 10.3402/jev.v3.23430.
87. Welton JL, Webber JP, Botos L-A, Jones M, Clayton A. Ready-made chromatography columns for extracellular vesicle isolation from plasma. J Extracell Vesicles. 2015;4.
88. Filipe V, Hawe A, Jiskoot W. Critical Evaluation of Nanoparticle Tracking Analysis (NTA) by NanoSight for the Measurement of Nanoparticles and Protein Aggregates Pharm Res. 2010;27(5):796–810.
89. Pol E, Coumans F, Varga Z, Krumrey M, Nieuwland R. Innovation in detection of microparticles and exosomes. J Thromb Haemost. 2013;11(s1):36–45.
90. Hoo C, Starostin N, West P, Mecartney M. A comparison of atomic force microscopy (AFM) and dynamic light scattering (DLS) methods to characterize nanoparticle size distributions. J Nanopart Res. 2008;10(1):89–96.
91. Momen-Heravi F, Balaj L, Alian S, Tigges J, Toxavidis V, Ericsson M, Distel RJ, Ivanov AR, Skog J, Kuo WP. Alternative methods for characterization of extracellular vesicles. Front Physiol. 2012;3:354.
92. Dragovic RA, Gardiner C, Brooks AS, Tannetta DS, Ferguson DJP, Hole P, Carr B, Redman CWG, Harris AL, Dobson PJ, Harrison P, Sargent IL. Sizing and phenotyping of cellular vesicles using nanoparticle tracking analysis. Nanomed Nanotech Biol Med. 2011;7(6): 780–8.
93. van der Pol E, Coumans FAW, Grootemaat AE, Gardiner C, Sargent IL, Harrison P, Sturk A, van Leeuwen TG, Nieuwland R. Particle size distribution of exosomes and microvesicles determined by transmission electron microscopy, flow cytometry, nanoparticle tracking analysis, and resistive pulse sensing. J Thromb Haemost. 2014;12(7):1182–92.
94. Yuana Y, Oosterkamp T, Bahatyrova S, Ashcroft B, Garcia Rodriguez P, Bertina R, Osanto S. Atomic force microscopy: a novel approach to the detection of nanosized blood microparticles. J Thromb Haemost. 2010;8(2):315–23.
95. Kanno T, Yamada T, Iwabuki H, Tanaka H, Kuroda SI, Tanizawa K, Kawai T. Size distribution measurement of vesicles by atomic force microscopy. Anal Biochem. 2002;309(2): 196–9.
96. Van Der Pol E, Hoekstra AG, Sturk A, Otto C, Van Leeuwen TG, Nieuwland R. Optical and non-optical methods for detection and characterization of microparticles and exosomes. J Thromb Haemost. 2010;8(12):2596–607.

97. Andriola Silva AK, Di Corato R, Gazeau F, Pellegrino T, Wilhelm C. Magnetophoresis at the nanoscale: tracking the magnetic targeting efficiency of nanovectors. Nanomedicine. 2012;7(11):1713–27.
98. Lacroix R, Robert S, Poncelet P, Kasthuri RS, Key NS, Dignat-George F. Standardization of platelet-derived microparticle enumeration by flow cytometry with calibrated beads: results of the International Society on Thrombosis and Haemostasis SSC Collaborative workshop. J Thromb Haemost. 2010;8(11):2571–4.
99. Arraud N, Gounou C, Linares R, Brisson AR. A simple flow cytometry method improves the detection of phosphatidylserine-exposing extracellular vesicles. J Thromb Haemost. 2015;13(2):237–47.
100. Palomo L, Casal E, Royo F, Cabrera D, van-Liempd S, Falcon-Perez JM. Considerations for applying metabolomics to the analysis of extracellular vesicles. Front Immunol. 2014;5:651.
101. Kreimer S, Belov AM, Ghiran I, Murthy SK, Frank DA, Ivanov AR. Mass spectrometry-based molecular characterization of extracellular vesicles: lipidomics and proteomics. J Proteome Res. 2015;14:2367–84.
102. Lötvall J, Hill AF, Hochberg F, Buzás EI, Di Vizio D, Gardiner C, Gho YS, Kurochkin IV, Mathivanan S, Quesenberry P. Minimal experimental requirements for definition of extracellular vesicles and their functions: a position statement from the International Society for Extracellular Vesicles. J Extracell Vesicles. 2014;3:10.3402.
103. Webber J, Clayton A. How pure are your vesicles? J Extracell Vesicles. 2013;2:10.3402
104. Al Faraj A, Gazeau F, Wilhelm C, Devue C, Guérin CL, Péchoux C, Paradis V, Clément O, Boulanger CM, Rautou P-E. Endothelial cell-derived microparticles loaded with iron oxide nanoparticles: feasibility of MR imaging monitoring in mice. Radiology. 2012;263(1):169–78.
105. Grange C, Tapparo M, Bruno S, Chatterjee D, Quesenberry PJ, Tetta C, Camussi G. Biodistribution of mesenchymal stem cell-derived extracellular vesicles in a model of acute kidney injury monitored by optical imaging. Int J Mol Med. 2014;33(5):1055–63.
106. Lai CP, Mardini O, Ericsson M, Prabhakar S, Maguire CA, Chen JW, Tannous BA, Breakefield XO. Dynamic Biodistribution of Extracellular Vesicles in Vivo Using a Multimodal Imaging Reporter ACS Nano. 2014;8(1):483–94.
107. Barenholz YC. Doxil®—the first FDA-approved nano-drug: lessons learned. J Control Release. 2012;160(2):117–34.
108. Wang R, Billone PS, Mullett WM. Nanomedicine in action: an overview of cancer nanomedicine on the market and in clinical trials. J Nanomater. 2013;2013:1.
109. Yoo J-W, Irvine DJ, Discher DE, Mitragotri S. Bio-inspired, bioengineered and biomimetic drug delivery carriers. Nat Rev Drug Discov. 2011;10(7):521–35.
110. Johnsen KB, Gudbergsson JM, Skov MN, Pilgaard L, Moos T, Duroux M. A comprehensive overview of exosomes as drug delivery vehicles—endogenous nanocarriers for targeted cancer therapy. Biochim Biophys Acta. 2014;1846(1):75–87.
111. Natasha G, Gundogan B, Tan A, Farhatnia Y, Wu W, Rajadas J, Seifalian AM. Exosomes as Immunotheranostic Nanoparticles Clin Ther. 2014;36(6):820–9.
112. Batrakova EV, Kim MS. Using exosomes, naturally-equipped nanocarriers, for drug delivery. J Control Release. 2015;219:396–405.
113. Yin W, Ouyang S, Li Y, Xiao B, Yang H. Immature Dendritic Cell-Derived Exosomes: a Promise Subcellular Vaccine for Autoimmunity Inflammation. 2013;36(1):232–40.
114. Tian Y, Li S, Song J, Ji T, Zhu M, Anderson GJ, Wei J, Nie G. A doxorubicin delivery platform using engineered natural membrane vesicle exosomes for targeted tumor therapy. Biomaterials. 2014;35(7):2383–90.
115. Katakowski M, Buller B, Zheng X, Lu Y, Rogers T, Osobamiro O, Shu W, Jiang F, Chopp M. Exosomes from marrow stromal cells expressing miR-146b inhibit glioma growth. Cancer Lett. 2013;335(1):201–4.

Magnetic Particle Imaging

Bo Zheng, Kuan Lu, Justin J. Konkle, Daniel W. Hensley, Paul Keselman, Ryan D. Orendorff, Zhi Wei Tay, Elaine Yu, Xinyi Y. Zhou, Mindy Bishop, Beliz Gunel, Laura Taylor, R. Matthew Ferguson, Amit P. Khandhar, Scott J. Kemp, Kannan M. Krishnan, Patrick W. Goodwill, and Steven M. Conolly

1 Unmet Biomedical Imaging Needs

The development of new clinical diagnosis and treatments, including safe ww, cancer diagnosis, and treatments, and stem cell-based therapies, requires quantitative, noninvasive in vivo imaging methods that can quantitatively assess the physiology and etiology of diseased tissues anywhere in the body. Ideally researchers would need the ability to perform dozens of serial scans over a period of several months.

Unfortunately, we currently lack an ideal in vivo imaging modality for many clinical and preclinical applications. Figure 1 compares several common preclinical imaging methods such as optical imaging, T_2*-MRI, and computed tomography (CT) to magnetic particle imaging (MPI), a new imaging technique which we describe in this chapter. As can be seen in the figure, optical imaging techniques are superb near the skin, but optical absorption and scatter restrict these methods to

B. Zheng (✉) • K. Lu • D.W. Hensley • P. Keselman • R.D. Orendorff • Z.W. Tay • E. Yu
X.Y. Zhou • M. Bishop • L. Taylor
Department of Bioengineering, UC Berkeley,
340 Hearst Memorial Mining Building, Berkeley, CA 94720, USA
e-mail: bozheng@berkeley.edu

J.J. Konkle
Magnetic Insight, Newark, CA, USA

B. Gunel
Department of Electrical Engineering and Computer Sciences, UC Berkeley, Berkeley, CA, USA

R.M. Ferguson • A.P. Khandhar • S.J. Kemp
LodeSpin Labs, Seattle, WA, USA

K.M. Krishnan
Materials Science and Engineering, University of Washington, Seattle, WA, USA
LodeSpin Labs, Seattle, WA, USA

J.W.M. Bulte, M.M.J. Modo (eds.), *Design and Applications of Nanoparticles in Biomedical Imaging*, DOI 10.1007/978-3-319-42169-8_4

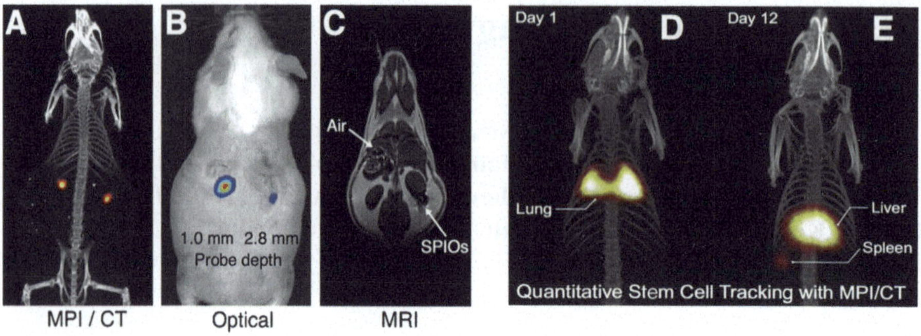

Fig. 1 MPI for high-contrast tracer imaging. (**a**) MPI scans with/CT overlay of two SPIO vials (at 1 and 2.8 mm deep) shows no signal change with depth. (**b**) Fluorescent scan of two implanted fluorescent probes at the same depths as (**a**) shows a dramatic signal loss with depth. (**c**) MRI (7T T_2*) images of two SPIO vials from (**a**). SPIOs induce areas with loss of signal, which are difficult to distinguish from naturally dark tissues such as air in the lungs or tendons. (**d**) MPI/CT overlay of IV injected stem cells trapped in the lungs on day 1, (**e**) with redistribution to and clearance from the liver on day 12

shallow depths. Also, many optical imaging methods rely on genetic editing (e.g., GFP), which is not clinically translatable [1]. Nuclear medicine methods are exquisitely sensitive for radiotracers, but are fundamentally limited by short tracer half-life and radioactive dose. For stem cell tracking applications, T_2*-MRI can detect even a single labeled cell when the SPIO-labeled cells are visualized against a uniform background of long-T_2 tissues. However, T_2*-MRI offers only "*negative contrast*," which can fail in regions of the body adjacent to tissues with naturally low signal, such as regions near the lungs, tendon, or bone (see Fig. 1c).

Unlike other techniques, MPI directly images the magnetization of SPIO tracers, so the MPI signal is not affected by any biological tissue. No biological tissue contains a magnetic signature similar to that of SPIOs, and hence there is no tissue background signal in MPI, akin to nuclear medicine techniques and fluorine MRI. Moreover, the low-frequency magnetic fields used in MPI are completely transparent through biological tissue, enabling imaging of deep anatomic

P.W. Goodwill
Department of Bioengineering, UC Berkeley,
340 Hearst Memorial Mining Building, Berkeley, CA 94720, USA

Magnetic Insight, Newark, CA, USA

S.M. Conolly
Department of Bioengineering, UC Berkeley,
340 Hearst Memorial Mining Building, Berkeley, CA 94720, USA

Magnetic Insight, Newark, CA, USA

Department of Electrical Engineering and Computer Sciences, UC Berkeley,
Berkeley, CA, USA

structures. These properties, combined with the linear quantitation in MPI for SPIO tracers and the safety of both the SPIO tracer and the nonionizing MPI scanning technique, make MPI ideal for a variety of sensitive clinical molecular imaging applications.

2 Introduction to Magnetic Particle Imaging

MPI was originally developed by Bernhard Gleich and Juergen Weizenecker at Philips Research in Hamburg [2]. The field of MPI research is rapidly expanding, and current efforts in MPI development are in progress both in industry at Philips research [2–6], Bruker [7], Lodespin Labs [8, 9], and Magnetic Insight [10], as well as in academia at UC Berkeley [11–22], University of Lubeck [23–25], Bilkent University [26], Physikalisch-Technische Bundesanstalt and Charité-Universitätsmedizin Berlin [27, 28], TU Braunschweig [29, 30], University of Wuerzburg [31], Osaka University [32, 33], University of Washington [9, 34–39], Harvard Massachusetts General Hospital, Case Western Reserve University [40], Johns Hopkins University [41], University of Twente [42, 43], Dartmouth University [44, 45], and TU Hamburg-Harburg [46, 47], among many others.

Since its development, MPI has shown promise for a variety of medical imaging applications, such as angiography and blood pool imaging [2–4, 11, 14–16, 21], cancer imaging [16], perfusion imaging, and stem cell tracking [13, 20, 41]. In MPI, we use SPIO tracers (also used in T_2*-MRI), but MPI scans *cannot* be obtained with an MRI scanner; the MPI hardware is completely distinct.

2.1 MPI Trade-Offs

MPI visualizes SPIO tracers by direct induction, so the MPI signal scales exactly linearly with the number of SPIOs in the imaging volume, meaning MPI is a "hotspot" imaging technique without background signal, similar to nuclear medicine techniques. For stem cell tracking, MPI provides quantitative cell counts anywhere in the body for up to months in the same animal [13]. Unlike optical imaging, ultrasound, X-ray, and other imaging methods, the MPI magnetic signal suffers zero attenuation with depth and there is no MPI signal from background tissues. The MPI signal also has excellent mass sensitivity because MPI detects the intense *electronic* magnetization of an SPIO, which is 22 million times more intense than the *nuclear* paramagnetism of water in 7T MRI. Hence, our prototype MPI scanner can detect even 2 nanograms of SPIOs within a single voxel. Theoretically, this translates to roughly 200 nanomolar concentration of iron, or to 200 cells with in vivo scans in state-of-the-art MPI scanners. Hence, the contrast and sensitivity of MPI rival nuclear medicine techniques like PET and SPECT without concerns about radiation dose or "half-life" constraints. Indeed, a few SPIO formulations have previously been approved by EU and FDA as diagnostic agents. MPI uses no ionizing radiation, and it can operate within the human magnetic safety limits so human translation will soon be feasible. Current disadvantages of MPI include relatively poor spatial resolution (roughly 1 mm in a mouse), and the lack of anatomical information implying that it cannot be used as a single imaging modality alone. Recently, a commercial murine MPI scanner was released in Germany (Bruker), and a preclinical scanner is in development at Magnetic Insight (California). A start-up in Seattle (LodeSpin Labs) has been formed around the synthesis of MPI-tailored SPIOs.

2.2 *How MPI Works*

The physics governing MPI is completely different from MRI, and it is not possible to perform MPI with an MRI scanner or vice versa. An overview of the physics, hardware, and reconstruction theory of MPI can be found in references [2–4, 11, 12, 14–17, 21, 48]. The SPIO tracers used in MPI obey steady-state Langevin physics as shown in Fig. 2. They achieve steady-state alignment shown here with applied fields in tens of microseconds. MPI scanners employ a strong gradient field to force all SPIOs outside of the gradient field origin, called the field-free point (FFP), to be saturated. Saturated SPIOs induce no signal in the MPI inductive receiver coil because they are static.

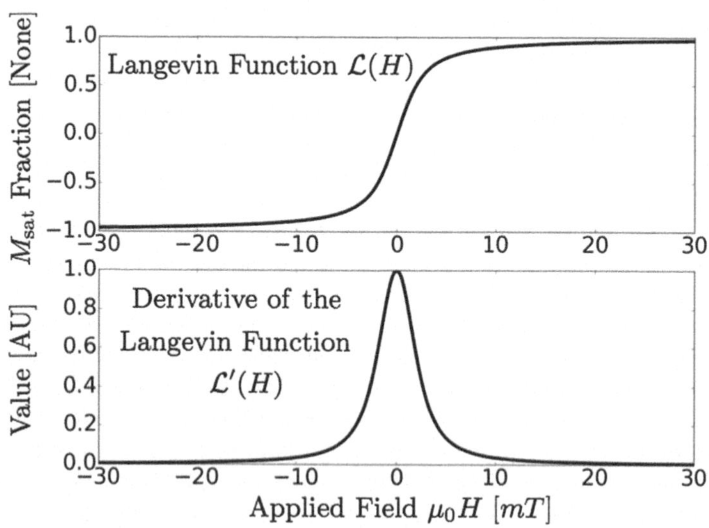

Fig. 2 Langevin theory describes the steady-state magnetization of an ensemble of SPIOs. (*Top*) SPIOs magnetize and quickly fully align, or saturate, in the presence of applied magnetic fields. (*Bottom*) We observe the SPIO density after it is blurred (or convolved) with the MPI point spread function (PSF), which is the derivative of the Langevin function

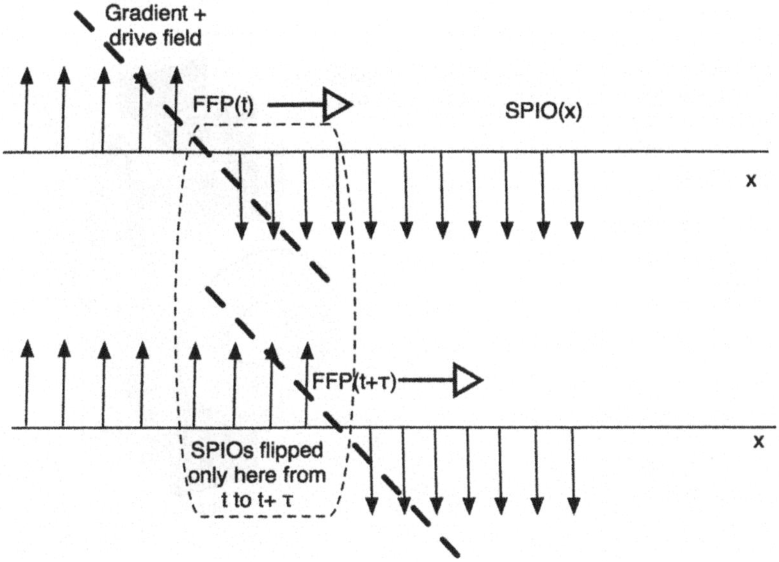

Fig. 3 In MPI, we raster a gradient field across the field of view, typically at 25 kHz. This is illustrated here in 1D but the principle is the same in multiple dimensions. Any SPIOs near the instantaneous FFP (or FFL) are flipped, inducing a voltage in the pickup coil that is linearly proportional to the mass of SPIOs at the FFP (or FFL) location. This allows for fast and robust x-space image reconstruction

All MPI scanners today scan the FFP across the FOV at roughly 25 kHz to create an MP image, as illustrated in Fig. 3. The SPIOs flip their orientation when the FFP sweeps past their location. The flip of even 2 nanograms of SPIOs in a voxel induces a detectable signal in a receiver coil. Importantly, the induced signal is linearly proportional to the SPIO mass at each location in space. Hence, the time axis of the MPI induction signal maps directly to spatial location in the SPIO image. This is the intuition behind the x-space MPI reconstruction algorithm [11, 12, 14–17, 48]. Multidimensional MPI is a simple extension of the 1D imaging process. Movement of the FFP in 3D is implemented by dynamically driving three spatially homogeneous "slow shift field" coils, one aligned with each direction (*x*, *y*, *z*). This effectively scans the FFP, or field-free line (FFL), in two or three dimensions, as shown in Fig. 4.

2.3 *The Direct Feed-Through Challenge*

Unfortunately, all MPI scanners must deal with a crucial technical challenge. The received MPI signal occurs at the same time as the drive field, which can be orders of magnitude stronger in amplitude. Hence, the receive coil picks up a massive interference from the time-varying drive fields. Fortunately, this direct

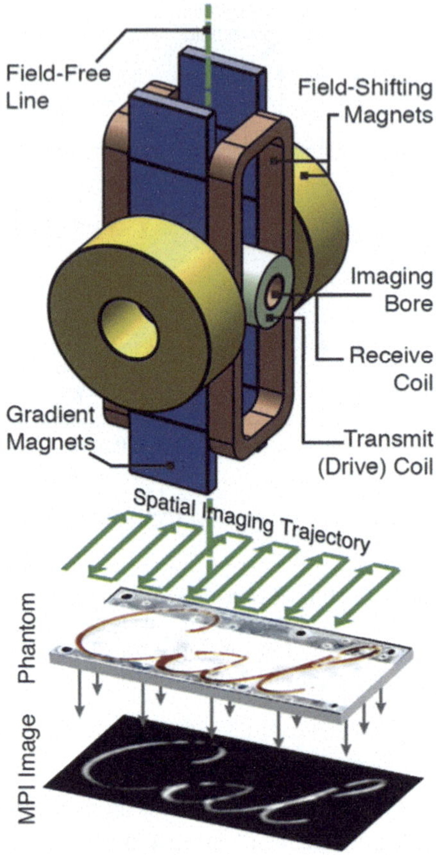

Fig. 4 Illustration of our prototype FFL scanner and 2D MPI scanning process. Three superposed magnets (field-shifting magnets, gradient magnets, and transmit magnet) produce and translate a magnetic field-free line (FFL) across an imaging volume. As the FFL is rastered across a distribution of SPIO particles using a spatial imaging trajectory, the particle ensemble magnetization changes in magnitude and orientation in response. The changing particle magnetization is detected via a detector coil and reconstructed to form an image

feed-through interference is localized (mostly) at the drive frequency, typically 25 kHz, by meticulous design of the transmit power electronics to ensure a low total harmonic distortion at 10–100 ppm. Hence, while the MPI signal is not contaminated by direct feed-through at higher harmonics, the signal at the fundamental frequency is swamped with direct feed-through interference. As a result, all MPI scanners reject the direct feed-through using a sharp passive filter, with more than 30,000-fold suppression. Unfortunately, this filter also removes the first harmonic content of the SPIO signal. The MPI image reconstruction challenge is to reconstruct a high-quality SPIO image using only the SPIO signal derived from higher harmonics of the 25 kHz drive frequency.

Both the system matrix reconstruction approach, advanced by Philips, and the x-space reconstruction approach, developed by UC Berkeley, have shown that the lost first harmonic information corresponded precisely to the average (baseline or DC) signal in a partial imaging volume [4, 11]. Hence, it is possible to design a robust image reconstruction algorithm that restores the lost DC information using excitation trajectories with overlapping partial imaging volumes, or field of views

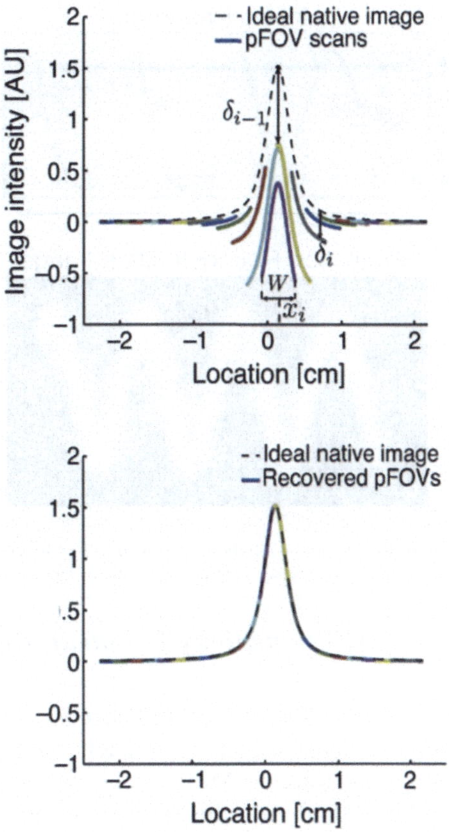

Fig. 5 Illustration of the x-space MPI DC recovery algorithm [11]. Overlapping partial FOVs lose all local baseline information due to the direct feed-through rejection filter. Using the *a priori* information that the MPI distribution is smooth, this continuity algorithm recovers the lost information to reconstruct a smooth 1D image

(pFOV). These are then stitched together using a fast continuity algorithm [11], which is demonstrated in Fig. 5. Importantly, Lu et al. and Zheng et al. proved experimentally and theoretically that MPI is a linear and shift invariant imaging system after the lost baseline information is restored [11, 13]. Several studies have also proven that the fundamental 1D point spread function (PSF) of MPI is simply a scaled version of the derivative of the SPIO's Langevin function, as shown in Fig. 2 [14, 15, 17]. Below we summarize the fundamental properties of MP images: *contrast, sensitivity, and resolution.*

2.4 *MPI Signal Is Independent of Depth*

Unlike in optical or ultrasound imaging, biological tissue does not attenuate the MPI signal. There is also no signal from living tissue in MPI as demonstrated experimentally in Fig. 6. This makes MPI uniquely suitable for linearly quantitative and radiation-free cell tracking at any depth in vivo.

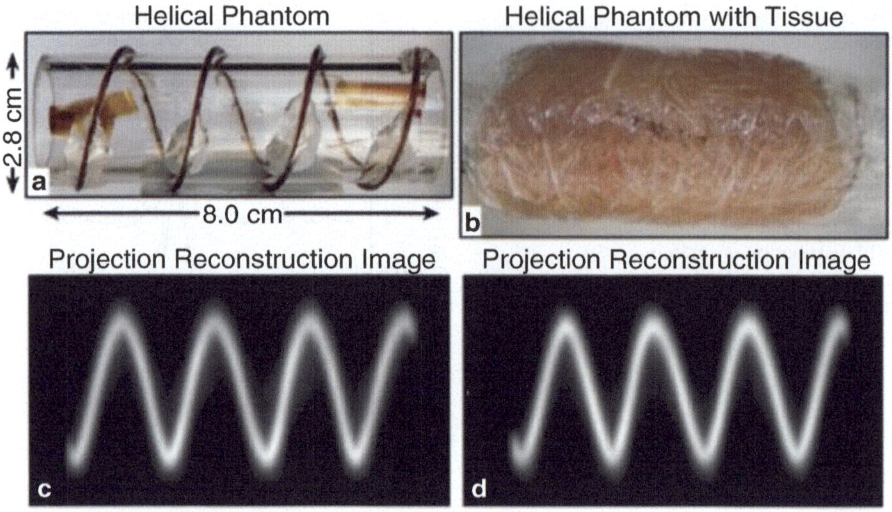

Fig. 6 MPI scans of a helical phantom (*left*) and a helical phantom embedded in animal tissue (*right*). This comparison makes clear that tissue creates no MPI signal and does not attenuate the image

2.5 MPI Sensitivity Is Linear, Robust, and Ideal

As shown in Fig. 7, the MPI image intensity signal scales linearly with the mass of SPIO particles at each pixel due to the linearity and shift invariance properties of MPI. Therefore, the MPI signal is perfectly linear and quantitative with cell count. MPI directly detects the SPIO electronic magnetization, which is 22 million times more intense than the nuclear paramagnetism imaged in 7 T high-field MRI. Direct detection of SPIO tracers enables "hotspot imaging" in MPI as opposed to the "negative contrast" when the same SPIOs are used in T_2*-MRI. We have experimentally demonstrated a linear detection limit of 200 labeled cells in vitro [13], with potential for significant further improvement. This linear sensitivity is superior to all other imaging modalities, the detection limits of which are compared in [1].

Importantly, although SPIOs respond nonlinearly to the applied magnetic field, the MPI signal induced from the rotating SPIOs in the receiver coil is perfectly linear with respect to the mass of SPIOs. Hence, there is absolutely no saturation of the MPI signal at higher cell counts. The details of the linearity and shift invariance of MPI are shown in [11, 14].

2.6 MPI Resolution

The fundamental spatial resolution of MPI depends only on the saturation characteristics of the SPIO particles and the strength of the magnetic gradient. Doubling the gradient strength improves resolution by twofold in every dimension. Currently

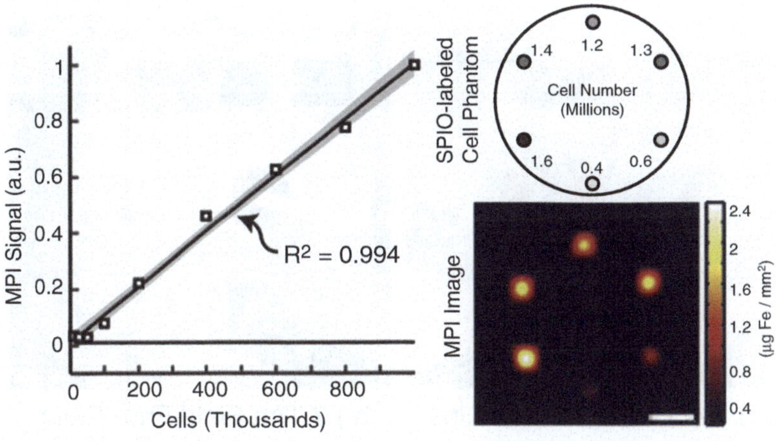

Fig. 7 (*Left*) Graph of MPI signal vs. number of labeled hESC-derived cells. The MPI signal is linear and quantitative (R^2=0.994). (*Right*) MPI of phantom of six SPIO-labeled cell populations ranging from 0.4 million to 1.6 million cells, with full image intensity scale at 2.4 μg Fe/mm^2 corresponding to the 1.6 million cell pellet

available SPIOs saturate at roughly plus or minus 3 mT, which translates to a spatial resolution of roughly 1 mm in a 6T/m gradient field. While this resolution is comparable to preclinical nuclear medicine, it is not yet competitive with CT or MRI small-animal imaging. To improve MPI's spatial resolution we and others have studied SPIOs with varying core diameters, since Langevin physics predicts a cubic resolution improvement with increasing SPIO core size. This has produced exciting results effectively reducing the FWHM resolution by a factor of two over Resovist [8]. Unfortunately, researchers have noted that improvements beyond roughly 25 nm core diameter seem to be limited, perhaps due to relaxation-induced blurring [18, 19]. This area remains a major active research area for tailoring nanoparticles to improve MPI performance.

2.7 Prototype Preclinical MPI Scanners

All of the imaging data shown in this chapter are obtained from the two scanners shown in Fig. 8. The top shows our 2.4,T/m FFL scanner, which allowed us to perform fast projection imaging as well as 3D projection reconstruction imaging, which is similar in scanning and reconstruction to X-ray computed tomography [48, 49]. Because we synthesize the 3D images from *N* projections, this volumetric imaging method improves image SNR by the square root of N over standard FFP imaging. We employed both the x-space reconstruction algorithm and the filtered back projection algorithm to reconstruct these images. The bottom scanner is our 7T/m FFP MPI scanner, which has produced some of the highest spatial resolution MP images reported thus far.

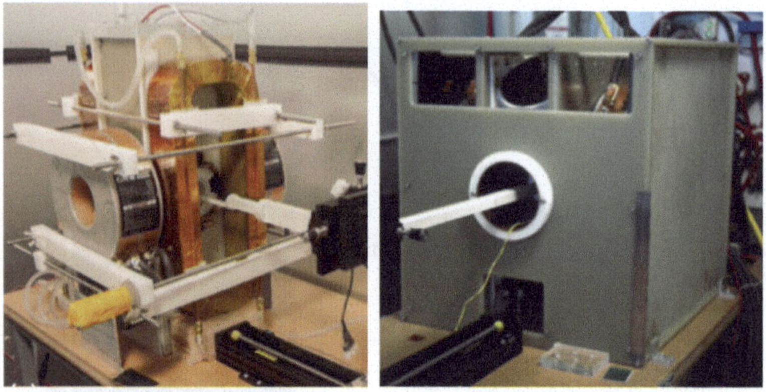

Fig. 8 Photos of two of our prototype MPI scanners. These scanners were used to acquire all of the MP images in this chapter. (**a**) 2.4T/m field-free line (FFL) scanner, able to perform magnetic computed tomography (akin to X-ray CT), or fast projection imaging. (**b**) High-gradient 7T/m field-free point (FFP) scanner

3 Clinical and Preclinical Applications of Magnetic Particle Imaging

3.1 Long-Term MPI Tracking of Neural Progenitor Cells

Recently, multiple initial studies have shown the successful use of MPI for tracking SPIO-labeled stem cells in vivo. Our group has demonstrated that MPI offers significant advantages for safe robust, noninvasive, long-term, and quantitative stem cell tracking [13, 20]. Others have shown that neural and mesenchymal stem cells implanted in mice can be detected as "hot spots" in a quantitative fashion as well [41]. In our first experiment [13], we tracked neural progenitor cell (NPC) grafts in the rodent brain using MPI. NPCs have shown regenerative properties for a variety of diseases, including Parkinson's disease, epilepsy, and ischemia [50–52].

In this study, we used MPI to image SPIO-labeled (Resovist) neural progenitor cell implants over an 87-day period. Our MP images (Fig. 9) show high image contrast and a persistent, quantifiable NPC signal throughout the 87-day duration of the study. Moreover, we were able to quantify the decay and clearance of the NPC grafts over time (Fig. 10), including the rapid loss of cells that were cleared by ventricular circulation, as validated via immunohistology and postmortem MRI [13]. The results from this study indicate that MPI is useful for long-term cell tracking studies.

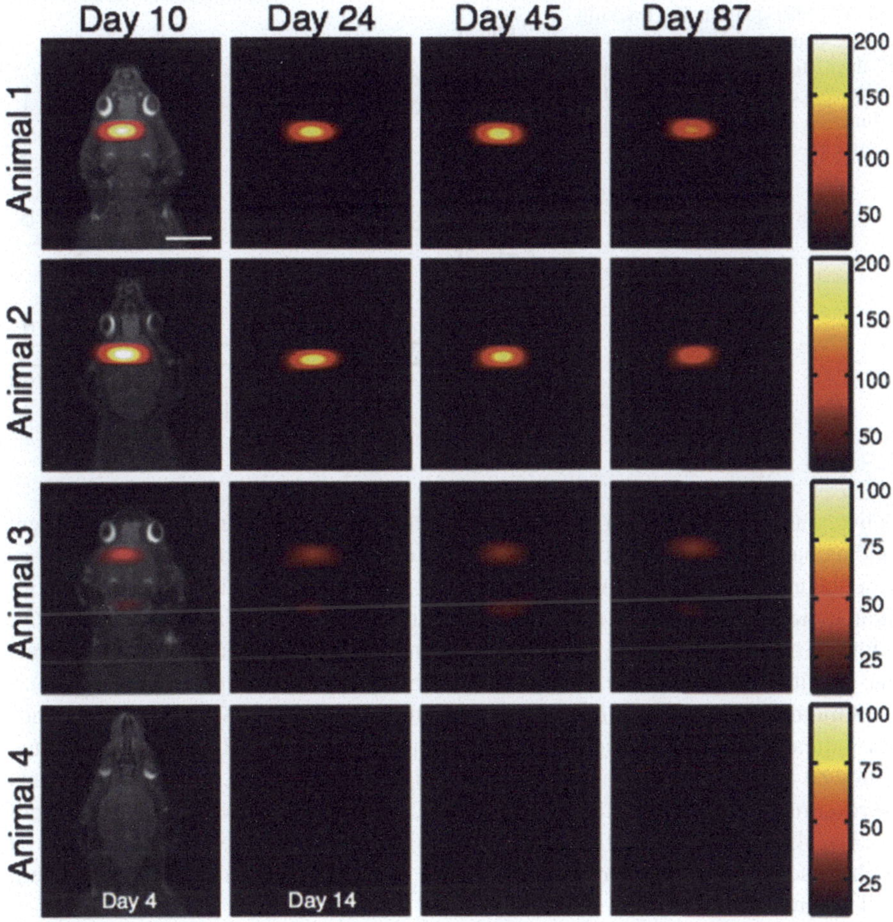

Fig. 9 Longitudinal MPI tracking of neural progenitor grafts from 10 to 87 days post-implantation. Labeled cells were implanted in the forebrain cortex (animals 1–2) and near-lateral ventricle (animal 3). Animal 4 shows injection of the SPIO tracer without cells, which quickly cleared from the body, as negative control

3.2 Dynamic, Systemic Stem Cell Tracking

Next, we evaluated MPI for dynamic monitoring of systemically administered cells, e.g., as mesenchymal stem cells (MSCs) [20]. MSCs are of particular therapeutic interest because they can control inflammation and modify the proliferation and cytokine production of immune cells [53]. MSC-based therapies have shown promise for treating diseases such as stroke, graft-versus-host disease, myocardial infarction, traumatic brain injury, and cancer [53, 54].

Intravenous injections are sometimes used to deliver MSCs in both animal models and clinical trials [55, 56]. However, it remains difficult to noninvasively monitor the

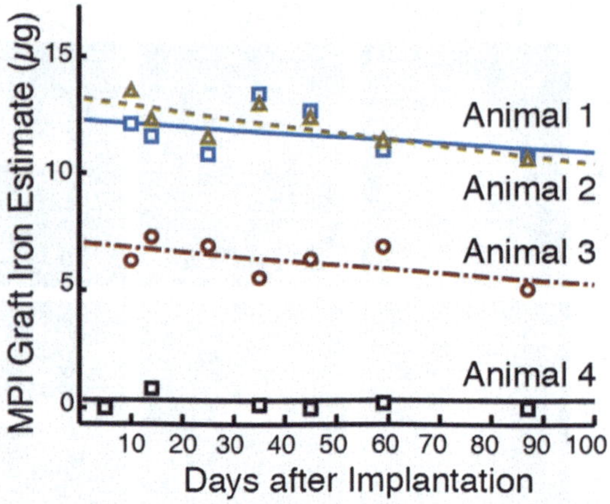

Fig. 10 MPI quantitative tracking of NPC neural implants in rats over 87 days. Total iron MPI estimates for *in vivo* cell grafts are plotted as a function of time using an exponential fit

delivery and biodistribution of administered cells into target organs [55, 57]. Recent studies have suggested that more than 80 % of MSCs are entrapped in pulmonary vasculature following intravenous injection [53, 57–60]. In this study, we demonstrated the first use of MPI to quantitatively track systemically administered MSCs, with CT coregistration. We intravenously administered SPIO-labeled MSCs in rats and monitored the dynamic distribution of MSCs with MPI over 12 days (Fig. 11).

Similar to previous studies that did not employ MPI [53, 58–61], our MPI studies indicated a rapid trafficking of intravenously injected MSCs to lung tissue, followed by clearance through the liver and spleen (Fig. 12). Our measurements were compared to gold standard ICP data, which showed excellent agreement ($R^2=0.943$). Hence, this MPI stem cell tracking study indicates that MPI could be highly useful for whole-body tracking of the biodistribution and clearance of stem cell therapies in real time.

3.3 MPI Angiography Enabled by Advanced Image Reconstruction

Iron oxides are metabolized through the liver [35, 62] and are not excreted by kidneys like gadolinium or iodine which are commonly used in MRI and X-ray studies. The latter agents are contraindicated for patients with damaged kidney function, including chronic kidney disease (CKD). Ferumoxytol is an USPIO iron supplement prescribed for certain types of CKD patients [63]. Hence, the use of CKD-safe iron oxides with MPI may present less risk for angiography and blood pool imaging in those patient populations.

X-ray fluoroscopy is a standard projection-imaging technique for angiography that exhibits both excellent spatial resolution and fast temporal resolution. Similar to X-ray fluoroscopy, projection-format MP imaging can be implemented using a

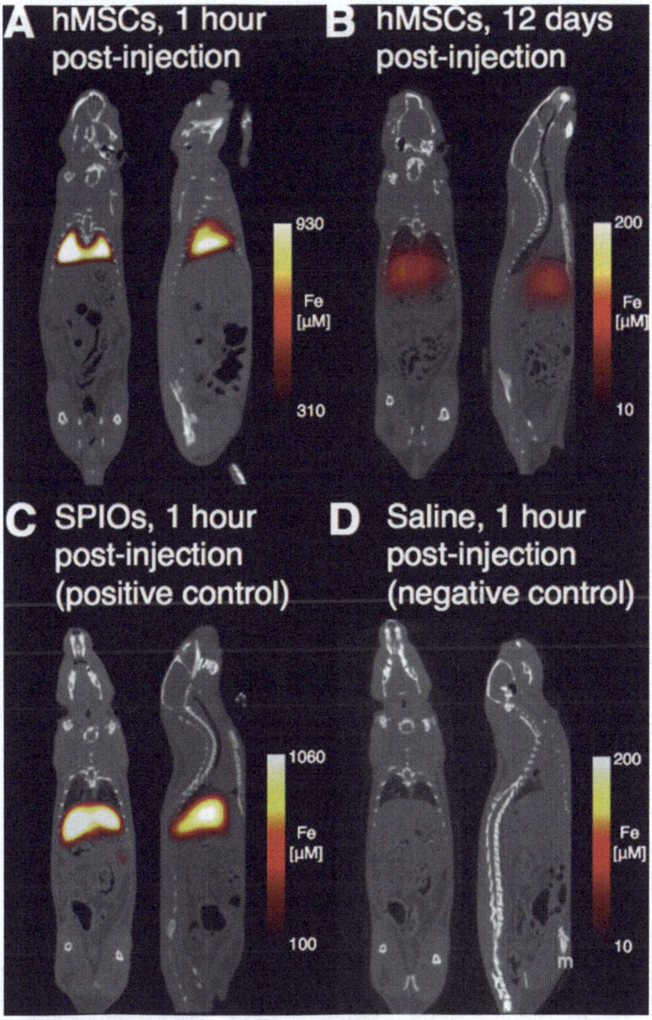

Fig. 11 MPI-CT imaging of intravenously injected MSCs. MPI shows immediate entrapment of intravenous MSC injections in the lungs (**a**), and redistribution to the liver by day 12 (**b**). SPIO-only injections are immediately redistributed to the liver (**c**), while saline injections show no detectable signal (**d**)

FFL magnetic architecture, which allows for rapidly acquired, 2D projection MPI as well as 3D magnetic computed tomography (via filtered back projection reconstruction) images. Figure 13 shows a phantom 3D projection reconstruction scan of a spiral phantom, as well as a 3D FFP scan of an angiography phantom [10].

Another major challenge for MPI angiography applications is the anisotropic, hazy nature of native MPI (seen in Fig. 14) [15]. Previously, it was shown that MPI is linear and shift invariant [11], but the 3D native MPI PSF is asymmetric; that is, the PSF is wider in directions orthogonal to the scan direction. Recently, we proved that adding

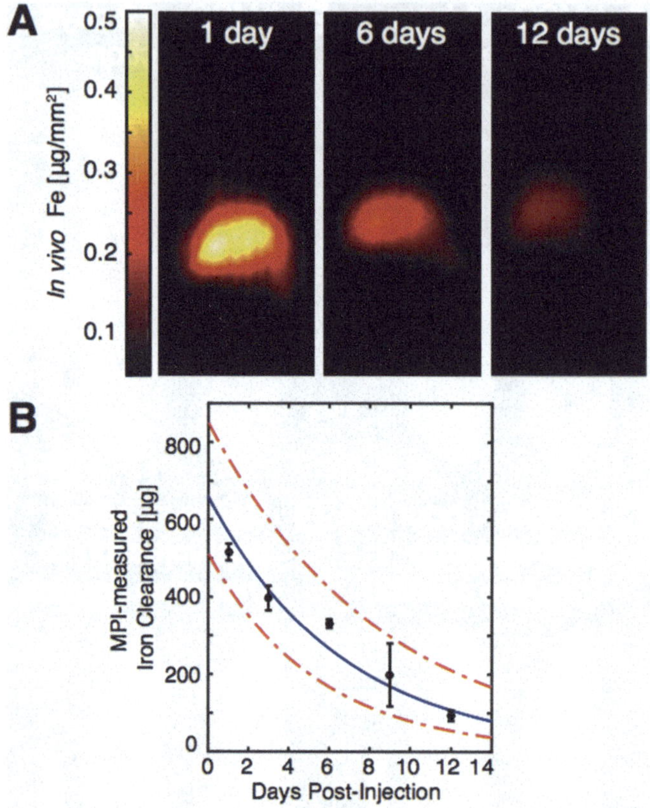

Fig. 12 (**a**) MPI monitoring of MSC clearance in liver from 1 to 12 days post-injection. (**b**) In vivo MPI measurements of MSC clearance from liver (time constant = 4.6 days)

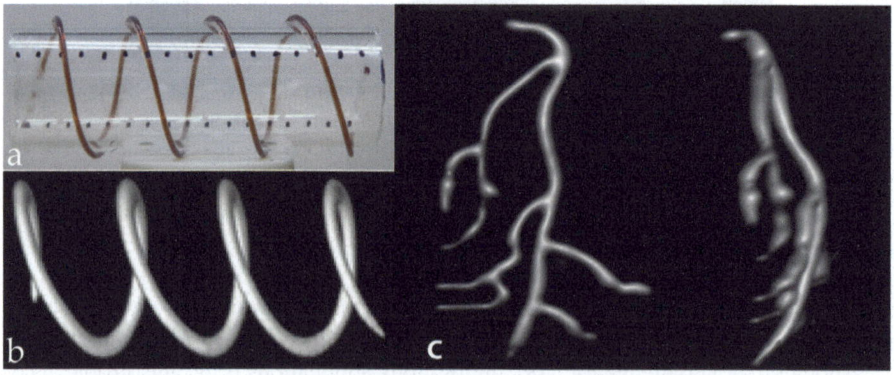

Fig. 13 3D MPI data using magnetic computed tomography. (**a**) Photograph of a spiral phantom and (**b**) 3D image using our FFL MPI scanner. (**c**) MPI of a 3D coronary artery phantom obtained with our FFP scanner and matrix-free optimized reconstruction

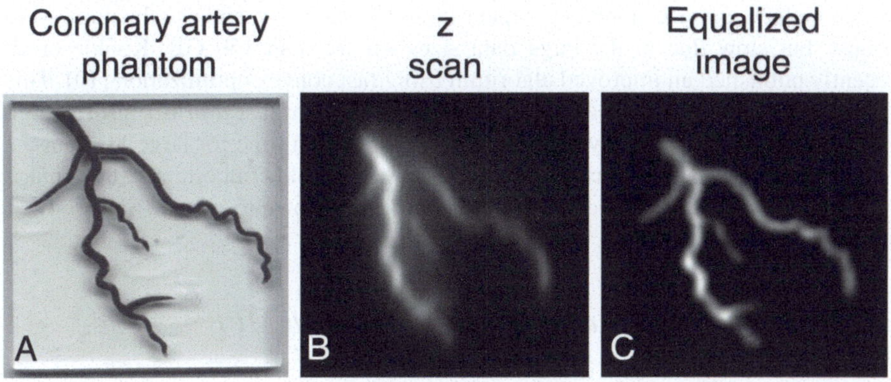

Fig. 14 MPI reconstruction advances. (**a**) Coronary artery SPIO phantom. (**b**) MP imaging along the z-direction shows asymmetric blur and image haze, arising from SPIO Langevin physics. (**c**) Combining y- and z-scans achieves isotropic resolution, and the additional use of an equalization filter sharpens MPI with minimal SNR loss

Fig. 15 MPI blood volume imaging in the rat brain. A 3D FFP image was acquired following injection of MPI-tailored SPIO particles (LS-008) [8]. Shown is a maximum intensity projection. Major blood vessels in the brain and head can be discerned, and the blood volume within capillaries is also appreciable, which is manifested as a background haze

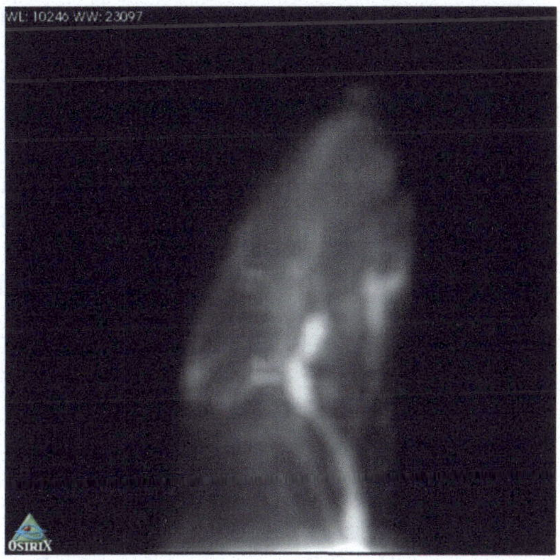

an orthogonal drive field and using a fast image reconstruction algorithm are sufficient to achieve isotropic resolution in MPI, from just two acquisitions, shown in Fig. 15. Further, MPI's PSF in 2D and 3D also has "long tails," which is clear from Langevin theory applied to higher dimensions and makes MP images appear hazy [15]. It is feasible to remove this haze, by compensating for the over-sampled density at lower spatial frequencies. The results are shown in Fig. 15. Most important, this improvement comes with only a minor loss of signal (<40 %) and zero noise gain.

Fast imaging applications such as angiography demand rapid scanning and image reconstruction times. Our first 3D x-space image reconstruction algorithms, which include x-space velocity compensation, DC recovery, and equalization, were robust, but slow due to the large data sizes—more than 150 GB. Konkle et al. recently published an improved algorithm using fast convex optimization [10]. This framework greatly reduces data size by incorporating robust *a priori* information, such as SPIO non-negativity. This reduced computation time for large 3D datasets to a few minutes and reduced memory use by 10^{11}-fold. Importantly, this image reconstruction performs no deconvolution, which is known to impede SNR [64].

3.4 Blood Volume and Tissue Perfusion with MPI

The use of MPI for vascular imaging is promising not only for angiography applications, but also for general blood volume imaging studies, such as blood perfusion imaging, which may be useful in the diagnosis and staging of stroke and cancer [65, 66]. Figure 15 shows a blood volume image of the brain and head region of a rat injected with MPI-tailored SPIO blood pool tracer LS-008 [8], acquired using our 7 T/m FFP MPI scanner. Both the major blood vessels in the brain and head, as well as blood within capillaries, can be appreciated in the image. The capillary blood volume is only grossly visible with nuclear medicine techniques today. In CT and MRI, the signal of the background tissue often dominates the capillary blood signal, which is typically about 5 % of the total tissue volume. This is a unique advantage of MPI. We are actively developing MPI perfusion methods, using wash-in and wash-out dynamics. MPI perfusion has the potential for brain and lung perfusion and may prove useful for quantitative determination of perfusion in diseases like stroke and cancer.

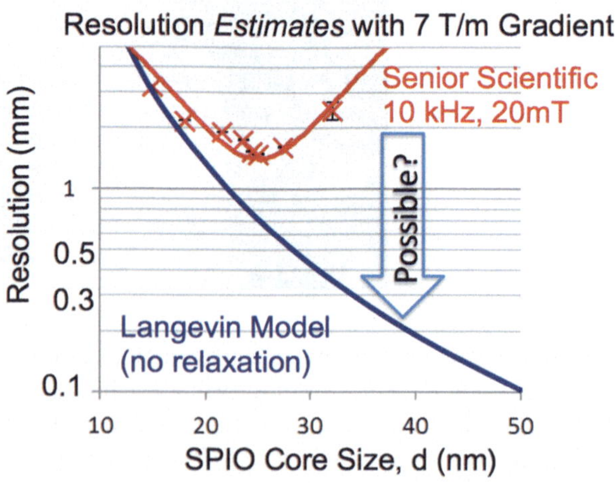

Fig. 16 Data from our relaxometer, showing the MPI point spread function (PSF) full-width half-max resolution as a function of SPIO core diameter for SPIOs. Steady-state Langevin theory [4, 14] predicts that spatial resolution should improve cubically with SPIO core diameter (*blue curve*). However, our data (*red curve*) clearly indicate that, above a certain size, the resolution actually begins to worsen, presumably due to the longer Brownian relaxation times for large SPIOs

4 Grand Challenges in MPI Nanoparticle Design

There are numerous open challenges in the nascent field of MPI. The field of MPI is currently in a stage of development like MRI was in 1980: no commercial scanners, no commercial vascular contrast agents (or tracers), and no agreement on image reconstruction methods, because the consensus on the k-space theory of MRI took several years after it was developed in 1983. Hence MPI is an exciting area of research. Below we discuss only two of these open challenges: how MPI could compete on spatial resolution with MRI and CT and how MPI might *revolutionize* cancer screening and diagnosis.

4.1 Competing with CT and MRI on Spatial Resolution

Perhaps the most urgent engineering priority for MPI-tailored nanoparticle research is to dramatically improve the spatial resolution of MPI. While today's 1 mm resolution is competitive with human-scale MRI, this is not competitive with murine MRI, where 100 μm resolution is considered routine. It would be an enormous contribution to improve MPI spatial resolution to match MRI or CT spatial resolution.

To improve MPI's spatial resolution we and others have researched SPIOs with varying core diameters, since Langevin steady-state physics predicts a *cubic* resolution improvement with increasing SPIO core size [4, 14]. This effort has already produced exciting results, effectively reducing the FWHM resolution by a factor of two over Resovist, using *both* the system matrix and the x-space reconstruction method [8].

Unfortunately, MPI researchers have noted that spatial resolution actually does not improve for particles beyond roughly 25 nm core diameter. Indeed, the spatial resolution actually worsens for particles larger than about 25 nm. The precise optimal SPIO diameter depends on the frequency and amplitude of the drive field, as well as the SPIO and coating properties, but this general trend is observed for all SPIOs. Figure 16 shows data obtained with our MPI relaxometer, using a range of SPIO sizes purchased from Senior Scientific, with identical coatings. The blue curve shows the predicted improvement in spatial resolution from steady state Langevin theory. Clearly, the steady-state Langevin theory fails for particles larger than 25 nm, which show a marked decrease in resolution (red curve).

Why does the steady-state Langevin theory fail for particles larger than roughly 25 nm? This remains an open research question. Both the system matrix and the x-space analytic models for MPI assume that relaxation is so fast that it can be neglected. Hence, we hypothesized that relaxation effects (Neel and Brownian relaxation) render the Langevin steady-state model inaccurate, and we recently showed experimentally that relaxation induces a *temporal* blur, which effectively causes a second *spatial* convolution over and above the Langevin steady-state PSF spatial blur [18, 19]. The data in Fig. 16 leads us to believe that the *dominant* blur for particles larger than the optimum is actually the Brownian relaxation blurring, although recent publications suggest that there is a Neel component to the relaxation that fits with standard models as well [67, 68]. It is well known that the Brownian relaxation time scales *cubically* with the SPIO hydrodynamic diameter and coating type [69], which is consistent with the worsening upward trend for particles larger than the optimum in Fig. 16. Hence, we believe that a dramatic improvement in MPI spatial resolution must obviate this so-called Brownian wall. Deconvolution approaches may help but they are known to reduce SNR dramatically [64]. Perhaps a nanoparticle specialist could devise an ingenious, high-resolution SPIO that has much faster relaxation times. Alternatively, MPI may require more innovative MPI data acquisition (both hardware and pulse sequences) that can mitigate relaxation blurring. This is an urgent area of research for all MPI researchers.

4.2 Cancer Imaging with MPI

One of the most promising potential clinical applications of MPI is cancer screening and diagnosis. Nanoparticles targeted to tumors often show excellent contrast via the enhanced permeability and retention (EPR) effect [70–72]. Our group recently performed a preclinical study to image human breast tumors within rats at 4 weeks after tumor implantation, using the blood pool tracer LS-008. Our results, shown in Fig. 17, indicate that MPI can be used to visualize tumors within small

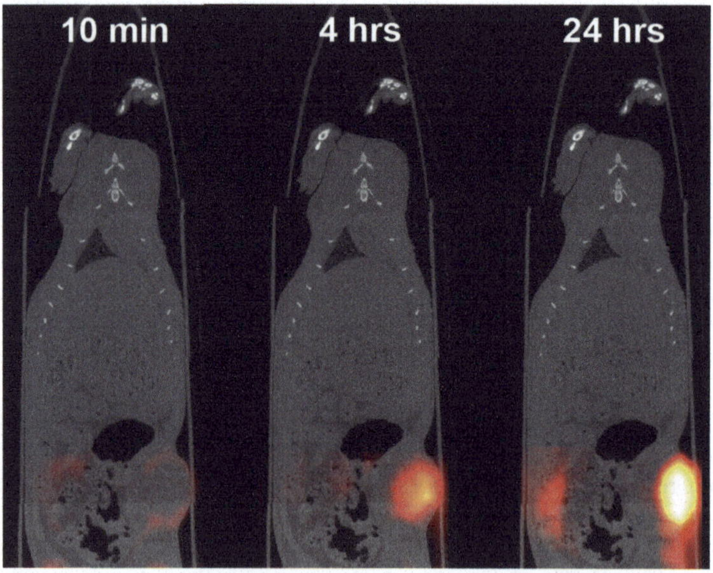

Fig. 17 *In vivo* MPI-CT image of a human breast tumor model. Non-targeted particles (LS-008) [8] were injected via tail vein 4 weeks after tumor implantation and monitored with MPI using a limited field of view over 24 h. A tumor vascular wash-in and wash-out can be appreciated. The physiologic image contrast is likely due to the enhanced permeability and retention effect

animals with excellent contrast and sensitivity. The contrast peaked at 24 h post-SPIO administration. The high image contrast for tumor imaging, using *non-targeted* SPIO particles (LS-008), is likely due to the EPR effect. These results are comparable in contrast to noise (CNR) to nuclear medicine scans, and this result strongly encourages further study.

A natural next step in MPI of cancer is to test tumor-targeted nanoparticle tracers to see if these improve the sensitivity and specificity of MP imaging. Given the rapid growth of cancer immunotherapy research, MPI techniques to image and track tumor-targeting immune cells may help accelerate the development of such therapies. Many targeting approaches aim to bind nanoparticles to the surface of blood vessels that specifically express vascular biomarkers of inflammation or angiogenesis (e.g., RGD, CREKA) [73]. This general approach may robust than tumor cell surface reporters, since particle extravasation from the blood is not necessary when targeting the endothelial wall. Notably, chemical binding can potentially generate significant changes in the MPI signal. This is because the SPIO must rotate nearly 180° as the FFP or FFL traverses an SPIO, and binding can significantly alter the Brownian relaxation time. It is important to note that chemical binding does not significantly alter T2*-MRI contrast. Our group and others have already experimentally demonstrated color-contrast MPI, as shown in Fig. 18 [5, 74]. Color MPI could possibly be used to distinguish SPIOs bound to the endothelium from SPIOs freely flowing in the blood.

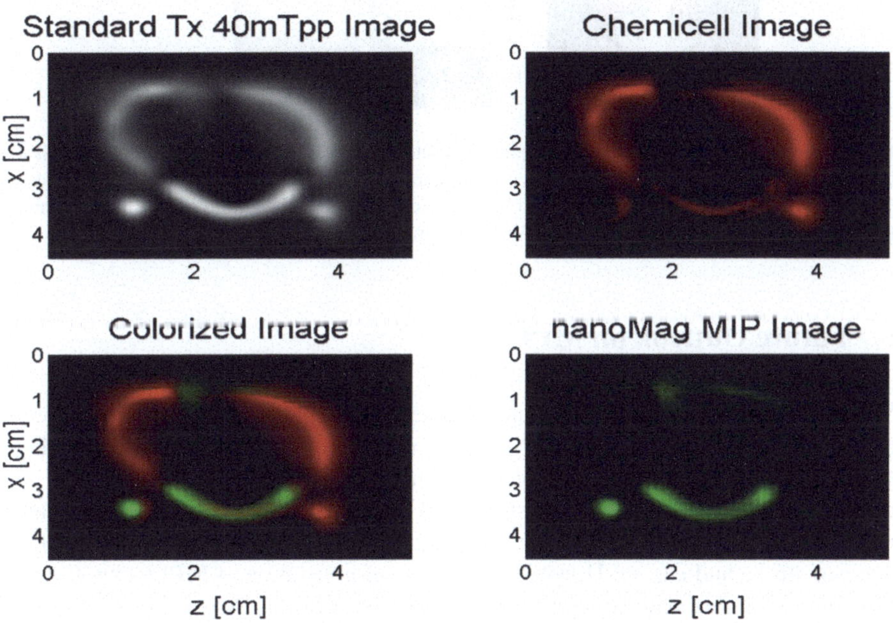

Fig. 18 "Color" MPI scans of phantom constructed using two MPI tracers with different relaxation properties. (*Top left*) Standard MPI. (*Top right*) Chemicell MPI tracer. (*Bottom right*) NanoMag MIP tracer. (*Bottom left*) Combined colorized image. Color MPI, when improved, may distinguish SPIOs in the blood from SPIOs bound to the endothelial wall

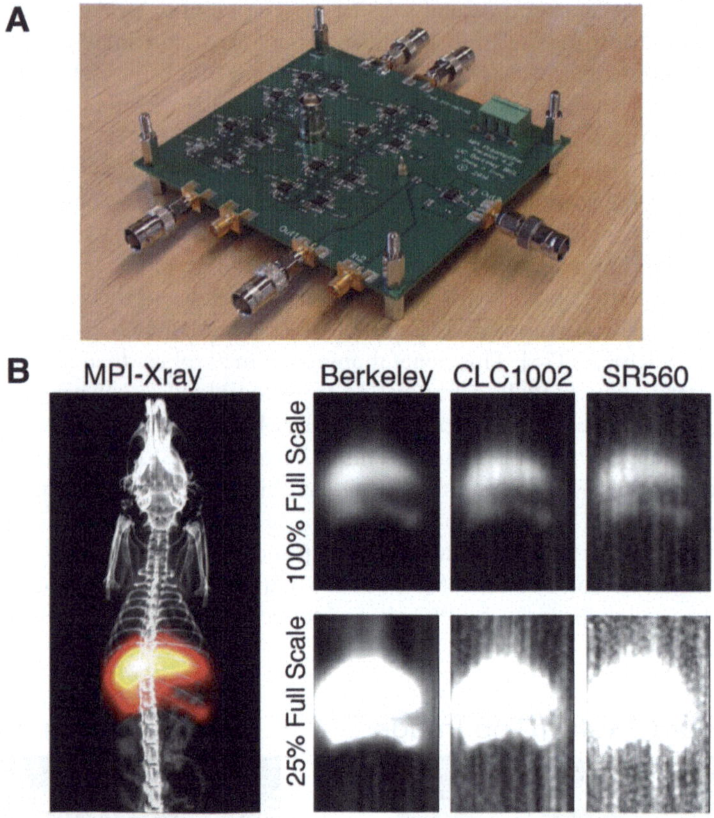

Fig. 19 (**a**) Photo of our custom noise-matched MPI preamp. (**b**) MP imaging following IV SPIO injections in a rat using the Stanford Research Systems SR560, Cadeka CLC1002, and the custom preamp. Our MPI preamp shows a tenfold SNR improvement over the commercial SR560 preamp. Images are shown for a 4×4×10 cm FOV and 8-min acquisition time

5 Future Outlook on MPI

With the improvement of SPIO nanoparticle tracers, as well as scanner hardware and reconstruction technology, MPI may join conventional nuclear medicine, MRI, or CT imaging techniques for molecular imaging applications. Already MPI has demonstrated the ability to track stem cells in small-animal models systemically and longitudinally, and to visualize brain and tumor vasculature for potential diagnosis of stroke and cancer. However, a successful translation of MPI technology to the clinic will also require significant engineering efforts. Below we discuss key challenges in MPI technology for clinical translation.

Our group recently showed that MPI has exquisite detection sensitivity, with detection and quantification of as few as 200 labeled cells in a voxel [13]. The true physics limit of detection sensitivity in MPI has not yet been achieved. This will be crucial for clinical translation and for fast, sensitive imaging of diseases like metastatic cancer.

Recently, methods to improve noise performance in MPI for enabling broadband noise matching to the detector preamplifier were identified, which guided the development of new low-noise preamplifier topologies with far improved broadband noise matching over commercial devices [75]. Figure 19 compares MP imaging performance using an MPI-optimized low-noise preamplifier from our lab to a commercial Stanford Research Systems SR560 preamplifier for imaging the liver and spleen via intravenous SPIO injections. Here, the MPI-optimized preamplifier showed a further tenfold SNR boost over the commercial device. While this detector system achieves detector coil-noise dominance, it still does not reach the fundamental physical noise limits, which is thermal noise arising from the patient. Hence, as MPI detector systems are further improved to reach patient noise dominance, configuring novel noise-matching schemes will be critical for optimizing image SNR in MPI.

Several groups have been advocating for using an FFL MPI scanner instead of an FFP scanner. FFL scanners offer greatly improved SNR and better resolution, with only a modest cost in scanning speed [17, 48, 49]. Hence, future MPI scanner architectures are likely to be projection-mode or 3D projection reconstruction FFL designs. As an example, using a 7.5T/m FFL MPI scanner, it is estimated that projection imaging of standard Resovist tracers can achieve 1.1 mm native MPI resolution, while with currently available MPI-optimized nanoparticles, this may be as low as 0.7 or 0.8 mm [8]. Such a high-gradient FFL MPI system is currently under development in our lab.

Further successful development of effective MPI tracers and robust MPI scanner architectures is needed to translate MPI technology to the clinic. The major remaining challenges in human-scale MPI operation include incorporating patient safety limits and scaling up the magnets in a cost-effective manner. MPI scanning parameters must be optimized to achieve high imaging speed and SNR while operating within FDA and EU limits for patient heating (SAR) and peripheral nerve stimulation (PNS) with magnetic fields [22, 26]. Akin to the evolution of MRI technology almost 35 years ago, it is likely that MPI FFL magnets will employ strong superconducting magnets. The technology should be similar in cost and complexity to existing low-field or mid-field MRI scanners.

In conclusion, while MPI remains in its technical infancy, prototype scanners built at our university appear to be competitive in sensitivity, contrast, safety, and robustness compared to professionally manufactured MRI and CT scanners for specific preclinical applications such as in vivo stem cell tracking, cancer imaging, and brain and lung perfusion. However, improvements in MPI tracer technologies are crucial in further advancing the MPI field for spatial resolution, sensitivity, and safety. Already several groups have made promising advances in mass sensitivity and resolution in MPI tracer technology [8, 28, 38]. However, much remains unknown about the physical mechanisms and trade-offs between particle magnetization and dynamic relaxation behavior. Further, the role of particle shape, size, and coating material on their MPI performance has not yet been extensively investigated. Hence, there is an urgent need for nanoparticle researchers to solve some of these important open challenges in the MPI field.

References

1. Nguyen PK, Riegler J, Wu JC. Stem cell imaging: from bench to bedside. Cell Stem Cell. 2014;14(4):431–44.
2. Gleich B, Weizenecker J. Tomographic imaging using the nonlinear response of magnetic particles. Nature. 2005;435(7046):1214–7.
3. Weizenecker J, Gleich B, Rahmer J, Dahnke H, Borgert J. Three-dimensional real-time in vivo magnetic particle imaging. Phys Med Biol. 2009;54(5):L1–10.
4. Rahmer J, Weizenecker J, Gleich B, Borgert J. Signal encoding in magnetic particle imaging: properties of the system function. BMC Med Imaging. 2009;9:4.
5. Rahmer J, Halkola A, Gleich B, Schmale I, Borgert J. First experimental evidence of the feasibility of multi-color magnetic particle imaging. Phys Med Biol. 2015;60(5):1775–91.
6. Rahmer J, Weizenecker J, Gleich B, Borgert J. Analysis of a 3-D system function measured for magnetic particle imaging. IEEE Trans Med Imaging. 2012;31(6):1289–99.
7. Franke J, Heinen U, Matthies L, Niemann V, Jaspard F, Heidenreich M, Buzug T. First hybrid MPI-MRI imaging system as integrated design for mice and rats: description of the instrumentation setup. In: Magnetic particle imaging (IWMPI), 2013 international workshop on 2013, p. 1–1.
8. Ferguson RM, Khandhar AP, Kemp SJ, Arami H, Saritas EU, Croft LR, Konkle J, Goodwill PW, Halkola A, Rahmer J, Borgert J, Conolly SM, Krishnan KM. Magnetic particle imaging with tailored iron oxide nanoparticle tracers. IEEE Trans Med Imaging. 2015;34(5):1077–84.
9. Ferguson RM, Khandhar AP, Krishnan KM. Tracer design for magnetic particle imaging (invited). J Appl Phys. 2012;111(7):7B318–7B3185.
10. Konkle JJ, Goodwill PW, Hensley DW, Orendorff RD, Lustig M, Conolly SM. A convex formulation for magnetic particle imaging X-space reconstruction. PLoS One. 2015;10(10): e0140137.
11. Lu K, Goodwill PW, Saritas EU, Zheng B, Conolly SM. Linearity and shift invariance for quantitative magnetic particle imaging. IEEE Trans Med Imaging. 2013;32(c):1565–75.
12. Goodwill PW, Lu K, Zheng B, Conolly SM. An x-space magnetic particle imaging scanner. Rev Sci Instrum. 2012;83(3):033708.
13. Zheng B, Vazin T, Goodwill PW, Conway A, Verma A, Ulku Saritas E, Schaffer D, Conolly SM. Magnetic particle Imaging tracks the long-term fate of in vivo neural cell implants with high image contrast. Sci Rep. 2015;5:14055.
14. Goodwill PW, Conolly SM. The X-space formulation of the magnetic particle imaging process: 1-D signal, resolution, bandwidth, SNR, SAR, and magnetostimulation. IEEE Trans Med Imaging. 2010;29(11):1851–9.
15. Goodwill PW, Conolly SM. Multidimensional x-space magnetic particle imaging. IEEE Trans Med Imaging. 2011;30(9):1581–90.
16. Goodwill PW, Saritas EU, Croft LR, Kim TN, Krishnan KM, Schaffer DV, Conolly SM. X-space MPI: magnetic nanoparticles for safe medical imaging. Adv Mater. 2012;24(28): 3870–7.
17. Goodwill PW, Konkle JJ, Zheng B, Saritas EU, Conolly SM. Projection x-space magnetic particle imaging. IEEE Trans Med Imaging. 2012;31(5):1076–85.
18. Croft LR, Goodwill PW, Conolly SM. Relaxation in x-space magnetic particle imaging. IEEE Trans Med Imaging. 2012;31(12):2335–42.
19. Croft LR, Goodwill PW, Konkle JJ, Arami H, Price DA, Li AX, Saritas EU, Conolly SM. Low drive field amplitude for improved image resolution in magnetic particle imaging. Med Phys. 2016;43(1):424–35.
20. Zheng B, von See MP, Yu E, Gunel B, Lu K, Vazin T, Schaffer DV, Goodwill PW, Conolly SM. Quantitative magnetic particle imaging monitors the transplantation, biodistribution, and clearance of stem cells in vivo. Theranostics. 2016;6(3):291–301.
21. Saritas EU, Goodwill PW, Croft LR, Konkle JJ, Lu K, Zheng B, Conolly SM. Magnetic particle imaging (MPI) for NMR and MRI researchers. J Magn Reson. 2013;229:116–26.

22. Saritas EU, Goodwill PW, Zhang GZ, Conolly SM. Magnetostimulation limits in magnetic particle imaging. IEEE Trans Med Imaging. 2013;32(9):1600–10.
23. Panagiotopoulos N, Duschka RL, Ahlborg M, Bringout G, Debbeler C, Graeser M, Kaethner C, Lüdtke-Buzug K, Medimagh H, Stelzner J, Buzug TM, Barkhausen J, Vogt FM, Haegele J. Magnetic particle imaging: current developments and future directions. Int J Nanomed. 2015; 10:3097–114.
24. Graeser M, Knopp T, Grüttner M, Sattel TF, Buzug TM. Analog receive signal processing for magnetic particle imaging. Med Phys. 2013;40(4):042303.
25. Erbe M, Knopp T, Sattel TF, Biederer S, Buzug TM. Experimental generation of an arbitrarily rotated field-free line for the use in magnetic particle imaging. Med Phys. 2011;38(9):5200–7.
26. Saritas EU, Goodwill PW, Conolly SM. Effects of pulse duration on magnetostimulation thresholds. Med Phys. 2015;42(6):3005–12.
27. Eberbeck D, Wiekhorst F, Wagner S, Trahms L. How the size distribution of magnetic nanoparticles determines their magnetic particle imaging performance. Appl Phys Lett. 2011;98(18):182502.
28. Eberbeck D, Dennis CL, Huls NF, Krycka KL, Gruttner C, Westphal F. Multicore magnetic nanoparticles for magnetic particle imaging. IEEE Trans Magn. 2013;49(1):269–74.
29. Wawrzik T, Kuhlmann C, Ludwig F, Schilling M. Scanner setup and reconstruction for three-dimensional magnetic particle imaging. SPIE Med Imaging. 2013;8672:86721B–86721B–8.
30. Lak A, Wawrzik T, Ludwig F, Schilling M. Synthesis of single-core iron oxide nanoparticles as a tracer for magnetic particle imaging. Int Workshop Magn Particle Imaging. 2012. pp. 93–97.
31. Vogel P, Ruckert MA, Klauer P, Kullmann WH, Jakob PM, Behr VC. Traveling wave magnetic particle imaging. IEEE Trans Med Imaging. 2014;33(2):400–7.
32. Murase K, Hiratsuka S, Song R, Takeuchi Y. Development of a system for magnetic particle imaging using neodymium magnets and gradiometer. Jpn J Appl Phys. 2014;53(6):067001.
33. Nishimoto K, Mimura A, Aoki M, Banura N, Murase K. Application of magnetic particle imaging to pulmonary imaging using nebulized magnetic nanoparticles. Open J Med Imaging. 2015;5(2):49.
34. Arami H, Krishnan KM. Intracellular performance of tailored nanoparticle tracers in magnetic particle imaging. J Appl Phys. 2014;115(17):17B306.
35. Arami H, Khandhar A, Liggitt D, Krishnan KM. In vivo delivery, pharmacokinetics, biodistribution and toxicity of iron oxide nanoparticles. Chem Soc Rev. 2015.
36. Arami H, Khandhar AP, Tomitaka A, Yu E, Goodwill PW, Conolly SM, Krishnan KM. In vivo multimodal magnetic particle imaging (MPI) with tailored magneto/optical contrast agents. Biomaterials. 2015;52:251–61.
37. Tomitaka A, Arami H, Gandhi S, Krishnan KM. Lactoferrin conjugated iron oxide nanoparticles for targeting brain glioma cells in magnetic particle imaging. Nanoscale. 2015.
38. Ferguson RM, Minard KR, Khandhar AP, Krishnan KM. Optimizing magnetite nanoparticles for mass sensitivity in magnetic particle imaging. Med Phys. 2011;38(3):1619.
39. Ferguson RM, Minard KR, Krishnan KM. Optimization of nanoparticle core size for magnetic particle imaging. J Magn Magn Mater. 2009;321:1548–51.
40. Bauer LM, Situ SF, Griswold MA, Samia ACS. Magnetic particle imaging tracers: state-of-the-art and future directions. J Phys Chem Lett. 2015;6(13):2509–17.
41. Bulte JWM, Walczak P, Janowski M, Krishnan KM, Arami H, Halkola A, Gleich B, Rahmer J. Quantitative "hot spot" imaging of transplanted stem cells using superparamagnetic tracers and magnetic particle imaging (MPI). Tomography. 2015;1(2):91–7.
42. Visscher M, Waanders S, Krooshoop HJG, ten Haken B. Selective detection of magnetic nanoparticles in biomedical applications using differential magnetometry. J Magn Magn Mater. 2014;365:31–9.
43. Visscher M, Waanders S, Pouw J, ten Haken B. Depth limitations for in vivo magnetic nanoparticle detection with a compact handheld device. J Magn Magn Mater. 2015;380:246–50.
44. Rauwerdink AM, Weaver JB. Viscous effects on nanoparticle magnetization harmonics. J Magn Magn Mater. 2010;322(6):609–13.
45. Weaver JB, Kuehlert E. Measurement of magnetic nanoparticle relaxation time. Med Phys. 2012;39(5):2765–70.

46. Knopp T, Them K, Kaul M, Gdaniec N. Joint reconstruction of non-overlapping magnetic particle imaging focus-field data. Phys Med Biol. 2015;60(8):L15–21.
47. Weber A, Werner F, Weizenecker J, Buzug TM, Knopp T. Artifact free reconstruction with the system matrix approach by overscanning the field-free-point trajectory in magnetic particle imaging. Phys Med Biol. 2015;61(2):475.
48. Konkle JJ, Goodwill PW, Carrasco-Zevallos OM, Conolly SM. Projection reconstruction magnetic particle imaging. IEEE Trans Med Imaging. 2013;32(2):338–47.
49. Konkle JJ, Goodwill PW, Saritas EU, Zheng B, Lu K, Conolly SM. Twenty-fold acceleration of 3D projection reconstruction MPI. Biomed Tech. 2013;58(6):565–76.
50. Chu K, Kim M, Jeong S-W, Kim SU, Yoon B-W. Human neural stem cells can migrate, differentiate, and integrate after intravenous transplantation in adult rats with transient forebrain ischemia. Neurosci Lett. 2003;343(2):129–33.
51. Kelly S, Bliss TM, Shah AK, Sun GH, Ma M, Foo WC, Masel J, Yenari MA, Weissman IL, Uchida N, Palmer T, Steinberg GK. Transplanted human fetal neural stem cells survive, migrate, and differentiate in ischemic rat cerebral cortex. Proc Natl Acad Sci U S A. 2004; 101(32):11839–44.
52. De Feo D, Merlini A, Laterza C, Martino G. Neural stem cell transplantation in central nervous system disorders. Curr Opin Neurol. 2012;25(3):322–33.
53. Eggenhofer E, Luk F, Dahlke MH, Hoogduijn MJ. The life and fate of mesenchymal stem cells. Front Immunol. 2014;5:1–6.
54. Hoogduijn MJ, Roemeling-van Rhijn M, Engela AU, Korevaar SS, Mensah FKF, Franquesa M, de Bruin RWF, Betjes MGH, Weimar W, Baan CC. Mesenchymal stem cells induce an inflammatory response after intravenous infusion. Stem Cells Dev. 2013;22(21):2825–35.
55. Harting MT, Jimenez F, Xue H, Fischer UM, Baumgartner J, Dash PK, Cox CS. Intravenous mesenchymal stem cell therapy for traumatic brain injury. J Neurosurg. 2009;110(6):1189–97.
56. Wu Y, Zhao RCH. The role of chemokines in mesenchymal stem cell homing to myocardium. Stem Cell Rev. 2012;8(1):243–50.
57. Ge J, Guo L, Wang S, Zhang Y, Cai T, Zhao RCH, Wu Y. The size of mesenchymal stem cells is a significant cause of vascular obstructions and stroke. Stem Cell Rev Rep. 2014;10:295–303.
58. Eggenhofer E, Benseler V, Kroemer A, Popp FC, Geissler EK, Schlitt HJ, Baan CC, Dahlke MH, Hoogduijn MJ. Mesenchymal stem cells are short-lived and do not migrate beyond the lungs after intravenous infusion. Front Immunol. 2012;3:297.
59. Fischer UM, Harting MT, Jimenez F, Monzon-Posadas WO, Xue H, Savitz SI, Laine GA, Cox CS. Pulmonary passage is a major obstacle for intravenous stem cell delivery: the pulmonary first-pass effect. Stem Cells Dev. 2009;18(5):683–92.
60. Gao J, Dennis JE, Muzic RF, Lundberg M, Caplan AI. The dynamic in vivo distribution of bone marrow-derived mesenchymal stem cells after infusion. Cells Tissues Organs. 2001;169:12–20.
61. Kraitchman DL, Tatsumi M, Gilson WD, Ishimori T, Kedziorek D, Walczak P, Segars WP, Chen HH, Fritzges D, Izbudak I, Young RG, Marcelino M, Pittenger MF, Solaiyappan M, Boston RC, Tsui BMW, Wahl RL, Bulte JWM. Dynamic imaging of allogeneic mesenchymal stem cells trafficking to myocardial infarction. Circulation. 2005;112(10):1451–61.
62. Arbab AS, Wilson LB, Ashari P, Jordan EK, Lewis BK, Frank JA. A model of lysosomal metabolism of dextran coated superparamagnetic iron oxide (SPIO) nanoparticles: implications for cellular magnetic resonance imaging. NMR Biomed. 2005;18(6):383–9.
63. Lu M, Cohen MH, Rieves D, Pazdur R. FDA report: ferumoxytol for intravenous iron therapy in adult patients with chronic kidney disease. Am J Hematol. 2010;85(5):315–9.
64. Shahram M, Milanfar P. Imaging below the diffraction limit: a statistical analysis. IEEE Trans Image Process. 2004;13(5):677–89.
65. Lui YW, Tang ER, Allmendinger AM, Spektor V. Evaluation of CT perfusion in the setting of cerebral ischemia: patterns and pitfalls. Am J Neuroradiol. 2010;31(9):1552–63.
66. Padhani AR, Liu G, Koh DM, Chenevert TL, Thoeny HC, Takahara T, Dzik-Jurasz A, Ross BD, Van Cauteren M, Collins D, Hammoud DA, Rustin GJS, Taouli B, Choyke PL. Diffusion-weighted magnetic resonance imaging as a cancer biomarker: consensus and recommendations. Neoplasia. 2009;11(2):102–25.

67. Shah SA, Reeves DB, Ferguson RM, Weaver JB, Krishnan KM. Mixed Brownian alignment and Néel rotations in superparamagnetic iron oxide nanoparticle suspensions driven by an ac field. Phys Rev B Condens Matter Mater Phys. 2015;92(9).
68. Shah SA, Ferguson RM, Krishnan KM. Slew-rate dependence of tracer magnetization response in magnetic particle imaging. J Appl Phys. 2014;116(16):163910.
69. Brown Jr WF. Thermal fluctuations of a single-domain particle. Phys Rev. 1963;130(5):1677.
70. Gao X, Cui Y, Levenson RM, Chung LWK, Nie S. In vivo cancer targeting and imaging with semiconductor quantum dots. Nat Biotechnol. 2004;22(8):969–76.
71. Loo C, Lowery A, Halas N, West J, Drezek R. Immunotargeted nanoshells for integrated cancer imaging and therapy. Nano Lett. 2005;5(4):709–11.
72. James ML, Gambhir SS. A molecular imaging primer: modalities, imaging agents, and applications. Physiol Rev. 2012;92(2):897–965.
73. Simberg D, Duza T, Park JH, Essler M, Pilch J, Zhang L, Derfus AM, Yang M, Hoffman RM, Bhatia S, Sailor MJ, Ruoslahti E. Biomimetic amplification of nanoparticle homing to tumors. Proc Natl Acad Sci U S A. 2007;104(3):932–6.
74. Hensley D, Goodwill P, Croft L, Conolly S. Preliminary experimental X-space color MPI. In: Magnetic particle imaging (IWMPI), 2015 5th international workshop on 2015, p. 1–1.
75. Zhang W, Zheng B, Goodwill P, Conolly S. A custom low-noise preamplifier for magnetic particle imaging. In: Magnetic particle imaging (IWMPI), 2015 5th international workshop on 2015, p. 1–1.

Multispectral MR Imaging and Sensing Using Shaped Nanoparticles

Gary Zabow

1 Introduction

The discovery of fluorescent proteins and their application in bioimaging have significantly advanced molecular and cellular biology [1–3]. Color-labeling with such fluorophores enables distinction between different cell types and different biomolecules, allows multiplexing and high-throughput bioassays, supports colorimetric sensing for real-time visualization of biomolecular processes and functions, and is critical to a host of new superresolution microscopy techniques [4, 5]. In addition to such molecular labels and probes, the bioimaging revolution that they started is now driven also by multicolor nanoparticle-based analogs. Prominent among these are semiconductor nanocrystal quantum dots [6–9], metallic plasmonic nanostructures [10, 11], and more recently nanodiamonds [12] and related carbon-based, low-toxicity fluorescent nanoparticles [13].

Despite unquestioned utility, however, optical probes still suffer from limited in vivo functionality, where optical access is more challenging or entirely precluded. The growing realization that optically accessible in vitro, or largely two-dimensional, testing does not accurately mimic more realistic, three-dimensional, in vivo physiological conditions, neither chemically nor mechanically [14], has spurred much research into extending optical penetration depths through biological tissues [15]. Among others, this includes confocal [16] and two- or multiphoton imaging schemes [17, 18], photoacoustic imaging [19], and adaptic-optic wavefront-shapings that can dynamically refocus light through turbid, scattering media [20, 21]. Concurrently, optical

G. Zabow (✉)
Applied Physics Division, National Institute of Standards and Technology (NIST), 325 Broadway, Boulder, CO 80305, USA
e-mail: zabow@boulder.nist.gov

J.W.M. Bulte, M.M.J. Modo (eds.), *Design and Applications of Nanoparticles in Biomedical Imaging*, DOI 10.1007/978-3-319-42169-8_5

nanoparticle probes are being reengineered with resonances shifted towards the more favorable near-infrared region of the spectrum where biological phototoxicity is reduced and intervening biological media offer less photon attenuation and less interfering background autofluorescence [22, 23]. Even so, optical signal intensities and spatial resolutions still fall rapidly with increasing depth below the surface.

For in vivo probing, magnetic resonance imaging (MRI) would seem an ideal alternative were it not for its largely monochromatic nature that forfeits many of the multicolor advantages that make multispectral optical probes so valuable in the first place. MRI is a powerful imaging technology that offers safe, deep in vivo imaging with excellent soft-tissue contrast and high spatial resolution compared to other radiological imaging modalities [24]. Not requiring any ionizing radiation, it has become one of the most widely accepted medical imaging and diagnostics tools. Operating in the radio-frequency range, MRI can noninvasively penetrate tissue without signal loss or distortion and is inherently immune to background autofluorescence, photobleaching, and phototoxicity issues that plague optical imaging. Moreover, MRI has the capability to discern not only anatomical form but also physiological function through many unique contrast mechanisms that provide sensitivity to such key variables as water diffusion [25–27], paramagnetic metal ion content (particularly iron [28]), and blood flow [29], perfusion [30], and oxygenation levels [31]. Even in vivo cellular and molecular level information is becoming accessible through an ever-growing array of targeted, and even genetically expressed, MRI contrast mechanisms [32]. These enable in vivo probing of many biomarkers and include early-stage disease detection through cell-type or epitope recognition [33–36] and imaging of gene expression and enzyme activity [37–41].

However, MRI has traditionally lacked the multiplexing advantages of multicolor optical imaging modalities. Images typically comprise amplitude (and/or phase) maps of the local water signal, determined by the local density of water and its movement and relaxation properties, which enable a rich array of different contrast mechanisms. Nevertheless, just as optical colors represent different wavelengths or optical frequencies, a color MRI mapping requires also some analogous form of frequency-based distinction. NMR spectroscopy already does just that, identifying different chemical molecules through their different NMR spectra and to some extent this chemical-shift information carries over to MRI through chemical-shift imaging and magnetic resonance spectroscopy (MRS). But in vivo MRS sensitivities are low, enabling high-resolution imaging of only those biomolecules that naturally occur in sufficiently high concentrations [42, 43].

This historical lack of multiplexed image contrast has not only impeded potential in vivo multiplexing applications, but also limited quantitative accuracy of MRI analyses (discussed below). Recognizing such weaknesses, several new multispectral approaches to MRI contrast have been advanced in recent years including agents based on chemical exchange and on ^{19}F, offering some of the first MRI multiplexing alternatives that sidestep optical access limitations of optical reporters. One of the most recent approaches—the focus of this chapter—uses magnetic nanoparticles whose special shapes enable multispectral MRI labeling and sensing. To place these new multispectral shaped contrast agents into better context, however, other multispectral, as well as more traditional, approaches to MRI contrast are first discussed.

Many excellent reviews already exist on the topic of MRI contrast agents [44, 45]; thus here only a brief overview is given.

2 MRI Contrast Agents

In addition to the many endogenous contrast schemes accessible to MRI, the clinical success of MRI owes much to the invention of exogenous contrast agents, which may be either molecular or nanoparticle based. Roughly a third of all clinical MRI exams currently employ an administered exogenous contrast agent, amounting to some 10 to 20 million contrast agent-enhanced exams annually. Such agents are regularly employed to highlight blood flow and detect abnormal, damaged tissue, perhaps most notably including detection of blood-brain barrier disruption and the presence of brain tumors [46].

2.1 *T_1 and T_2 Agents*

By far the most common clinical MRI contrast agents are molecular paramagnetic agents based on chelates of the Gd^{3+} ion [47, 48], although agents based on Mn^{2+}, which can directly enter cells, are also used [49–51]. With seven unpaired electrons, the relatively large, fluctuating magnetic fields around Gd^{3+} ions can significantly affect both the longitudinal (T_1) and transverse (T_2) relaxation rates of hydrogen protons in closely neighboring water molecules. Because their relative effect on T_1 is larger, however, Gd-chelates are generally referred to as T_1 contrast agents and used with T_1 image-weighting schemes. By rapidly repeating image acquisitions, these T_1-weighted images exploit the locally reduced longitudinal relaxation time, T_1, of the nearby water protons to increase image brightness or signal (positive contrast) near the contrast agent as compared to the weaker signal accrued from more slowly relaxing water protons further away.

Many nanoparticle-based contrast agents also exist. These include nanoparticles of manganese oxide [52, 53], of gadolinium oxide [54, 55], and, most prominently, of superparamagnetic iron oxide (SPIO) that benefit from many already-developed synthesis methods [56, 57]. SPIO nanoparticles contain magnetite (Fe_3O_4) or maghemite (γ-Fe_2O_3), which offer much greater magnetic moments than chelated paramagnetic ions. Therefore, they can affect water relaxivity out to relatively large distances, much greater than the contrast agent particle sizes, which themselves span a large range from a few nanometers up to a few micrometers. They include nanometer-sized monocrystalline iron oxide nanoparticles (MIONs) [58] and ultrasmall SPIO (USPIO) [59] agents (often dextran coated for biocompatibility and solubility), as well as assorted agglomerations thereof that include magnetodendrimers [60] a few tens of nanometers in size, larger composite SPIO agents [61, 62] up to a few hundred nanometers in diameter, and even micrometer-scale particles of iron oxide (MPIOs) that may be as large as several micrometers across [63] and that can be individually detected [64]. While these particulate-based agents also affect both T_1 and T_2, they are referred to as T_2 contrast agents since their effect on T_2 is far greater, especially as particle sizes increase. For T_2 agents, the inhomogeneous magnetic fields surrounding the nanoparticles cause a transverse dephasing of surrounding hydrogen proton spins that increases over time. Independent of any longitudinal relaxation, this loss of transverse spin coherence locally reduces signal intensity or image brightness, yielding "negative contrast" near the nanoparticles (although schemes do exist to provide also positive contrast [65, 66]). Distinct from T_1-weighted imaging, T_2-weighted imaging therefore seeks to reduce image repetition rates, using longer repeat and echo times to enhance the amount of dephasing, accentuating the local image darkening in the contrast agent vicinity. Although negative T_2 contrast agents are clinically less common than positive T_1 agents, they have been used for in vivo liver imaging [67, 68]. They are also widely used in preclinical studies, in particular in MRI cell tracking [69–73], for which their magnetic field disturbances are large enough to enable single-cell detection in vitro [74–76] and even in vivo [77]. Additionally, superparamagnetic nanoparticles are employed in magnetic resonance-based bioassays through relaxation switch sensors that detect molecular binding events [78], and thereby analyte concentrations [79], through particle aggregation-dependent changes in T_2 relaxation rates [80].

Regardless of whether contrast agents are molecular or nanoparticle based, however, the net result of T_1 and of T_2 relaxation enhancers is a change in image brightness or amplitude. Such signal amplitude changes are analogous to brightness or intensity changes of optical imaging probes; what these T_1 and T_2 relaxation schemes lack is some equivalent to the different color, or, effectively, different optical frequency information, available to optical probes. Being solely amplitude based complicates distinction between different types of MRI contrast agents that may be simultaneously present; to some extent distinguishing between a T_1 and a T_2 agent is possible [81], but this falls short of the multiplexing capabilities of colored probes accessible to optical imaging modalities. This lack of distinction between different agent types also hinders quantitative signal analyses. Without a priori knowledge of the exact amount of contrast agent present, changes in signal intensity may result from a change in contrast agent concentration rather than in the quantity or functioning of some targeted biological variable under study. Distinguishable agents, on the other hand, may allow for ratiometric measures that eliminate concentration dependences, provided that the distinguishable agents are themselves present in some known concentration ratio to each other and are similar enough that they are tracked the same way through the body.

2.2 *Heteronuclear Agents*

Besides relatively low-sensitivity MRS, one spectroscopic approach to distinguishing between different agents is to base those agents on MRI-detectable nuclei other than the hydrogen protons used in conventional ^1H MRI. Fluorinated compounds, for example, which are inert, nontoxic, and based on highly NMR-sensitive ^{19}F

nuclei (comparable to ^{1}H nuclei sensitivity), have long been used as contrast agents or, better formulated, tracers [82, 83], easily predating all Gd^{3+} and magnetic nanoparticle T_1 and T_2 agents by several years [84]. Substrates labeled with hyperpolarized ^{13}C [85, 86] also find much use, particularly in metabolic imaging. Other common NMR-active nuclei include ^{15}N, ^{31}P, and noble gases like ^{129}Xe. And Si nanoparticles containing hyperpolarized ^{29}Si have also been recently introduced as possible MRI tracers [87]. Different nuclei have different gyromagnetic ratios. Therefore, unlike T_1 and T_2 agents, which alter water relaxivities, heteronuclear agents are detected at completely different resonance frequencies, well separated from those of water. They can therefore be distinguished from each other and from conventional contrast agents and, for cases like ^{19}F where there is negligible naturally occurring free fluorine in the body, offer the advantage of background-free detection and even a form of ^{19}F-based multispectral MRI [88].

Still, compared to available water (with hydrogen nuclei concentration up to 110 M), exogenous heteronuclear agents, even hyperpolarized and/or background free, would need to be administered in very high concentrations to match the signal-to-noise ratios achievable with ^{1}H MRI. This is not only because other nuclei may be less NMR sensitive than hydrogen, but also because for heteronuclear agents it is the agent itself that is being detected, whereas for ^{1}H contrast agents it is not the contrast agent, but instead the nearby water that is being detected. That is, T_1 and T_2 contrast agents benefit from interacting with large amounts of water to significantly boost their effective signals, albeit at the expense of no frequency discrimination between these signals.

2.3 Chemical Exchange Agents

An alternative approach that has enjoyed rapid growth over the past decade is the field of chemical exchange saturation transfer (CEST) agents [89] and their paramagnetic equivalents (PARACEST) [90]. Like T_2 contrast agents, CEST agents interact with the surrounding water to decrease the ^{1}H signal, but CEST agents are able to provide signals at different offset frequencies from the background water. They do this by exploiting proton exchange, a natural chemical exchange process whereby weakly bound protons on certain molecules continually exchange with free (unbound) protons in the surrounding bulk water. While attached, the protons experience a chemical shift due to the molecule that they are (temporarily) a part of. Irradiation at that particular chemically shifted offset frequency allows the magnetization of the bound protons to be selectively "saturated out" while leaving the unbound water untouched. When the bound protons subsequently exchange with the surrounding free water protons, they reduce the total magnetization then detected in the unbound water signal. A full magnetization saturation spectrum, referred to as a z-spectrum [91], can then be acquired by repeating the presaturing irradiation pulses over a series of different offset frequencies. Depending on how rapidly chemical exchange recurs, the off-resonance irradiation process can be repeated multiple times within the free water's T_1 relaxation time, multiplying the number of protons that can be magnetically saturated out and boosting the resulting difference in on-resonance signal magnitude. Thus the CEST effect provides an indirect, but clever way to amplify the frequency-shifted signatures of certain molecules that might otherwise yield only weak signal or be undetectable with magnetic resonance spectroscopy imaging. CEST imaging does still require high concentrations of exchanging protons but, fortuitously, some biomolecules with the necessary labile protons do naturally occur at high concentrations and are therefore amenable to such CEST imaging protocols [92]. This gives CEST the advantage of being applicable to both endogenous and exogenous contrast enhancement schemes.

One route to reducing required CEST concentrations is to use molecules with higher proton exchange rates, thus amplifying the signal per molecule. Proton exchange does broaden the shifted linewidths, however, limiting the useful frequency of exchange to of order the chemical frequency shift itself, the so-called "slow exchange" limit [93]. CEST molecules with as large a chemical shift as possible are therefore sought, but many candidate endogenous CEST molecules are diamagnetic offering relatively small shifts. These shifts are proportional to the magnetizing field of the MRI scanner and can thus to some extent be increased by working at higher fields, but this puts many potential applications out of clinical reach. Another workaround is to use exogenous CEST agents that include paramagnetic ions. These PARACEST agents generate larger frequency shifts that allow for more rapid proton exchange and aid in discriminating signals from the background water and from each other.

Whether CEST or PARACEST, their ability to generate signals at different frequency offsets allows for selectively addressable contrast and simultaneous use of more than one agent type, enabling a form of multispectral image contrast [94–96]. Many versions of PARACEST agents are currently being explored, including polymer and supramolecular versions [97, 98], liposome-based LIPOCEST agents [99], and even hyperpolarized (HYPERCEST) systems [100]. Promising as these CEST agents are, all are ultimately limited by their chemical shifts and by the number of exchangeable protons on the molecules used. As recently shown [101], however, NMR frequency shifts need not be restricted to only those from molecular chemical shifts or differing nuclear gyromagnetic ratios; NMR frequency shifts can also be engineered through a new class of contrast agents based on specially shaped magnetic nano- and microparticles.

3 Shaped Nanoparticles

An overwhelming majority of nanoparticles are spherical. This is not surprising. Without deliberate intention otherwise, energy minimization automatically renders many chemically synthesized particles roughly spherical in form. For many applications, it is also more size than shape that matters anyway. Often size is the fundamental determinant of novel nanoparticle functionality due to quantum effects that increasingly dominate over classical ones as particle sizes shrink. In other cases, size dominates simply because the decreasing footprints of nanoparticles make them less obtrusive, a key requirement in, for example, many biomedical applications.

But as the fields of nanotechnology mature—particularly the subfields of nanoparticle synthesis and related colloid and self-assembly research—appreciation of the unique functionalities enabled by nonspherical particles and new ways to synthesize them are growing [102, 103]. Nanoparticle shapes now comprise a veritable zoo of nanocreatures including such species as nanorods, nanotubes, nanorings, nanoshells, nanocubes, nanoellipsoids, nanopyramids, nanodisks, nanobelts, nanocylinders, nanoribbons, nanodumbbells, nanodiscoballs, nanopeanuts, nanostars, nanoscrews, nanocoils, nanosprings, nanotetrapods, nano-octapods, nanohexagons, nanobarrels, nanocages, nanomushrooms, nanoflowers, nanocrescents, nanoworms, and more besides including even such curiously sounding recent proposals as drug-delivering nanovolcanoes [104]. To be sure, many have yet to find any actual use. But among those that have, the applications enabled specifically by nonspherical particles are many and varied. They span a range of fields too broad to cover here, from enhanced platinum nanoparticle catalysis [105] to self-assembled materials with unique anisotropic optical and mechanical properties [106–108]. Even limited to biomedical fields, applications are numerous. Nanoparticle shape—not just size or surface charge—has been found to influence cell uptake rates [109–111], prolong in vivo circulation times and improve tumor targeting efficiency [112], and enhance preferential targeting of particular cell/tissue types [113]. Optical resonances, local electric field enhancements, and heating properties of metallic plasmonic nanoparticles [114–116], which find use in biomedical imaging, sensing, and potential photothermal therapies, are directly determined by both nanoparticle size and shape. And shaped magnetic particles are also attracting attention. Whereas spherical magnetic particles can only translate in a magnetic field, the shape anisotropy of nonspherical particles allows them to be also rotated. This is advantageous because translational magnetic forces, which depend on gradients in the magnetic field, often require close proximity to the source magnet and may be weaker than magnetic torques, which depend on the magnetic field strength itself. Thus asymmetric and corkscrew-shaped magnetic microparticles underpin much of the new field of so-called magnetic swimmers [117], which transduce externally applied magnetic field rotation into mechanical translation with a goal of magnetically guided drug delivery. New ways to attack cancer cells have also been proposed using targeted thin disk-shaped particles whose magnetic shape anisotropy allows them to be rapidly oscillated back and forth to destroy the cells to which they are attached [118]. As a further illustration, increased anisotropy of cube-shaped magnetic nanoparticles has been argued to enhance heat absorption over equivalent spherical particles for alternative magnetic hyperthermia approaches to treatment [119].

3.1 *Shaping NMR Relaxivity*

In MRI, deliberately shaped nanoparticles are less conspicuous, but not entirely absent either. To be sure, for T_1 and T_2 contrast agents, size is the more important qualifier than shape: smaller, molecular scale agents make better T_1 relaxers, while larger nano- and microparticles with higher magnetic moments excel at T_2 contrast generation. Changing size not only changes relaxation efficiency, but can even switch between predominantly T_1 and T_2 relaxation. For example, T_2 MnO particulate contrast agents transform into T_1 agents by releasing free manganese ions when dissolved in acidic cellular endosomes/lysosomes [120]. Conversely, T_1 Gd chelates become better T_2 relaxers when locally concentrated together.

But shape itself can also play a role. One recent example claims increased T_2 contrast through the use of octapod-shaped iron oxide nanoparticles [121], where the more pointed structure gives a more disperse distribution of material with a larger fraction of the magnetic material positioned further from the particle center than would be the case for a solid sphere. It is argued that the more open structures and effectively increased particle diameters allow the nanoparticles to interact with more water, increasing the transverse relaxivity. A comparable case for T_1 contrast agents may be recent Gd(III)-DNA-coated gold nanostars [122] showing increased T_1 contrast (compared to Gd(III)-DNA-coated spheres, for example), believed due at least in part to improved water access resulting from their nonspherical star shapes. Although contrast mechanisms differ, in both cases moving away from a spherical shape to one that is more open with higher curvature surfaces appears to increase water access and relaxivity. Gadolinium-containing ultrashort carbon nanotubes [123] represent another relatively new, and patently nonspherical, set of MRI contrast structures under development. They show significantly higher MR relaxivities than traditional agents, thought due to nanoscale clustering of Gd^{3+} ions confined within the nanotubes and, again, higher water access [124]. This is not to say that relaxivities are necessarily optimized through particles that are specifically octapod-, star-, or tube shaped, but it does show that nanoscale shape can be important.

4 Geometrically Based Multispectral MRI Contrast

The above examples increase image contrast; to add distinguishing spectral content a different approach dependent on more specifically designed, shaped magnetic microstructures was recently introduced [101]. Here the spectral information was a direct consequence of the precise magnetic geometries used, enabling controlled NMR frequency shifting, multiple uniquely identifiable contrast agents, and the prospect of larger scale multiplexing with MRI. Unlike the above nanostructure agents, these multispectral contrast agent structures do not necessarily need to be nanoscale; their operation, while highly dependent on particle shape, is to a large extent size independent.

4.1 Using Shape to Control Frequency

Because all magnetic particles add T_2 contrast, regardless of shape, it helps revisiting how that contrast is generated to better understand how particle geometries can also encode distinct spectral signatures. In a pure water sample free of any magnetic particles, the hydrogen proton spins all precess at the same Larmor frequency, which is proportional to the magnitude of the magnetic field of the MRI scanner. Once a magnetic particle is added to the water its own surrounding magnetic field adds to the MRI field, altering the precession frequencies of protons nearby. Because the particle field is different in different positions around the particle, precession frequencies differ between neighboring water molecules. Thus proton spins accrue different transverse phases with respect to one another, leading to destructive signal interference that locally darkens the image around the particle location. Or, spectrally, the NMR line is broadened since signal from around the particle introduces a continuous spread in precession frequencies.

Such contrast is largely independent of particle shape. What makes magnetic microparticles such powerful T_2 relaxers is the large distances out from the particle over which the particle magnetic field can cause appreciable transverse dephasing. Fields from a single micrometer-sized particle, for example, may be sufficiently strong to significantly dephase hydrogen protons out to of order 100 μm away from that particle [125], encompassing a volume of water many orders of magnitude larger than that of the particle itself. Therefore, unlike T_1 contrast that only relaxes water close to the agent—in the first or second coordination sphere of the paramagnetic ion used—for larger agents much T_2 contrast accrues from water relatively far from the particle. At such distances, any higher spatial frequency components, or equivalently higher multipole moments, in the surrounding particle field distribution have decayed away. Transverse dephasing due to microparticles is therefore predominantly due to spatially decaying dipole fields that, while proportional to net particle magnetic moment and/or size, retain no other information about particle shape.

Avoiding such dephasing, which blurs any potential distinguishing spectral content, requires instead a spatially extended region about the magnetic particle over which the magnetic field does not change. Within such a uniform field region, the precession frequencies of water protons would all be the same, locally keeping all proton spins in phase. This would locally avoid NMR line broadening, but would frequency-shift the water line proportionally to the shifted field magnitude within that region. Such a field setup might initially seem unlikely but, even though all magnetized particles project dipolar far fields, this does not preclude generating the necessary uniform field in the near-field regions of specially designed magnetic microstructures. Further, since signals from water within such a near-field region are spectrally distinct from water further away, selective signal amplification becomes possible. Thus large signals can still be acquired from the engineered uniform near-field volumes even though those water volumes may be much smaller than their dipole-field-dominated far-field counterparts.

4.2 *Frequency-Shifting Micromagnetic Structures*

As a concrete example, Fig. 1 shows one possible microstructure that can discretely shift the NMR frequency of nearby water. It consists of two magnetizable disks, spaced a distance apart roughly equal to the disk radii. When magnetically saturated

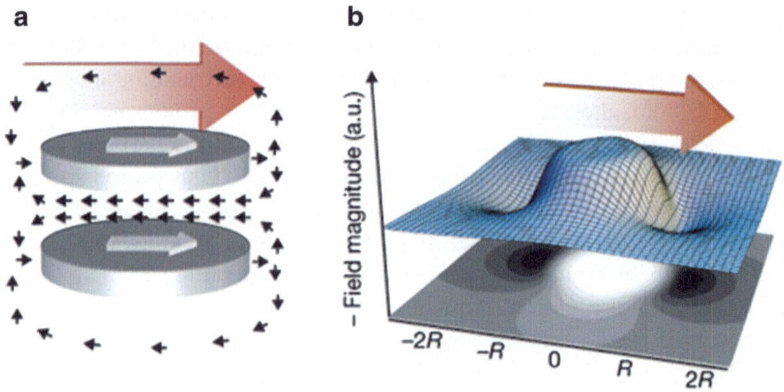

Fig. 1 Double-disk magnetic structure and field diagrams. (**a**) Schematic of the field (small *black arrows*) from two parallel disks magnetized to saturation by the background field of an MRI scanner (large *red arrow*). Nonmagnetic spacer elements are omitted for clarity. (**b**) Calculated (negative) field magnitude in the mid-plane through a typical magnetized disk set, contrasting its homogeneous nature between the disks with its rapid external decay (reproduced in part from [101], with permission, Nature Publishing Group)

in a typical MRI scanner, this particle geometry yields an extended spatial region between the disks where the field is roughly uniform and of a different magnitude to the fields far away from the particle. Note that this does not eliminate the dipolar field decay external to the structure. The dipole field still exists, dephasing surrounding water and broadening the NMR water line just like any other T_2 agent. But within the inner uniform field region, there is a volume of water where all proton spins precess at the same offset frequency, different to the resonance frequency of the bulk water signal. The structure therefore adds a distinct frequency-shifted peak to the acquired NMR spectra. The frequency shift, $\Delta\omega$, is proportional to the difference in magnitudes of the field between the disks and the field far away where water is unperturbed by their presence. The uniform field generated by the disks can be approximated analytically from the field at the center point of the structure. For magnetically saturated disks of thickness h, radius r, center-to-center separation $2S$, made from material with a saturation magnetic polarization density J_S, and immersed in a medium of gyromagnetic ratio γ, the resulting NMR frequency shift is [101]

$$\Delta\omega \approx -\gamma\, J_S\left(\frac{h\,r^2}{2(r^2+s^2)^{3/2}}\right)$$

where thin disks with $h << 2S \approx R$ are assumed for simplicity. Different frequency offsets can therefore be engineered by changing the magnitude of the field between the disks. This can be done by changing the spacing between the disks, the disk radii, the disk thicknesses, or the magnetic material from which the disks are made. All of these variables can be controlled when microfabricating such structures, allowing tailored frequency shifts over large frequency ranges. Thus families of structures can be created, each capable of offsetting the water NMR frequency by a different amount. These frequencies lie in the NMR radio-frequency (RF) range, but if each different frequency offset is associated with a different color (as in the optical spectrum) multispectral, or "RF color," contrast becomes possible. Although operational principles are different, in many ways such shaped magnetic particles become RF analogues to optical quantum dots, or plasmonic nanoparticles. All have resonances that can be engineered by controlling either the size of the quantum dots, the size and shape of the plasmonic nanoparticles, or the geometrical aspect ratios of the magnetic nano- or microparticle structures.

Double-disks are not the only structures that can frequency-shift the NMR signal. As mentioned, the main requirement is generating a uniform offset field region and this can be done in more than one way. For example, short, hollow cylindrical tubes with a length approximately equal to their diameter can also generate the necessary water-accessible, spatially extended regions of uniform field offset within their interior [126]. Figure 2 shows an example of the fields of such a structure which give frequency shifts of [126]

$$\Delta\omega \approx -\gamma J_S \left(\frac{4L\rho t}{\left(L^2 + 4\rho^2\right)^{3/2}} \right)$$

Here, γ and J_S are defined as previously but now L represents the tube length, 2ρ is its diameter, and t is its wall thickness. Again for simplicity a thin-walled structure with $t << L \approx 2\rho$ has been assumed. Analogously to the double-disk structures, the field magnitude can be controlled by changing the tube length, the tube diameter, the tube wall thickness, or the magnetic material from which the tubes are made.

Interestingly, uniform field regions can also be created by asymmetric structures. An example of this is open elliptical shell structures that have their inner and outer boundaries defined by ellipsoids of differing eccentricities [127]. As surfaces of second degree, ellipsoids offer truly uniform internal magnetizations [128]. For

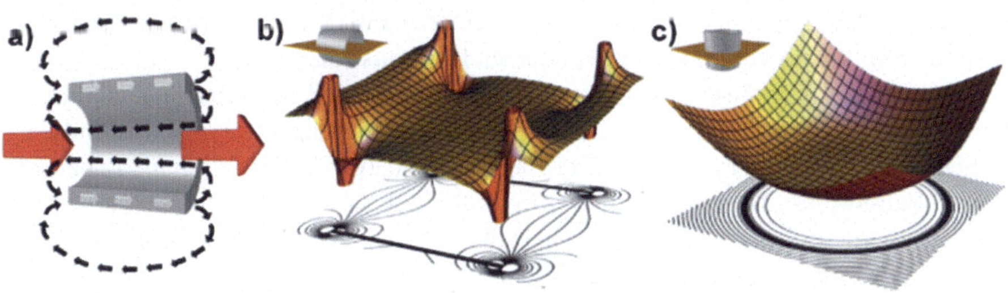

Fig. 2 Schematic of the fields generated by a magnetized hollow cylinder. (**a**) Cut-away schematic of the field (*black arrows*) of a hollow cylinder magnetized to saturation by that background MRI field (larger *red arrows*). (**b**) Calculated magnetic field magnitude profile with underlying field magnitude contour plot in a mid-plane through a magnetized hollow cylinder. Plane orientation shown in upper left corner. (**c**) As for (**b**) but for perpendicularly oriented mid-plane (reproduced in part from [126], with permission, IOP Publishing)

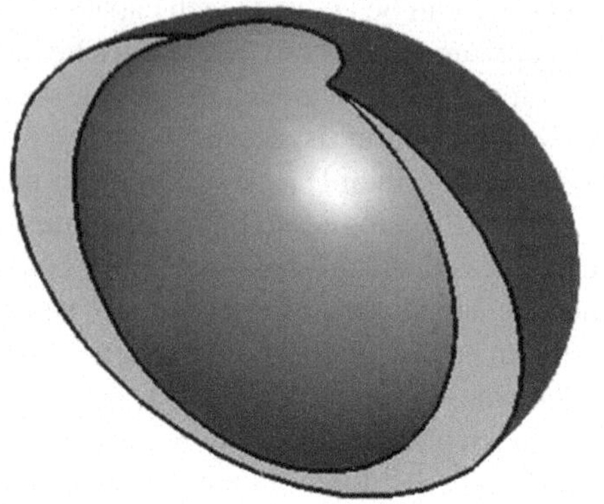

Fig. 3 Schematic of open, asymmetric ellipsoidal shell structure. Cross section through an open shell geometry defined by an oblate ellipsoid of revolution (oblate spheroid) after a vertically offset spherical volume has been removed

magnetically saturated ellipsoids, such uniform magnetization can be shown to result also for geometries that represent one ellipsoidal volume removed from within another, no matter whether the ellipsoidal volumes share a common center [127]. Thus various counterintuitive asymmetrical structures become possible, an example schematic being shown in Fig. 3. For these elliptical shell agents, the internal uniform field magnitudes and resulting frequency shifts depend on the difference in aspect ratios, or eccentricities, of the ellipsoidal shapes that define the structures' physical boundaries. A relatively simple case is a hollow shell structure formed by subtracting a spherical volume (of any radius r) from within an ellipsoid of revolution with semi-axes $r(1+\varepsilon_a)$ and $r(1+\varepsilon_b)$; frequency shifts are described by [127]

$$\Delta\omega \approx -\gamma J_S\left(\frac{4(\varepsilon_a-\varepsilon_b)}{15}\right)$$

where for simplicity again, a thin shell structure has been chosen with ε_a and ε_b much less than unity. Here, field magnitudes and NMR frequency shifts can be controlled by changing boundary ellipticities, by changing the size of either outer or inner bounding ellipsoid (thus changing shell "thickness"), or by changing the magnetic material from which the ellipsoidal shell is constructed.

As magnetic particles, all of these structures double as T_2 contrast agents. Indeed, one can imagine transforming a T_2 agent into a multispectral agent simply by redistributing its material, reforming the spherical (or randomly shaped) particle into a double-disk, hollow cylinder, elliptical shell, or any other shape capable of generating the necessary magnetic field profiles. No new magnetic material needs to be added or removed in the process. Since microparticle T_2 relaxation is dominated by transverse dephasing in the far field where particle shape is irrelevant, a multispectral microparticle agent created in this way would possess the same relaxing ability of the original microparticle agent. But by adding a local uniform field region, the material redistribution adds shape-identifying spectral information that enables particles that would otherwise appear identical in an MRI scan to be distinguished from one another.

Such particles therefore make interesting candidate labels for MRI-based cell tracking, a growing MRI application that may be of considerable value to new cell-based medical therapies [129, 130]. If surface functionalized such that different microstructure geometries target different cell types, labeled cells could be tracked using regular T_2*-weighted gradient echo imaging protocols, but also distinguished from one another through the geometrically encoded spectral content of their magnetic labels. Added spectral content could also distinguish hypointense image regions due to the administered contrast agent from those due to natural image darkenings or signal voids arising from air bubbles or other magnetic field inhomogeneities. Such "color" particle distinction can be seen in Fig. 4, which shows different resulting frequency shifts from double-disk structures with different disk thicknesses.

4.3 Frequency-Shifting Properties

Figure 5 displays several scanning electron micrograph (SEM) images of microfabricated magnetic double-disk, hollow cylinder, and ellipsoidal shell sample structures. Although geometrically distinct from one another, the shapes are unified through their frequency shifts that all reduce to

$$\Delta\omega \approx -\gamma J_S X$$

where X represents the bracketed portions in the equations above. As can be seen, in all cases X is a dimensionless function of solely the structure geometry. That is, at least as far as the magnetics are concerned (although not necessarily as far as the dynamics are concerned), shape-based frequency shifting is in principle scale invariant. Isotropically expanding or contracting any of the structures does not change the frequency shift, allowing agents to be produced over a large size range for different potential applications. Being independent of any chemical exchange processes and with material dependences appearing only through the gyromagnetic ratio and magnetic saturation polarization density, structures can also be used with any NMR-active medium and made from any magnetic material. This allows agents to be made from less toxic materials than the lanthanide ions used in PARACEST and clinical T_1 agents, which require powerful chelating ligands to protect the body from direct exposure [131]. To date, shaped particle agents have been made from nickel [101, 126, 127] as well as from more biocompatible materials including iron [132] and iron oxides [133]. They have also been shown to operate equally in water and in deuterium oxide [101], and have been produced with sizes ranging from milli- to micro- to nanoscale [101, 126, 134].

Lacking spherical symmetry, different particle orientations relative to the applied field can lead to different surrounding magnetic field profiles. These would change the resulting field shifts, were it not for the particles' self-aligning properties. Much

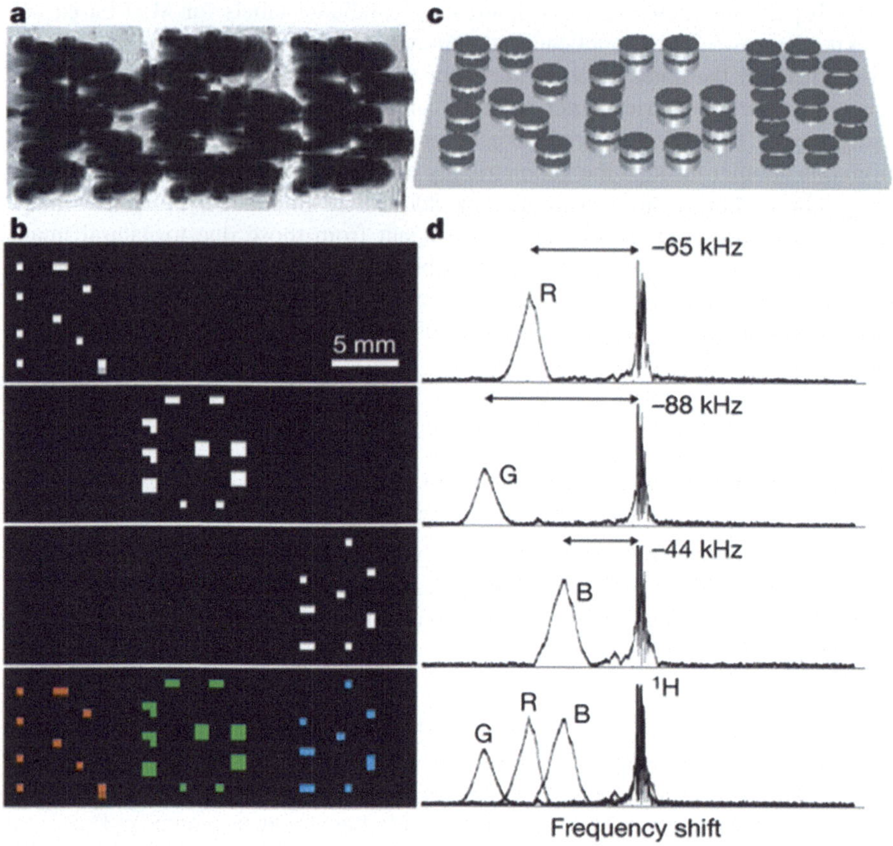

Fig. 4 Multispectral MRI. (**a–d**) Chemical shift imaging of demonstration 1.25 mm diameter particles magnetized by background MRI field. Particle frequency was varied by changing the thickness of electroplated nickel layers that formed the magnetizable disk pairs. As with normal magnetic particle detection, magnetic dephasing due to the particles' external fields enables the spatial imaging shown in the gradient-echo MRI (**a**). However, comparison between (**a**) and the chemical-shift images (**b**) shows that the additional spectral information both differentiates between particle types and improves particle localization. The particles are shown schematically (not to scale) in (**c**). With particle spectra (**d**), to the right of the corresponding chemical-shift images in (**b**) shifted well clear of the water proton line, different planes in the chemical-shift imaging map isolate different particle types for unambiguous color coding with minimal background interference (**b**, *bottom panel*). (Although still visible in the gradient-echo image, the top-corner particle of the letter "B" was damaged, causing its shifted frequency peak to vanish) (reproduced from [101], with permission, Nature Publishing Group)

like a compass needle in the earth's field, the particles' magnetic shape anisotropy causes strong magnetic torques that automatically align the particles parallel to one another when placed into the large fields of an MRI scanner [101]. This avoids random spread in resulting resonance shifts that might otherwise hinder the ability to distinguish different particle geometries.

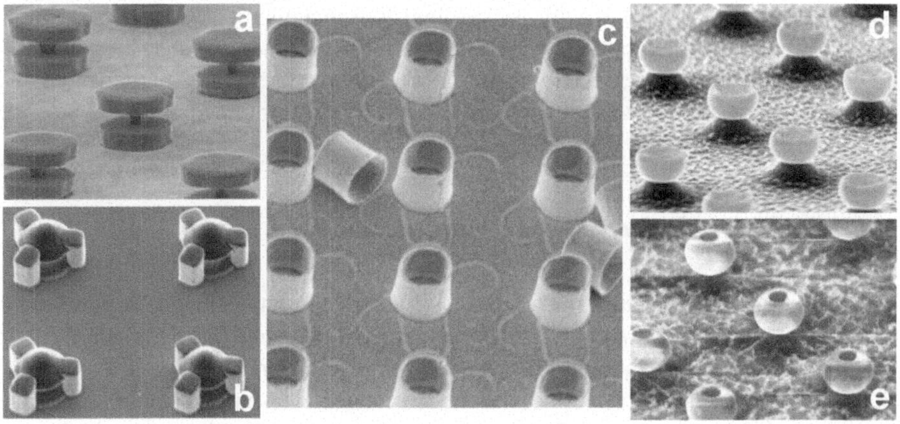

Fig. 5 Scanning electron micrographs (SEM) of microfabricated contrast agent microstructures. (**a, b**) SEM of magnetic double-disk structures separated by nonmagnetic internal or external spacing posts, respectively. (**c**) SEM of hollow magnetic cylinders. (**d, e**) SEM of open oblate and prolate ellipsoidal magnetic shells, respectively. For scale, all structures are a few micrometer in total size

Because particles can be made with ferromagnetic materials, which have far greater magnetic permeabilities than para- or diamagnetic materials, the range of accessible frequency shifts is large. Iron has a J_S value of over 2 T, for example, allowing NMR shifts for water protons to be engineered anywhere from zero up to tens of MHz. (In theory, shifting up to 100 MHz is possible, albeit only for unwieldly structures with very thick magnetic layers.) But even at 1 MHz or a fraction thereof, shifts easily exceed anything yet achieved with molecular agents.

Unlike paramagnetic (or diamagnetic) CEST agents, shifts generated by microparticle structures can also be field independent because typical MRI field strengths magnetically saturate the constituent ferromagnetic materials. Large field-independent shifts are advantageous because they raise the possibility of using such agents at clinical field strengths, rather than at the higher field strengths used to increase the shifts of molecular based (PARA)CEST agents. Whereas most chemical shifts are proportional to the MRI field and thus reported in relative terms of parts per million (ppm), for these shaped particle agents NMR shifts are absolute. In conventional, relative terms their shifts will therefore appear to change based on the applied field but, to aid comparison, for a 1.5 T clinical field strength MRI scanner, a MHz shift corresponds to around 15,000 ppm. Chemical shifts in NMR spectroscopy, by contrast, typically measure just a few ppm. Frequency shifting far from the background bulk water relaxes bandwidth constraints on the off-resonance radiation pulses used to address these agents. With little chance of any RF power leaking over to excite the background water, it enables virtually background-free imaging. That is, even though the signal still comes from water surrounding the magnetic particle rather than from the particle itself, the large shifts enable a form of "hotspot" imaging [135], different in mechanism, but not unlike that reported with highly shifted proton

(HSP) MR imaging [136] or with background-free imaging of ^{19}F labels. Based on magnetic nano- or microparticles that interact with surrounding water, these agents do however offer higher sensitivity than HSP or ^{19}F imaging, in large part due to their large frequency shifts that also enable significant signal amplification.

4.4 Diffusion-Driven Signal Amplification

In some cases, the frequency-shifted signals encoded by shaped magnetic nano- and microstructures can be directly imaged via chemical-shift imaging, a standard MRI protocol that spatially maps NMR frequencies. An instance of this was already shown in Fig. 4. Just like any other magnetic particles the shaped structures appear indistinguishable in a T_2-weighted gradient echo image. But in a chemical-shift image they are clearly distinguished from one another (as well as from the background water) through their ability to locally frequency-shift the nearby water by different amounts. Also clear, however, is that the frequency-shifted signals cover smaller areas than do the T_2 contrast signals. This is because the frequency shifting occurs within the particles' homogeneous field regions, which are small compared to the far-field volume over which water is transversely dephased. This means that shaped particles can be better spatially localized through their added spectral content, but it also means that their spectrally distinguishing signals are not as strong as their associated T_2 contrast signals.

To boost spectral signals, therefore, a variation of magnetization transfer imaging [137] is used, not unlike that employed for amplifying CEST-based signals. Here though, proton exchange does not imply chemical exchange. Instead exchange is driven by water self-diffusion that randomly moves water molecules into and out of the homogeneous field region. The particles' open structure design exploits this natural water diffusion to continually refresh protons contained with the field-shifted region. This yields an effective water volume from which signal can be acquired that may be several orders of magnitude larger than the homogeneous field region itself. That is, signal can be acquired not just from water that happens to be in the field-shifted region at one point in time, but from water passing through that region over an extended time period (of order the bulk-water longitudinal relaxation time, T_1).

The signal acquisition protocol is similar to that of CEST agents. A series of presaturating RF pulses, applied at a specific offset frequency from the background water resonance, are followed by a single on-resonance pulse and the resulting free induction decay (FID) signal used to infer the resulting reduction in the bulk water signal. This process can then be repeated for different offset frequencies, building up a z-spectrum that reveals the particle-induced shifted resonance through a spectrally localized dip in the remaining water signal. (Examples of such NMR z-spectra can be found later in Fig. 6.) Alternatively, knowing contrast particles' specific offset resonance frequencies allows for selectively addressable contrast that can be turned off and on.

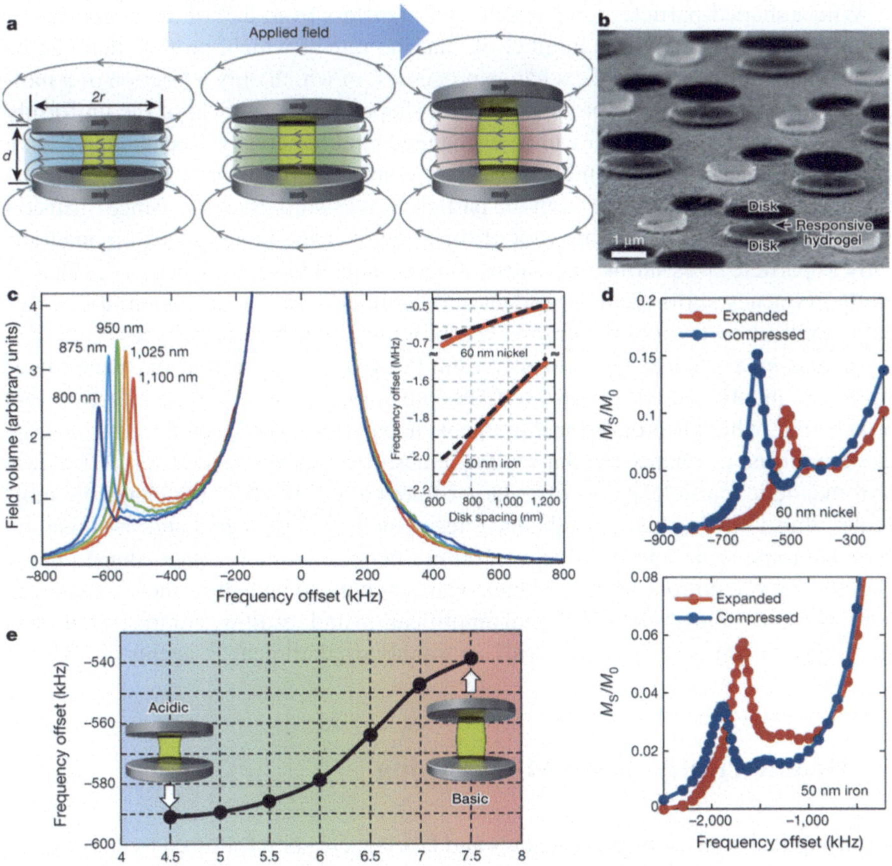

Fig. 6 Principles of shape-changing RF colorimetric sensors. (**a**) Schematic of sensor assemblies comprising two parallel disks magnetized by applied MRI field (*blue arrow*) and separated by stimuli-responsive hydrogel spacers (*yellow*). Resulting magnetic fields (*grey curves*) are uniform between the disks and locally shifted NMR frequencies of water passing through proportionally to the field magnitude, which depends on disk spacing *d*. Different frequency shifts represent different effective RF "colors." (**b**) Scanning electron micrograph of sensors. (Interspersed features are nonmagnetic residual topography from the microfabrication process.) (**c**) Theoretical precession frequency (or equivalently, field) histograms, mimicking NMR spectra, for 60 nm thick, 1 μm radius nickel disk pairs with disk spacing indicated. Background water appears at zero offset; shifted peaks result from uniform field regions between the disks. Inset: frequency offset versus disk spacing for nickel and iron disks with thicknesses shown. *Dashed black curves* are analytic approximations (see equation (1) in ref. [132]); *red curves* result from numerical field simulations. (**d**) Experimental NMR z-spectra for nickel (*top*) and iron (*bottom*) sensors with hydrogel spacers in compressed and expanded states. Magnetization saturated out M_S is normalized to the initial water magnetization M_0. (**e**) Experimental pH-dependent NMR shifts for nickel-based sensors containing pH-sensitive hydrogel spacers designed to shrink (expand) at low (high) pH with peak sensitivity in the physiological pH range (reproduced from [132], with permission, Nature Publishing Group)

Where shaped-particle signal amplification differs from that of molecular-based chemical exchange is in the number of, and the rate at which, protons that can be exchanged. Unlike chemical exchange processes, in which only a fraction of a molecule's protons are able to participate, with shaped particles most of the uniformly field-shifted volume of water can be exchanged simultaneously. Exchange rates are also no longer dependent on chemical rate constants; they depend on the time it takes water to self-diffuse through the particle's field-shifted region. Since diffusion distances scale with the square root of time, the exchange rate speeds up quadratically as particle sizes shrink, increasing the total signal that can be acquired. That is, while frequency shifts are independent of overall particle size, diffusion-driven signal amplification favors smaller (and less biologically invasive) particles.

Increasing exchange rates do increasingly broaden the shifted resonance linewidth. As mentioned with regard to CEST imaging, for agents to be effective, this broadening, which is proportional to the exchange rate, should not become so large that the shifted resonance overlaps with the unshifted background water [93]. Being ferromagnetic, particle agents offer larger frequency shifts than paramagnetic molecules, thus allowing higher exchange rates, which increase signal and should allow particles to be scaled down to below 100 nm in size. Taken together, shaped particles allow more protons to be simultaneously exchanged and allow those exchanges to recur more rapidly, yielding signal amplification and resulting contrast agent sensitivities that compare favorably with commonly used clinical T_1 agents [132].

5 Geometrically Based MRI Sensing

Just as optically based fluorescent tags and labels were soon followed by fluorescent sensors, MRI contrast agents have since expanded to include responsive MRI probes [138]. Such agents change surrounding image contrast in response to some chosen biomarker, which may include various biologically significant metal ions, biomolecules, or surrounding environmental conditions such as temperature or pH. Examples include changes in T_1 due to modified water access to the Gd ions [139], switches in T_2 due to induced aggregation of magnetic nanoparticles [78], changes in CEST contrast due to modified proton exchange rates [140], and signal switching in shaped particle structures by (un)blocking their homogeneous shifted field regions [101].

Whether based on T_1, T_2, CEST, or shaped particles, these sensing examples all amount to changes in the amplitude of the contrast agent signal. As mentioned, however, such signal changes cannot necessarily be distinguished from more mundane changes in the contrast agent concentration. For in vitro tests, concentrations may be well controlled, but this is not always true for in vivo studies where precise trafficking and pharmacokinetics of the administered contrast agents may be poorly known. An option is to administer two different contrast agents [141]. By assuming that at least the ratio of their concentrations remains constant, unwanted concentration dependences can be eliminated. But such ratiometric correction may require larger overall amounts of exogenous agent and may still fail if there is any difference in agent pharmacokinetics.

One way to avoid ambiguity is through a responsive probe that reports via changes in NMR signal frequency rather than amplitude. Unlike signal amplitudes, NMR frequencies need not depend on contrast agent concentrations, allowing for more quantitative measurements. With resonant frequency shifts that can be directly engineered through particle geometry, multispectral shaped-particle contrast agents are well suited to the task. Converting such contrast agents into sensing agents is conceptually simple, requiring only a modified structure whose shape is no longer static, but can dynamically vary in an appropriate way in response to the chosen biomarker. As the structure changes shape, its surrounding magnetic field profile changes, in turn changing the local frequency shift imparted on the surrounding water signal. Given the large frequency shifts of shaped-particle agents, resulting responsive changes in those frequency shifts can be similarly large, in principle enabling even small biomarker changes to be detected.

The first such shape-changing, frequency-based, MRI sensor agents have only recently been published [132]. Referred to as Geometrically Encoded Magnetic (GEM) sensors, they employ stimulus-responsive polymer gels that effect the necessary shape changes and that enable continuous and reversible operation. (Interestingly, Paul Lauterbur, one of the original inventors of MRI, first proposed using a gel-based agent over two decades ago [142]. Except, without specific control over their magnetic fields, those agents yielded particle aggregation-dependent changes in relaxation, or signal amplitude, much like relaxation switch sensors [78].) The first examples of sensors that controllably change fields and therefore frequencies, however, used acid-sensitized hydrogels [143] to transduce local pH levels into NMR-readable frequency shifts. These sensors borrowed from the double-disk multispectral MRI agent geometry, using this time nanoscopic hydrogel pillars as spacing elements between the two magnetic disks. As these spacers swell or shrink in response to the pH of the surrounding solution, the attached disks move further apart or closer together. The resulting change in magnetic field magnitude between the disks then manifests as a changed offset resonance frequency for water passing between the disks, as detailed in Fig. 6. Since hydrogels can expand or contract by large amounts and since frequency changes can be shown to be proportional to length changes in the hydrogels separating the disks [132], large stimulus-induced spectral shifts are possible. Although stimuli-reponsive hydrogels, which rely on solute diffusion through the gel, offer notoriously slow macroscopic response times, for the nanoscopic hydrogel elements incorporated into the GEM sensors diffusion times are rapid, enabling sub-second response times.

As their name suggests, GEM sensors depend on geometry. Like the originating multispectral shaped-particle MRI contrast agents, GEM sensor signals are an intrinsic function of the shapes of the nanostructures involved. They afford not just another example of how particle shape enables new functionality, but of how the shape itself can function as a reporter of local conditions, which may extend beyond pH to include reporting on many other conditions and/or biomolecules of interest. Many different hydrogel formulations—many of them biocompatible—have already been developed in other fields and there exists considerable literature demonstrating the ability to sensitize them to many variables of interest [144]. (Nor are hydrogels the only possible polymers that could be used to effect the necessary shape changes.) Gel sensitization techniques include molecular imprinting of gels [145], the incorporation of catalytic enzymes [146] or enzyme-cleavable substrates [147], and the inclusion of specific receptor-ligand-type recognition bondings [148]. By using hydrogels with responses tailored to different targets, the same geometrically based sensing platform should be adaptable to measure a variety of biomarkers. The multispectral nature of these agents suggests also that multiple different biomarkers might be measured simultaneously by using multiple different GEM sensors. Provided that these sensors are engineered with different initial frequency offsets, different sensor signals can be spectrally isolated from one another even if spatially co-located. This may allow for different sensors to be calibrated against each other and/or for panels of biomarkers to be measured simultaneously to better discriminate between pathologies.

6 Particle Synthesis

A distinguishing feature of shaped particle agents is their synthesis. MRI contrast agents have always been produced by bottom-up chemical synthesis routes; shaped agents have instead leveraged top-down microfabrication techniques [149]. They are produced using the same technology that underpins the electronics revolution and the ever-present, but ever-shrinking, integrated circuit. Supported by an indomitable semiconductor industry, investment in micro- and nanofabrication research over many years has resulted in powerful sets of tools, able to pattern materials with features that are now reaching less than a hundred atoms on a side. Although not used before for contrast agent synthesis, the extraordinary control that such tools offer over both structure shape and composition makes them a good choice for a contrast agent whose function depends on its geometry and material makeup. Of course, an MRI contrast agent is quite different from an integrated circuit and their microfabrication does require adaptation of traditional microfabrication protocols, but several routes have already been proven [126, 127, 132, 133, 150].

Top-down fabrication is not without its limitations, however. As the billions of transistors in each of today's billions of smartphones abundantly prove, top-down microfabrication is well suited to the creation of very small, very precise structures in a massively parallel manner. But throughput still cannot match that possible through bottom-up chemical synthesis, which may lack the precision, but which can produce large volumes of particles at a time. And even though resulting microfabricated components need not be expensive, the necessary microfabrication tools themselves are sometimes out of reach of academic labs, limiting the number of researchers that might otherwise be able to further develop particle-based multispectral contrast agent technology. Ironically then, while microfabrication is a key, enabling technology for proof-of-principle demonstrations of new structures and new functionalities, a current handicap of such structures may be precisely their dependence on such technology.

An open question remains whether it might be possible to chemically synthesize all or some of the above shaped contrast agent structures with sufficient geometrical accuracy and size monodispersity. Having resonance frequencies determined by structure shapes allows differently "colored" agents to be engineered; but it also means that any errors in particle geometry, or variations in shape across a batch of particles, can reduce or blur their distinguishing spectral signals. Precise shape control can be difficult through bottom-up synthesis and monodispersity often deteriorates as particle sizes increase beyond a few tens of nanometer in size. But this is not to say that it is impossible either. Given the particles' large frequency-shifted signals, some trade-off between simplicity of particle synthesis and resulting spectral resolution may be acceptable. As the above particle zoo indicates, skill in the bottom-up chemical control of shapes is also quickly growing. Or perhaps hybrid template-based syntheses might offer simpler approaches. Although lower throughput than bottom-up chemical synthesis, they might nonetheless yield particles in reasonable quantity, particularly if compatible with some form of roll-to-roll processing [151].

7 Conclusion

In some ways, shaped agents can be regarded as a mix between particle-based T_2 and molecular-based (PARA)CEST agents, borrowing advantages from both. Based on magnetic particles, they provide T_2 contrast while remaining selectively addressable through frequency-shifted signals similar to those of CEST agents. In principle they can be made from exactly the same material as a T_2 agent, just reshaped to add identifying spectral content. And they can be imaged through similar gradient-echo T_2-weighted pulse sequences while their spectral information is acquired via similar signal-amplifying magnetization transfer protocols used with CEST agents, just driven now by diffusion, rather than chemical exchange.

But there are differences too. Deliberately shaped nano- and microparticles represent a considerable departure from traditional MRI contrast agents, an approach that may still be too new for its potential, or its pitfalls, to be fully appreciated yet. Their more controlled synthesis through top-down microfabrication has enabled new particle functionality and allowed precise tuning of desired contrast properties. But less accessible fabrication equipment has thus far also limited uptake in the community, retarding what might otherwise be more rapid development of the technology and fuller exploration of its potential. Currently, at around a micrometer in size, the shaped agents are also still relatively large compared to most (though not all) other MRI contrast agents. This need not necessarily be a detriment [152]. Their size may already be adequate for cell tracking studies—indeed, cell viability and tracking have already been tested using larger chemically synthesized spherical particles [63]. But to increase biological utility and reduce size-related biological delivery issues, sizes will ultimately need to be reduced further. This is not impossible: in theory, agents should continue to function down to below 100 nm size scales, but such agents must still be developed and proven in practice.

Exactly how and where such new shaped agents may find their largest impact is therefore still unclear. Possibly new classes of structures with new properties still await discovery. Viewing them as RF analogs to quantum dot or plasmonic nanoparticles suggests a variety of potential applications within biology and beyond. What does seem already clear, however, is that magnetic particle shape, not just size, enables novel imaging and sensing functionalities, offering new avenues to explore in the burgeoning field of nanoparticle-based biomedical imaging.

References

1. Tsien RY. The green fluorescent protein. Ann Rev Biochem. 1998;67:509–44.
2. Giepmans BNG, Adams SR, Ellisman MH, Tsien RY. Review—the fluorescent toolbox for assessing protein location and function. Science. 2006;312:217–24.
3. Chalfie M, Tu Y, Euskirchen G, Ward WW, Prasher DC. Green fluorescent protein as a marker for gene-expression. Science. 1994;263:802–5.
4. Betzig E, et al. Imaging intracellular fluorescent proteins at nanometer resolution. Science. 2006;313:1642–5.
5. Huang B, Bates M, Zhuang X. Super-resolution fluorescence microscopy. Ann Rev Biochem. 2009;78:993–1016.
6. Bruchez Jr M, Moronne M, Gin P, Weiss S, Alivisatos AP. Semiconductor nanocrystals as fluorescent biological labels. Science. 1998;281:2013–6.
7. Alivisatos P. The use of nanocrystals in biological detection. Nat Biotechnol. 2004;22:47–52.
8. Chan WCW, Nie S. Quantum dot bioconjugates for ultrasensitive nonisotopic detection. Science. 1998;281:2016–8.
9. Michalet X, et al. Quantum dots for live cells, in vivo imaging, and diagnostics. Science. 2005;307:538.
10. Anker JN, et al. Biosensing with plasmonic nanosensors. Nat Mater. 2008;7:442.
11. Hu M, et al. Gold nanostructures: engineering their plasmonic properties for biomedical applications. Chem Soc Rev. 2006;35:1084–94.
12. Fu CC, et al. Characterization and application of single fluorescent nanodiamonds as cellular biomarkers. Proc Natl Acad Sci U S A. 2007;104:727–32.
13. Lim SY, Shen W, Gao Z. Carbon quantum dots and their application. Chem Soc Rev. 2015;44:362–81.
14. Pederson JA, Swartz MA. Mechanobiology in the third dimension. Ann Biomed Eng. 2005;33:1469–90.
15. Ntziachristos V. Going deeper than microscopy: the optical imaging frontier in biology. Nat Methods. 2010;7:603–14.
16. Webb RH. Confocal optical microscopy. Rep Prog Phys. 1996;59:427–71.
17. Helmchen F, Denk W. Deep-tissue two-photon microscopy. Nat Methods. 2005;2:932–40.
18. Zipfel WR, Williams RM, Webb WW. Nonlinear magic: multiphoton microscopy in the biosciences. Nat Biotechnol. 2003;21:1368–76.
19. Ntziachristos V, Ripoll J, Wang LHV, Weissleder R. Looking and listening to light: the evolution of whole-body photonic imaging. Nat Biotechnol. 2005;23:313–20.
20. Mosk AP, Lagendijk A, Lerosey G, Fink M. Controlling waves in space and time for imaging and focusing in complex media. Nat Photonics. 2012;6:283–92.

21. Katz O, Small E, Guan Y, Silberberg Y. Noninvasive nonlinear imaging through strongly-scattering turbid layers. Optica. 2014;3:170–4.
22. Hilderbrand SA, Weissleder R. Near-infrared fluorescence: application to in vivo molecular imaging. Curr Opin Chem Bio. 2010;14:71.
23. Guo ZQ, Park S, Yoon J, Shin I. Recent progress in the development of near-infrared fluorescent probes for bioimaging. Chem Soc Rev. 2014;43:16–29.
24. Callaghan PT. Principles of nuclear magnetic resonance microscopy. New York: Oxford Univ. Press; 1991.
25. Moseley ME, et al. Diffusion-weighted MR imaging of anisotropic water diffusion in cat central-nervous-system. Radiology. 1990;176:439–45.
26. Basser PJ. Inferring microstructural features and the physiological state of tissues from diffusion-weighted images. NMR Biomed. 1995;8:333–44.
27. LeBihan D, et al. Diffusion tensor imaging: concepts and applications. J Magn Reson Imag. 2001;13:534–46.
28. Haacke EM, Xu YB, Cheng YCN, Reichenbach JR. Susceptibility weighted imaging (SWI). Magn Reson Med. 2004;52:612–8.
29. Calamante F, Thomas DL, Pell GS, Wiersma J, Turner R. Measuring cerebral blood flow using magnetic resonance imaging techniques. J Cereb Blood Flow Metab. 1999;19:701–35.
30. Detre JA, Leigh JS, Williams DS, Koretsky AP. Perfusion imaging. Magn Reson Med. 1992;23:37–45.
31. Ogawa S, Lee TM, Nayak AS, Glynn P. Oxygenation-sensitive contrast in magnetic-resonance image of rodent brain at high magnetic fields. Magn Reson Med. 1990;14:68–78.
32. Sosnovik DE, Weissleder R. Emerging concepts in molecular MRI. Curr Opin Biotechnol. 2006;18:4–10.
33. Sipkins DA, Cheresh DA, Kazemi MR, Nevin LM, Bednarski MD, Li KC. Detection of tumor angiogenesis in vivo by alpha$_V$beta$_3$-targeted magnetic resonance imaging. Nat Med. 1998;4:623–6.
34. Yu X, et al. High-resolution MRI characterization of human thrombus using a novel fibrin-targeted paramagnetic nanoparticle contrast agent. Magn Reson Med. 2000;44:867–72.
35. Flacke S, et al. Novel MRI contrast agent for molecular imaging of fibrin: implications for detecting vulnerable plaques. Circulation. 2001;104:1280–5.
36. Weissleder R, Reimer R, Lee AS, Wittenberg J, Brady TJ. MR receptor imaging—ultrasmall iron-oxide particles targeted to asialoglycoprotein receptors. Am J Roentgenology. 1990;155:1161–7.
37. Louie AY, et al. In vivo visualization of gene expression using magnetic resonance imaging. Nat Biotechnol. 2000;18:321–5.
38. Weissleder R, et al. In vivo magnetic resonance imagine of transgene expression. Nat Med. 2000;6:351–4.
39. Genove G, DeMarco U, Xu H, Goins WF, Ahrens ET. A new transgene reporter for in vivo magnetic resonance imaging. Nat Med. 2005;11:450–4.
40. Gilad AA, et al. Artificial reporter gene providing MRI contrast based on proton exchange. Nat Biotechnol. 2007;25:217–9.
41. Gilad AA, Ziv K, McMahon MT, van Zijl PCM, Neeman M, Bulte JWM. MRI reporter genes. J Nucl Med. 2008;49:1905–8.
42. Glunde K, Artemov D, Penet M-F, Jacobs MA, Bhujwalla ZM. Magnetic resonance spectroscopy in metabolic and molecular imaging and diagnosis of cancer. Chem Rev. 2010;110:3043.
43. Mountford CE, Stanwell P, Lin A, Ramadan S, Ross B. Neurospectroscopy: the past, present and future. Chem Rev. 2010;110:3060–86.
44. Nelson KL, Runge VM. Basic principles of MR contrast. Topics Magn Reson Imaging. 1995;7:124–36.
45. Merbach A, Helm H, Tóth E, editors. The chemistry of contrast agents in medical magnetic resonance imaging. 2nd ed. West Sussex, UK: Wiley; 2013.
46. Watanabe M, Tanaka R, Takeda N. Magnetic-resonance-imaging and histopathology of cerebral gliomas. Neuroradiology. 1992;34:463–9.

47. Caravan P, Ellison JJ, McMurry TJ, Lauffer RB. Gadolinium(III) chelates as MRI contrast agents: structure, dynamics, and applications. Chem Rev. 1999;99:2293–352.
48. Bottrill M, Kwok L, Long NJ. Lanthanides in magnetic resonance imaging. Chem Soc Rev. 2006;35:557–71.
49. Rocklage SM, Cacheris WP, Quay SC, Hahn FE, Raymond KN. Manganese(II) n, n′-dipyrid oxylethylenediamine-n, n′-diacetate 5,5′-bis(phosphate)—synthesis and characterization of a paramagnetic chelate for magnetic-resonance imaging enhancement. Inorg Chem. 1989;28: 477–85.
50. Koretsky AP, Silva AC. Manganese-enhance magnetic resonance imaging. NMR Biomed. 2004;17:527–31.
51. Silva AC, Lee JH, Aoki I, Koretsky AP. Manganese-enhanced magnetic resonance imaging (MEMRI): methodological and practical considerations. NMR Biomed. 2004;17:532–43.
52. Na HB, et al. Development of a T-1 contrast agent for magnetic resonance imaging using MnO nanoparticles. Angew Chem Int Ed. 2007;46:5397–401.
53. Gilad AA, et al. MR tracking of transplanted cells with "positive contrast" using manganese oxide nanoparticles. Magn Reson Med. 2008;60:1–7.
54. Park JY, et al. Paramagnetic ultrasmall gadolinium oxide nanoparticles as advanced T-1 MR1 contrast agent: account for large longitudinal relaxivity, optimal particle diameter, and in vivo T-1 MR images. ACS Nano. 2009;3:3663–9.
55. Engstrom M, Klasson A, Pedersen H, Vahlberg C, Kall PO, Uvdal K. High proton relaxivity for gadolinium oxide nanoparticles. Magn Reson Mat Phys Bio Med. 2006;19:180–6.
56. Hyeon T. Chemical synthesis of magnetic nanoparticles. Chem Commun. 2003;8:927–34.
57. Gupta AK, Gupta M. Synthesis and surface engineering of iron oxide nanoparticles for biomedical applications. Biomaterials. 2005;26:3995–4021.
58. Shen T, Weissleder R, Papisov M, Bogdanov Jr A, Brady TJ. Monocrystalline iron oxide nanocompounds (MION): physicochemical properties. Magn Reson Med. 1993;29:599–604.
59. Weissleder R, Elizondo G, Wittenberg J, Rabito CA, Bengele HH, Josephson L. Ultrasmall superparamagnetic iron oxide: characterization of a new class of contrast agents for MR imaging. Radiology. 1990;175:489–93.
60. Bulte JWM, et al. Magnetodendrimers allow endosomal magnetic labeling and in vivo tracking of stem cells. Nat Biotech. 2001;19:1141–7.
61. Jung CW, Jacobs P. Physical and chemical properties of superparamagnetic iron oxide MR contrast agents: ferumoxides, ferumoxtran, ferumoxsil. Magn Reson Imaging. 1995;13: 661–74.
62. Wang YX, Hussain SM, Krestin GP. Superparamagnetic iron oxide contrast agents: physicochemical characteristics and applications in MR imaging. Eur Radiol. 2001;11:2319–31.
63. Shapiro EM, Skrtic S, Koretsky AP. Sizing it up: cellular MRI using micron-sized iron oxide particles. Magn Reson Med. 2005;53:329–38.
64. Shapiro EM, Skrtic S, Sharer K, Hill JM, Dunbar CE, Koretsky AP. MRI detection of single particles for cellular imaging. Proc Natl Acad Sci. 2004;101:10901–6.
65. Seppenwoolde J-H, Viergever MA, Bakker CJG. Passive tracking exploiting local signal conservation: the white marker phenomenon. Magn Reson Med. 2003;50:784–90.
66. Cunningham CH, Arai T, Yang PC, McConnell MV, Pauly JM, Conolly SM. Positive contrast magnetic resonance imaging of cells labeled with magnetic nanoparticles. Magn Reson Med. 2005;53:999–1005.
67. Bellin MF, Zaim S, Auberton E, Sarfati G, Duron JJ, Khayat D, Grellet J. Liver metastase—safety and efficacy of detection with superparamagnetic iron-oxide in MR-imaging. Radiology. 1994;193:657–63.
68. Weinmann HJ, Ebert W, Misselwitz B, Schmitt-Willich H. Tissue-specific MR contrast agents. Eur J Radiol. 2003;46:33–44.
69. Frank JA, et al. Clinically applicable labeling of mammalian and stem cells by combining superparamagnetic iron oxides and transfection agents. Radiology. 2003;228:480–7.

70. Bulte JWM, Kraitchman DL. Iron oxide MR contrast agents for molecular and cellular imaging. NMR Biomed. 2004;17:484–99.
71. Modo M, Hoehn M, Bulte JWM. Cellular MR imaging. Mol Imaging. 2005;4:143–64.
72. Wu YL, Ye Q, Foley LM, Hitchens TK, Sato K, Williams JB, Ho C. In situ labeling of immune cells with iron oxide particles: an approach to detect organ rejection by cellular MRI. Proc Natl Acad Sci. 2006;103:1852–7.
73. Shapiro EM, Gonzalez-Perez O, Garcia-Verdugo JM, Alvarez-Buylla A, Koretsky AP. Magnetic resonance imaging of the migration of neuronal precursors generated in the adult rodent brain. Neuroimage. 2006;32:1150–7.
74. Dodd SJ, Williams M, Suhan JP, Williams DS, Koretsky AP, Ho C. Detection of single mammalian cells by high-resolution magnetic resonance imaging. Biophys J. 1999;76:103–9.
75. Hinds KA, et al. Highly efficient endosomal labeling of progenitor stem cells with large magnetic particles allows magnetic resonance imaging of single cells. Blood. 2003;102: 867–72.
76. Foster-Gareau P, Heyn C, Alejski A, Rutt BK. Imaging single mammalian cells with a 1.5 T clinical MRI scanner. Magn Reson Med. 2003;49:968–71.
77. Shapiro EM, Sharer K, Skrtic S, Koretsky AP. In vivo detection of single cells by MRI. Magn Reson Med. 2006;55:242–9.
78. Perez JM, Josephson L, O'Loughlin T, Högemann D, Weissleder R. Magnetic relaxation switches capable of sensing molecular interactions. Nat Biotechnol. 2002;20:816–20.
79. Sun EY, Weissleder R, Josephson L. Continuous analyte sensing with magnetic nanoswitches. Small. 2006;2:1144–7.
80. Tanimoto A, Pouliquen D, Kreft BP, Stark DD. Effects of spatial distribution on proton relaxation enhancement by particulate iron oxide. J Magn Reson Imaging. 1994;4:653–7.
81. Gilad AA, et al. MR tracking of transplanted cells with "positive contrast" using manganese oxide nanoparticles. Magn Reson Med. 2008;60:1–7.
82. Ruiz-Cabello J, Barnett BP, Bottomley PA, Bulte JWM. Fluorine (F-19) MRS and MRI in biomedicine. NMR Biomed. 2011;24:114–29.
83. Chen JJ, Lanza GM, Wickline SA. Quantitative magnetic resonance fluorine imaging: today and tomorrow. Wiley Interdiscip Rev Nanomed Nanobiotechnol. 2010;2:431–40.
84. Holland GN, Bottomley PA, Hinshaw WS. F-19 magnetic-resonance imaging. J Magn Reson. 1977;28:133–6.
85. Gallagher FA, et al. Magnetic resonance imaging of pH in vivo using hyperpolarized (13) C-labelled bicarbonate. Nature. 2008;453:940–3.
86. Golman K, Petersson JS. Metabolic imaging and other applications of hyperpolarized C-13. Acad Radiol. 2006;13:932–42.
87. Cassidy MC, Chan HR, Ross BD, Bhattacharya PK, Marcus CM. In vivo magnetic resonance imaging of hyperpolarized silicon particles. Nat Nanotech. 2013;8:363–8.
88. Partlow KC, et al. F-19 magnetic resonance imaging for stem/progenitor cell tracking with multiple unique perfluorocarbon nanobeacons. FASEB J. 2007;21:1647–54.
89. Ward KM, Aletras AH, Balaban RS. A new class of contrast agents for MRI based on proton chemical exchange dependent saturation transfer (CEST). J Magn Reson. 2000;143:79–87.
90. Zhang S, Merritt M, Woessner DE, Lenkinski RE, Sherry AD. PARACEST agents: modulating MRI contrast via water proton exchange. Acc Chem Res. 2003;36:783–90.
91. Grad J, Bryant RG. Nuclear magnetic cross-relaxation spectroscopy. J Magn Reson. 1990;90:1.
92. Zhou JY, Payen JF, Wilson DA, Traystman RJ, van Zijl PCM. Using the amide proton signals of intracellular proteins and peptides to detect pH effects in MRI. Nat Medicine. 2003;9:1085–90.
93. Woods M, Woessner DE, Sherry AD. Paramagnetic lanthanide complexes as PARACEST agents for medical imaging. Chem Soc Rev. 2006;35:500–11.
94. Aime S, Carrera C, Delli Castelli D, Crich SG, Terreno E. Tunable imaging of cells labeled with MRI-PARACEST agents. Angew Chem Int Ed. 2005;44:1813–5.

95. McMahon MT, Gilad AA, DeLiso MA, Cromer Berman SM, Bulte JWM, van Zijl PCM. New "multicolor" polypeptide diamagnetic chemical exchange saturation transfer (DIACEST) contrast agents for MRI. Magn Reson Med. 2008;60:803–12.
96. Nicholls FJ, Ling W, Ferrauto G, Aime S, Modo M. Simultaneous MR imaging for tissue engineering in a rat model of stroke. Sci Rep. 2015. doi: 10.1038/srep14597.
97. Aime S, Delli Castelli D, Terreno E. Supramolecular adducts between poly-L-arginine and [Tm^{III}dotp]: a route to sensitivity-enhanced magnetic resonance imaging-chemical exchange saturation transfer agents. Angew Chem Int Ed. 2003;42:4527.
98. Wu Y, et al. Polymeric PARACEST agents for enhancing MRI contrast sensitivity. J Am Chem Soc. 2008;130:13854.
99. Aime S, Delli Castelli D, Terreno E. Highly sensitive MRI chemical exchange saturation transfer agents using liposomes. Angew Chem Int Ed. 2005;44:5513–5.
100. Schröder L, Lowery TJ, Hilty C, Wemmer DE, Pines A. Molecular imaging using a targeted magnetic resonance hyperpolarized biosensor. Science. 2006;314:446–9.
101. Zabow G, Dodd S, Moreland J, Koretsky A. Micro-engineered local field control for high-sensitivity multispectral MRI. Nature. 2008;453:1058–63.
102. Sau TK, Rogach AL, editors. Complex-shaped metal nanoparticles: bottom-up syntheses and applications. Weinheim: Wiley; 2012.
103. Champion JA, Katare YK, Mitragotri S. Making polymeric micro- and nanoparticles of complex shapes. Proc Natl Acad Sci. 2007;104:11901–4.
104. Zhang XA, Elek J, Chang C-H. Three-dimensional nanolithography using light scattering from colloidal particles. ACS Nano. 2013;7:6212–8.
105. Narayanan R, El-Sayed MA. Shape-dependent catalytic activity of platinum nanoparticles in colloidal solution. Nano Lett. 2004;4:1343–8.
106. Grzelczak M, Vermant J, Furst EM, Liz-Marzan LM. Directed self-assembly of nanoparticles. ACS Nano. 2010;4:3591.
107. Ding T, Song K, Clays K, Tung C. Fabrication of 3D photonic crystals of ellipsoids: convective self-assembly in magnetic field. Adv Mater. 2009;21:1936.
108. Mittal M, Furst EM. Electric field-directed convective assembly of ellipsoidal colloidal particles to create optically and mechanically anisotropic thin films. Adv Funct Mater. 2009;19:3271.
109. Gratton SEA, et al. The effect of particle design on cellular internalization pathways. Proc Natl Acad Sci U S A. 2008;105:11613–8.
110. Champion JA. Mitragotri. Role of target geometry in phagocytosis. Proc Natl Acad Sci U S A. 2006;103:4930–4.
111. Chithrani BD, Ghazini AA, Chan WCW. Determining the size and shape dependence of gold nanoparticle uptake into mammalian cells. Nano Lett. 2006;6:662–8.
112. Park JH, et al. Systematic surface engineering of magnetic nanoworms for in vivo tumor targeting. Small. 2009;5:694–700.
113. Kolhar P, et al. Using shape effects to target antibody-coated nanoparticles to lung and brain endothelium. Proc Natl Acad Sci U S A. 2013;110:10753–8.
114. Kelly KL, et al. The optical properties of metal nanoparticles: the influence of size, shape, and dielectric environment. J Phys Chem B. 2003;107:668–77.
115. Wang YC, et al. Comparison study of gold nanohexapods, nanorods, and nanocages for photothermal cancer treatment. ACS Nano. 2013;7:2068–77.
116. Cole JR, Mirin NA, Knight MW, Goodrich GP, Halas NJ. Photothermal efficiencies of nanoshells and nanorods for clinical therapeutic applications. J Phys Chem C. 2009;113:12090–4.
117. Tottori S, et al. Magneti helical micromachines: fabrication, controlled swimming, and cargo transport. Adv Mater. 2012;24:811–6.
118. Kim DH, et al. Biofunctionalized magnetic-vortex microdiscs for targeted cancer-cell destruction. Nat Mater. 2010;9:165–71.
119. Martinez-Boubeta C, et al. Learning from nature to improve the heat generation of iron-oxide nanoparticles for magnetic hyperthermia applications. Sci Rep. 2013. Doi 10.1038/srep01652.

120. Shapiro EM, Koretsky AP. Convertible manganese contrast for molecular and cellular MRI. Magn Reson Med. 2008;60:265–9.
121. Zhao ZH, et al. Octapod iron oxide nanoparticles as high-performance T-2 contrast agents for magnetic resonance imaging. Nat Comm. 2013. doi 10.1038/ncomms3266.
122. Rotz MW, et al. High relaxivity Gd(III)-DNA gold nanostars: investigation of shape effects on proton relaxation. ACS Nano. 2015;9:3385–96.
123. Sitharaman B, et al. Superparamagnetic gadonanotubes are high-performance MRI contrast agents. Chem Comm. 2005;31:3915–7.
124. Sethi R, Mackeyev Y, Wilson LJ. The gadonanotubes revisited: a new frontier in MRI contrast agent design. Inorg Chim Acta. 2012;393:165–72.
125. Zabow G, Dodd SJ, Shapiro E, Moreland J, Koretsky AP. Microfabricated high-moment micrometer-sized MRI contrast agents. Magn Reson Med. 2011;65:645–55.
126. Zabow G, Dodd SJ, Moreland J, Koretsky AP. The fabrication of uniform cylindrical nanoshells and their use as spectrally tunable MRI contrast agents. Nanotechnology. 2009;20:385301.
127. Zabow G, Dodd SJ, Koretsky AP. Ellipsoidal microcavities: electromagnetic properties, fabrication, and use as multispectral MRI agents. Small. 2014;10:1902–7.
128. Maxwell JC. A treatise on electricity and magnetism, vol. 2. 3rd ed. Oxford: Clarendon; 1904.
129. Long CM, Bulte JWM. In vivo tracking of cellular therapeutics using magnetic resonance imaging. Expert Opin Biol Ther. 2009;9:293–306.
130. Ahrens ET, Bulte JWM. Tracking immune cells in vivo using magnetic resonance imaging. Nat Rev Immunol. 2013;13:755–63.
131. Hao DP, Ai T, Goerner F, Hu XM, Runge VM, Tweedle M. MRI contrast agents: basic chemistry and safety. J Magn Reson Imag. 2012;36:1060–71.
132. Zabow G, Dodd SJ, Koretsky AP. Shape-changing magnetic assemblies as high-sensitivity NMR-readable nanoprobes. Nature. 2015;520:73–7.
133. Wang X, Wang C, Anderson S, Zhang X. Microfabricated iron oxide particles for tunable, multispectral magnetic resonance imaging. Mater Lett. 2013;110:122–6.
134. Wang C, Wang X, Anderson S, Zhang X. Biocompatible, micro- and nanofabricated magnetic cylinders for potential use as contrast agents for magnetic resonance imaging. Sens Actuators B Chem. 2014;196:670–5.
135. Bulte JWM. Hot spot MRI emerges from the background. Nat Biotechnol. 2005;23:945–6.
136. Schmidt R, et al. Highly shifted proton MR imaging: cell tracking by using direct detection of paramagnetic compounds. Radiology. 2014;272:785–95.
137. Henkelman RM, Stanisz GJ, Graham SJ. Magnetization transfer in MRI: a review. NMR Biomed. 2001;14:57.
138. Yoo B, Pagel MD. An overview of responsive MRI contrast agents for molecular imaging. Front Biosci. 2008;13:1733–52.
139. Moats RA, Fraser SE, Meade TJ. A "smart" magnetic resonance imaging agent that reports on specific enzymatic activity. Angew Chem Int Ed. 1997;36:726–8.
140. Ward KM, Balaban RS. Determination of pH using water protons and chemical exchange dependent saturation transfer (CEST). Magn Reson Med. 2000;44:799–802.
141. Martinez GV, et al. Imaging the extracellular pH of tumors by MRI after injection of a single cocktail of T1 and T2 contrast agents. NMR Biomed. 2011;24:1380–91.
142. Frank S, Lauterbur PC. Voltage-sensitive magnetic gels as magnetic resonance monitoring agents. Nature. 1993;363:334–6.
143. Peppas NA, Hilt JZ, Khademhosseini A, Langer R. Hyrogels in biology and medicine: from molecular principles to bionanotechnology. Adv Mater. 2006;18:1345–60.
144. Ulijn RV, et al. Bioresponsive hydrogels. Mater Today. 2007;10:40.
145. Byrne ME, Park K, Peppas NA. Molecular imprinting within hydrogels. Adv Drug Deliv Rev. 2002;54:149–61.
146. Fischel-Ghodsian F, Brown L, Mathiowitz E, Brandenburg D, Langer R. Enzymatically controlled drug delivery. Proc Natl Acad Sci U S A. 1988;85:2403–6.

147. Plunkett KN, Berkowski KL, Moore JS. Chymotrypsin responsive hydrogel: application of a disulfide exchange protocol for the preparation of methacrylamide containing peptides. Biomacromolecules. 2005;6:632–7.
148. Miyata T, Asami N, Uragami T. A reversibly antigen-responsive hydrogel. Nature. 1999;399:766–9.
149. Madou MJ. Fundamentals of microfabrication and nanotechnology. 3rd ed. Boca Raton, FL: CRC Press; 2011.
150. Zabow G, Koretsky AP, Moreland J. Design and fabrication of a micromachined multispectral magnetic resonance imaging agent. J Micromech Microeng. 2009;19:025020.
151. Perry JL, Herlihy KP, Napier ME, Desimone JM. PRINT: a novel platform toward shape and size specific nanoparticle theranostics. Acc Chem Res. 2011;44:990–8.
152. Whitesides GM. The “right” size in nanobiotechnology. Nat Biotech. 2003;21:1161.

Magnet-Targeted Delivery and Imaging

P. Stephen Patrick, Quentin A. Pankhurst, Christopher Payne, Tammy L. Kalber*, and Mark F. Lythgoe*

1 Introduction

Magnetic nanoparticles, in combination with applied magnetic fields, can non-invasively focus delivery of small-molecule drugs and human cells to specific regions of the anatomy. This emerging technology could solve one of the main challenges in therapy development: delivery of a high concentration of the therapeutic agent to the target organ or tissue whilst reducing systemic dosing and off-target side effects. Several challenges, however, must be met before this technology can be applied either effectively or safely in the clinic to augment therapies.

Multiple nanoparticle features interact to influence the efficiency of magnet-targeted delivery, and so their design will have a large influence on the success of therapeutic targeting. For example iron oxide core size and composition affect the type (superparamagnetism/ferrimagnetism/antiferromagnetism) and strength of magnetism, and thus the amount of force that can be applied by an external magnetic field. Furthermore, particle behaviour within biological systems can be affected by particle size and coating, including their cell uptake, extravasation rate, circulation time, clearance, aggregation, and degradation.

To assess the impact of these factors on particle biodistribution and success of delivery, it is useful to be able to image nanoparticles non-invasively with a clinically

*Author contributed equally with all other contributors.

P.S. Patrick • C. Payne • T.L. Kalber • M.F. Lythgoe (✉)
Division of Medicine, UCL Centre of Advanced Biomedical Imaging, University College London, Paul O'Gorman Building, 72 Huntley Street, London WC1E 6DD, UK
e-mail: m.lythgoe@ucl.ac.uk

Q.A. Pankhurst
UCL Healthcare Biomagnetics Laboratory, University College London, 21 Albemarle Street, London W1S 4BS, UK

J.W.M. Bulte, M.M.J. Modo (eds.), *Design and Applications of Nanoparticles in Biomedical Imaging*, DOI 10.1007/978-3-319-42169-8_6

available imaging modality. One solution to this is the use of magnetic resonance imaging (MRI), which is sensitive to the presence of magnetic particles, such as iron oxide, in most biological tissue. This can give high-resolution anatomical information on particle location, providing a translatable method to confirm delivery success. However, MRI lacks the quantitative ability to assess whole-body biodistribution, and so the incorporation of other imaging agents, such as radionuclides, into particles could also be beneficial (see Chap. 10).

Preclinical researchers have investigated the use of magnetic targeting-based therapies across a wide range of conditions, and positive results have been reported [1, 2]. Furthermore, improved drug delivery to tumours has been demonstrated in the small number of clinical trials completed to date [3, 4]. Other potential cancer therapy applications include embolization and the delivery of radiotherapy agents, as well as the heating of targeted magnetic particles using alternating current magnetic fields for hyperthermia-based treatment or the controlled release of drugs [5]. For regenerative medicine, stem and progenitor cells have been targeted to specific organs such as the heart and brain to restore function to damaged tissue in mice and rats [6]. In the emerging field of "micro-robotics," bacteria, magnetic particles, and microsurgical robots have been directed through the vasculature or organs in live animals or mazes using computer-interfaced magnetic devices and real-time MRI to follow their progress [7–10].

In this chapter, we provide a basic introduction to the physical principles behind magnetic targeting technology, relevant design features of nanoparticles and magnetic targeting devices, an overview of preclinical and clinical applications, and an introduction to imaging magnetic particles in vivo.

2 General Physical Principles

The magnitude and direction of force that can be exerted on a magnetic particle will determine its potential to be targeted. In general terms, this force depends only on the magnetisation of the particle (*M*), and the strength and spatial distribution of the magnetic field (*B*), as described by (1)

$$F_m = (M \times \nabla) B \tag{1}$$

In this, and the following equations, it is important to note that we are dealing with three-dimensional vectors that describe the resulting magnetic force, as well as the magnetic field and gradient; so for example $\nabla = \left(\frac{\partial}{\partial x}, \frac{\partial}{\partial y}, \frac{\partial}{\partial z} \right)$ is the vector gradient operator, and $(M \times \nabla)$ is a (3×3) matrix with elements $M_x \frac{\partial}{\partial x}$, etc. In small magnetic fields, where ***M*** is linearly proportional to the applied magnetic field, Eq. (1) may be written as [1]

$$F_m = \frac{\Delta \chi V_m}{\mu_o} (B \times \nabla) B \tag{2}$$

where μ_0 is the magnetic permeability constant of free space. The other variables are explained in more detail below:

$$\Delta\chi$$

The relative magnetic susceptibility of the particle: This can be understood as a ratio of the induced magnetisation to the strength of the inducing field, and is characteristic for each type of magnetic particle in a given arrangement. During targeting experiments, the amount of force that can be exerted on the particle is proportional to $\Delta\chi$, which is the difference between the magnetic susceptibility of the particle and that of the external medium. As the susceptibility of normal biological tissue and fluids, such as the blood, is usually negligible in comparison to the particle, $\Delta\chi$ is typically equivalent to the magnetic susceptibility χ of the particle.

$$V_m$$

The volume of the magnetic particle: (Note that we use "particle" to denote the magnetic entity that is being actuated, and that this may be a multicomponent entity, e.g. one in which many inorganic magnetic cores may be encapsulated in an organic matrix.) In cases where a cell or magnetoliposome is loaded with multiple magnetic particles and an aggregation of magnetic particles exists, the particles will typically behave as a single magnetic particle. For simplification the volume of the particles in this situation can be summed to estimate the force acting upon them:

$$\left(B\times\nabla\right)B$$

The external magnetic field—both its strength, $B=|B|$, and its spatial variation (commonly referred to as the "magnetic field gradient"), as expressed by the vector product with the gradient operator: The amount of force that can be applied to a magnetic particle is proportional to the field strength, up to field strengths that approach the saturation magnetisation (Fig. 1a) for a given particle (typically $0.5<B<1.0$ T). Field strength depends on the strength of the magnetic device and the position of the particle relative to the magnetic field created by the device. The magnetic field gradient depends on the shape of the magnetic field that is produced by the magnetic targeting device, and the position of the particle within the field. The direction of the gradient is the main controllable parameter that will determine the direction of particle movement. In a uniform magnetic field lacking a gradient, the directional magnetic force acting upon a particle will be negligible.

As the magnetic force acting on a particle is only linearly dependent on magnetic field strength over a certain range, Eq. (2) is not applicable for higher magnetic field environments, where the particle approaches its saturation magnetisation, M_s. This can be visualised on the hysteresis curve (Fig. 1), which shows the induced magnetism of a magnetic particle on the y-axis in response to increasing external magnetic field strength along the x-axis. At higher field strengths, another approximation can be used to estimate the force [11]:

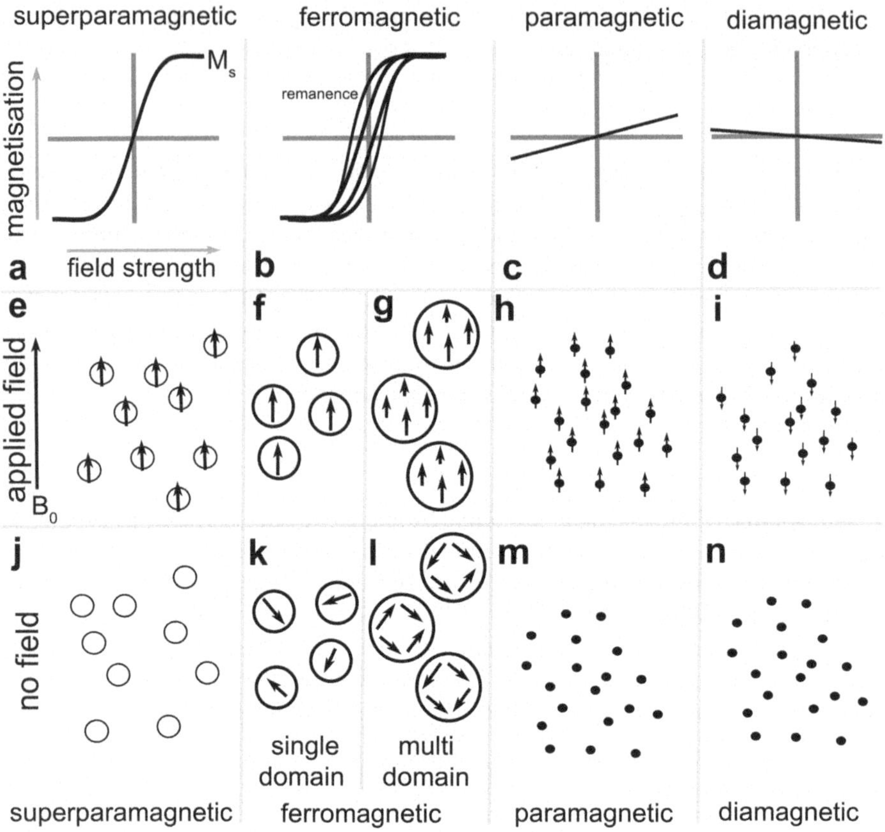

Fig. 1 Hysteresis curves showing net particle magnetisation induced by varying external field strength, for (**a**) superparamagnetic, (**b**) ferromagnetic (or ferrimagnetic), (**c**) paramagnetic, and (**d**) diamagnetic materials. Particle magnetisation in an applied field for (**e**) superparamagnetic, (**f**) single-domain ferromagnetic (or ferrimagnetic), (**g**) multi-domain ferromagnetic (or ferrimagnetic), (**h**) paramagnetic, and (**i**) diamagnetic materials. Net magnetisation in the absence of a magnetic field for (**j**) superparamagnetic, (**k**) single-domain ferromagnetic, (**l**) multi-domain ferromagnetic, (**m**) paramagnetic, and (**n**) diamagnetic materials. Further explanations on the different forms of magnetism can be found in Sect. 3.1

$$F_m = \frac{V_m M_s L(x)}{\mu_o B}(B \times \nabla)B \tag{3}$$

where $L(x) = \coth(x) - 1/x$ is the Langevin function for $x = \mu_o V_m M_s B / k_B T$, k_B is Boltzmann's constant, and T is temperature. By incorporating the Langevin function into this equation, it becomes possible to model the sigmoidal induction of magnetisation in response to increasing field strength seen on the hysteresis curve (Fig. 1), overcoming the assumption of linearity in Eq. (2). This allows Eq. (3) to be used for any field strength. Please see Riegler et al. for further details [11].

Table 1 Bulk saturation magnetisation for selected minerals and metals/alloys at room temperature [12]

Composition	Magnetic order	M_s (emu/g)
Fe_3O_4	Ferrimagnetic	90–92
γ-Fe_2O_3	Ferrimagnetic	70–80
α-Fe_2O_3	Canted-antiferromagnetic	0.4
$CoFe_2O_4$	Ferrimagnetic	80
$MnFe_2O_4$	Ferrimagnetic	77
$NiFe_2O_4$	Ferrimagnetic	51
Co	Ferromagnetic	161
CoFe	Ferromagnetic	235
Fe	Ferromagnetic	218
Ni	Ferromagnetic	55
Ni_3Fe	Ferromagnetic	120

(Note that size and surface effects may lead to lower M_s values in nanoparticulate materials than those listed here.) Most magnetic particles used for biomedical applications are iron oxide based, and consist of either maghemite or magnetite. Nanoparticles made from metal alloys such as CoFe would have greater saturation magnetisation than iron oxides, potentially providing more force at high field strengths

As the saturation magnetisation becomes relevant at higher field strength, this might influence the choice of magnetic material used. Of the two most common types of iron oxide used in magnetic particles, magnetite (Fe_3O_4) typically has higher saturation magnetism than maghemite (γ-Fe_2O_3), though higher values are found in some metals and alloys (Table 1).

Counteracting the magnetic force, targeting efficiency will be affected by additional physical parameters specific to the precise anatomical context, including fluid velocity (v_w) and viscosity (η), difference in particle or cell velocity and fluid velocity (Δv), and particle or cell radius (R_m). These can be used to calculate the drag force acting on the particle and against the magnetic force that is exerted on a particle within a given fluid [1]:

$$F_d = 6\pi\eta R_m \Delta v$$

As drag force is proportional to the surface area of a particle, smaller particles will experience greater drag per unit volume than larger particles or particle-loaded cells or liposomes. Therefore the resulting force will be proportionally greater the larger the object to be targeted, assuming that it is well loaded with magnetic particles. Though in general this means that larger magnetic particles and particle-loaded entities will be targeted more effectively in the vasculature, in some circumstances this might be offset to a degree by certain advantages of smaller particles, such as increased extravasation and diffusion rate through certain tissues, which is discussed in part 3.4 of this chapter. Mathematical modelling of the forces affecting magnetic particle targeting in biological tissues has been further explored by Nacev et al. [12].

3 Magnetic Particles

3.1 Particle Size and Magnetism

Core size and material affect the type and potential strength of magnetism that a particle displays (Fig. 2), which in turn influences how it behaves in magnetic targeting applications. (Temperature also has a significant effect, but in this chapter it should be taken as read that the ambient temperature is in the room/body temperature range.) In the low nanometre range (up to 2–30 nm depending on the material [13]) most iron-, nickel-, and cobalt-based magnetic materials are superparamagnetic—they are magnetic only in the presence of an external magnetic field, and their magnetism increases with external magnetic field strength up to their saturation magnetisation (Fig. 1 a, e, j).

Up to a higher size limit (50–100 nm, again dependent on the material [13]) the same materials will possess a single magnetic domain (Fig. 1 b, f, k). When a particle is in the single-domain state, its internal magnetisation is aligned in the same direction, whereas in multi-domain particles each domain may point in a different direction. In the application of a magnetic field, individual single-domain magnetic particles do not increase their magnetisation, which remains at saturation (M_s); however, multiple neighbouring particles that are initially magnetised in varying directions (along their easy axis of magnetisation) will align with the magnetic field in chains (if in solution), or against their easy axis if they are not free to rotate. This increases the net magnetisation of the group and therefore the amount of directional magnetic force that can be applied to the particles by a field gradient.

At larger sizes, the magnetic cores will split into multiple non-aligned domains to reduce the internal energy of the particle (Fig. 1g, l). These are known as multi-domain magnetic particles. Domain wall motion can be induced by an external field

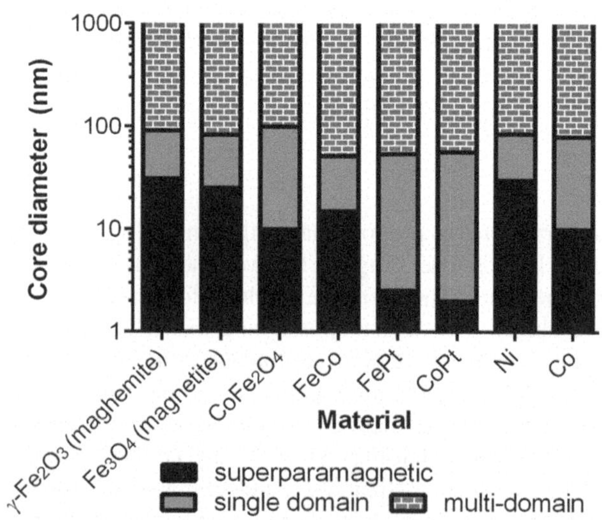

Fig. 2 Particle core size and material determine whether particles display superparamagnetism, single-domain or multi-domain ferromagnetism/ferrimagnetism [13]

to increase the net magnetisation, with domains that are magnetised in directions favoured by the applied field growing at the expense of unfavourably oriented domains, and saturation magnetisation may be reached if a sufficiently strong external field is applied. The field strength at which this is achieved, and the retention of magnetic domain alignment after removal of an external magnetic field (known as remanence), can be determined from a hysteresis curve for the particles (Fig. 1). Though the same net effect is seen in groups of single-domain particles due to magnetic anisotropy, this is due to different underlying reasons as illustrated in Fig. 1.

The magnetic susceptibility of a given assembly of particles is influenced by the above factors, such as the composition and magnetic structure of the particle core. Magnetic susceptibility is important for targeting applications, as it is proportional to the amount of magnetic force that can be applied below field strengths at which the magnetisation is saturated. This can also be visualised on a hysteresis curve (Fig. 1). This allows us to compare magnetic properties for several potential candidate magnetic particles to determine which will be most suitable for a specific magnetic targeting application. This is particularly important when considering nanoparticles originally made for use as MRI contrast agents (superparamagnetic iron oxide particles—SPIOs), which were not designed for the effective production of force, but for their ability to produce local magnetic field inhomogeneities resulting in areas of hypo-intensity in T_2-weighted MR images (due to de-phasing of the surrounding ^1H proton spins). For this reason, some MRI contrast agents are not well suited for certain magnetic targeting applications due to their small core size, especially the ultra-small SPIOs or USPIOs.

Despite this potential mismatch in required design features between the two applications, some SPIOs designed as MRI contrast agents do have suitable properties for particular magnetic targeting applications. Among these are some of the SPIO particles approved for human use by the FDA, Endorem (Feridex), and Ferucarbotran (Resovist), which have both been used successfully for a number of preclinical magnetic targeting studies [2] (Table 2). Furthermore, most commercially available SPIOs have been designed for low toxicity and good cell uptake and stability, which also suits them to both cell tracking with MRI and magnetic targeting applications that use cells.

3.2 *Particle Aggregation*

Particle aggregation can be promoted by both biological and magnetic factors, and might be beneficial or detrimental according to the type of magnetic targeting application and the duration and scale of aggregation. As superparamagnetic particles are magnetic only in the presence of a magnetic field, they will not magnetically aggregate in zero field. However, temporary chains of superparamagnetic particles can form at sufficiently high concentration and field strength as a result of attractive dipole-dipole interactions [14]. Permanent magnetic particles however will cluster together in the absence of an external magnetic field, and will form chains parallel

Table 2 Selection of particles used for magnetic targeting of cells

Particle name	Hydrodynamic diameter	Organs targeted	Cell types used	Core type	Reference(s)
Feridex/Endorem	80–150 nm	Brain, knee joint cartilage, liver, spinal cord, vasculature	hNSC, MSC, BMSC, EPC, MNC	Multiple (5–6 nm) superparamagnetic cores	[15–20]
Resovist/Ferucarbotran	62 nm	Femur, heart, skeletal muscle	MSC	Multiple 3–5 nm superparamagnetic cores	[21–23]
Feraheme/ferumoxytol	17–31 nm	Heart	CDC	6–8 nm Superparamagnetic core	[24]
FluidMag	50, 100, and 200 nm	Retina, vasculature	MSC	Multiple superparamagnetic cores	[25, 26]
Biomag/superparamagnetic microspheres	0.5–2 μm	Arteries, heart, vasculature	EC, CDC, MNC,	Multi-domain permanent magnetic core (0.5–2 μm)	[20, 27]
Magnetospirillum sp.-derived particles	~40 nm	Hind limb	EPC	Single-domain permanent magnetic core (~40 nm)	[28]
Multifunctional upconversion nanoparticles	240 nm	Skin wound	MSC	Multiple 5 nm superparamagnetic cores	[29]

Clinically approved particles listed in bold type, commercially available particles shown in italics. *hNSC* human neural stem cell, *MSC* mesenchymal stem cell, *BMSC* bone marrow stromal cells, *EPC* endothelial progenitor cell, *MNC* mononuclear cell, *EC* endothelial cells, *CDC* cardiosphere-derived cells to an applied external field. Clustering and the formation of chains in particles and particle-loaded cells are also dependent on their concentration, surface charge and chemistry, and viscosity and flow rate of the solution. This can be modelled for a given scenario to predict the optimal parameters to use [30, 31].

For some applications requiring the production of vascular embolisms, particle aggregation is desirable, and larger magnetic particles into the micron range or above have been used for this purpose [32, 33]. For other applications such as drug or cell delivery that require even distribution throughout a tissue and efficient extravasation, aggregation will impede delivery and so the use of smaller single-domain or superparamagnetic particles might be more effective [34]. However, the tendency of particles to form temporary chains or small aggregations during delivery will increase the amount of force that can be applied by an external field and can increase targeting efficiency, so this must be balanced against the probability of embolization to achieve optimum delivery. In addition to tuning particle size, the balance between the application of sufficient force and the possibility of embolization can be achieved by assessing a range of targeting device field strengths for a given application [22].

Recent work has also suggested that microfabricated particles comprising multilayer stacks of alternating magnetic and non-magnetic materials can be used to avoid aggregation. These can be produced with diameters on the scale of multidomain ferromagnetic particles (2 μm), but with zero remanence, and with an efficient switch to saturation magnetisation upon application of a given field strength [35]. This avoids some of the limitations associated with the aggregation of larger particles while retaining many of the benefits including the higher amount of force that can be applied per particle and increased circulation times (Sect. 3.4). The construction of larger composite particles from multiple small superparamagnetic cores can also avoid the magnetic aggregation typical of larger permanently magnetic particles while retaining the benefit of higher magnetic force per surface area and thus increased overall targeting efficiency. Several other methods can achieve similar effects, including the creation of "nanoworms" from strings of particles [36, 37], or embedding multiple superparamagnetic cores within dextran [38] or silica coating [39], hydrogel, [40], vesicles [41], or magnetoliposomes [42].

3.3 Cell Uptake and Labelling

Particle size also affects cell uptake, and therefore the resulting amount of force that can be applied to a particle-loaded cell will be determined not only by the particle core size, but also by how many particles the cell has taken up. Around ten different biological mechanisms of particle uptake have been discovered in cells, and are active to varying degrees depending on the specific cell type and its environment, and the size, charge, and coating of the particle [43, 44]. These include non-specific uptake mechanisms such as macropinocytosis (particles around 1 μm) and caveolar-mediated endocytosis (for particles ~60 nm), as well as receptor-dependent mechanisms such as clathrin-mediated endocytosis (~120 nm) and clathrin-and caveolin-independent endocytosis (~90 nm) [45]. Due to the complexity of cell uptake mechanisms, it is difficult therefore to predict with accuracy the range of particle sizes that will be taken up effectively by a given cell type, except for cells that have already been investigated with similar particles. Generally, particles between 20 and 200 nm in diameter are more effectively taken up by most non-phagocytic cell types than particles either side of that range [46–48], though increased particle charge will improve the uptake of larger particles into the micron scale [48]. In addition to this, other interacting variables such as particle shape, stiffness, and surface chemistry will affect the uptake. A complete discussion of this topic falls outside the scope of this chapter and the reader is referred to a recent comprehensive review [44].

As most commercially available and FDA-approved magnetic particles designed for imaging applications fall within this range of 20–200 nm, and have been optimised for cell uptake by size and surface coating, these have found successful use in several magnetic cell-targeting studies [2] and can be used for reference in designing particles for this purpose (Table 2).

When labelling adherent cells with magnetic particles, uptake can sometimes be promoted by application of a magnetic field. This has been demonstrated with a range of particle and cell types [49, 50]. Additionally, this technique can be used to simultaneously introduce DNA to the cell (known as magnetofection), which might be useful if the cells are to be used for gene therapy [51].

Addition of antibodies to the particle surface can also promote uptake or attachment to specific cell types, according to the choice of antibody. In this way, cells can be labelled either prior to implantation or in vivo. For example, antibody-labelled magnetic particles administered intravenously can label specific cell types within the circulation, allowing these cells to be magnetically targeted to specific locations within the body for therapeutic purposes. This strategy has been demonstrated to improve repair of infarct myocardium using circulating stem cells [6]. Here a dual mechanism of targeting was demonstrated, in which cells were targeted to the heart using a magnetic field, and also by the particles that displayed antibodies for both the stem cell and for the damaged cardiac tissue. As this approach required only the intravenous administration of the particles, translation would be more straightforward than in therapies in which cells must be harvested, isolated, and labelled before being given to the patient. This strategy might also prove effective for magnetic targeting of antibody-based drugs, or together with the delivery of other small-molecule drugs or therapeutics such as siRNA (see Chap. 7) contained in magnetoliposomes or adsorbed to the particle surface.

3.4 Targeting Particles for Drug and Therapeutic Delivery

The required design features of particles that are to be used directly for therapy differ from those that are to be used to label cells, as do the constraints on these. This is due to several biological and physical factors.

For example, particle size affects extravasation rate and diffusion though tissue, and will affect the ability to target regions distant to the vasculature. This may be a particular concern for certain applications such as drug delivery to poorly vascularised metastases, or the targeting of particles for hyperthermia therapy. The effect of particle size on diffusion and extravasation can be modelled in a number of ways, and used to predict the range of particle sizes that might benefit from magnetic targeting, or which would be effective in a given tissue. In cases where particle size falls outside this range (20–400 nm, based on a simulation of tumour tissue), delivery by endogenous diffusion mechanisms might prove superior to targeting strategies [34].

Clearance of particles by the reticuloendothelial system (RES) and renal filtration are also dependent on size. Renal filtration will quickly clear the majority of circulating particles under 5 nm, but barely any that are above 9 nm [52]. Larger particles can be cleared by immune cells including Kupffer cells in the liver, monocytes, spleen, and macrophages, which engulf nanometre- to micron-sized particles, decreasing delivery to the target organ. Targeting efficiency will therefore be improved the faster the particles can be attracted to the target site, before clearance has occurred, though this may be challenging when most particles are cleared on the first-pass circulation post-injection. The rate at which particles are scavenged in this way is affected by their coating and size. It is outside the scope of this chapter to review this topic in detail, but several modifications can reduce RES uptake such as polyethylene glycan (PEG) coating, which has been demonstrated to enhance the circulation time of particles and improve magnetic targeting efficiency [53]. The utility of PEGylation in reducing RES uptake is further illustrated by the results achieved with Doxil—a liposomal formulation of doxorubicin, and the first nano-encapsulated drug to obtain clinical approval [54].

Another strategy in reducing RES uptake utilises the properties of red blood cells, which naturally evade clearance and achieve long circulation times through a combination of their size, surface composition, and charge. For this reason they have been used in magnetic targeting applications for drug delivery [55, 56]. In one study red blood cells were loaded with a drug for photodynamic therapy and targeted to the tumour, resulting in a decrease in tumour volume following phototherapy compared to non-magnetically targeted control conditions [56]. In an adaptation of this concept, nanoparticles can be coated with extracted cell membrane of red blood cells, thereby "cloaking" the underlying material. This provides a strategy to utilise the unique binding properties of red blood cells to diseased tissue and pathogens [57], and could facilitate particle-based drug delivery to sites of inflammation in combination with magnetic targeting.

On the other hand, several iron oxide contrast agents for MRI, including those that have been clinically approved, were originally designed with size and surface coatings to optimise their uptake by the RES, for the purpose of imaging uptake in the liver. These might not therefore be optimal for magnet-targeted delivery of drugs due to their increased clearance rate.

To bypass normal clearance mechanisms, several magnetic targeting studies have opted for local delivery to the target organ using a directed injection [4, 26, 58], instead of a systemic injection. For example in a clinical trial targeting hepatocellular carcinoma, particles were injected via a catheter inserted into the hepatic artery, which could be moved intraoperatively between the acquisition of magnetic resonance images to ensure that particles were effectively delivered to the entire tumour (Fig. 3) [3]. This strategy can reduce the number of particles or cells that can be cleared by the RES prior to reaching the site of interest, increasing the efficiency of delivery. Though this approach increases the invasiveness of the technique, it may be required in several situations to achieve effective delivery. However, some difficult-to-target organs such as the brain have been efficiently targeted using a systemic intravenous injection of particles [38], though this may depend on blood-brain-barrier breakdown related to specific disease states. Together, these results suggest that both systemic and local delivery can be effective.

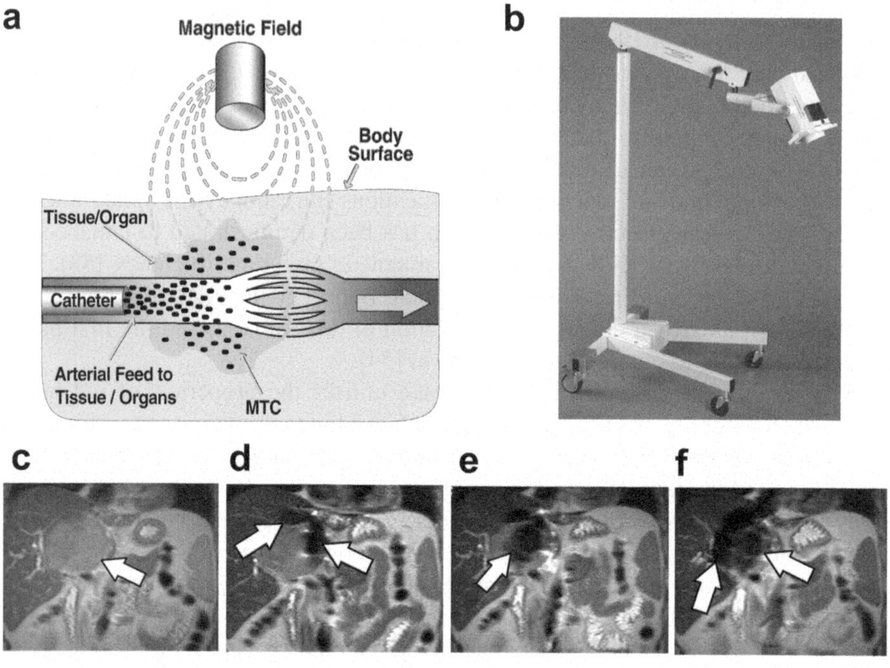

Fig. 3 (**a**) Diagram depicts the mode of action of magnetic targeted therapy. After leaving the intra-arterial catheter, magnetically targeted carrier (MTC) doxorubicin is drawn out of the artery into surrounding tumour and/or liver tissue by the influence of the local magnetic field (image courtesy of FeRx). (**b**) 5 kG portable magnet used for magnetic targeted therapy. The overall height of the magnet and holding apparatus is 1.4 m. (**c–f**) Coronal single-shot spin-echo MR images (839/80, one signal acquired, 6 mm section thickness) obtained, **c**, before MTC-DOX administration and, **d**, after the first, **e**, the second, and, **f**, the third dose of MTC-DOX. The selective hepatic arterial catheter was repositioned between each dose [3]. Figure reproduced from Wilson et al. Hepatocellular carcinoma: Regional therapy with a magnetic targeted carrier bound to Doxorubicin in a dual MR imaging/conventional angiography suite—Initial experience with four patients. Radiology 2004;230:287–293, with permission from the Radiological Society of North America

Particle size also affects the surface area-to-volume ratio, which is relevant to the adsorption of drugs to the particle surface. However, there is a trade-off between the motivations of using smaller particles, and the increased force that can be gained from using larger particles, and so a range of particle sizes can be considered. For example, magnet-targeted drug delivery in the clinic has been trialled across a scale of particle sizes: 100 nm particles coated with glucose and bound to epirubicin [4], and 0.5–5 μm sized particles with an activated carbon coating bound to doxorubicin [3]. Even larger metallic objects in the millimetre scale have been used for targeting and hyperthermia therapy, which does not require the same extent of extravasation and distribution of particles, due to transfer of heat across tissue [7].

For targeting tumours, particle retention can be improved by the enhanced permeability and retention (EPR) effect, which has been well documented to improve delivery [59]. As this effect relies on leaky vasculature and diffusion of particles through the extracellular space, it is a size-specific phenomenon, with the enhancement of particle retention decreasing with particle diameters above 30 nm in one tumour model [60], though this will differ to some degree depending on tumour type and nature of its vascularisation. Though enhanced nanoparticle delivery is achievable in difficult-to-access sites, such as brain tumours with magnetic targeting [61], one limitation of magnetic targeting for cancer treatment is the issue of metastases, which can be remote, disperse, and difficult to identify. These factors would limit the ability for targeting to be used, as they might require a separate targeting field to be applied for each site that could be identified.

For drug delivery it is also necessary to consider the attachment of cargo to the particle. In the simplest form this can rely on absorption of the drug onto the particle surface [3, 4]. However controlled release of the drug can be achieved if it is trapped in a solid coating, such as silica, around the magnetic particle. By heating the particle using a focused alternating field around the target tissue, as during a hyperthermia treatment, the coating can be made to melt, thereby releasing the drug [5, 41]. Similar strategies can be employed to release drugs from magnetoliposomes [62].

4 Magnetic Devices for In Vivo Targeting

Appropriate selection or design of magnetic targeting devices depends largely on the anatomical region to be targeted. Other considerations include the sort of particle that will be targeted (small superparamagnetic particles will typically require a stronger magnetic field and field gradient to exert the same force on them than larger or permanent single-magnetic-domain particles), as well as the type of steering and/or imaging that is required for the application. Several broad categories of magnetic targeting device will now be discussed, including their benefits, limitations, and applications to which they are suited.

4.1 Permanent Magnets

Small permanent magnets have found wide use in preclinical magnetic targeting studies, as they are readily available and provide high field strengths and gradients over a relatively small area, without the larger hardware size or cooling required for electromagnets. While suitable for tissue depths of a few centimetres, the limited spatial extent of such fields would restrict their external use in the clinic to targeting superficial tissue [3, 4], unless the particles are delivered to the internal site by direct injection (Fig. 3). Furthermore, a single permanent magnet cannot produce a local maximum of field gradient within the patient (unless surgically implanted—see below), meaning that for practical purposes that magnetic particles will be attracted from the inside of the patient to the surface of the magnet to produce a focal accumulation towards the skin, rather than vice versa. This makes it difficult to target an internal location non-invasively or achieve an even distribution of particles throughout a target organ. One solution to this is to change the location of the magnet during the targeting procedure, for example from one side of the liver to the other, so that the targeted particles are more evenly spread out [34]. This was demonstrated ex vivo to increase the concentration of magnetic particles within poorly perfused liver metastases by 1.8-fold above diffusion alone, and was predicted to be effective for particles in the 30–500 nm diameter range, with normal diffusion acting more effectively than targeting for smaller particles [34]. Another option is the implantation [18] or stereotactic insertion of magnets [32, 33]. While this can achieve effective targeting to internal organs, it does not completely avoid the invasive procedures associated with more traditional surgical implantation methods. Alternatively, magnetic stents can be manoeuvred to a specific point of the vasculature to achieve targeting of cells or particles [63], but again this is invasive and only useful for targeting areas that are accessible in this way.

4.2 Permanent Magnet Arrays

By superimposing two or more magnetic fields in an array of permanent magnets, some of the limitations of single permanent magnets can be overcome to produce more complex field shapes and gradients. With certain configurations this can allow the device to remain external to the target tissue and steer magnetic particles away from the magnet itself. A simple example is two bar magnets that have been arranged so that their fields overlap to create a remote point from which magnetic particles are pushed away from the magnet (Fig. 4). This has been posited as a solution for magnetically "injecting" nanomedicines into the inner ear [64], and has been validated in a rat model over distances relevant to use in the human ear [65]. This is important as the ear is behind the blood-brain barrier and can be a difficult organ in which to achieve drug delivery. It is also protected by the tympanic, round window, and oval membranes, and magnetic targeting using a "pulling" magnetic field from the opposite side of the head would require an unfeasibly strong magnetic field to achieve the same amount of force [64].

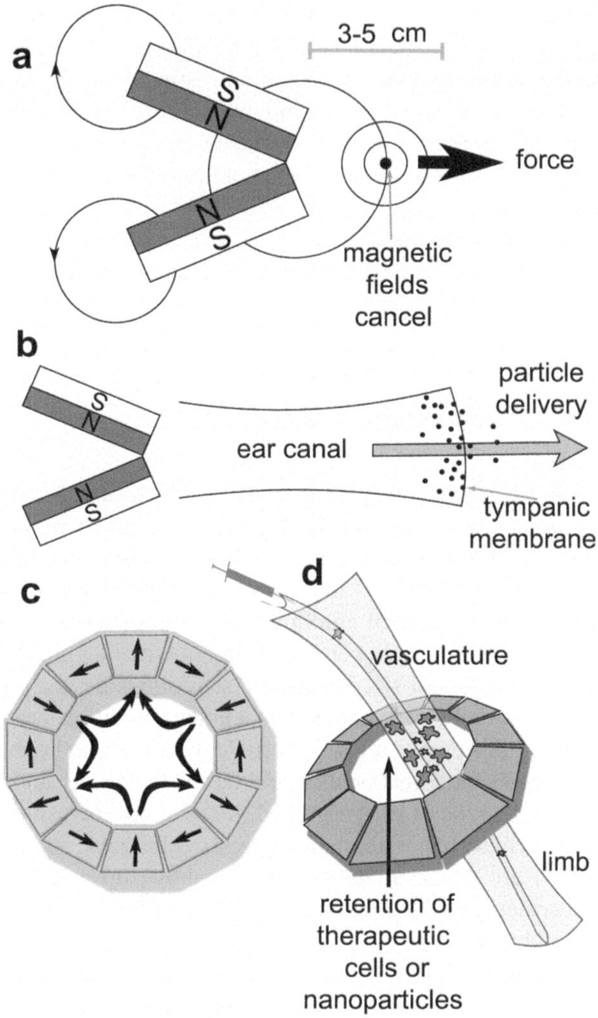

Fig. 4 (**a**) Two permanent magnets arranged at an angle can create a remote focal point about which magnetic particles may be repelled. (**b**) This magnet configuration has been proposed as a way to deliver magnetic particles to deep-tissue locations such as the inner ear [63, 64]. (**c**) Cylindrical Halbach array of permanent magnets [26], where the *arrows* show the direction of force. (**d**) The Halbach array pictured in (**c**) can be used to retain therapeutic cells within the vasculature, providing potential regenerative effects in cases of peripheral limb ischemia [11, 26]

Using the same principles an optimal 2D Halbach array of 36 magnets was computationally designed to create a pushing force at 10 cm distance with nine times the force achievable with only two magnets of the same overall volume and field strength [66]. These results hold promise for the development of relevant clinical devices, as one of the challenges in translating preclinical results will be scaling up the technology for use with greater tissue depths.

As well as increasing the magnetic pushing force that can be applied at a specific point, 2D and 3D Halbach arrays can also be designed to increase the attractive force from permanent magnets [66]. A relevant example is the Halbach cylinder [26], which is a circular array of permanent magnets that can be built to fit around a limb and target cells or magnetic particles for retention within a specific portion of the vasculature (Fig. 4c, d). This has been suggested as a good solution to achieve targeting of therapeutic cells for peripheral limb ischemia [11, 26]—currently one of the largest disease areas being investigated in clinical cell therapy trials.

Though arrays of permanent magnets are a promising solution to several applications due to their relatively low cost, size, and ease of operation, several limitations apply. One of the most important of these is predicted in Earnshaw's theorem, which holds that for most situations relevant here, no arrangement of static magnetic fields will create a remote focal area in the magnetic field in which magnetic particles will accumulate. The next section will discuss electromagnets, which can be used to work around this limitation.

4.3 *Electromagnet-Based Arrays*

Using computer-interfaced control, electromagnet-based systems can produce switchable and tuneable field gradients to direct magnetic particles along a three-dimensional route controlled in real time. This can be illustrated by several working examples, including a "magnetic stereotaxis" system based on four electromagnets which was capable of guiding a magnetic pellet along a defined route within a live dog brain to enable hyperthermia treatment of a glioma [7]. In a development of this system, a clinical sized magnetic stereotaxis device based on six superconducting coils was built, and validated on a gelatin brain phantom [67]. X-ray imaging capability in this device allowed particle movement to be monitored intraoperatively, so that feedback adjustments could be made to steer and improve navigational accuracy. This type of image-guided, focused non-invasive tumour ablation is currently only achievable with radiosurgical techniques such as "gamma-knife" therapy, which are prone to damage the healthy surrounding tissue, whereas a targeted hyperthermia-based treatment might avoid this due to the higher heat tolerance of normal tissue.

Computer-interfaced electromagnetic devices can also control much smaller magnetic devices on a fine scale, for example for steering micro-robots within the eye [9]. This has been demonstrated with control of both robot orientation and direction of movement, and is proposed as a solution to enable microsurgery and drug delivery in what is another difficult organ on which to operate. This was demonstrated during testing through the control of a magnet attached to a hypodermic needle, which was used to pierce a target blood vessel in vivo [9].

For control of multiple particles at once, as might be beneficial in a nanoparticle drug delivery system, an array of eight electromagnets was used together with a sequence of directional magnetic pulses to cause accumulation at an external focal site [68]. This technique made use of the ability of ferromagnetic rods to be rotated and moved by the application of external fields. More examples on particle delivery with electromagnets will be given in the next section, which will discuss the use of MRI scanners for targeting applications.

4.4 Targeting Using MRI Systems

Though designed for imaging and not particle manipulation, MRI scanners share several features with the purpose-built electromagnet-based targeting systems described in the previous section. Their high magnetic field strength and controllable field gradients have been used for a handful of targeting applications, while their imaging capability can provide feedback on particle location during delivery.

Proof-of-concept studies show in vitro that human cells loaded with magnetic particles can be targeted using unmodified preclinical MRI scanner gradients in vascular flow models [20]. Further work expanded upon this to steer oncolytic virus-producing, iron oxide-loaded macrophages to sites of pulmonary and prostate tumours in vivo, resulting in enhanced tumour necrosis [69]. Simultaneous targeting and imaging have also been achieved using an MRI scanner to control a millimetre-sized magnetic sphere within live pig vasculature [70], and to direct magnetotactic bacteria around a maze [8]. Magnetic drug-encapsulating particles have also been directed in this way to either the right or the left lobe of the rabbit liver, providing a potential treatment for tumour metastases, this time with the aid of additional gradients optimised for particle steering [58]. The use of MRI hardware for magnetic targeting is advantageous in that high-resolution imaging can be performed with the same equipment at intervals during the targeting experiment [8, 58, 70]. In this way, particle accumulation can be seen in near-real time, allowing feedback adjustments on the field gradient to improve targeting efficiency. One limitation to this approach is that the field gradients present in clinical MRI scanners typically have maximum values of 40–70 mT/m, which is an order of magnitude or two below those produced in some dedicated magnetic targeting devices [7, 67]. A solution to this is the addition of "propulsion" gradient sets to the MRI hardware without permanent modification to improve targeting ability [58], which can increase the amount of force that can be applied. In theory this would facilitate the translation of this technology to the clinic, as the majority of the necessary hardware is already in place.

5 Applications

To provide a general overview of what has been previously achieved and possible areas of future development or translation, in this section we will introduce some results reported in both clinical and preclinical magnetic targeting studies.

5.1 Magnet-Assisted Embolization for Aneurysm and Tumour Treatment

In the first clinical use of magnetic targeting (1966), metallic thrombi were produced to block ruptured aneurysms in the brain [32, 33]. Large carbonyl iron particles injected into the carotid artery or site of aneurysm were targeted to two magnets placed around the aneurysm to create a focal magnetic field (62 mm diameter, 375–750 mm length), using a stereotactic frame and X-ray guidance. Thrombus formation upon particle aggregation within the aneurysm was confirmed by X-ray imaging. This involved less invasive surgery than established alternatives, as the magnetic probes were inserted through small-diameter burr holes in the skull instead of exposing the brain to the site of aneurysm. Importantly, remote aneurysms that were too far from the skull to be otherwise operable could also be targeted, and dozens of patients were successfully treated. In a later refinement to the technique, magnetic particles were mixed with a rapidly polymerising acrylic to stabilise the particle aggregation post-injection and prevent the original metallic thrombus fragmenting [71]. Following this work, the same group tested the idea of using magnetic emboli to occlude tumour vasculature, though this did not progress to clinical testing [72]. Interest in this technique has been recurrent, with recent work combining magnetic tumour embolization with the delivery of radiotherapy using yttrium-90-conjugated magnetic particles [73]. In another example, a clinical MRI scanner equipped with an additional "propulsion gradient" set was used to target the formation of magnetic particle emboli in either the right or the left lobes of rabbit liver [58]. This suggests that non-invasive targeting of emboli might be possible in the clinic with little modification to currently available hardware, and without the need for invasive probe insertion.

5.2 *Cancer Hyperthermia*

The low heat resistance of tumour cells compared to healthy normal tissue can facilitate their selective killing when the target cancerous area is heated [74]. This can be achieved using a number of methods including whole-body heating, heating of inserted probes, and heating of targeted magnetic nanoparticles using an alternating current magnetic field or radiofrequency (RF) field. The first clinical demonstration was described in 1957 in gastrointestinal tract lymph node metastases [75], in which magnetic particles were injected into the lymphatic system, and then heated using an RF field. Interest in hyperthermia continues, with a number of recent clinical studies investigating iron oxide nanoparticle hyperthermia with alternating current magnetic fields in prostate tumours [76] and glioblastoma [77]. These trials have relied on direct injection of the particles without magnetic targeting, and though preclinical work has suggested the combination of these two procedures, few convincing demonstrations have been performed to date [78]. One study of note, however, combined the three-dimensional magnetic steering of a small (mm scale) magnetic pellet through the brain of a dog, guided by magnetic resonance imaging to a specific location, followed by RF-induced heating of the pellet once it reached the target site [7]. This was suggested as a possible future treatment for malignant gliomas and for combination with radiotherapy delivery. Though this was achieved using a pellet several millimetres in size, histological analysis showed that the particle did not cause haemorrhaging along its path. Adaptation for use with smaller particles might be possible and remove the need for direct brain implantation of the particle. The suitability of different magnetic particle types for hyperthermia has been systematically evaluated here [79].

5.3 *Cell Targeting for Regenerative Medicine*

Stem, progenitor, and differentiated cells have the potential to treat a range of currently incurable and degenerative diseases by initiating regenerative or immune-modulatory effects within the host tissue. A handful of cell therapies have gained approval for clinical use during the last 10 years, and many thousands more are currently being assessed in clinical trials [80]. It has therefore been suggested that cell therapy could become the third main class of therapeutic alongside small-molecule drugs and biological macromolecules, such as antibodies [81]; however such progress will rely on developments within the field and in associated technologies.

For many conditions, cell therapies require efficient delivery of cells to the afflicted organ or tissue, and this may be challenging without invasive surgical techniques or repeated implantations. The combination of magnetic targeting with cell therapy offers a solution to this by improving cell delivery and retention, and has increased treatment response compared to non-targeted controls in several preclinical disease models including myocardial infarction, brain ischemia, damaged skeletal muscle, skin wounds, bone fracture, spinal cord damage, and restenosis [2] (Table 2).

It is unclear in many cases if stem cells contribute to repair by differentiating into the cell types needed to reconstitute healthy tissue, or if their effect is mediated by paracrine signalling factors or immune modulation. It is likely that at least in some cases a combination of these effects is required, and so the ability of cells to maintain a therapeutic phenotype after magnetic particle labelling should be considered, including their ability to migrate and differentiate. To this end, it has been shown that it is possible to load mesenchymal stem cells with enough iron oxide particles to achieve magnetic targeting while retaining their differentiation capacity for multiple lineages [26]. As the potential toxic effects of magnetic particles are dependent both on particle and cell type, as well as the concentration of the particles used to label the cells, this would need to be established for each specific case.

This, however, assumes that the isolated therapeutic cells are to be loaded with magnetic nanoparticles and then transplanted into the patient. Though this is the dominant paradigm for magnet-targeted cell therapy [2], another strategy has recently been demonstrated to avoid many of the associated challenges. Here, iron oxide particles were labelled with antibodies specific for circulating stem cells, and then injected intravenously. These bound to stem cells within the blood supply, which were then targeted to the heart with an external magnetic field [6]. As the magnetic particles also had antibodies for damaged heart tissue, the targeting was achieved with both molecular and spatial specificity, and did not require the time-consuming pre-isolation and labelling of stem cells *ex vivo*. Furthermore, a significant increase in regeneration of ischemic heart tissue was achieved, and tracking of the particles to the heart was demonstrated using MRI [6]. By changing the antibodies for cell type or tissue type, as well as altering the location of the external magnetic field, this strategy could be easily altered for repair of other tissues.

To replicate this success in the clinic, several challenges will need to be met, including the production of magnetic fields on the scale of the human anatomy, optimisation of field strength [22], and development of suitable magnetic particles for labelling cells (Sect. 3.3). As a handful of magnetic particle types have already been approved for clinical use as MRI contrast agents, the latter of these barriers to translation is somewhat reduced (Table 2).

5.4 Drug Delivery

Drug delivery is currently one of the largest areas of magnetic targeting research [82], and the only one within which more recent efforts have been made in the clinic [3, 4]. It is also well suited for combination therapy, having the potential for use with multiple drugs and therapy types such as hyperthermia [78], radiation therapy, photodynamic therapy [56], and embolization [10]. The idea of magnetically targeting drugs emerged in the 1970s, with the suggestion of using magnetic particle-loaded erythrocytes [55] or magnetic microspheres as carriers [83]. The first preclinical demonstration of drug delivery showed that drug-carrying microspheres could be targeted to a specific section of the rat tail [83]. In this case, a 100-fold lower systemic dose of the targeted drug was needed as compared to the untargeted drug, to achieve the same local concentration in the targeted tissue. Early application of this technique to tumour treatment resulted in total remission in the targeted group, while nearly all of the animals receiving a non-targeted treatment died [84]. By reducing dosing in off-target areas, this could in theory reduce many of the side effects associated with chemotherapy while allowing higher doses to be achieved in the tumour for increased therapeutic efficiency.

The first of the two clinical assessments of drug delivery to tumours was a phase I trial which gave 14 patients varying intravenous doses of 100 nm magnetic particles bound to epirubicin, with effective magnetic fields of 0.5–0.8 T applied to tumours for 60–120 min. In four of these patients, delivery of magnetic particles was visible as a discolouration of the tissue exposed to the magnetic field, and in all but five of the patients the presence of the particles was detectable with T_2-weighted MRI [4]. All patients tolerated the therapy well, with no indications of toxicity and reduced side effects and plasma drug concentration compared to dosing with free epirubicin. However, the study concluded that the treatment of further patients would be needed to prove efficacy.

In the more recent clinical study that used doxorubicin bound to 0.5–5 μm sized particles injected into the tumour, four patients with hepatocellular carcinoma were treated in a phase I/II trial. In this study, tumours were characterised using angiography to identify the hepatic artery branch that supplied the tumour and for catheter guidance. MRI was also used pre- and post-delivery of treatment to reveal the coverage of the tumour with magnetic particles (Fig. 3). When particles were not delivered to the whole of the tumour, MRI was able to indicate this, allowing repositioning of the catheter and a second or third dosage to achieve greater coverage (Fig. 3c–f). This intraoperative assessment is a major advantage to using magnetic particle-bound drugs, as this non-invasive assessment of drug delivery is not possible with most traditional therapies.

Aside from cancer treatment, magnetic targeting has also been suggested as a method of delivery for anaesthetics such as lidocaine to the spine [85], and magnetic aerosols to the lung to treat a number of respiratory complaints [86], though progress in non-cancer applications has so far been limited.

5.5 Bacterial Targeting and Therapy

The concept of treating tumours with bacteria has been around for a number of decades, with effective oncolysis demonstrated in several human studies with non-pathogenic Clostridium species [87, 88]. Recent work indicates a renewed interest in bacterial therapies in general [89], including the use of genetically modified drug-synthesising bacteria [90], and magnetotactic bacteria [91–93]. Magnetotactic bacteria synthesise multiple aligned nanoscale magnetic particles along their axis in order to align themselves with geomagnetic fields (Fig. 5). This increases their ability to overcome Brownian motion and move in a consistent direction. In environments such as lakes, this is often combined with hypoxia sensing, allowing migration of the bacteria to low-oxygen environments in which they preferentially grow. This hypoxia seeking has been exploited in the development of potential therapies, as it promotes tumour infiltration by the bacteria, which can be detected using MRI [92].

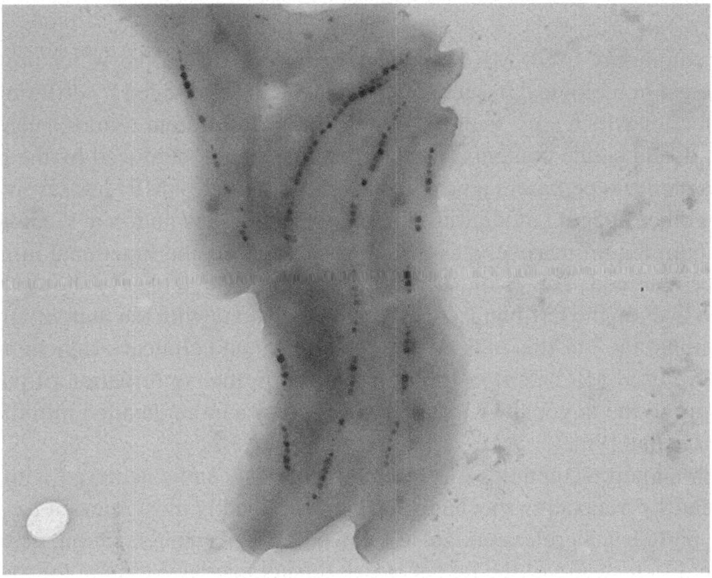

Fig. 5 Transmission electron micrograph showing aligned magnetotactic *Magnetospirillum* sp. bacteria. Individual magnetosomes are made of solid magnetite and appear as dark structures ~40 nm in diameter

Further work has demonstrated that bacteria can be attached to larger molecules such as nanoliposomes or microbeads, which might be useful for tumour drug delivery [94, 95]. One of the advantages of using bacteria to deliver drugs is that a certain amount of self-propulsion can be achieved from the flagellum, which can act together with the directional steering achieved by external magnetic fields. This has been demonstrated using an MRI scanner to steer magnetic bacteria around a maze [8].

Aside from the use of whole magnetotactic bacteria, the magnetic particles (magnetosomes) isolated from these have also been used for a number of applications due to the high quality of their magnetic material. This includes the magnetic targeting of human cells [28], hyperthermia [93], as well as their use as contrast agents for MRI.

6 Imaging Magnetic Particles

A handful of techniques exist for imaging magnetic particles. The most widely available of these is MRI [96], though the more recently developed technique known as magnetic particle imaging (MPI) does offer some advantages such as increased quantification and sensitivity [97], despite the lack of anatomical information that it provides.

6.1 Imaging with MRI

As most commonly used, MRI produces images based on the water proton (^{1}H) signal present in biological tissue. Anatomical detail is provided by differing proton relaxation rates which vary with the composition of different tissues, such as their water, lipid, and solute contents. Further contrast can be produced by the presence of paramagnetic, superparamagnetic, or magnetic particles. MRI has several advantages over other imaging modalities, including the high resolution of the images that can be acquired non-invasively, as well as the structural and functional information that can be obtained. This is illustrated in Fig. 3c–f, which clearly shows the structure of the liver on the left-hand side of the image, along with the tumour, intestines, and abdominal fat. For this reason, both preclinical and clinical magnetic targeting studies have used MRI, demonstrating its utility in the confirmation of particle or cell delivery to the target site [3, 8, 42, 61, 73]. For a more detailed introduction to MRI, see Gadian [96].

Though numerous methods of producing contrast can be achieved with MRI, it is the T_2 and T_2* relaxation mechanisms that are typically most relevant for imaging magnetic particles. T_2 relaxation, also known as transverse relaxation, or spin-spin relaxation, is a measure of the rate at which the net magnetisation in the *xy*-plane is lost following an RF pulse that flips the net magnetisation into the *xy*-plane from the *z*-axis. Specifically, it is the time taken for the signal to decay to 37 % of its initial value following the RF pulse.

Transverse relaxation is influenced by several factors which can be separated out into T_2 and T_2* effects. Transverse relaxation (T_2) is shorter in environments with high macromolecular content (which appear darker), and longer in tissues with higher water content (which appear brighter), though other factors also affect T_2 relaxation. Magnetic particles appear as hypo-intense (dark) areas on the image, though this results mainly from a T_2*effect. T_2* effects, such as magnetic field inhomogeneities, can alter the transverse relaxation rate by increasing the rate at which the de-phasing of spins occurs, and can be separated out from T_2 effects with certain pulse sequences. To reduce the effect of these T_2* factors, and produce a more T_2-weighted image, a specific type of pulse sequence called a spin echo can be used. To produce a T_2*-weighted image, which is more sensitive to the detection of superparamagnetic species such as iron oxides, a gradient-echo sequence can be used in which the de-phasing resulting from T_2 factors is reversed by a magnetic field gradient. These T_2*-weighted images, however, can be more sensitive to image artefacts caused by non-uniform magnetic fields present in the bore of the magnet.

T_2 contrast agents, such as magnetic particles, shorten the T_2 and T_2* time of nearby water molecules by de-phasing their spins with the magnetic field inhomogeneities around the particle, thus reducing the amount of signal collected from the surrounding water when the echo is recorded. This results in a hypo-intense or dark area on the image, which indicates the presence of magnetic particles in the tissue (Fig. 3).

The detection of magnetic particles with MRI can be very sensitive, allowing hundreds or fewer to be tracked in optimal conditions [98], and can be further improved by using long imaging times. However, in most settings this is limited by practical considerations such as how long the patient or animal can be kept within the scanner.

The ability to detect magnetic particles can also be complicated and unreliable in certain areas, such as the lungs or haemorrhages, where natural signal voids already occur. Another limitation is the quantification of magnetic particle number, which can also be challenging due to the saturation of contrast produced by focal accumulations of magnetic particles, and reduced detection sensitivity in areas with endogenous signal voids. If quantitative biodistribution data on particle location is required, it is recommended that particles are radiolabelled with isotopes for positron emission tomography (PET), which provides complementary information to MRI in many circumstances. Further information on imaging nanoparticles with other modalities can be found in the relevant chapters in this book.

6.2 *Magnetic Particle Imaging*

MPI differs from MRI in that the signal is obtained directly from the magnetic particles [97], and not from the effect they have on the spin of nearby water protons. For this reason it has several potential advantages, including the ability to quantify magnetic particle concentration within the host tissue, increased accuracy of identifying regions containing magnetic particles, increased sensitivity, and possibility of performing "multi-color" imaging using magnetic particles. Despite these factors, however, MPI does not on its own produce anatomical data, and so is best performed alongside CT or MRI imaging, has lower spatial resolution, and is also not yet widely available. A recent review covering the physical principles and nanoparticle tracers for magnetic particle imaging can be found here [99], and also in Chap. 4 of this book by Connolly.

6.3 Multimodality Imaging

Though MRI can enable the non-invasive detection of magnetic particles in the clinic, it does not provide some types of information that would be useful to the development of magnetic targeting techniques, such as whole-body quantitative biodistribution data. Radionuclide imaging (positron emission tomography (PET), and single-photon emission computed tomography (SPECT)) has many of the features also associated with MRI, including their clinical availability and non-invasiveness, and though they provide lower resolution images there are several complementary features. Radionuclide imaging relies on the detection of gamma rays emitted following radioactive decay of isotopes used to label the molecule, particle, or cell of interest. Unlike MRI, these techniques can provide quantitative whole-body detection of labelled molecules, allowing the non-invasive assessment of particle biodistribution to quantify the effects of magnetic targeting procedures. Further information on the design of magnetic particles for radionuclide imaging can be found in Chaps. 10–12. For tracking cells with PET or SPECT, the choice of isotope depends on how long the cells need to be tracked. One option is the use of ^{111}In-oxine (half-life of 2.8 days) and SPECT imaging, which has been used for tracking transplanted cells in the clinic for over 30 years. A recent modification to the technique uses ^{89}Zr-oxine (half-life of 3.3 days) for PET imaging, which could potentially provide better sensitivity and quantification. Both of these techniques can allow cells to be tracked for over a week due to their long half-lives, while the use of other isotopes such as ^{99m}Tc (half-life of 6 h) can be used for cell tracking over shorter periods of time.

7 Conclusions

Despite the numerous challenges involved in implementing magnetic targeting successfully and safely, it remains an exciting and active area of preclinical research due to the wide range of potential clinical applications. In theory, these potentially cover the majority of drug- and cell-based therapies, but in practice this scope will be limited by several physical and biological considerations that have yet to be determined and will be specific to each case. Magnetically targeted delivery offers the ability to simultaneously increase local concentrations of the active agent while reducing systemic dosing and off-target side effects. The inability to balance these so that the dose in the target organ is sufficient to act therapeutically, while the systemic dose is low enough not to cause unwanted side effects, is the reason why many potential drug candidates fail, and why many approved drugs are limited in their efficacy. For cell therapies, the field is yet to become well established, but the improved treatment response achieved in several preclinical studies from using magnetic targeting demonstrates the potential for its clinical use in the future [2].

Acknowledgments ML receives funding from Medical Research Council (MR/J013110/1); the King's College London and UCL Comprehensive Cancer Imaging Centre CR-UK & EPSRC, in association with the MRC and DoH (England); the National Centre for the Replacement, Reduction and Refinement of Animal in Research (NC3Rs); UK Regenerative Medicine Platform Safety Hub (MRC: MR/K026739/1); and Eli Lilly and Company. PS Patrick is funded by UK Regenerative Medicine Platform. TL Kalber is funded by an EPSRC Early Career Fellowship (EP/L006472/1).

References

1. Pankhurst QA, Connolly J, Jones SK, Dobson J. Applications of magnetic nanoparticles in biomedicine. J Phys D Appl Phys. 2003;36(13):R167–81.
2. Connell JJ, Patrick PS, Yu Y, Lythgoe MF, Kalber TL. Advanced cell therapies: targeting, tracking and actuation of cells with magnetic particles. Regen Med. 2015;10(6):757–72.
3. Wilson MW, Kerlan Jr RK, Fidelman NA, Venook AP, LaBerge JM, Koda J, et al. Hepatocellular carcinoma: regional therapy with a magnetic targeted carrier bound to doxorubicin in a dual MR imaging/conventional angiography suite—initial experience with four patients. Radiology. 2004;230(1):287–93.
4. Lubbe AS, Bergemann C, Riess H, Schriever F, Reichardt P, Possinger K, et al. Clinical experiences with magnetic drug targeting: a phase I study with 4′-epidoxorubicin in 14 patients with advanced solid tumors. Cancer Res. 1996;56(20):4686–93.
5. Thomas CR, Ferris DP, Lee JH, Choi E, Cho MH, Kim ES, et al. Noninvasive remote-controlled release of drug molecules in vitro using magnetic actuation of mechanized nanoparticles. J Am Chem Soc. 2010;132(31):10623–5.
6. Cheng K, Shen D, Hensley MT, Middleton R, Sun B, Liu W, et al. Magnetic antibody-linked nanomatchmakers for therapeutic cell targeting. Nat Commun. 2014;5:4880.
7. Grady MS, Howard 3rd MA, Broaddus WC, Molloy JA, Ritter RC, Quate EG, et al. Magnetic stereotaxis: a technique to deliver stereotactic hyperthermia. Neurosurgery. 1990;27(6):1010–5; discussion 5-6.
8. Martel S, Mohammadi M, Felfoul O, Lu Z, Pouponneau P. Flagellated magnetotactic bacteria as controlled MRI-trackable propulsion and steering systems for medical nanorobots operating in the human microvasculature. Int J Robotics Res. 2009;28(4):571–82.
9. Kummer MP, Abbott JJ, Kratochvil BE, Borer R, Sengul A, Nelson BJ. OctoMag: an electromagnetic system for 5-DOF wireless micromanipulation. IEEE Trans Robot. 2010;26(6):1006–17.
10. Pouponneau P, Soulez G, Beaudoin G, Leroux JC, Martel S. MR imaging of therapeutic magnetic microcarriers guided by magnetic resonance navigation for targeted liver chemoembolization. Cardiovasc Intervent Radiol. 2014;37(3):784–90.
11. Riegler J, Lau KD, Garcia-Prieto A, Price AN, Richards T, Pankhurst QA, et al. Magnetic cell delivery for peripheral arterial disease: a theoretical framework. Med Phys. 2011;38(7):3932–43.
12. Nacev A, Beni C, Bruno O, Shapiro B. The behaviors of ferromagnetic nano-particles in and around blood vessels under applied magnetic fields. J Magn Magn Mater. 2011;323(6):651–68.

13. Krishnan KM. Biomedical nanomagnetics: a spin through possibilities in imaging, diagnostics, and therapy. IEEE Trans Magn. 2010;46(7):2523–58.
14. Faraudo J, Andreu JS, Camacho J. Understanding diluted dispersions of superparamagnetic particles under strong magnetic fields: a review of concepts, theory and simulations. Soft Matter. 2013;9(29):6654–64.
15. Song M, Kim YJ, Kim YH, Roh J, Kim SU, Yoon BW. Using a neodymium magnet to target delivery of ferumoxidelabeled human neural stem cells in a rat model of focal cerebral ischemia. Hum Gene Ther. 2010;21(5):603–10.
16. Kobayashi T, Ochi M, Yanada S, Ishikawa M, Adachi N, Deie M, et al. A novel cell delivery system using magnetically labeled mesenchymal stem cells and an external magnetic device for clinical cartilage repair. Arthroscopy. 2008;24(1):69–76.
17. Arbab AS, Jordan EK, Wilson LB, Yocum GT, Lewis BK, Frank JA. In vivo trafficking and targeted delivery of magnetically labeled stem cells. Hum Gene Ther. 2004;15(4):351–60.
18. Sasaki H, Tanaka N, Nakanishi K, Nishida K, Hamasaki T, Yamada K, et al. Therapeutic effects with magnetic targeting of bone marrow stromal cells in a rat spinal cord injury model. Spine. 2011;36(12):933–8.
19. Kyrtatos PG, Lehtolainen P, Junemann-Ramirez M, Garcia-Prieto A, Price AN, Martin JF, et al. Magnetic tagging increases delivery of circulating progenitors in vascular injury. J Am Coll Cardiol Intv. 2009;2(8):794–802.
20. Riegler J, Wells JA, Kyrtatos PG, Price AN, Pankhurst QA, Lythgoe MF. Targeted magnetic delivery and tracking of cells using a magnetic resonance imaging system. Biomaterials. 2010;31(20):5366–71.
21. Kodama A, Kamei N, Kamei G, Kongcharoensombat W, Ohkawa S, Nakabayashi A, et al. In vivo bioluminescence imaging of transplanted bone marrow mesenchymal stromal cells using a magnetic delivery system in a rat fracture model. J Bone Joint Surg Br Vol. 2012;94(7):998–1006.
22. Shen Y, Liu X, Huang Z, Pei N, Xu J, Li Z, et al. Comparison of magnetic intensities for mesenchymal stem cell targeting therapy on ischemic myocardial repair: high magnetic intensity improves cell retention but has no additional functional benefit. Cell Transplant. 2014;24(10):1981–97.
23. Oshima S, Kamei N, Nakasa T, Yasunaga Y, Ochi M. Enhancement of muscle repair using human mesenchymal stem cells with a magnetic targeting system in a subchronic muscle injury model. J Orthop Sci. 2014;19(3):478–88.
24. Vandergriff AC, Hensley TM, Henry ET, Shen D, Anthony S, Zhang J, et al. Magnetic targeting of cardiosphere-derived stem cells with ferumoxytol nanoparticles for treating rats with myocardial infarction. Biomaterials. 2014;35(30):8528–39.
25. Yanai A, Hafeli UO, Metcalfe AL, Soema P, Addo L, Gregory-Evans CY, et al. Focused magnetic stem cell targeting to the retina using superparamagnetic iron oxide nanoparticles. Cell Transplant. 2012;21(6):1137–48.
26. Riegler J, Liew A, Hynes SO, Ortega D, O'Brien T, Day RM, et al. Superparamagnetic iron oxide nanoparticle targeting of MSCs in vascular injury. Biomaterials. 2013;34(8):1987–94.
27. Pislaru SV, Harbuzariu A, Gulati R, Witt T, Sandhu NP, Simari RD, et al. Magnetically targeted endothelial cell localization in stented vessels. J Am Coll Cardiol. 2006;48(9):1839–45.
28. Kang HJ, Kim JY, Lee HJ, Kim KH, Kim TY, Lee CS, et al. Magnetic bionanoparticle enhances homing of endothelial progenitor cells in mouse hindlimb ischemia. Korean Circ J. 2012;42(6):390–6.
29. Cheng L, Wang C, Ma X, Wang Q, Cheng Y, Wang H, et al. Multifunctional upconversion nanoparticles for dual-modal imaging-guided stem cell therapy under remote magnetic control. Adv Funct Mater. 2013;23(3):272–80.
30. Riegler J, Allain B, Cook RJ, Lythgoe MF, Pankhurst QA. Magnetically assisted delivery of cells using a magnetic resonance imaging system. J Phys D: Appl Phys. 2011;44(5).
31. Mathieu JB, Martel S. Aggregation of magnetic microparticles in the context of targeted therapies actuated by a magnetic resonance imaging system. J Appl Phys. 2009;106(4):44904.

32. Alksne JF. Stereotactic thrombosis of intracranial aneurysms. N Engl J Med. 1971;284(4): 171–4.
33. Alksne JF, Fingerhut AG, Rand RW. Magnetic probe for the stereotactic thrombosis of intracranial aneurysms. J Neurol Neurosurg Psychiatry. 1967;30(2):159–62.
34. Nacev A, Kim SH, Rodriguez-Canales J, Tangrea MA, Shapiro B, Emmert-Buck MR. A dynamic magnetic shift method to increase nanoparticle concentration in cancer metastases: a feasibility study using simulations on autopsy specimens. Int J Nanomedicine. 2011;6: 2907–23.
35. Vemulkar T, Mansell R, Petit D, Cowburn RP, Lesniak MS. Highly tunable perpendicularly magnetized synthetic antiferromagnets for biotechnology applications. Appl Phys Lett. 2015;107(1):12403.
36. Park JH, von Maltzahn G, Zhang LL, Schwartz MP, Ruoslahti E, Bhatia SN, et al. Magnetic iron oxide nanoworms for tumor targeting and imaging. Adv Mater. 2008;20(9):1630.
37. Wang GK, Inturi S, Serkova NJ, Merkulov S, McCrae K, Russek SE, et al. High-relaxivity superparamagnetic iron oxide nanoworms with decreased immune recognition and long-circulating properties. ACS Nano. 2014;8(12):12437–49.
38. Yu F, Zhang L, Huang Y, Sun K, David AE, Yang VC. The magnetophoretic mobility and superparamagnetism of core-shell iron oxide nanoparticles with dual targeting and imaging functionality. Biomaterials. 2010;31(22):5842–8.
39. Yoon TJ, Lee H, Shao HL, Hilderbrand SA, Weissleder R. Multicore assemblies potentiate magnetic properties of biomagnetic nanoparticles. Adv Mater. 2011;23(41):4793.
40. Paquet C, de Haan HW, Leek DM, Lin HY, Xiang B, Tian GH, et al. Clusters of superparamagnetic iron oxide nanoparticles encapsulated in a hydrogel: a particle architecture generating a synergistic enhancement of the T(2) relaxation. ACS Nano. 2011;5(4):3104–12.
41. Sanson C, Diou O, Thevenot J, Ibarboure E, Soum A, Brulet A, et al. Doxorubicin loaded magnetic polymersomes: theranostic nanocarriers for mr imaging and magneto-chemotherapy. ACS Nano. 2011;5(2):1122–40.
42. Fortin-Ripoche JP, Martina MS, Gazeau F, Menager C, Wilhelm C, Bacri JC, et al. Magnetic targeting of magnetoliposomes to solid tumors with MR imaging monitoring in mice: feasibility. Radiology. 2006;239(2):415–24.
43. Doherty GJ, McMahon HT. Mechanisms of endocytosis. Annu Rev Biochem. 2009;78: 857–902.
44. Zhang S, Gao H, Bao G. Physical principles of nanoparticle cellular endocytosis. ACS Nano. 2015;9:8655–71.
45. Petros RA, DeSimone JM. Strategies in the design of nanoparticles for therapeutic applications. Nat Rev Drug Discov. 2010;9(8):615–27.
46. Gratton SEA, Ropp PA, Pohlhaus PD, Luft JC, Madden VJ, Napier ME, et al. The effect of particle design on cellular internalization pathways. Proc Natl Acad Sci U S A. 2008;105(33):11613–8.
47. Thorek DLJ, Tsourkas A. Size, charge and concentration dependent uptake of iron oxide particles by non-phagocytic cells. Biomaterials. 2008;29(26):3583–90.
48. Foged C, Brodin B, Frokjaer S, Sundblad A. Particle size and surface charge affect particle uptake by human dendritic cells in an in vitro model. Int J Pharm. 2005;298(2):315–22.
49. Liu Q, Zhang J, Xia W, Gu H. Magnetic field enhanced cell uptake efficiency of magnetic silica mesoporous nanoparticles. Nanoscale. 2012;4(11):3415–21.
50. Child HW, Del Pino PA, De La Fuente JM, Hursthouse AS, Stirling D, Mullen M, et al. Working together: the combined application of a magnetic field and penetratin for the delivery of magnetic nanoparticles to cells in 3D. ACS Nano. 2011;5(10):7910–9.
51. Dobson J. Gene therapy progress and prospects: magnetic nanoparticle-based gene delivery. Gene Ther. 2006;13(4):283–7.
52. Longmire M, Choyke PL, Kobayashi H. Clearance properties of nano-sized particles and molecules as imaging agents: considerations and caveats. Nanomedicine. 2008;3(5): 703–17.

53. Cole AJ, David AE, Wang JX, Galban CJ, Hill HL, Yang VC. Polyethylene glycol modified, cross-linked starchcoated iron oxide nanoparticles for enhanced magnetic tumor targeting. Biomaterials. 2011;32(8):2183–93.
54. Barenholz Y. Doxil(R)-the first FDA-approved nano-drug: lessons learned. J Contr Release. 2012;160(2):117–34.
55. Zimmermann U, Pilwat G. Organ specific application of drugs by means of cellular capsule systems (author's transl). Z Naturforsch C. 1976;31(11–12):732–6.
56. Wang C, Sun XQ, Cheng L, Yin SN, Yang GB, Li YG, et al. Multifunctional theranostic red blood cells for magnetic-field-enhanced in vivo combination therapy of cancer. Adv Mater. 2014;26(28):4794.
57. Hu CJ, Fang RH, Wang KC, Luk BT, Thamphiwatana S, Dehaini D, et al. Nanoparticle biointerfacing by platelet membrane cloaking. Nature. 2015.
58. Pouponneau P, Bringout G, Martel S. Therapeutic magnetic microcarriers guided by magnetic resonance navigation for enhanced liver chemoembilization: a design review. Ann Biomed Eng. 2014;42(5):929–39.
59. Fang J, Nakamura H, Maeda H. The EPR effect: Unique features of tumor blood vessels for drug delivery, factors involved, and limitations and augmentation of the effect. Adv Drug Deliv Rev. 2011;63(3):136–51.
60. Cabral H, Matsumoto Y, Mizuno K, Chen Q, Murakami M, Kimura M, et al. Accumulation of sub-100 nm polymeric micelles in poorly permeable tumours depends on size. Nat Nanotechnol. 2011;6(12):815–23.
61. Chertok B, Moffat BA, David AE, Yu FQ, Bergemann C, Ross BD, et al. Iron oxide nanoparticles as a drug delivery vehicle for MRI monitored magnetic targeting of brain tumors. Biomaterials. 2008;29(4):487–96.
62. Podaru G, Ogden S, Baxter A, Shrestha T, Ren S, Thapa P, et al. Pulsed magnetic field induced fast drug release from magneto liposomes via ultrasound generation. J Phys Chem B. 2014;118(40):11715–22.
63. Polyak B, Fishbein I, Chorny M, Alferiev I, Williams D, Yellen B, et al. High field gradient targeting of magnetic nanoparticle-loaded endothelial cells to the surfaces of steel stents. Proc Natl Acad Sci U S A. 2008;105(2):698–703.
64. Shapiro B, Dormer K, Rutel IB. A two-magnet system to push therapeutic nanoparticles. AIP Conf Proc. 2010;1311(1):77–88.
65. Sarwar A, Lee R, Depireux DA, Shapiro B. Magnetic injection of nanoparticles into rat inner ears at a human head working distance. Ieee T Magn. 2013;49(1):440–52.
66. Sarwar A, Nemirovski A, Shapiro B. Optimal Halbach permanent magnet designs for maximally pulling and pushing nanoparticles. J Magn Magn Mater. 2012;324(5):742–54.
67. McNeil RG, Ritter RC, Wang B, Lawson MA, Gillies GT, Wika KG, et al. Functional design features and initial performance characteristics of a magnetic-implant guidance system for stereotactic neurosurgery. IEEE Trans Biomed Eng. 1995;42(8):793–801.
68. Nacev A, Weinberg IN, Stepanov PY, Kupfer S, Mair LO, Urdaneta MG, et al. Dynamic inversion enables external magnets to concentrate ferromagnetic rods to a central target. Nano Lett. 2015;15(1):359–64.
69. Muthana M, Kennerley AJ, Hughes R, Fagnano E, Richardson J, Paul M, et al. Directing cell therapy to anatomic target sites in vivo with magnetic resonance targeting. Nat Commun. 2015;6:8009.
70. Martel S, Mathieu JB, Felfoul O, Chanu A, Aboussouan E, Tamaz S, et al. Automatic navigation of an untethered device in the artery of a living animal using a conventional clinical magnetic resonance imaging system. Appl Phys Lett. 2007;90(11):14105.
71. Alksne JF, Smith RW. Iron-acrylic compound for steriotaxic aneurysm thrombosis. J Neurosurg. 1977;47(2):137–41.
72. Turner RD, Rand RW, Bentson JR, Mosso JA. Ferromagnetic silicone necrosis of hypernephromas by selective vascular occlusion to the tumor: a new technique. J Urol. 1975; 113(4):455–9.

73. Kobeiter H, Georgiades CS, Leakakos T, Torbenson M, Hong K, Geschwind JF. Targeted transarterial therapy of Vx-2 rabbit liver tumor with Yttrium-90 labeled ferromagnetic particles using an external magnetic field. Anticancer Res. 2007;27(2):755–60.
74. Chu KF, Dupuy DE. Thermal ablation of tumours: biological mechanisms and advances in therapy. Nat Rev Cancer. 2014;14(3):199–208.
75. Gilchrist RK, Medal R, Shorey WD, Hanselman RC, Parrott JC, Taylor CB. Selective inductive heating of lymph nodes. Ann Surg. 1957;146(4):596–606.
76. Johannsen M, Thiesen B, Wust P, Jordan A. Magnetic nanoparticle hyperthermia for prostate cancer. Int J hyperther. 2010;26(8):790–5.
77. Maier-Hauff K, Ulrich F, Nestler D, Niehoff H, Wust P, Thiesen B, et al. Efficacy and safety of intratumoral thermotherapy using magnetic iron-oxide nanoparticles combined with external beam radiotherapy on patients with recurrent glioblastoma multiforme. J Neuro-Oncol. 2011;103(2):317–24.
78. Bealle G, Di Corato R, Kolosnjaj-Tabi J, Dupuis V, Clement O, Gazeau F, et al. Ultra magnetic liposomes for MR imaging, targeting, and hyperthermia. Langmuir. 2012;28(32): 11834–42.
79. Jordan A, Wust P, Fahling H, John W, Hinz A, Felix R. Inductive heating of ferrimagnetic particles and magnetic fluids: physical evaluation of their potential for hyperthermia. 1993. Int J Hyperther. 2009;25(7):499–511.
80. Heathman TR, Nienow AW, McCall MJ, Coopman K, Kara B, Hewitt CJ. The translation of cell-based therapies: clinical landscape and manufacturing challenges. Regen Med. 2015;10(1):49–64.
81. Fischbach MA, Bluestone JA, Lim WA. Cell-based therapeutics: the next pillar of medicine. Science translational medicine. 2013;5(179):179ps7.
82. Estelrich J, Escribano E, Queralt J, Busquets MA. Iron oxide nanoparticles for magnetically-guided and magnetically-responsive drug delivery. Int J Mol Sci. 2015;16(4):8070–101.
83. Senyei AE, Reich SD, Gonczy C, Widder KJ. In vivo kinetics of magnetically targeted low-dose doxorubicin. J Pharm Sci. 1981;70(4):389–91.
84. Widder KJ, Morris RM, Poore GA, Howard DP, Senyei AE. Selective targeting of magnetic albumin microspheres containing low-dose doxorubicin: total remission in Yoshida sarcoma-bearing rats. Eur J Canc Clin Oncol. 1983;19(1):135–9.
85. Thiele RH, Colquhoun DA, Gillies GT, Tiouririne M. Manipulation of hyperbaric lidocaine using a weak magnetic field: a pilot study. Anesth Analg. 2012;114(6):1365–7.
86. Dames P, Gleich B, Flemmer A, Hajek K, Seidl N, Wiekhorst F, et al. Targeted delivery of magnetic aerosol droplets to the lung. Nat Nanotechnol. 2007;2(8):495–9.
87. Heppner F, Mose JR. The liquefaction (oncolysis) of malignant gliomas by a non pathogenic Clostridium. Acta Neurochir. 1978;42(1–2):123–5.
88. Schmidt W, Fabricius EM, Schneeweiss U. The tumour-Clostridium phenomenon: 50 years of developmental research (Review). Int J Oncol. 2006;29(6):1479–92.
89. Forbes NS. Engineering the perfect (bacterial) cancer therapy. Nat Rev Cancer. 2010;10(11): 784–93.
90. Jiang Y, Sigmund F, Reber J, Dean-Ben XL, Glasl S, Kneipp M, et al. Violacein as a genetically-controlled, enzymatically amplified and photobleaching-resistant chromophore for optoacoustic bacterial imaging. Sci Rep. 2015;5:11048.
91. Martel S, Felfoul O, Mohammadi M, Mathieu JB. Interventional procedure based on nanorobots propelled and steered by flagellated magnetotactic bacteria for direct targeting of tumors in the human body. Conf Proc IEEE Eng Med Biol Soc. 2008;2008:2497–500.
92. Benoit MR, Mayer D, Barak Y, Chen IY, Hu W, Cheng Z, et al. Visualizing implanted tumors in mice with magnetic resonance imaging using magnetotactic bacteria. Clin Canc Res. 2009;15(16):5170–7.
93. Alphandery E, Faure S, Seksek O, Guyot F, Chebbi I. Chains of magnetosomes extracted from AMB-1 magnetotactic bacteria for application in alternative magnetic field cancer therapy. ACS Nano. 2011;5(8):6279–96.

94. Taherkhani S, Mohammadi M, Daoud J, Martel S, Tabrizian M. Covalent binding of nanoliposomes to the surface of magnetotactic bacteria for the synthesis of self-propelled therapeutic agents. ACS Nano. 2014;8(5):5049–60.
95. Park SJ, Park SH, Cho S, Kim DM, Lee Y, Ko SY, et al. New paradigm for tumor theranostic methodology using bacteria-based microrobot. Sci Rep. 2013;3:3394.
96. Gadian DG. NMR and its applications to living systems, 2nd Edition. Oxford Science Publications. 1996.
97. Gleich B, Weizenecker R. Tomographic imaging using the nonlinear response of magnetic particles. Nature. 2005;435(7046):1214–7.
98. Heyn C, Ronald JA, Ramadan SS, Snir JA, Barry AM, MacKenzie LT, et al. In vivo MRI of cancer cell fate at the single-cell level in a mouse model of breast cancer metastasis to the brain. Magn Reson Med. 2006;56(5):1001–10.
99. Bauer LM, Situ SF, Griswold MA, Samia AC. Magnetic particle imaging tracers: state-of-the-art and future directions. J Phys Chem Lett. 2015;6(13):2509–17.

Spinning to a Different Beat: ^{19}F Agents for "Hot-Spot" Cellular MR Imaging

Edyta Swider and Mangala Srinivas

1 Introduction

The imaging of cellular therapeutics plays a significant and vital role in the development, optimization, and personalization of such therapy, as it can provide essential information on cell localization, number, and functionality (Fig. 1a), thus allowing dosage and delivery optimization, among others. Magnetic resonance imaging (MRI) is an anatomic and functional imaging technique which excels at soft-tissue imaging, without limitations on penetration depth. MRI is one of the commonly used imaging modalities due to its excellent safety, noninvasive nature, high resolution, lack of ionizing radiation, and intrinsic anatomic contrast. Standard (i.e. most commonly used) MRI focuses on the detection of differences in relaxation rates of the protons (^1H) in mobile water molecules in various tissues. However, it is normally not possible to distinguish specific therapeutic cells from their surrounding cellular milieu without the use of a specific label for the relevant cells. There have been several approaches to develop contrast agents (CAs) to improve contrast, e.g., gadolinium or superparamagnetic iron oxide (SPIO). However, metal CAs can be toxic, may not allow proper cell quantification, and often require "before" and "after" images to detect changes in contrast. Only SPIO-based agents have been used in the clinic for cell tracking [1] and these agents are no longer commercially available.

E. Swider • M. Srinivas (✉)
Department of Tumor Immunology, Radboud Institute for Molecular Life Sciences, Radboud University Medical Center, Nijmegen, The Netherlands
e-mail: mangala.srinivas@radboudumc.nl

J.W.M. Bulte, M.M.J. Modo (eds.), *Design and Applications of Nanoparticles in Biomedical Imaging*, DOI 10.1007/978-3-319-42169-8_7

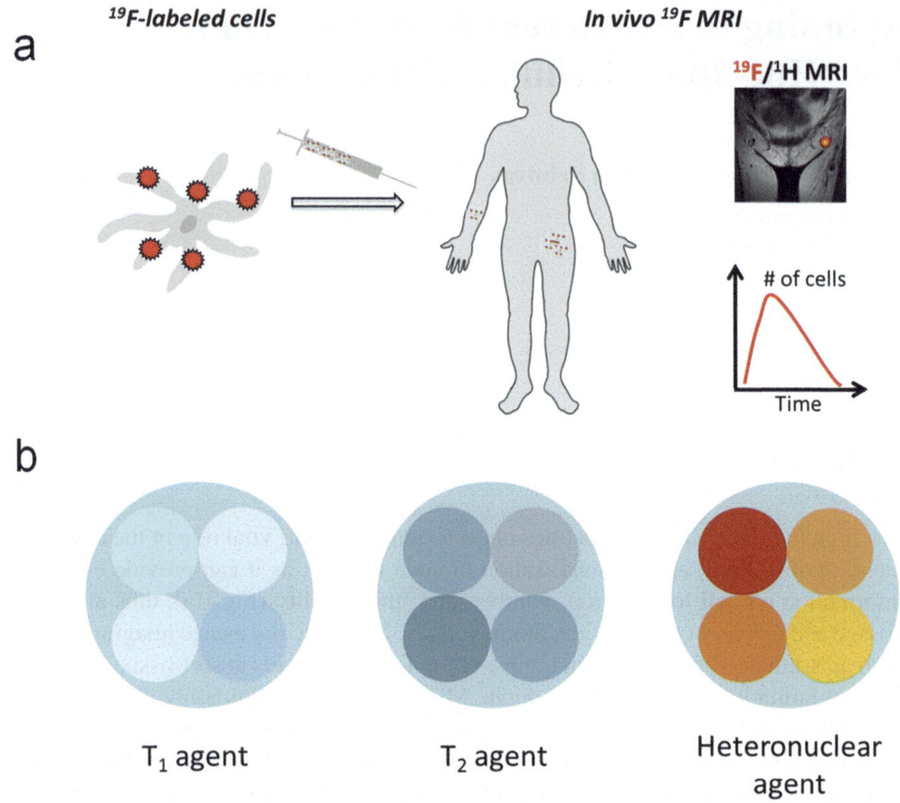

Fig. 1 General ^{19}F MRI cell labeling. (**a**) A general protocol for ^{19}F MRI of labeled cells involves ex vivo labeling of the cells, transfer to the subject, and then imaging. A smaller number of studies use direct in vivo labeling, typically macrophages. (**b**) "Hot-spot" imaging. The figure demonstrates the images typically obtained with T_1, T_2 contrast agents, and a ^{19}F label. The contrast agents change the contrast in the anatomic (^{1}H) scan, while the ^{19}F image must be acquired separately, rendered in false color, and is then typically superimposed over the standard anatomic scan (in *gray* scale). Thus, the detection of ^{19}F signal is less ambiguous, and, more importantly, the signal intensity directly relates to agent concentration, given sufficient signal to noise. In practice, this means that the cell numbers can be estimated directly from image data, one of the main advantages of the technique

In the last decade, "hot-spot" imaging has gained a lot of attention. Hot-spot imaging is hetero-nuclear MRI using ^{13}C, ^{23}Na, ^{31}P, or ^{19}F in addition to ^{1}H, which enables direct detection rather than indirect, as is the case for CAs (Fig. 1b). Out of these four elements, ^{19}F has been the main focus as it is essentially absent in biological tissues; thus there is no endogenous background signal, which makes it more attractive for hot-spot imaging [2]. One clear example of the advantage of the lack of background signal is a recent paper using ^{19}F MRI to track inflammatory macrophages and monocytes in a murine model of multiple sclerosis [3]. Here the inflammatory cells were found not only in the diseased spinal cord, as expected, but also in neighboring bone marrow tissue. This ability to locate ^{19}F signal even in unexpected locations without comparison to a "before" image, given a sufficient signal-to-noise ratio, is one of the key advantages of the technique.

^{19}F MRI has been applied as a powerful imaging tool for diverse applications, such as cell tracking [4], drug metabolism, plaque detection, and monitoring inflammation. Cell tracking using ^{19}F MRI has been applied to various cell types, such as dendritic cells (DCs), T cells, stem cells, and macrophages [5]. In order to distinguish relevant cells from the background, they need to be labeled with fluorinated labels, described in the following section.

2 Fluorocarbons

2.1 General Properties of Fluorine

Using fluorine in MRI has several advantages as ^{19}F is a spin ½ nucleus, which means that it has an odd number of protons and neutrons and spherical charge distribution, and a gyromagnetic ratio close to the proton. Thus, for a given magnetic field value, the spin processes at almost the same frequency, permitting the use of a conventional proton spectrometer to detect fluorine [6]. Fluorine has 100 % natural abundance and an extremely broad chemical shift and is not naturally present in biological tissues and therefore has no biological background signal; the traces of fluorine in bone and teeth exhibit a very short spin-spin relaxation time (T_2) and are not visible to conventional MRI methods [2]. In all, fluorine MRI gives a very high contrast-to-noise ratio and specificity, in a situation where a fluorinated element is introduced as a contrast agent. Such agent requires a very high density of ^{19}F nuclei on the molecule in addition to a high tissue concentration in order to produce an image quality such as the one of ^{1}H MRI. However, it is important to realize that the administered levels of ^{19}F will still be many orders of magnitude lower than the amount of ^{1}H present in biological tissues, and thus, despite all its advantages in terms of imaging properties, the low concentrations of ^{19}F frequently result in images that hover at the detection limit with low signal-to-noise ratios (SNRs). Finally, some fluorocarbons show a T_1 sensitivity to oxygen tension, which may allow its use as an in vivo sensor [2]; alternatively, this T_1 variability can be a nuisance, hindering accurate quantification.

2.2 Fluorinated Compounds/Molecules

An example of fluorinated compounds includes fluorinated peptides, polymers, and also fluorinated small molecules, such as sugars; however these result in much lower cell loading and, as a result, a lower detection sensitivity. Small fluorinated molecules have lower ^{19}F content than larger fluorocarbons, such as perfluorocarbons (PFCs); however their advantage is that they can be used to study cell functionality [5]. Despite this benefit, the insufficient sensitivity of the small molecules makes them less attractive for in vivo use. Recent developments in small fluorinated molecules include fluorinated albumin [7] and self-assembling nanoparticles sensitive to their environment [8], although neither of these have been applied to cell tracking. Some examples are listed in Table 1.

There have been limited studies using fluorinated drugs (see Table 1). The limit here is the high reactivity and electronegativity of fluorine, which when incorporated into a drug can significantly alter drug chemistry and subsequent metabolism. Chemical shift artifacts can also be an issue. For all these reasons, such small molecules are seldom used for cell tracking.

2.3 *Perfluorocarbons*

Perfluorocarbons (PFCs) are synthetic molecules mainly composed of carbon and fluorine atoms (see Table 2 for a summary of their physical properties). Depending on their chemical structure and molecular weight, PFCs exist in a gaseous, liquid, or even solid form. The compounds are extremely stable, generally have low vapor pressures, and are inert. PFCs do not react with living tissues due to their carbon-fluorine bonds, which cannot be metabolized in vivo, and they have low intermolecular forces, and hence, the surface tension of the liquid form of PFCs is extremely low. Most PFCs have a surface tension of 14–18 dyne/cm [9]. As gases, PFCs show some toxicity, while in a liquid form they are relatively safe [10]. Clearance of PFC agents from the body generally occurs via the mononuclear phagocyte system (MPS) uptake (phagocytic cells primarily in lymph nodes, spleen, and liver) and finally lung exhalation. Some PFCs can also bind and dissolve oxygen. This led to their early use as blood substitutes, although the idea has been commercially abandoned. PFCs are still used clinically, for example for partial liquid ventilation [LiquiVent, Alliance Pharmaceutical Corporation], eye surgery [Perfluoron™, Alcon® Inc., USA], and as ultrasound contrast agents as gas bubbles [Sonovue®, Bracco International B.V.] (Fig. 2a).

PFCs have now found their use as imaging labels for ultrasonography, computed tomography, and ^{19}F MRI. Their use in the latter imaging method is especially beneficial, as PFCs have a high payload of fluorine atoms, which makes ^{19}F MRI a promising modality for in vivo quantitative imaging, including in vivo cell tracking. Here, liquid PFCs are more suitable, as they result in the most favorable imaging characteristics.

For imaging purposes, PFCs offer a high ^{19}F density per molecule, and thus high sensitivity, provided that multiple fluorine resonances do not occur. To avoid such chemical shift artifacts, PFCs such as ring ethers or longer linear ether chains are frequently used [5, 11]. The most frequently used ^{19}F MRI labels consist of perfluorooctyl

Table 1 Overview of different ^{19}F labels, cell types, and cell loading

Major ^{19}F component	Additives	Hydrodynamic diameter (nm)	Cell type	Average cell loading ($\times 10^{12}$ ^{19}F atoms/cell)	No. of (dominant) resonances	Ref.
Various PFCs	Atto 647	200–300	Primary human DCs	30	1	[22]
PFCE PFOB		233 224	Primary human stem/ progenitor cells	20 0.5	1 Multiple	[33]
PFCE	Dye		Rodent glioma	0.2	1	[34]
PFOB	Drug	225	Primary human umbilical cord smooth muscle cells		Multiple	[35]
Various PFCs	Fluorinated quantum dots	200	Murine macrophage Human carcinoma		Multiple	[36]
PFCE	Targeting peptide	170	Human umbilical cord vein Endothelial cells		1	[37]
PFOB	Targeting antibody Transfection agent		Human cancer cell lines		Multiple	[38, 39]
PFCE	Dye	145	Macrophages		1	[40]
Hyperbranched fluorinated polymer	Drug	20–30	Human glioblastoma cell line	0.0005	1	[41]
PERFECTA	n/a	140–220	Dendritic cells	19.2	1	[14]
Fluorinated colchicine derivatives	n/a	n/a	Human leukemia cell line		Multiple	[42]
Fluorinated reporter molecules	n/a	n/a	Various, including bacteria and human cancer cell lines			[43–45]
Fluorinated prodrug			Human cancer cell lines			[46]

Reproduced with permission from Ref. [5]

Table 2 General physical properties of various PFCs used in cell tracking

Perfluorocarbon	Formula	Molecular weight	Density (g/mL)	Boiling point (°C)
Perfluoropentane	C_5F_{12}	288	1.62	26–36
Perfluorohexane	C_6F_{14}	338	1.76	58–60
Perfluorooctyl bromide	$C_8F_{17}Br$	499	1.93	140.5
Perfluorononane	C_9F_{20}	488	1.80	126
Perfluorodecalin	$C_{10}F_{18}$	462	1.99	144
Perfluoro-15-crown-5-ether	$C_{10}F_{20}O_5$	580	1.78	145
PERFECTA	$C_{21}H_8F_{36}O_4$	1008	2.09	ND

bromide (PFOB), perfluoro-15-crown-5-ether (PFCE), and a linear PFPE polymer mixture, which is commercialized as Cell Sense (Celsense Inc., Pittsburgh PA, USA). Characteristics of examples of these PFCs are shown in Table 1, while Fig. 2b shows the structure of PFCE, PFOB, and PFPE.

PFCs have a unique and sometimes undesirable characteristic, in that they are fluorophilic, i.e., simultaneously lipophobic and hydrophobic (Fig. 2b). Consequently, PFCs separate when placed in an aqueous environment and are immiscible in cell membranes. Because of this, they require emulsification, to form nanoemulsions, or entrapment into particles to achieve stability in biological environments. In such forms, PFCs can also be used to label cells for in vivo cell tracking.

3 ^{19}F MRI Labels

The ideal ^{19}F label needs to have certain characteristics, described in more detail elsewhere [2]. Briefly, such an agent has to be biologically inert and chemically stable. Fluorinated compounds should be nontoxic for the host and have minimal impact on labeled cell function. The label needs to be suitable for cell uptake, with smaller droplets or particles for intracellular labeling and potentially larger ones for blood pool imaging. What is more, cells need to take up sufficient amounts of the label in order to be detected (ranging from 0.5 to 30×10^{12} ^{19}F atoms/cell) [5]. The "sufficient amount" is based on the number of cells per voxel and T_1 and T_2 characteristics of the compound. Last, but not least, the ^{19}F compounds must have beneficial relaxation parameters, with a short T_1 and a long T_2 [2]. Typical labels are composed of a fluorinated molecule and a coating or surfactant which stabilizes the fluorocarbon to make it more suitable for cell labeling and cell uptake (Fig. 3).

In general, high signal density per molecule is necessary. Therefore, most labels consist of fluorocarbon or even liquid PFCs such as perfluoropolyethers (PFPE), typically administered as nanoemulsions with droplet sizes around 100–300 nm. Due to their poor miscibility, PFCs need to be "packaged" to stabilize them in aqueous biological environments. These packaging techniques are discussed in the following section.

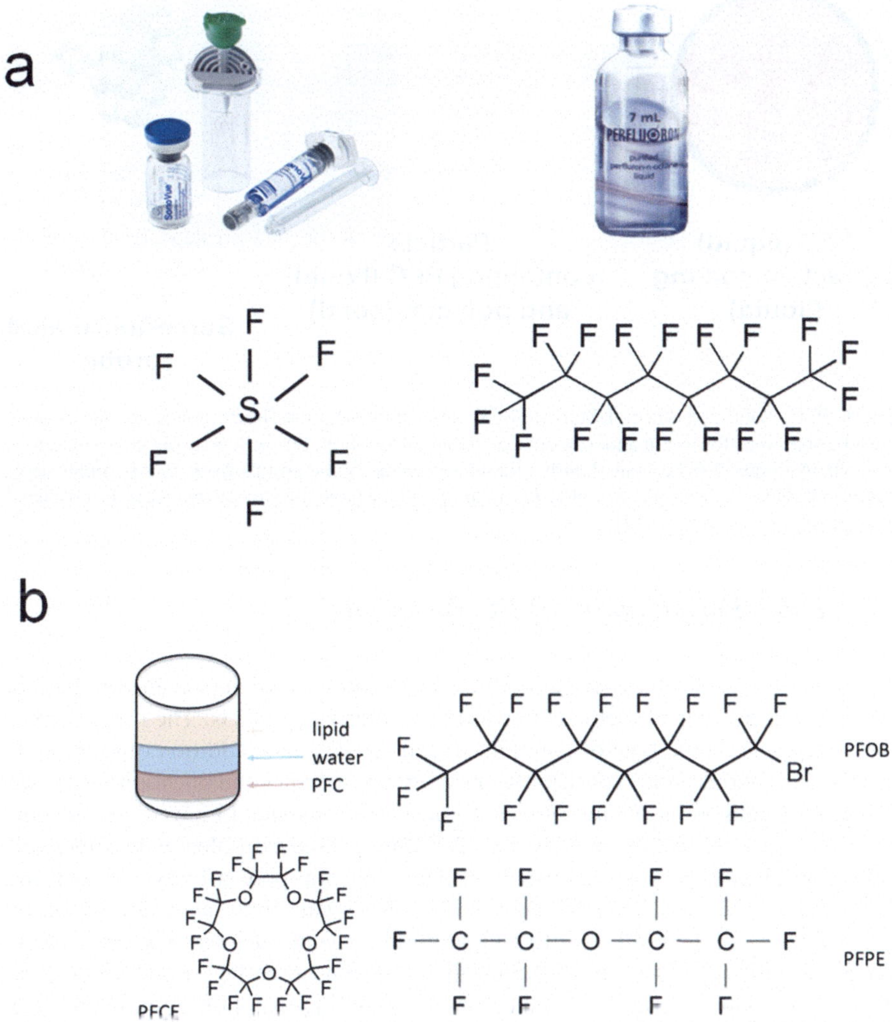

Fig. 2 Perfluorocarbons. (**a**) PFCs are currently in clinical use, as both gases and liquids. Shown is an ultrasound microbubble contrast agent containing sulfur hexafluoride gas (Sonovue®) and a perfluorooctane liquid used in eye surgery (Perfluoron™). (**b**) Most cell labels are perfluorocarbons (PFCs), polymers of carbon and fluorine. These compounds have unique properties; in particular they mix with neither aqueous nor organic solvents. Some of the more commonly used PFCs for cell tracking are PFOB, PFCE, and PFPE, shown here

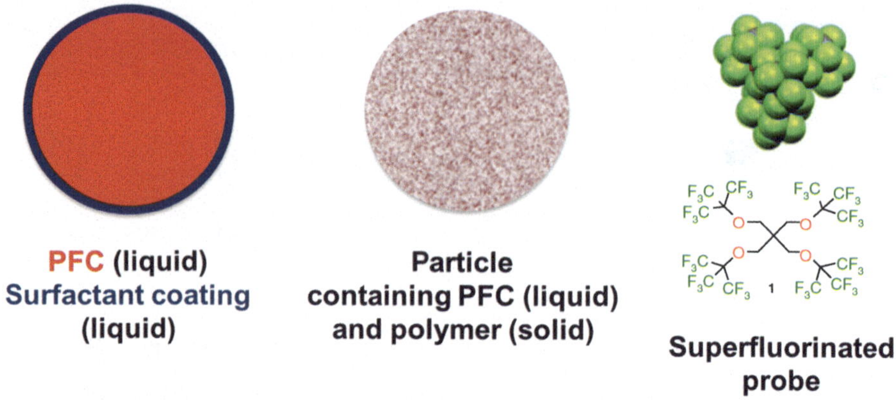

Fig. 3 PFCs stabilized for cell labeling. PFCs, being simultaneously hydrophobic and lipophobic, must be stabilized for use in aqueous environments. Most often, this is done using emulsion droplets, where a liquid PFC is coated with a liquid surfactant. Some groups work with particles, containing a liquid PFC in a solid polymer. The final option is a large fluorinated molecule. Reproduced with permission from Ref. [14]

3.1 Label Formulation: "PFC Packaging"

Cell tracking applications require PFCs to be formulated into a biocompatible label. There are several methods to formulate fluorinated labels. The route of synthesis depends on the type of label, its size, properties, and addition of extra molecules, e.g., fluorescent dyes. Emulsions are prepared via emulsification processes using, e.g., sonication or microfluidization. Emulsions should have a small, uniform droplet size, ideally smaller than 200 nm, as larger droplets may affect cell activation phenotype post-labeling [11]. However, the poor affinity between the PFC, surfactant, and aqueous continuous phase can limit emulsion stability. Furthermore, it can be difficult to add additional agents such as fluorescent dyes, drugs, or antibodies without changing the emulsion properties. Furthermore, the addition of extra moieties to an emulsion can result in separation of those moieties from PFC in the cell or in vivo, with no assurance of their localization. Recently, nanoparticles have received a lot of interest [5, 12], due to their enhanced stability. A key advantage of using particles is that the components of the particles can be modified without altering the formulation protocol.

3.2 Emulsions

Emulsions are the simple product of vigorously mixing oil and water. More specifically, an emulsion is a suspension of two immiscible liquids. In the literature, emulsion droplets are frequently referred to as "particles" but to avoid confusion, here particles are defined as solids which can be isolated from and added back to the continuous phase (water or buffer) without undergoing changes other than hydration, with the particles being able to be stored as powder. Emulsion droplets cannot be isolated in such fashion without losing stability.

Emulsions are the most commonly used cell labels due to their high content of PFCs per droplet and simplicity of production. Emulsions consist of an immiscible compound that is surrounded by a surfactant. The role of the surfactant is to stabilize the formed droplets in the continuous phase (water) and delay the PFC from settling out and coalescing to form a separate layer. PFC emulsions for MRI generally exhibit poor stability due to the fact that the surfactant used cannot be miscible with both the PFC and water, unless it is a fluorinated surfactant. Fluorinated surfactants are avoided, due to the potential for toxicity and the chemical shift artifacts that can result from the added fluorine. PFC emulsions are described in more detail elsewhere [13].

Recently a new compound called PERFECTA has been developed as a ^{19}F MRI tracer. PERFECTA is a superfluorinated molecular probe suitable for in vivo cell tracking. It has high detection sensitivity, a single, sharp resonance peak, and attractive T_1 and T_2 values [14]. PERFECTA is not fully fluorinated, but has a hydrocarbon polar core and four ether bonds, which might undergo enzymatic degradation in vivo. Most other recent studies have focused on conventional, existing PFCs [12, 15–17]. Figure 3 shows a representation of an emulsion droplet, a nanoparticle, and a large fluorinated molecule.

3.3 *Micelles*

Micelles are lipid-based compounds composed of amphiphilic molecules that undergo spontaneous self-assembly in an aqueous solution. This spontaneous self-assembly is due to the amphipathic nature of the molecule. To form a micelle, the amphiphilic molecules arrange in a way where the hydrophobic groups create a core, while the hydrophilic groups are on the outside and have contact with water [18]. The size of micelles ranges from 2 to 20 nm, and it depends on their composition, modifications, and concentration. There have been several studies incorporating a hydrophilic Gd(III) chelate into micelles for in vivo imaging [19–21] to enhance the T_1 MRI contrast. Micelles can be further modified with groups that bind to specific targets.

3.4 *Dendrimers*

Dendrimers are nanoscale macromolecules that have well-defined polymeric chains which branch off from a central core. They can be functionalized with polyfluorinated groups and targeting moieties. Dendrimers modified with Gd(III) chelate show increase in rotational correlation times which leads to strong enhancement in T_1 relaxation rates when compared to lower molecular weight chelates [18]. Dendrimers typically present prolonged retention times within the vascular system in comparison to lower molecular weight chelates, which usually undergo a fast renal clearance.

Dendrimers are very appealing in ^{19}F MRI label design due to large number of surface functional groups which can be reacted with polyfluorinated precursors, leading to materials with potentially high concentrations of fluorine nuclei [18]. However, it is possible that dendrimers may be metabolized in vivo leading to potentially toxic compounds. In general, less is known about these extremely large molecules and their in vivo behavior.

3.5 PFC Nanoparticles

Particles are an attractive alternative for ^{19}F MRI compared to conventional liquid emulsion droplets, as they are more stable and can be frozen as a dry powder for easy storage and transport [5]. PFC nanoparticles typically consist of a PFC entrapped in a solid polymer. We previously reported on stable encapsulation of PFC in poly(D,L-lactide-*co*-glycolide acid) (PLGA) particles [22]. The PFC-PLGA particles were prepared using single-emulsion solvent-evaporation method. In this method, the polymer is dissolved in an organic solvent, and the PFCE is added. Next, this organic phase is added to an aqueous solution containing surfactant under sonication. The role of surfactant is to stabilize the cells and prevent Ostwald ripening (which is very common in emulsions, including PFC emulsions) [23]. Sonication is then followed by solvent evaporation, and washing and freeze-drying of particles. After the last step, particles can be stored in a powder form for extended periods of time. PLGA-PFC particles can be easily modified with targeting ligands, and fluorescent dyes, and they can also be used to encapsulate drugs.

There are several methods to characterize PFC nanoparticles, for example dynamic light scattering, electron microscopy, and nuclear magnetic resonance (NMR). The first two methods are used to study diameter, while ^{19}F NMR is used to measure fluorine content. PFC particles prepared with solvent evaporation method typically have a diameter of 200 nm, comparable to many PFC emulsion droplets.

4 Cell Labeling

Most of the studies performed thus far in this field have used naturally phagocytic cells, such as macrophages or DCs. These types of cells readily phagocytose label droplets or particles, generally resulting in labeling in the order of 10^{10}–10^{13} fluorine atoms per cell (see Table 1).

A general cell labeling protocol is shown in Fig. 4. Key labeling steps include the length of time the cells are incubated with the label, the concentration of the label, and removal of excess label. Label uptake can be encouraged when necessary through the use of positive changes on the labels, such as through transfection agents [24]. It is important to show that the label is incorporated in a stable manner within the cell, and will remain within the relevant cell in vivo. In addition, it needs to be ensured that any excess label is removed through careful washing steps before transfer to the subject.

It is also necessary to balance the uptake of label with any effects on cell functionality and cytotoxicity. In particular, it is vital to assess any effect of the label on cell migration and other functions that are necessary in vivo. It has been shown that iron oxide labeling can alter cell migration in neural stem cells [25], although another study that exhaustively compared various labels including iron oxide and ^{19}F did not find any effect in mesenchymal stem cells [26]. Thus, the effect of labeling appears to be cell dependent and label dependent at the very least. Furthermore, with some fluorine labels, cell migration is known to be unaffected, and ^{19}F MRI has in fact been suggested as a cell migration assay for larger cell numbers [27]. However, one cannot assume that any particular label will not affect migration, or any particular function, in a different cell type, i.e., each cell type-label pair needs to be individually optimized and thoroughly characterized.

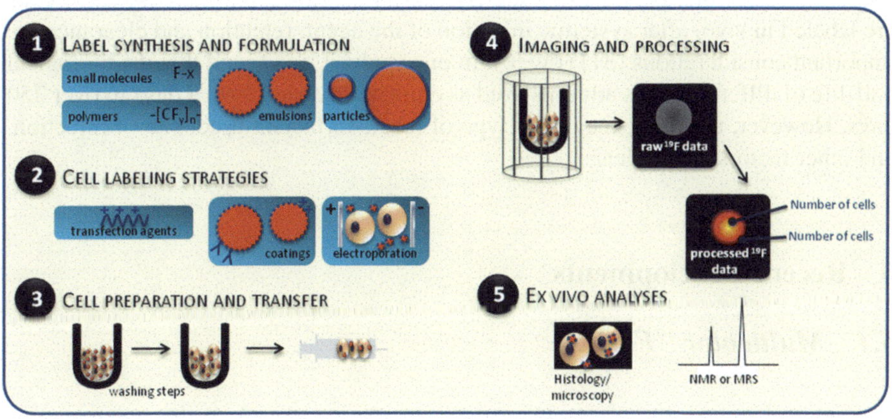

Fig. 4 Typical ex vivo cell labeling protocol. Key steps in cell labeling for ^{19}F MRI. The figure shows the most commonly used protocols for each step; variations proposed in the literature are discussed in the text. Appropriate selection of label and labeling protocol is crucial for success of the experiment (Step 1). Cell labeling (Step 2) may require enhancement through the use of coatings or transfection agents. After suitable preparation (Step 3) the cells can be imaged. Post-processing leads to quantification (Step 4). Finally, image post-processing can be carried out (Step 5) to corroborate the in vivo data. Reproduced with permission from Ref. [5]

In all cases, quantification of cell numbers is much more accurate when cells are labeled ex vivo, as the average ^{19}F content per cell can be readily determined. When the label is injected systemically [28] in vivo, and subsequently taken up by target cells (typically macrophages), quantification becomes more complex, as the label uptake per cell cannot be determined. Even with ex vivo-labeled cells, the label can be lost or transferred to other cell types, particularly macrophages present within the surrounding host tissue. For example, labeled mesenchymal stem cells lost their ^{19}F agent to bystander macrophages by 17 days after transfer [29]. Thus, the signal detected could be falsely interpreted at later time points. Such label transfer may vary depending on cell type, label type, transfer site, division status of the cells, and various other factors. Similar results were also obtained in another study [30], where transgenic neural stem cells expressing a bioluminescence marker were also labeled with a fluorinated agent. Here, the ^{19}F signal was retained for at least several days after the bioluminescence signal was lost, and indicator of cell death (Fig. 5). This slow loss of ^{19}F signal even after cell death could be a result of various factors including the particular properties of the label or cell, or simply slow clearance from the brain. Regardless, these studies highlight that the interpretation of cell tracking can become more complex after its initial stages.

The fate of the agent, both in cells and in vivo, is an important consideration, particularly in terms of any toxicity and the clearance rate and route. This may be less of an issue if cells are prelabeled ex vivo before injection to the subject, as then only very small amounts of PFCs are administered. However, in cases where cells are labeled in vivo, after systemic injection of the agent, retention and clearance are important considerations [31] (Fig. 6). In one study, it was found that the biological half-life of different PFCs administered as emulsions varied from 9 days to over 250 days. However, the influence of the type of label, surfactant used, site of injection, and other factors is not clear.

5 Recent Developments

5.1 *Multicolor ^{19}F MRI*

The broad chemical shift of ^{19}F allows the possibility of imaging well-separated resonances. Multiple resonances are most often a drawback, as they represent "lost" ^{19}Fs that do not contribute to the signal, or that create artifacts making the image "fuzzy" (Fig. 7). However, in some cases this drawback can be harnessed to image multiple cell lines at different resonance frequencies. Recently, this has been applied to different therapeutic cell subsets, which were labeled with different PFCs (PFO and PFCE, respectively) [12] (Fig. 8). A standard gradient-echo sequence was used for imaging, where the excitation peak was sufficiently broad so as to excite both the main

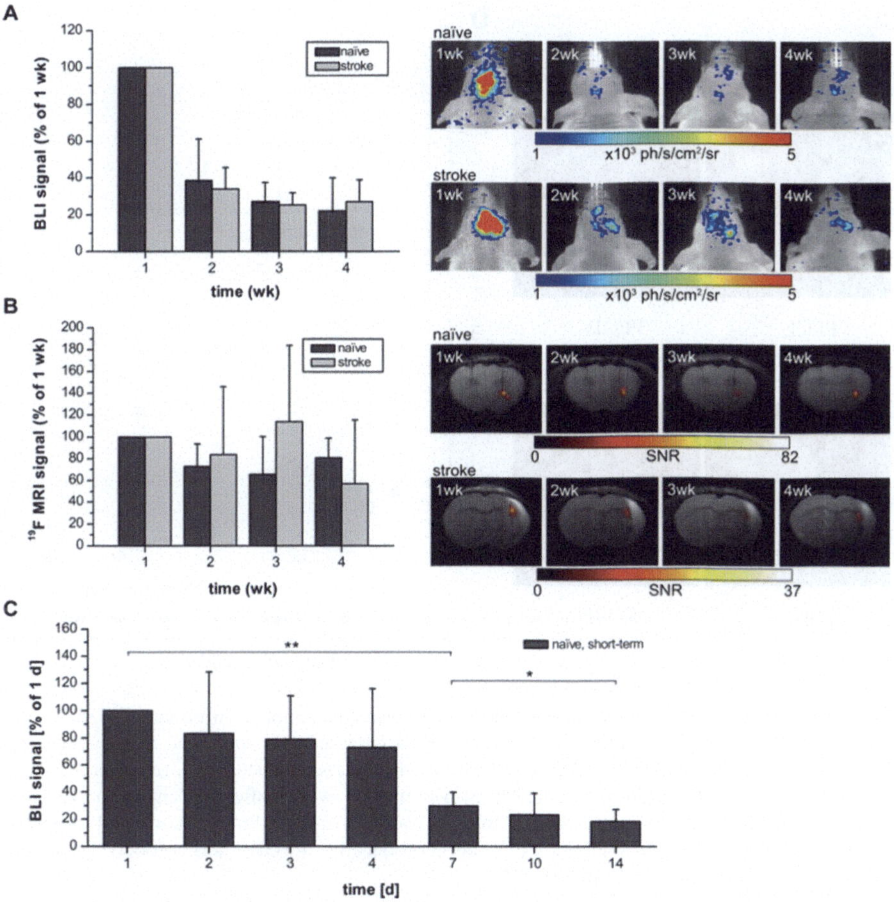

Fig. 5 ^{19}F MRI signal can persist even after cell death in some cases. BLI and ^{19}F MRI signal after implantation of neural stem cells into naïve mice and stroked animals that underwent middle cerebral artery occlusion (MCAO) 48 h earlier. (**a**) BLI signal in naïve ($n=5$) or stroked ($n=4$) animals decreased rapidly over 4 weeks indicating impaired graft survival. (**b**) ^{19}F SNR in naïve and stroked animals persisted, with more scatter in the stroke group. (**c**) To better resolve the decrease in BLI signal, a separate group of naïve animals ($n=6$) underwent BLI one day after implantation and up to 2 weeks after. The onset of decreased graft survival was found at 7 days. */**: Significance level $p \leq 0.05/0.01$, all values presented as mean ± standard deviation. Reproduced with permission from Ref. [30]

resonance of the PFO and the single resonance of the PFCE. Both were detected simultaneously, but due to the known bandwidth per voxel, the signals could be separated. The technique clearly has limitations, for example, in applications where the cell distribution is very complex. Still, it allows imaging of two cell populations simultaneously. Such work has also been carried out with stem cells, using spectroscopic imaging [32].

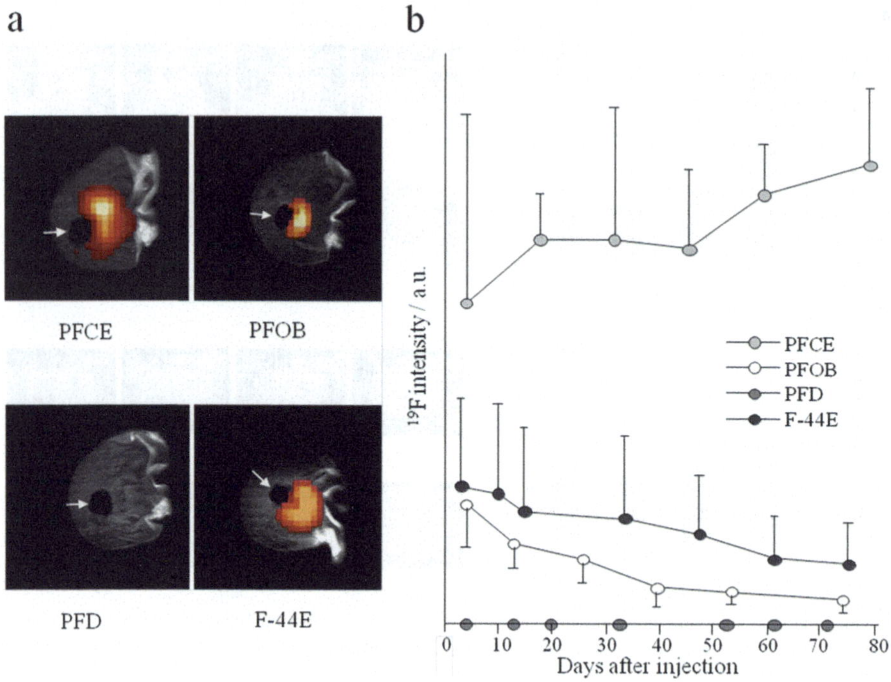

Fig. 6 Retention and sensitivity of different PFCs. In vivo ^{19}F chemical shift imaging signal in the ear after induction of inflammation by ear clipping (holes marked by *arrows*) and subsequent injection of 500 μL perfluorocarbon (PFC) emulsion ($n=4$ animals for each PFC). (**a**) Overlay of ^{19}F signal and anatomical scan for the initial MR measurement 3–4 days after PFC injection. (**b**) Time course of the fluorine signals for the different PFCs. F-44E, trans-bis-perfluorobutyl ethylene; PFCE, perfluoro-15-crown-5 ether; PFD, perfluorodecalin; PFOB, perfluorooctyl bromide. Reproduced with permission from Ref. [31]

5.2 Multimodal Imaging

No single imaging modality is perfect and able to provide all the necessary answers with certainty. In fact, nowadays the majority of ^{19}F cell tracking imaging studies use multimodal imaging, most often combining MRI with fluorescence, either in vivo or ex vivo using histopathology. The advantages of multimodal cell tracking include verification of data obtained using a new imaging agent or technique with an established one, and the ability to gather complementary information from a single experiment.

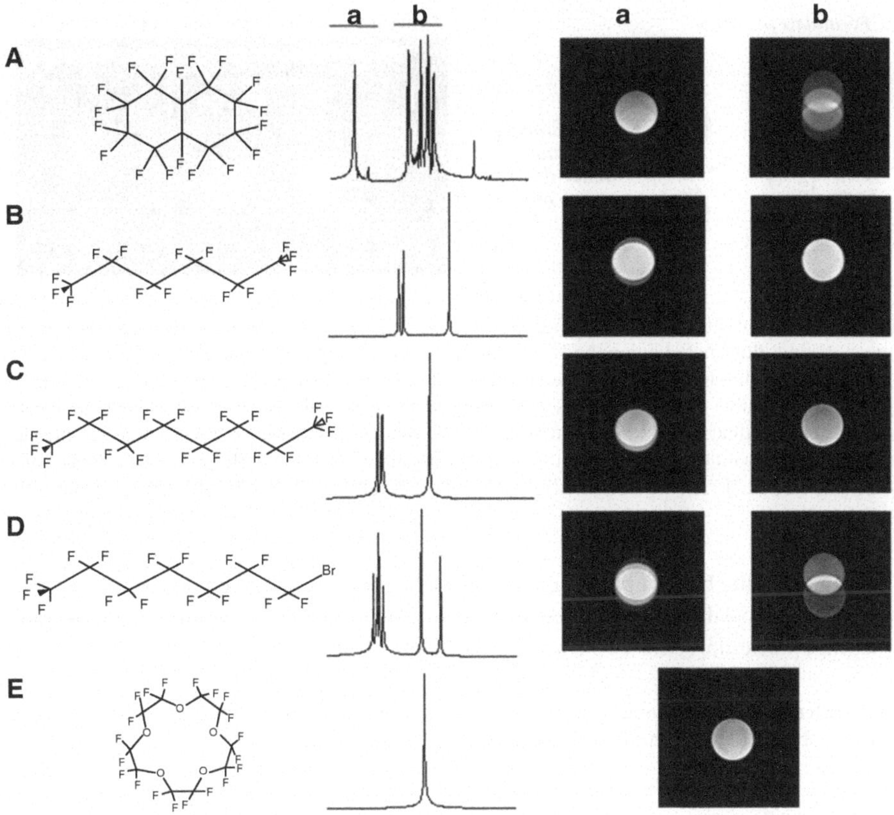

Fig. 7 Chemical shift artifact. The ^{19}F NMR spectrum, chemical structure, and the ^{19}F MR images are shown. The acquisition frequency of the MR images was centered on each peak or group of peaks, and the corresponding image is shown (indicated by **a** and **b**). (**a**) Perfluorodecalin ($C_{10}F_{18}$) has several peaks in the ^{19}F spectrum. It is used in commercial oxygen carriers and blood pool agents. (**b**) Perfluorohexane (C_6F_{14}) is also used as a contrast agent for ultrasound imaging. (**c**) Perfluorooctane (C_8F_{18}) is also used as an oxygen carrier. (**d**) Perfluorooctylbromide (C_8BrF_{17}), or PFOB, is a blood substitute that has also been used for ^{19}F MRI. However, the presence of multiple ^{19}F resonances complicates the imaging process. (**e**) Perfluoro-15-crown ether ($C_{10}F_{20}O_5$) has 20 equivalent ^{19}F atoms, resulting in a single resonance peak. The compound has been used for ^{19}F MRI for cell tracking and as an oxygen sensor. Reproduced with permission from Ref. [22]

6 Conclusions and Future Outlook

^{19}F MRI for cell tracking has recently been demonstrated in the clinic [15]; thus its clinical applicability is no longer in dispute. There is still no consensus on which PFC, if any, is the best for in vivo application, particularly in terms of balancing imaging favorability with retention times, and new and more "ideal" compounds are continually being generated. Furthermore, the relatively long imaging times necessary and limited availability of suitable hardware to do ^{19}F MRI are still a

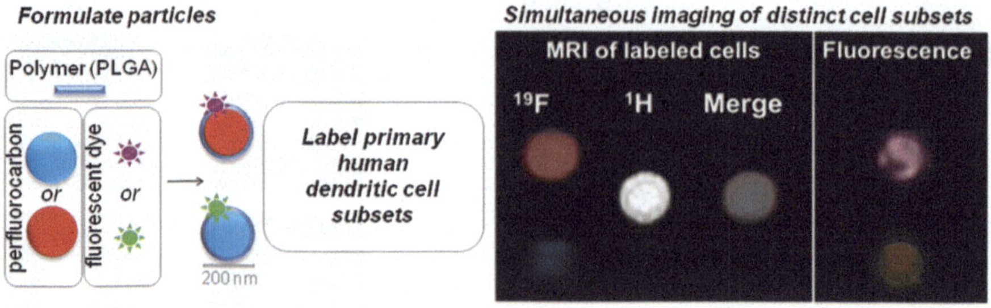

Fig. 8 Imaging multiple resonance frequencies. In some cases, the chemical shift can be used to image distinct fluorinated compounds separately. In this example, PLGA nanoparticles were formulated with different PFCs (here shown in *blue* or *red*), and different fluorescent dyes (*purple* or *green*). The resultant particles were used to label primary human therapeutic cells, which were then imaged. The ^{19}F image shows two signals, due to the two resonance frequencies, which can then be resolved in line with the ^{1}H image. Fluorescence images are also shown. Reproduced with permission from Ref. [12]

hindrance for the broad application of ^{19}F MRI. Regardless, given all the advantages and unique possibilities outlined briefly in this chapter, the future of "hot-spot" imaging, particularly for cell tracking, is getting hotter.

Acknowledgment This work was supported by the European Research Council (ERC) Starting Grant (CoNQUeST Grant no. 336454) to MS.

References

1. de Vries IJ et al. Magnetic resonance tracking of dendritic cells in melanoma patients for monitoring of cellular therapy. Nat Biotechnol. 2005;23(11):1407–13.
2. Srinivas M, Heerschap A, Ahrens ET, Figdor CG, de Vries IJ. 19F MRI for quantitative in vivo cell tracking. Trends Biotechnol. 2010;28.
3. Zhong J, Narsinh K, Morel PA, Xu H, Ahrens ET. In vivo quantification of inflammation in experimental autoimmune encephalomyelitis rats using fluorine-19 magnetic resonance imaging reveals immune cell recruitment outside the nervous system. PLoS One. 2015;10(10).
4. Ahrens ET, Flores R, Xu H, Morel PA. In vivo imaging platform for tracking immunotherapeutic cells. Nat Biotechnol. 2005;23.
5. Srinivas M, Boehm-Sturm P, Figdor CG, de Vries IJ, Hoehn M. Labelling cells for in vivo cell tracking using 19F MRI. Biomaterials. 2012;33.
6. Diou O, Tsapis N, Giraudeau C, Valette J, Gueutin C, Bourasset F, et al. Long-circulating perfluorooctyl bromide nanocapsules for tumor imaging by 19FMRI. Biomaterials. 2012;33.
7. Chubarov AS, Zakharova OD, Koval OA, Romaschenko AV, Akulov AE, Zavjalov EL, et al. Design of protein homocystamides with enhanced tumor uptake properties for 19F magnetic resonance imaging. Bioorg Med Chem. 2015;23(21).
8. Yuan Y, Ge S, Sun H, Dong X, Zhao H, An L, et al. Intracellular self-assembly and disassembly of 19F nanoparticles confer respective "Off" and "On" 19F NMR/MRI signals for Legumain activity detection in zebrafish. ACS Nano. 2015;9(5).

9. Sarkar S, Paswan A, Prakas S. Liquid ventilation. Anesth Essays Res. 2014;8(3).
10. Cosco D, Fattal E, Fresta M, Tsapis N. Perfluorocarbon-loaded micro and nanosystems for medical imaging: A state of the art. J Fluorine Chem. 2015;171.
11. Ahrens ET, Zhong J. In vivo MRI cell tracking using perfluorocarbon probes and fluorine-19 detection. NMR Biomed. 2013;26(7).
12. Srinivas M, Tel J, Schreibelt G, Bonetto F, Cruz LJ, Amiri H, et al. PLGA-encapsulated perfluorocarbon nanoparticles for simultaneous visualization of distinct cell populations by 19F MRI. Nanomedicine. 2015;10(15).
13. Janjic JM, Ahrens ET. Fluorine-containing nanoemulsions for MRI cell tracking. Adv Rev. 2009;1(5).
14. Ilaria Tirotta, et al. A superfluorinated molecular probe for highly sensitive in vivo19F-MRI. J Am Chem Soc. 2014;136.
15. Ahrens ET, Helfer BM, O'Hanlon CF, Schirda C. Clinical cell therapy imaging using a perfluorocarbon tracer and fluorine-19 MRI. Magn Reson Med. 2014;72(6).
16. Kadayakkara DK, Damodaran K, Hitchens TK, Bulte JW, Ahrens ET. (19)F spin-lattice relaxation of perfluoropolyethers: Dependence on temperature and magnetic field strength (7.0-14.1T). J Magn Reson. 2014;242.
17. de Vries A, Moonen R, Yildirim M, Langereis S, Lamerichs R, Pikkemaat JA, et al. Relaxometric studies of gadolinium-functionalized perfluorocarbon nanoparticles for MR imaging. Contrast Media Mol Imaging. 2014;9(1).
18. Knight JC, Edwards PG, Paisey SJ. Fluorinated contrast agents for magnetic resonance imaging; a review of recent developments. RSC Adv. 2011;1.
19. Zhang G, Zhang R, Wen X, Li L, Li C. Micelles based on biodegradable poly(L-glutamic acid)-b-polylactide with paramagnetic Gd ions chelated to the shell layer as a potential nanoscale MRI-visible delivery system. Biomacromolecules. 2008;9(1).
20. Grogna M, Cloots R, Luxen A, Jérôme C, Passirani C, Lautram N, et al. Polymer micelles decorated by gadolinium complexes as MRI blood contrast agents: design, synthesis and properties. Polymer Chem. 2010;1.
21. Mouffouk F, Simão T, Dornelles DF, Lopes AD, Sau P, Martins J, et al. Self-assembled polymeric nanoparticles as new, smart contrast agents for cancer early detection using magnetic resonance imaging. Int J Nanomed. 2015;10.
22. Srinivas M, Cruz LJ, Bonetto F, Heerschap A, Figdor CG, de Vries IJ. Customizable, multifunctional fluorocarbon nanoparticles for quantitative in vivo imaging using 19F MRI and optical imaging. Biomaterials. 2010;31(27).
23. Patel SK, Williams J, Janjic JM. Cell Labeling for 19F MRI: new and improved approach to perfluorocarbon nanoemulsion design. Biosensors. 2013;3(3).
24. Srinivas M, Turner MS, Janjic JM, Morel PA, Laidlaw DH, Ahrens ET. In vivo cytometry of antigen-specific t cells using 19F MRI. Magn Reson Med. 2009;62(3).
25. Cromer Berman SM, Kshitiz, Wang CJ, Orukari I, Levchenko A, Bulte JW, Walczak P. Cell motility of neural stem cells is reduced after SPIO-labeling, which is mitigated after exocytosis. Magn Reson Med. 2013;69(1).
26. Muhammad G, Jablonska A, Rose L, Walczak P, Janowski M. Effect of MRI tags: SPIO nanoparticles and 19F nanoemulsion on various populations of mouse mesenchymal stem cells. Acta Neurobiol Exp. 2015;75(2).
27. Bonetto F, Srinivas M, Weigelin B, Cruz LJ, Heerschap A, Friedl P, et al. A large-scale (19)F MRI-based cell migration assay to optimize cell therapy. NMR Biomed. 2012;25(9).
28. Bönner F, Merx MW, Klingel K, Begovatz P, Flögel U, Sager M, et al. Monocyte imaging after myocardial infarction with 19F MRI at 3 T: a pilot study in explanted porcine hearts. Eur Heart J Cardiovasc Imaging. 2015;16(6).
29. Gaudet JM, Ribot EJ, Chen Y, Gilbert KM, Foster PJ. Tracking the fate of stem cell implants with fluorine-19 MRI. PLoS One. 2015;10(3).
30. Boehm-Sturm P, Aswendt M, Minassian A, Michalk S, Mengler L, Adamczak J, et al. A multimodality platform to image stem cell graft survival in the naïve and stroke-damaged mouse brain. Biomaterials. 2014;35(7).

31. Jacoby C, Temme S, Mayenfels F, Benoit N, Krafft MP, Schubert R, et al. Probing different perfluorocarbons for in vivo inflammation imaging by 19F MRI: image reconstruction, biological half-lives and sensitivity. NMR Biomed. 2013;27(3).
32. Partlow KC, Chen J, Brant JA, Neubauer AM, Meyerrose TE, Creer MH, et al. 19F magnetic resonance imaging for stem/progenitor cell tracking with multiple unique perfluorocarbon nanobeacons. FASEB J. 2007;21(8).
33. Mizukami S, Takikawa R, Sugihara F, Hori Y, Tochio H, Wälchli M, et al. Paramagnetic relaxation-based 19f MRI probe to detect protease activity. J Am Chem Soc. 2008;130(3).
34. Boehm-Sturm P, Mengler L, Wecker S, Hoehn M, Kallur T. In vivo tracking of human neural stem cells with 19F magnetic resonance imaging. PLoS One. 2011;6(12).
35. Liu L, Ye Q, Wu Y, Hsieh WY, Chen CL, Shen HH, et al. Tracking T-cells in vivo with a new nano-sized MRI contrast agent. Nanomedicine. 2012;8(8).
36. Waiczies H, Lepore S, Janitzek N, Hagen U, Seifert F, Ittermann B, et al. Perfluorocarbon particle size influences magnetic resonance signal and immunological properties of dendritic cells. PLoS One. 2011;6(7).
37. Thorek DLJ, Tsourkas A. Size, charge and concentration dependent uptake of iron oxide particles by non-phagocytic cells. Biomaterials. 2009;29(26).
38. Jain AK, Das M, Swarnakar NK, Jain S. Engineered PLGA nanoparticles: an emerging delivery tool in cancer therapeutics. Critic Rev Therapeut Drug Carrier Syst. 2011;28(1).
39. Luo R, Neu B, Venkatraman SS. Surface functionalization of nanoparticles to control cell interactions and drug release. Small. 2012;8(16).
40. Flaim SF. Pharmacokinetics and side effects of perfluorocarbon-based blood substitutes. Artif Cells Blood Subst Biotechnol. 1994;22(4).
41. Du W, Xu Z, Nyström AM, Zhang K, Leonard JR, Wooley KL. 19F- and fluorescently labeled micelles as nanoscopic assemblies for chemotherapeutic delivery. Bioconjugate Chem. 2008;19(12).
42. Hitchens TK, Ye Q, Eytan DF, Janjic JM, Ahrens ET, Ho C. 19F MRI detection of acute allograft rejection with in vivo perfluorocarbon labeling of immune cells. Magn Reson Med. 2011;65(4).
43. Ruiz-Cabello J, Barnett BP, Bottomley PA, Bulte JW. Fluorine (19F) MRS and MRI in biomedicine. NMR Biomed. 2011;24(2).
44. Maki J, Masuda C, Morikawa S, Morita M, Inubushi T, Matsusue Y, et al. The MR tracking of transplanted ATDC5 cells using fluorinated poly-l-lysine-CF3. Biomaterials. 2007;28(3).
45. Masuda C, Maki Z, Morikawa S, Morita M, Inubushi T, Matsusue Y, et al. MR tracking of transplanted glial cells using poly-l-lysine-CF3. Neurosci Res. 2006;56(2).
46. Bommerich U, Trantzschel T, Mulla-Osman S, Buntkowsky G, Bargon J, Bernarding J. Hyperpolarized 19F-MRI: parahydrogen-induced polarization and field variation enable 19F-MRI at low spin density. Phys Chem Chem Phys. 2010;12(35).

Nanoformulations for Pharmacological siRNA Delivery in Cancer

Byunghee Yoo and Zdravka Medarova

1 Introduction

The phenomenon of gene silencing by double-stranded RNA, known as small interfering RNA (siRNA), was discovered in the nematode worm (*Caenorhabditis elegans*) in 1998 [1]. The application of siRNA for posttranscriptional gene silencing was validated in mammalian cells in 2001 [2] and mice in 2002 [3]. Since then, RNA interference (RNAi) has attracted great interest because of its potential to knock down specific genes that regulate protein expression in diverse human diseases. Because of its higher specificity, lower systemic toxicity, and flexibility for target gene selection, siRNA-mediated RNAi has advantages as a therapeutic approach over conventional chemical inhibitors.

RNAi is an endogenous pathway for posttranscriptional gene silencing. It can be activated by dsRNAs, which include small interfering RNAs (siRNA), short-hairpin RNAs (shRNA), long noncoding RNAs (lncRNA), and microRNAs (miRNA). siRNA of 21–23 nucleotides consists of a passenger (sense) strand and a guide (antisense) strand. The sequence of the guide strand is designed to perfectly match the target messenger RNA (mRNA) to avoid erroneous gene silencing and is modified at the 5'-end with linkers when it needs further conjugation with nanoparticles. In principle, siRNA is exogenous and synthesized to target specific mRNA and initiate RNAi when it is delivered in the cytosol. First, siRNA is processed by Dicer (an RNase III endonuclease) and subsequently interacts with Argonaute-2, a multifunctional protein (Ago2) that is incorporated into a multiprotein complex known as the RNA-induced silencing complex (RISC). During this process, the passenger (sense) strand of siRNA is

B. Yoo (✉) • Z. Medarova (✉)
Molecular Imaging Laboratory, Department of Radiology, MGH/MIT/HMS Athinoula A. Martinos Center for Biomedical Imaging, Massachusetts General Hospital and Harvard Medical School, 13th Street, Boston, MA 02129, USA
e-mail: byunghee@nmr.mgh.harvard.edu; zmedarova@partners.org

J.W.M. Bulte, M.M.J. Modo (eds.), *Design and Applications of Nanoparticles in Biomedical Imaging*, DOI 10.1007/978-3-319-42169-8_8

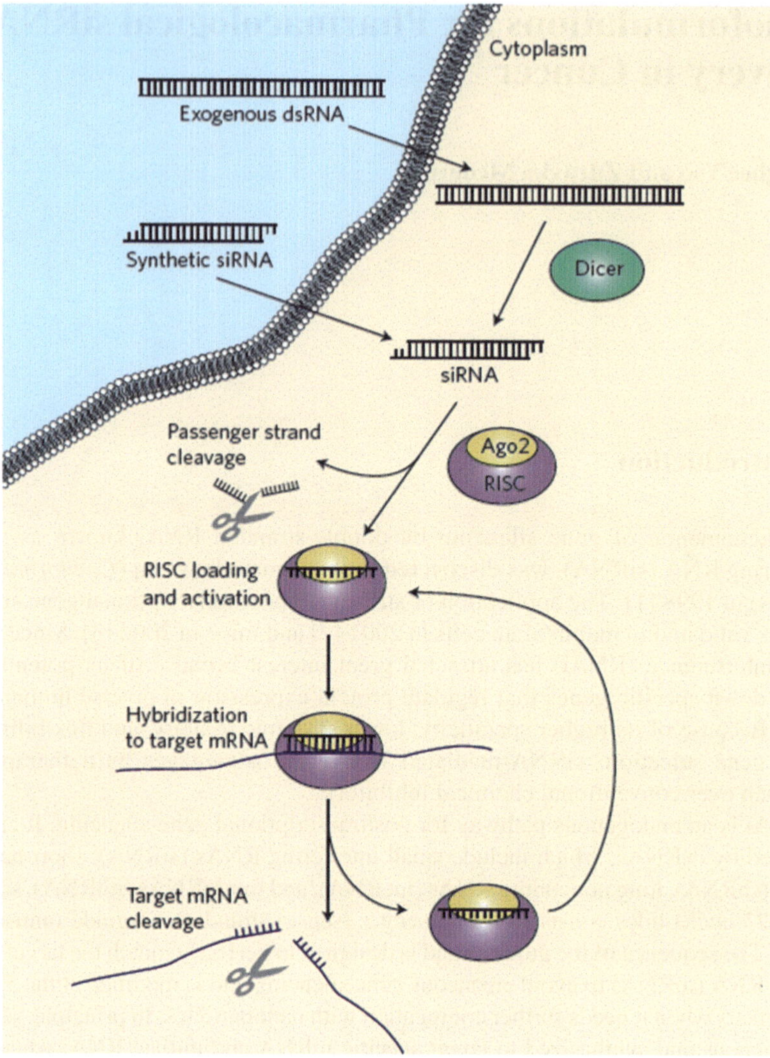

Fig. 1 Mechanism of RNA interference. Long dsRNA introduced into the cytoplasm is processed by the enzyme Dicer into 22-nt pieces with 2-nt single-stranded overhangs on the 3′ ends. The structure of synthetic siRNA mimics that of Dicer products. The siRNA guide strand is loaded into the RNA-induced silencing complex (RISC), and the passenger strand is cleaved by Argonaute-2 (Ago2). The activated RISC–guide-strand complex identifies and cleaves mRNA that is complementary to the guide strand, preventing translation and thereby silencing gene expression. Reproduced with permission from Ref. [4]

released and degraded, whereas the guide (antisense) strand binds to Ago2 and guides RISC to its complementary target mRNA. The complex of siRNA-RISC cleaves the target mRNA to silence gene expression and repress translation of mRNA. Importantly, the role of the RISC is catalytic and the complex is recycled (Fig. 1) [4].

Despite its promise, siRNA has not realized its potential as a therapeutic because of obstacles related to its delivery in vivo. After intravascular administration, siRNA is susceptible to degradation by RNase A-type nucleases or is aggregated by serum proteins in the plasma, and is rapidly eliminated by the kidneys, resulting in very short intravascular circulating half-lives in the range of seconds to minutes [5]. After entering the circulation, siRNA has to pass across vascular endothelial walls and diffuse through the extracellular matrix (ECM) in the interstitium. This represents a significant delivery hurdle since the ECM forms a dense network structure consisting of polysaccharides and fibrous proteins that accommodate macrophages. Finally, siRNA delivery is limited by the negatively charged membranes of target cells mainly because the negative charges on the phosphodiester backbone and the large molecular weight of siRNA (~13 kDa) hinder uptake by mammalian cells. In addition, siRNA could induce an innate immune response that is mediated by type I interferon and proinflammatory cytokines, further limiting the feasibility of delivering unmodified naked siRNA in vivo [6, 7].

To overcome these obstacles, a wide range of nano-sized delivery vehicles (nanodrugs) has been investigated. The nano-delivery agents can be classified into three types: lipid-based, polymer-based, and inorganic nanodrugs. Considering the obstacles listed above, an optimal nanodrug should be equipped with the physicochemical properties of protecting siRNA from degradation, elongating blood circulation, localizing in desired loci in the body, facilitating the cellular uptake, and releasing siRNA within the cytosol to initiate the RNAi process. Also, nanodrugs are expected to satisfy requirements related to their physiological safety, including lack of immunogenicity, noncoagulation with serum proteins, and low nonspecific uptake by normal tissues or cells.

To accommodate the requirements for efficient siRNA delivery, nanodrugs are mainly optimized in terms of surface charge and particle size. The outer surface of nanodrugs is usually charged with cations (positive charges) for enhanced uptake by cell membranes. However, cationic surface charges increase the chances of aggregate formation of negatively charged plasma proteins that are entrapped in the pulmonary capillary bed or taken up by the mononuclear phagocyte system. The introduction of polyethylene glycol (PEG) on the surface can neutralize the surface charge and impart a "stealth function" that avoids the interactions with plasma proteins, resulting in the elongation of intravascular circulation time and faster diffusion in the extracellular matrix [8]. In contrast, PEGylation interferes with spontaneous cellular uptake through opsonization and requires an additional moiety for receptor-mediated endocytosis. With respect to nanoparticle size, nanodrugs are limited by the requirement that the size of nanodrugs should be bigger than the pore size of the glomerular filtering system (>7 nm) to avoid renal clearance. It is also beneficial to keep the size smaller than 100 nm because the discontinuous endothelia in solid tumors are also found in some other organs including liver, spleen, and bone marrow, where nanoparticles bigger than 100 nm in diameter are entrapped [9–12].

The conventional size of nanodrugs is adjusted in the range of 10–100 nm in diameter to allow delivery to tissues through the enhanced permeation and retention (EPR) effect. This is particularly relevant for solid tumors with leaky vasculature [9, 13, 14]. Even in poorly permeable tumors, nanodrugs smaller than 50 nm can penetrate the capillary endothelium, whereas micellar nanodrugs of 70 nm are retained in the vasculature [15]. Based on these observations, the optimal size of nanodrugs can be narrowed to 10–50 nm, especially when the goal is passive delivery for cancer treatment.

In this review, we will focus on the formulation of nanodrugs for pharmacological siRNA delivery according to the categories classified above: lipid-based, polymer-based, and inorganic nanodrugs.

2 Polymeric Nanoparticles

Polymer-based nanoparticles have numerous advantages as a template for siRNA nanodrugs including thermodynamic stability, high drug loading efficiency, high cellular uptake, adjustable drug release rate, and the feasibility of surface modification. However, polymeric nanodrugs have potential issues originating from the properties of synthetic or natural polymers, such as biodegradability, biocompatibility, toxicity, and immunogenicity.

Polyethyleneimine (PEI) is one of the most commonly used polymers. PEI-based nanoparticles show high cellular uptake due to their cationic surface charges. PEI exhibits very different compatibility profiles depending on its incorporation of specific molecular structures and the choice of branched (obtained from acid-catalyzed polymerization of aziridine) vs. linear forms (obtained via polyisoxazoline precursors) [16, 17]. With an acceptable biocompatibility profile, nanodrugs formulated with linear PEI show no significant increase of proinflammatory cytokines or hepatic enzymes after systemic administration [18, 19]. Branched PEI is recognized as more toxic than linear PEI, activating inflammatory responses and susceptible to macrophage uptake and rapid clearance. Chemical modification can improve the toxicity profile of branched PEI. PEGylation of branched PEI prevented the induction of mutations and oxidative DNA damage (8-OH-dG) in FE1 lung epithelial cells, whereas alkylation showed low toxicity and enhanced knockdown efficiency due to stabilization of encapsulated siRNA [16, 20, 21].

In addition to PEGylation and alkylation, PEI has been complexed with lipids, polymers, surfactants, and inorganic materials to enhance the properties of nanodrugs [22–27]. For example, low-molecular-weight polyamine (PEI600) was utilized for the synthesis of nanoparticles by reacting with C_{15} epoxide-terminated lipids at a 14:1 molar ratio, and formulating the nanoparticles with $C_{14}PEG_{2000}$. The resulting nanodrugs formed multilamellar vesicles rather than the periodic aqueous compartments containing siRNA that make up stable nucleic-acid lipid particle formulations. The resulting nanodrugs had a diameter between 35 and 60 nm, and were delivered into endothelial cells without any target moieties. The surface charge of 7C1 nanodrugs was electrically neutral at pH 7.4 (in the blood stream), but its pK_a was 5.0. Unlike lipid and lipidoid-based nanodrugs, 7C1 nanodrugs showed high specificity of transfection to endothelial cells in vivo even at low doses, and did not significantly reduce gene expression in nontargeted cells, such as hepatocytes, peritoneal immune cells, pulmonary epithelial cells, or pulmonary immune cells (Fig. 2) [27].

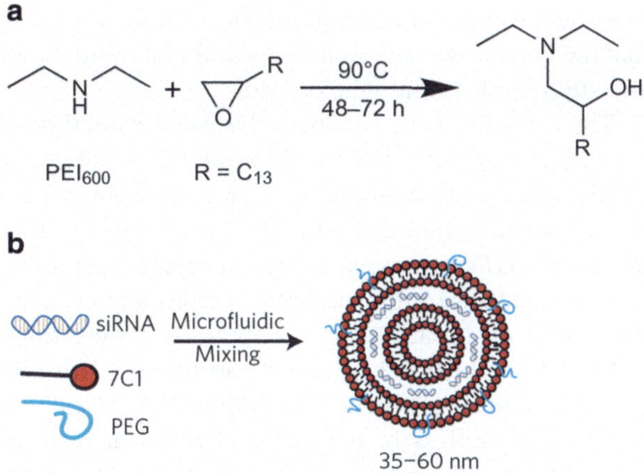

Fig. 2 Preparation of polymeric nanoparticles with low molecular weight. (**a**) 7C1 synthesis scheme. (**b**) 7C1 formulation scheme. 7C1 nanoparticles were mixed with C14PEG2000 and siRNA in a high-throughput microfluidic chamber. Reproduced with permission from Ref. [27]

Co-delivery of two siRNAs and paclitaxel was investigated by employing a two-amphiphilic polymer system of polyethyleneimine-block-poly[(1,4-butanediol)-diacrylate-β-5-hydroxyamylamine] (PEI-PDHA) and polyehtylene glycol-block-poly[(1,4-butanediol)-diacrylate-β-5-hydroxyamylamine] (PEG-PDHA) by self-assembly. PDHA is acid responsive and releases drugs in an acidic environment. Two siRNAs and paclitaxel were encapsulated in polymeric particles to form PEI-PDHA/PEG-PDHA/PTX/siSna/siTwi (PPSTs) complex nanodrugs. The size of PPSTs nanodrugs (80–140 nm) was small enough to expect passive targeting through EPR effects. PPSTs nanodrugs were administered intravenously, and the accumulation of two siRNAs and paclitaxel was confirmed in metastatic 4 T1 breast tumors in a mouse model. The administration of dual siRNAs resulted in suppression of breast cancer by preventing cancer cell migration and invasion and the effect was more significant than the administration of single siRNA [28].

An abundance of polymers has been explored for the delivery of anticancer therapeutics. Some examples include poly(lactide-co-glycolide) (PLGA), polylactide (PLA), polyglycolide, polycaprolactone (PCL), and poly(D,L-lactide). These polymers have also been chemically modified with lipids, other polymers, surfactants, and inorganic materials to improve the physicochemical properties and optimize the formulations for siRNA delivery [29–34]. For the treatment of bladder cancer, for example, PLGA nanoparticles were modified with positively charged mucoadhesive polysaccharide chitosan chains (2.5 or 20 kDa) to enhance transurothelial penetration and delivery of survivin siRNA. Surface modification of PLGA nanoparticles with short chain chitosan (NP-siSUR-2.5, 137 ± 51 nm) increased the release of siRNA by ten times compared to long chain chitosan (NP-siSUR-20, 130 ± 56 nm). Nanodrugs were delivered using passive targeting, since no targeting moieties were incorporated into the design. The higher molecular weight (long chain) chitosan entangled and trapped the negatively charged siRNA more tightly than the lower molecular weight (short chain) chitosan [34, 35]. This result implies that the balanced surface charge should be optimized for enhanced cellular uptake and siRNA release into the cytosol.

The formulation of nanoparticles using natural polymers has also been explored for the delivery of siRNAs. Representative natural polymers are chitosan, hyaluronic acids, sodium alginate, gelatin, and polypeptides [36–41]. Glycol chitosan was chemically modified with 5β-cholanic acid to form a hydrophobic core where doxorubicin (Dox) was encapsulated in self-assembling nanoparticles. Also, siRNA was conjugated to a thiolated glycol chitosan via a cleavable disulfide for release in the reductive intracellular environment. Interestingly, Dox-encapsulated CNPs (290 ± 4.5 nm) or Bcl-2 siRNA-encapsulated CNPs (301 ± 9.3 nm) exhibited similar physicochemical properties, including size, surface properties, and pH sensitive behavior, regardless of the different physical features of Dox and Bcl-2 siRNA. Also, the two different nanodrugs showed similar patterns of in vivo biodistribution and pharmacokinetics in PC3 tumor bearing mice, and mediated a dose-dependent therapeutic effect following sequential administration [41].

Dendrimers represent another common formulation for siRNA delivery. The most widely used dendrimers consist of polyamidoamine (PAMAM, Starburst™), poly(propylenemine) (PPI or DAB, Astramol™), and PEG polyester dendrons, and are functionalized with a $-NH_2$ (for oligonucleotide delivery), a –COOH (for dendrimer platinates), and an –OH (for dendrimer-derived magnetic resonance imaging contrast agents). The size of dendrimers can be regulated by stepwise growth and adjustable branching patterns. Their hydrodynamic diameter is well defined according to the degree of generations, such as generation 3 (G3, 3.1 nm), generation 4 (G4, 4.0 nm), and generation 5 (G5, 5.3 nm). The surface functional groups allow for the adjustment of surface charges and permit functionalization with siRNA, targeting probes and surface coating materials [42–47]. However, dendrimers, especially cationic dendrimers, suffer from low biocompatibility and toxicity limiting their clinical translation [48, 49].

Dendrimer-inspired nanodrugs were developed by combining PAMAM or poly(propylenimine) dendrimers of increasing generations with alkyl epoxides of various carbon chain lengths. In this scenario, the positively charged dendrimer core entrapped negatively charged siRNAs for Tie2 gene knockdown in immortal and primary endothelial cells. 1,2-dimyristoyl-sn-glycero-3-phosphoethanolamine-N-$mPEG_{2000}$ and cholesterol were incorporated with the dendrimers to generate nanodrugs with sizes in the range of 50–100 nm. Contrary to low molecular weight PEI-based polymer that was capable of targeting both lung endothelial cells and epithelial lung tumors, this formulation of Tie2 siRNA nanodrugs was specifically taken up by lung endothelial cells and not by epithelial lung tumor cells, whereas the integration of cholesterol improved gene knockdown efficiency. However, this formulation was not completely free from off-target effects in the endothelial cells of other organs because some knockdown was still observed at high doses [47].

3 Lipid-Based Nanoparticles

Lipid based nanodrugs include liposomes, solid lipid nanoparticles, stable nucleic acid-lipid nanoparticles (SNALPs), and lipoid nanoparticles. Liposomes are the most popular and extensively investigated lipid-based nanoparticles and are employed to deliver a variety of payloads, such as anti-cancer drugs, oligonucleotides, DNAs, RNAs, antigens, and proteins. The advantages of liposomes for siRNA delivery include the prevention of degradation, accumulation in tumor tissues by passive targeting, feasibility in surface modification for active targeting, and high biocompatibility for systemic delivery in animals and humans. Described in "polymeric nanodrugs," liposomes are coated with polyethylene glycol (PEG) to avoid rapid clearance by the reticuloendothelial system (RES) in liver, spleen, lungs, and bone marrow for a longer circulation half-life. Also, liposomes are formulated with polymers, lipidoids, inorganic materials, and polypeptides to improve physicochemical properties for efficient siRNA delivery [50–55].

Layer-by-layer (LbL) nanoparticles are prepared by sequential deposition of oppositely charged polymers on top of nanoparticles to build a highly stable multilayer film. This permits precise control at the nanometer-scale level to adapt a range of polycationic materials with siRNA loading and releasing, film stability, transfection efficiency, and cytotoxicity. LbL nanodrugs were developed for co-delivery of MRP1 siRNA and anticancer therapeutics by stacking siRNA-loaded LbL films atop of a doxorubicin-loaded liposome with an exterior negatively charged phospholipid membrane. Cationic poly-L-arginine was deposited in the inner most layer to reverse the surface charge of the liposome from negative to positive which can drive negative siRNA molecules to form an outer shell in the LbL film layer. The size of the nanodrugs was determined to be approximately 120 nm in diameter with a zeta potential of 55 mV and the hydrodynamic diameter increased by 5 nm with the addition of each layer. The number of siRNA molecules per nanoparticle was determined to be 3500. The LbL nanodrugs were finalized with a hyaluronic acid (2000 kDa) coat for prolonged circulations with a serum half-life of up to 28 h [55].

In addition to liposomes, solid lipid-based nanodrugs were also developed for systemic siRNA delivery with the benefit of using biodegradable and nontoxic lipids that form solid nanoparticles at a physiological body temperature. These nanodrugs include "stable nucleic acid-lipid particles (SNALPs)" and "solid-lipid nanoparticles (SLNs)." SNALPs consist of a lipid bilayer containing a mixture of cationic and fusogenic lipids that enable the cellular uptake and endosomal release of siRNA. SLNs are composed of cholesteryl ester, triglyceride, cholesterol, dioleyl phosphatidyl ethanolamine (DOPE), and 3-β-[*N*-(*N*′,*N*′-dimethylamino ethane)carbamoyl]-cholesterol (DC-cholesterol). The formulations of SNALPs and SLNs are also modified with PEG groups, lipids, surfactants, polymers, and human serum proteins, and are adjusted for the purpose of siRNA delivery [53, 56–62].

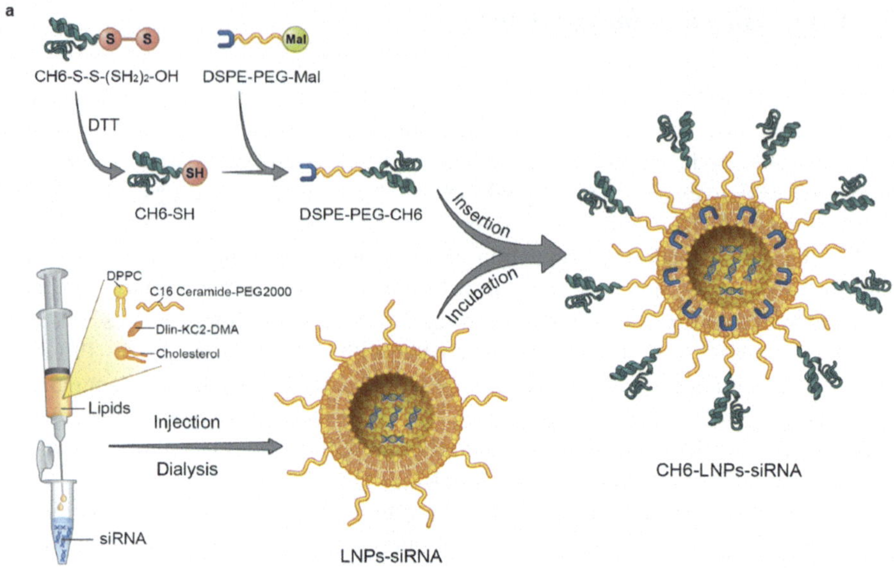

Fig. 3 Preparation of aptamer-functionalized lipid nanoparticles. LNPs are first prepared by spontaneous vesicle formation after a lipid/ethanol solution is slowly injected into siRNA buffer solution, followed by dialysis. The 3′ thiol and 2′-*O*-methyl-modified CH6 aptamer are then activated followed by conjugation to DSPE-PEG2000-Mal to form CH6-PEG2000-DSPE. Finally, CH6-PEG2000-DSPE in the form of micelles is inserted into the surface of LNPs. Reproduced with permission from Ref. [62]

Lipidoid nanoparticles are lipid-like delivery molecules conjugated with cholesterol and formulated with PEG-coated lipids for delivery of siRNAs [63–66]. For the delivery of osteogenic *Plekho1* siRNA, lipid nanoparticles (size: <90 nm) were formulated to encapsulate siRNAs by use of DPPC, C16 celarmide-PEG2000, and cholesterol, and modified with the osteoblast-specific aptamer CH6 on the surface. The siRNA encapsulating lipid nanoparticles showed in vitro osteoblast selective uptake via macropinocytosis, and down-regulated in vivo osteoblast-specific *Plekho1* gene expression, which resulted in promoted bone formation, improved bone microstructure, increased bone mass, and enhanced mechanical properties in both osteopenic and healthy rodents (Fig. 3) [62].

4 Inorganic Nanoparticles

Inorganic nanoparticles can be classified as metallic nanoparticles and nonmetallic nanoparticles. Metallic (or metal-hybridized) nanoparticles include gold [67–70], silver [71], copper [72], manganese oxide [73], and iron oxide [74, 75], as well as quantum dots [76–78]. Nonmetallic nanoparticles include silica (silicate) and calcium phosphate (hydroxyapatite) [79–84]. These inorganic nanoparticles can be modified with PEG groups, polymers, lipids, and surfactants for the enhancement of drug loading, blood circulation time, and specificity to the target biomarkers.

Magnetic nanoparticles (MNs) have been used for the purpose of contrast enhancement in magnetic resonance imaging (MRI) and have also been employed for the delivery of siRNAs. In terms of diagnostic capability and delivery of siRNA therapeutics, MNs are considered a principle template for theranostic imaging. For the investigation of drug resistance in glioma against DNA-methylating agents, amine-derivatized dextran coated MN was loaded with siRNA for silencing O^6-methylguanine methyltransferase (MGMT), which reverses the anti-glioma effects of DNA methylating anticancer drugs. The MNs were also conjugated with a chloride ion specific peptide, Chlorotoxin, for effective uptake by glioma cells. By suppressing MGMT gene expression in an intracranial glioma model, the therapeutic effects of Temozolomide (DNA-alkylating anti-glioma agent) were significantly increased (Fig. 4) [74].

Nanodrugs loaded with siRNA are taken up by cells through endocytic pathways, and are prone to entrapment within subcellular compartments, which requires an increased dosage of siRNA. Polycationic nanodrugs can enhance the cellular uptake and facilitate the escape of siRNA from endosomes by taking advantage of the "proton sponge effect" [85], but can also cause cytotoxicity. For the optimal delivery of siRNA into the cytosol, nanoparticle-stabilized nanocapsules (NPSCs) were designed based on supramolecular guanidine–carboxylate interactions between the arginine-functionalized gold nanoparticles (Arg-AuNPs) of the shell and the hydrophobic fatty acid "oil" components in the core. NPSCs were successfully shown to be effective siRNA delivery vehicles into the cytosol by their ability to silence green fluorescent protein (deGFP) and polo-like kinase (siPLK1) with concomitant cytotoxicity [86].

Mature dendritic cells (DCs) can capture tumor antigens and induce potent antigen-specific antitumor immunity in tumor-draining lymph nodes by presenting the antigens to T cells and secreting proinflammatory cytokines that enhance T cell activation. Activation of STAT3 in DCs inhibits the expression of numerous immunostimulatory molecules triggered by TLR ligands that regulate T cell activation. For the suppression of STAT3, amine-modified quantum dots were conjugated with STAT3 siRNAs and encapsulated in a PLGA matrix. After intratumoral injection, STAT3 mRNA expression was downregulated by up to 73 % and the amount of proinflammatory cytokines increased significantly [87].

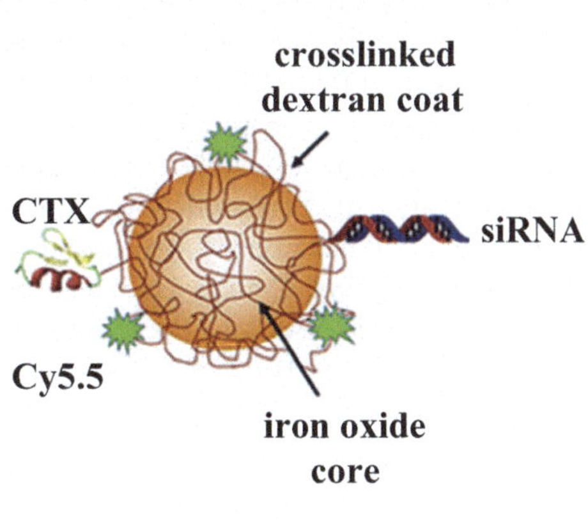

Fig. 4 Schematic structure of a nanodrug for targeted delivery of siRNA. Reproduced with permission from Ref. [74]

Porous silicon nanoparticles (pSiNPs) are actively investigated as nanoparticle-based drug delivery vehicles and have advantages in terms of biocompatibility, biodegradability, and high payload capacity [88, 89]. The pore sizes of pSiNPs can be controlled and provide criteria to classify SiNPs, such as micropores (smaller than 2 nm) through mesopores (between 2 and 50 nm), and macropores (larger than 50 nm) [90]. pSiNPs loaded with MRP1-siRNA showed efficient cellular uptake by T98G glioblastoma cells (even after a 30 min incubation) and successfully downregulated MRP1 mRNA (40 %) and protein (30 %) [91].

The delivery efficiency of silicon particles inside the body is affected by the size, shape, and surface physical and chemical properties of the nanoparticles. Discoidal silicon particles are more effectively taken up by tumor cells than spherical or cylindrical silicon particles. In addition, systemic administration of the discoidal SiNPs does not cause acute or subacute toxicity in wild-type mice [92, 93]. Polycation-functionalized porous silicon (PCPS) was prepared for the delivery siRNA. PCPS was fabricated by oxidizing the surface of the porous silicon to introduce a hydroxyl group that was used to conjugate arginine and PEI through a linker (3-aminopropyl-triethoxysilane). Loading of STAT3 siRNA was achieved through electrostatic interaction between the positively charged Arg-PEI on PCPS and the negatively charged siRNA. After uptake by breast tumor cells, STAT3 siRNA was released in the process of PCPS degradation to induce a significant 91 % knockdown of the target gene [94].

5 Future Outlook

RNA interference (RNAi) holds great potential as a therapeutic strategy for the treatment of many diseases, including cancers. RNA interference can be accomplished by two routes: the "chop-up" or "zip-up" route. In the zip-up mechanism, unnatural synthetic RNA oligonucleotides are employed to form a thermodynamically and physicochemically stable complex with target RNAs, which are not able to participate in the gene translation process. Following the chop-up mechanism, siRNAs are synthesized by the use of natural RNA building blocks and are utilized for the interference of target gene expression by binding the target gene to form a partial duplex in the sequence that is prone to RISC digestion. Therefore, siRNAs consist of natural RNA bases and are easily degraded by nucleases in sera. The role of nanoparticles is to protect siRNAs during circulation in blood vessels and diffusion in interstitium, to safely deliver siRNAs to the desired loci, and to release siRNA for silencing target mRNAs. So far, a wide range of organic, inorganic, and hybrid materials have been investigated for the purpose of siRNA delivery to the desired loci in vivo. The key challenge, however, is to translate the application of pharmacologic siRNAs into the clinic. This capability can have a transformative impact on healthcare because siRNAs provide incomparable specificity, including the capability to target single-nucleotide mutations.

In addition to their high level of specificity, siRNAs also have a broad applicability. Theoretically, we can target any genes using a modular strategy. This is important because practically every disease is influenced by genetic predisposition, including diseases that are caused directly by environmental factors. So far, a very limited number of therapeutic siRNAs have been evaluated in clinical trials for the treatment of diverse diseases, such as diabetic macular edema, immunotherapy of melanoma, myeloid leukemia, metastatic solid tumors, and liver cancer. Twenty clinical studies have been completed already and eight clinical studies are planned for the evaluation of siRNA in patients (clinicaltrials.gov). Considering the increasing number of preclinical investigations, it is reasonable to expect that the near future will see many more siRNA-based pharmacologic agents tested as monotherapies or combination therapies. However, before siRNAs can fully realize their potential in a clinical setting, it is imperative to develop optimized nanodrug platforms that are both safe and efficient.

References

1. Fire A, Xu S, Montgomery MK, Kostas SA, Driver SE, Mello CC. Potent and specific genetic interference by double-stranded RNA in Caenorhabditis elegans. Nature. 1998;391(6669): 806–11.
2. Elbashir SM, Harborth J, Lendeckel W, Yalcin A, Weber K, Tuschl T. Duplexes of 21-nucleotide RNAs mediate RNA interference in cultured mammalian cells. Nature. 2001;411(6836): 494–8.
3. McCaffrey AP, Meuse L, Pham T-TT, Conklin DS, Hannon GJ, Kay MA. Gene expression: RNA interference in adult mice. Nature. 2002;418(6893):38–9.
4. Kanasty R, Dorkin JR, Vegas A, Anderson D. Delivery materials for siRNA therapeutics. Nat Mater. 2013;12(11):967–77.
5. Soutschek J, Akinc A, Bramlage B, Charisse K, Constien R, Donoghue M, et al. Therapeutic silencing of an endogenous gene by systemic administration of modified siRNAs. Nature. 2004;432(7014):173–8.
6. Judge AD, Sood V, Shaw JR, Fang D, McClintock K, MacLachlan I. Sequence-dependent stimulation of the mammalian innate immune response by synthetic siRNA. Nat Biotechnol. 2005;23(4):457–62.
7. Hornung V, Guenthner-Biller M, Bourquin C, Ablasser A, Schlee M, Uematsu S, et al. Sequence-specific potent induction of IFN-α by short interfering RNA in plasmacytoid dendritic cells through TLR7. Nat Med. 2005;11(3):263–70.
8. Stylianopoulos T, Poh M-Z, Insin N, Bawendi MG, Fukumura D, Munn LL, et al. Diffusion of particles in the extracellular matrix: the effect of repulsive electrostatic interactions. Biophys J. 2010;99(5):1342–9.
9. Lin Q, Chen J, Zhang Z, Zheng G. Lipid-based nanoparticles in the systemic delivery of siRNA. Nanomedicine. 2014;9(1):105–20.
10. Kanasty RL, Whitehead KA, Vegas AJ, Anderson DG. Action and reaction: the biological response to siRNA and its delivery vehicles. Mol Ther. 2012;20(3):513–24.
11. Aird WC. Phenotypic heterogeneity of the endothelium I. Structure, function, and mechanisms. Circ Res. 2007;100(2):158–73.
12. Wisse E, Jacobs F, Topal B, Frederik P, De Geest B. The size of endothelial fenestrae in human liver sinusoids: implications for hepatocyte-directed gene transfer. Gene Ther. 2008;15(17): 1193–9.

13. Choi HS, Liu W, Misra P, Tanaka E, Zimmer JP, Ipe BI, et al. Renal clearance of quantum dots. Nat Biotechnol. 2007;25(10):1165–70.
14. Singh S, Sharma A, Robertson GP. Realizing the clinical potential of cancer nanotechnology by minimizing toxicologic and targeted delivery concerns. Cancer Res. 2012;72(22):5663–8.
15. Cabral H, Matsumoto Y, Mizuno K, Chen Q, Murakami M, Kimura M, et al. Accumulation of sub-100 nm polymeric micelles in poorly permeable tumours depends on size. Nat Nanotechnol. 2011;6(12):815–23.
16. Zintchenko A, Philipp A, Dehshahri A, Wagner E. Simple modifications of branched PEI lead to highly efficient siRNA carriers with low toxicity. Bioconjug Chem. 2008;19(7):1448–55.
17. Jäger M, Schubert S, Ochrimenko S, Fischer D, Schubert US. Branched and linear poly (ethylene imine)-based conjugates: synthetic modification, characterization, and application. Chem Soc Rev. 2012;41(13):4755–67.
18. Bonnet M-E, Erbacher P, Bolcato-Bellemin A-L. Systemic delivery of DNA or siRNA mediated by linear polyethylenimine (L-PEI) does not induce an inflammatory response. Pharm Res. 2008;25(12):2972–82.
19. Shim MS, Kwon YJ. Acid-responsive linear polyethylenimine for efficient, specific, and biocompatible siRNA delivery. Bioconjug Chem. 2009;20(3):488–99.
20. Beyerle A, Long AS, White PA, Kissel T, Stoeger T. Poly (ethylene imine) nanocarriers do not induce mutations nor oxidative DNA damage in vitro in MutaMouse FE1 cells. Mol Pharm. 2011;8(3):976–81.
21. Thomas M, Klibanov AM. Enhancing polyethylenimine's delivery of plasmid DNA into mammalian cells. Proc Natl Acad Sci. 2002;99(23):14640–5.
22. Alshamsan A, Hamdy S, Samuel J, El-Kadi AO, Lavasanifar A, Uludağ H. The induction of tumor apoptosis in B16 melanoma following STAT3 siRNA delivery with a lipid-substituted polyethylenimine. Biomaterials. 2010;31(6):1420–8.
23. Ghonaim HM, Li S, Blagbrough IS. Very long chain N 4, N 9-diacyl spermines: non-viral lipopolyamine vectors for efficient plasmid DNA and siRNA delivery. Pharm Res. 2009;26(1): 19–31.
24. Shen J, Yin Q, Chen L, Zhang Z, Li Y. Co-delivery of paclitaxel and survivin shRNA by pluronic P85-PEI/TPGS complex nanoparticles to overcome drug resistance in lung cancer. Biomaterials. 2012;33(33):8613–24.
25. Xiao J, Duan X, Yin Q, Miao Z, Yu H, Chen C, et al. The inhibition of metastasis and growth of breast cancer by blocking the NF-kB signaling pathway using bioreducible PEI-based/p65 shRNA complex nanoparticles. Biomaterials. 2013;34(21):5381–90.
26. Huang H, Yu H, Tang G, Wang Q, Li J. Low molecular weight polyethylenimine cross-linked by 2-hydroxypropyl-γ-cyclodextrin coupled to peptide targeting HER2 as a gene delivery vector. Biomaterials. 2010;31(7):1830–8.
27. Dahlman JE, Barnes C, Khan OF, Thiriot A, Jhunjunwala S, Shaw TE, et al. In vivo endothelial siRNA delivery using polymeric nanoparticles with low molecular weight. Nat Nanotechnol. 2014;9(8):648–55.
28. Tang S, Yin Q, Su J, Sun H, Meng Q, Chen Y, et al. Inhibition of metastasis and growth of breast cancer by pH-sensitive poly (β-amino ester) nanoparticles co-delivering two siRNA and paclitaxel. Biomaterials. 2015;48:1–15.
29. Luo G, Jin C, Long J, Fu D, Yang F, Xu J, et al. RNA interference of MBD1 in BxPC-3 human pancreatic cancer cells delivered by PLGA-poloxamer nanoparticles. Cancer Biol Ther. 2009; 8(7):594–8.
30. Pan X, Zhu Q, Sun Y, Li L, Zhu Y, Zhao Z, et al. PLGA/poloxamer nanoparticles loaded with EPAS1 siRNA for the treatment of pancreatic cancer in vitro and in vivo. Int J Mol Med. 2015;35(4):995–1002.
31. Liu P, Yu H, Sun Y, Zhu M, Duan Y. A mPEG-PLGA-b-PLL copolymer carrier for adriamycin and siRNA delivery. Biomaterials. 2012;33(17):4403–12.
32. Liu X-Q, Xiong M-H, Shu X-T, Tang R-Z, Wang J. Therapeutic delivery of siRNA silencing HIF-1 alpha with micellar nanoparticles inhibits hypoxic tumor growth. Mol Pharm. 2012; 9(10):2863–74.

33. Stigliano C, Aryal S, de Tullio MD, Nicchia GP, Pascazio G, Svelto M, et al. siRNA-Chitosan complexes in poly (lactic-co-glycolic acid) nanoparticles for the silencing of aquaporin-1 in cancer cells. Mol Pharm. 2013;10(8):3186–94.
34. Martin DT, Steinbach JM, Liu J, Shimizu S, Kaimakliotis HZ, Wheeler MA, et al. Surface-modified nanoparticles enhance transurothelial penetration and delivery of survivin siRNA in treating bladder cancer. Mol Cancer Ther. 2014;13(1):71–81.
35. Techaarpornkul S, Wongkupasert S, Opanasopit P, Apirakaramwong A, Nunthanid J, Ruktanonchai U. Chitosan-mediated siRNA delivery in vitro: effect of polymer molecular weight, concentration and salt forms. Aaps Pharmscitech. 2010;11(1):64–72.
36. Christie RJ, Matsumoto Y, Miyata K, Nomoto T, Fukushima S, Osada K, et al. Targeted polymeric micelles for siRNA treatment of experimental cancer by intravenous injection. ACS Nano. 2012;6(6):5174–89.
37. Abdelghany SM, Schmid D, Deacon J, Jaworski J, Fay F, McLaughlin KM, et al. Enhanced antitumor activity of the photosensitizer meso-tetra (N-methyl-4-pyridyl) porphine tetra tosylate through encapsulation in antibody-targeted chitosan/alginate nanoparticles. Biomacromolecules. 2013;14(2):302–10.
38. Zheng C, Zheng M, Gong P, Deng J, Yi H, Zhang P, et al. Polypeptide cationic micelles mediated co-delivery of docetaxel and siRNA for synergistic tumor therapy. Biomaterials. 2013;34(13):3431–8.
39. Kim E, Yang J, Kim H-O, An Y, Lim E-K, Lee G, et al. Hyaluronic acid receptor-targetable imidazolized nanovectors for induction of gastric cancer cell death by RNA interference. Biomaterials. 2013;34(17):4327–38.
40. Lee SJ, Yhee JY, Kim SH, Kwon IC, Kim K. Biocompatible gelatin nanoparticles for tumor-targeted delivery of polymerized siRNA in tumor-bearing mice. J Control Release. 2013;172(1):358–66.
41. Yoon HY, Son S, Lee SJ, You DG, Yhee JY, Park JH, et al. Glycol chitosan nanoparticles as specialized cancer therapeutic vehicles: Sequential delivery of doxorubicin and Bcl-2 siRNA. Sci Rep. 2014;4.
42. Taratula O, Garbuzenko OB, Kirkpatrick P, Pandya I, Savla R, Pozharov VP, et al. Surface-engineered targeted PPI dendrimer for efficient intracellular and intratumoral siRNA delivery. J Control Release. 2009;140(3):284–93.
43. Agrawal A, Min D-H, Singh N, Zhu H, Birjiniuk A, Von Maltzahn G, et al. Functional delivery of siRNA in mice using dendriworms. ACS Nano. 2009;3(9):2495–504.
44. Liu X, Liu C, Laurini E, Posocco P, Pricl S, Qu F, et al. Efficient delivery of sticky siRNA and potent gene silencing in a prostate cancer model using a generation 5 triethanolamine-core PAMAM dendrimer. Mol Pharm. 2012;9(3):470–81.
45. Pérez-Martínez FC, Carrión B, Lucío MI, Rubio N, Herrero MA, Vázquez E, et al. Enhanced docetaxel-mediated cytotoxicity in human prostate cancer cells through knockdown of cofilin-1 by carbon nanohorn delivered siRNA. Biomaterials. 2012;33(32):8152–9.
46. Cho SK, Pedram A, Levin ER, Kwon YJ. Acid-degradable core-shell nanoparticles for reversed tamoxifen-resistance in breast cancer by silencing manganese superoxide dismutase (MnSOD). Biomaterials. 2013;34(38):10228–37.
47. Khan OF, Zaia EW, Jhunjhunwala S, Xue W, Cai W, Yun DS, et al. Dendrimer-inspired nanomaterials for the in vivo delivery of siRNA to lung vasculature. Nano Lett. 2015.
48. Duncan R, Izzo L. Dendrimer biocompatibility and toxicity. Adv Drug Deliv Rev. 2005; 57(15):2215–37.
49. Tomalia DA, Naylor AM, Goddard WA. Starburst dendrimers: molecular-level control of size, shape, surface chemistry, topology, and flexibility from atoms to macroscopic matter. Angewandte Chemie Int Ed English. 1990;29(2):138–75.
50. Pirollo KF, Rait A, Zhou Q, Hwang SH, Dagata JA, Zon G, et al. Materializing the potential of small interfering RNA via a tumor-targeting nanodelivery system. Cancer Res. 2007; 67(7):2938–43.
51. Peer D, Park EJ, Morishita Y, Carman CV, Shimaoka M. Systemic leukocyte-directed siRNA delivery revealing cyclin D1 as an anti-inflammatory target. Science. 2008;319(5863): 627–30.

52. Yagi N, Manabe I, Tottori T, Ishihara A, Ogata F, Kim JH, et al. A nanoparticle system specifically designed to deliver short interfering RNA inhibits tumor growth in vivo. Cancer Res. 2009;69(16):6531–8.
53. Chen Y, Sen J, Bathula SR, Yang Q, Fittipaldi R, Huang L. Novel cationic lipid that delivers siRNA and enhances therapeutic effect in lung cancer cells. Mol Pharm. 2009;6(3):696–705.
54. Wang Y, Zhang L, Guo S, Hatefi A, Huang L. Incorporation of histone derived recombinant protein for enhanced disassembly of core-membrane structured liposomal nanoparticles for efficient siRNA delivery. J Control Release. 2013;172(1):179–89.
55. Deng ZJ, Morton SW, Ben-Akiva E, Dreaden EC, Shopsowitz KE, Hammond PT. Layer-by-layer nanoparticles for systemic codelivery of an anticancer drug and siRNA for potential triple-negative breast cancer treatment. ACS Nano. 2013;7(11):9571–84.
56. Resnier P, David S, Lautram N, Delcroix GJ-R, Clavreul A, Benoit J-P, et al. EGFR siRNA lipid nanocapsules efficiently transfect glioma cells in vitro. Int J Pharm. 2013;454(2): 748–55.
57. Lee JB, Zhang K, Tam YYC, Tam YK, Belliveau NM, Sung VY, et al. Lipid nanoparticle siRNA systems for silencing the androgen receptor in human prostate cancer in vivo. Int J Cancer. 2012;131(5):E781–90.
58. Yang X-Z, Dou S, Wang Y-C, Long H-Y, Xiong M-H, Mao C-Q, et al. Single-step assembly of cationic lipid–polymer hybrid nanoparticles for systemic delivery of siRNA. ACS Nano. 2012;6(6):4955–65.
59. Hatakeyama H, Akita H, Ito E, Hayashi Y, Oishi M, Nagasaki Y, et al. Systemic delivery of siRNA to tumors using a lipid nanoparticle containing a tumor-specific cleavable PEG-lipid. Biomaterials. 2011;32(18):4306–16.
60. Li L, Wang R, Wilcox D, Zhao X, Song J, Lin X, et al. Tumor vasculature is a key determinant for the efficiency of nanoparticle-mediated siRNA delivery. Gene Ther. 2012;19(7):775–80.
61. Sahay G, Querbes W, Alabi C, Eltoukhy A, Sarkar S, Zurenko C, et al. Efficiency of siRNA delivery by lipid nanoparticles is limited by endocytic recycling. Nat Biotechnol. 2013;31(7): 653–8.
62. Liang C, Guo B, Wu H, Shao N, Li D, Liu J, et al. Aptamer-functionalized lipid nanoparticles targeting osteoblasts as a novel RNA interference-based bone anabolic strategy. Nat Med. 2015;21(3):288–94.
63. Wang M, Alberti K, Varone A, Pouli D, Georgakoudi I, Xu Q. Enhanced intracellular siRNA delivery using bioreducible lipid-like nanoparticles. Adv Healthcare Mater. 2014;3(9): 1398–403.
64. Brock A, Krause S, Li H, Kowalski M, Goldberg MS, Collins JJ, et al. Silencing HoxA1 by intraductal injection of siRNA lipidoid nanoparticles prevents mammary tumor progression in mice. Sci Transl Med. 2014;6(217):217ra2.
65. Whitehead KA, Dorkin JR, Vegas AJ, Chang PH, Veiseh O, Matthews J, et al. Degradable lipid nanoparticles with predictable in vivo siRNA delivery activity. Nat Commun. 2014;5.
66. Goldberg MS, Xing D, Ren Y, Orsulic S, Bhatia SN, Sharp PA. Nanoparticle-mediated delivery of siRNA targeting Parp1 extends survival of mice bearing tumors derived from Brca1-deficient ovarian cancer cells. Proc Natl Acad Sci. 2011;108(2):745–50.
67. Kim HJ, Takemoto H, Yi Y, Zheng M, Maeda Y, Chaya H, et al. Precise engineering of siRNA delivery vehicles to tumors using polyion complexes and gold nanoparticles. ACS Nano. 2014;8(9):8979–91.
68. DeLong RK, Akhtar U, Sallee M, Parker B, Barber S, Zhang J, et al. Characterization and performance of nucleic acid nanoparticles combined with protamine and gold. Biomaterials. 2009;30(32):6451–9.
69. Chen AM, Taratula O, Wei D, Yen H-I, Thomas T, Thomas T, et al. Labile catalytic packaging of DNA/siRNA: control of gold nanoparticles "out" of DNA/siRNA complexes. ACS Nano. 2010;4(7):3679–88.
70. Huschka R, Barhoumi A, Liu Q, Roth JA, Ji L, Halas NJ. Gene silencing by gold nanoshell-mediated delivery and laser-triggered release of antisense oligonucleotide and siRNA. ACS Nano. 2012;6(9):7681–91.

71. Lin J, Huang Z, Wu H, Zhou W, Jin P, Wei P, et al. Inhibition of autophagy enhances the anticancer activity of silver nanoparticles. Autophagy. 2014;10(11):2006–20.
72. Laha D, Pramanik A, Maity J, Mukherjee A, Pramanik P, Laskar A, et al. Interplay between autophagy and apoptosis mediated by copper oxide nanoparticles in human breast cancer cells MCF7. Biochim Biophys Acta. 2014;1840(1):1–9.
73. Bae KH, Lee K, Kim C, Park TG. Surface functionalized hollow manganese oxide nanoparticles for cancer targeted siRNA delivery and magnetic resonance imaging. Biomaterials. 2011; 32(1):176–84.
74. Yoo B, Ifediba MA, Ghosh S, Medarova Z, Moore A. Combination treatment with theranostic nanoparticles for glioblastoma sensitization to TMZ. Mol Imaging Biol. 2014;16(5):680–9.
75. Ghosh SK, Yigit MV, Uchida M, Ross AW, Barteneva N, Moore A, et al. Sequence-dependent combination therapy with doxorubicin and a survivin-specific small interfering RNA nanodrug demonstrates efficacy in models of adenocarcinoma. Int J Cancer. 2014;134(7):1758–66.
76. Derfus AM, Chen AA, Min D-H, Ruoslahti E, Bhatia SN. Targeted quantum dot conjugates for siRNA delivery. Bioconjug Chem. 2007;18(5):1391–6.
77. Yezhelyev MV, Qi L, O'Regan RM, Nie S, Gao X. Proton-sponge coated quantum dots for siRNA delivery and intracellular imaging. J Am Chem Soc. 2008;130(28):9006–12.
78. Jung J, Solanki A, Memoli KA, Kamei K, Kim H, Drahl MA, et al. Selective inhibition of human brain tumor cells through multifunctional quantum-dot-based siRNA delivery. Angew Chem. 2010;122(1):107–11.
79. Xia T, Kovochich M, Liong M, Meng H, Kabehie S, George S, et al. Polyethyleneimine coating enhances the cellular uptake of mesoporous silica nanoparticles and allows safe delivery of siRNA and DNA constructs. ACS Nano. 2009;3(10):3273–86.
80. Tanaka T, Mangala LS, Vivas-Mejia PE, Nieves-Alicea R, Mann AP, Mora E, et al. Sustained small interfering RNA delivery by mesoporous silicon particles. Cancer Res. 2010;70(9): 3687–96.
81. Giger EV, Castagner B, Räikkönen J, Mönkkönen J, Leroux JC. siRNA transfection with calcium phosphate nanoparticles stabilized with PEGylated chelators. Adv Healthcare Mater. 2013;2(1):134–44.
82. Tseng Y-C, Xu Z, Guley K, Yuan H, Huang L. Lipid–calcium phosphate nanoparticles for delivery to the lymphatic system and SPECT/CT imaging of lymph node metastases. Biomaterials. 2014;35(16):4688–98.
83. Xie Y, Qiao H, Su Z, Chen M, Ping Q, Sun M. PEGylated carboxymethyl chitosan/calcium phosphate hybrid anionic nanoparticles mediated hTERT siRNA delivery for anticancer therapy. Biomaterials. 2014;35(27):7978–91.
84. Parodi A, Haddix SG, Taghipour N, Scaria S, Taraballi F, Cevenini A, et al. Bromelain surface modification increases the diffusion of silica nanoparticles in the tumor extracellular matrix. ACS Nano. 2014;8(10):9874–83.
85. Boussif O, Lezoualc'h F, Zanta MA, Mergny MD, Scherman D, Demeneix B, et al. A versatile vector for gene and oligonucleotide transfer into cells in culture and in vivo: polyethylenimine. Proc Natl Acad Sci. 1995;92(16):7297–301.
86. Jiang Y, Tang R, Duncan B, Jiang Z, Yan B, Mout R, et al. Direct cytosolic delivery of siRNA using nanoparticle-stabilized nanocapsules. Angew Chem Int Ed. 2015;54(2):506–10.
87. Kim JH, Noh YW, Heo MB, Cho MY, Lim YT. Multifunctional hybrid nanoconjugates for efficient in vivo delivery of immunomodulating oligonucleotides and enhanced antitumor immunity. Angewandte Chemie. 2012;124(38):9808–11.
88. Tasciotti E, Liu X, Bhavane R, Plant K, Leonard AD, Price BK, et al. Mesoporous silicon particles as a multistage delivery system for imaging and therapeutic applications. Nat Nanotechnol. 2008;3(3):151–7.
89. Bimbo LM, Sarparanta M, Mäkilä E, Laaksonen T, Laaksonen P, Salonen J, et al. Cellular interactions of surface modified nanoporous silicon particles. Nanoscale. 2012;4(10): 3184–92.
90. Anglin EJ, Cheng L, Freeman WR, Sailor MJ. Porous silicon in drug delivery devices and materials. Adv Drug Deliv Rev. 2008;60(11):1266–77.

91. Wan Y, Apostolou S, Dronov R, Kuss B, Voelcker NH. Cancer-targeting siRNA delivery from porous silicon nanoparticles. Nanomedicine. 2014;9(15):2309–21.
92. Decuzzi P, Godin B, Tanaka T, Lee S-Y, Chiappini C, Liu X, et al. Size and shape effects in the biodistribution of intravascularly injected particles. J Control Release. 2010;141(3):320–7.
93. Xu R, Huang Y, Mai J, Zhang G, Guo X, Xia X, et al. Multistage vectored siRNA targeting ataxia-telangiectasia mutated for breast cancer therapy. Small. 2013;9(9-10):1799–808.
94. Shen J, Xu R, Mai J, Kim H-C, Guo X, Qin G, et al. High capacity nanoporous silicon carrier for systemic delivery of gene silencing therapeutics. ACS Nano. 2013;7(11):9867–80.

Forming Magnetosome-Like Nanoparticles in Mammalian Cells for Molecular MRI

Donna E. Goldhawk, Neil Gelman, R. Terry Thompson, and Frank S. Prato

1 Introduction

As an understanding of the molecular basis of disease becomes crucial for treatment and diagnosis, there is a growing need to noninvasively image these processes in preclinical animal models and human clinical trials. However, in order to track molecular activity effectively using current medical imaging platforms, increased sensitivity of detection as well as improved spatial and temporal resolution are needed. With optical imaging technology, such as bioluminescence and fluorescence, the use of reporter genes has been very successful for interrogating molecular activity in cells and very small animal models [1–3]. As optical imaging is limited due to light scatter and attenuation, in vivo reporter gene imaging of larger animals currently relies on nuclear medicine and magnetic resonance (MR) reporter genes. In addition, there is growing evidence that results from small animal imaging cannot be easily translated to humans. This difficulty relates not only to species-specific differences [4, 5] but also to differences in imaging platform, wherein the

D.E. Goldhawk (✉) • F.S. Prato
Imaging Program, Lawson Health Research Institute,
268 Grosvenor Street, London, ON, Canada, N6A 4V2

Medical Biophysics, Western University, London, ON, Canada

Collaborative Graduate Program in Molecular Imaging, Western University,
London, ON, Canada
e-mail: dgoldhawk@lawsonimaging.ca

N. Gelman • R.T. Thompson
Imaging Program, Lawson Health Research Institute,
268 Grosvenor Street, London, ON, Canada, N6A 4V2

Medical Biophysics, Western University, London, ON, Canada

J.W.M. Bulte, M.M.J. Modo (eds.), *Design and Applications of Nanoparticles in Biomedical Imaging*, DOI 10.1007/978-3-319-42169-8_9

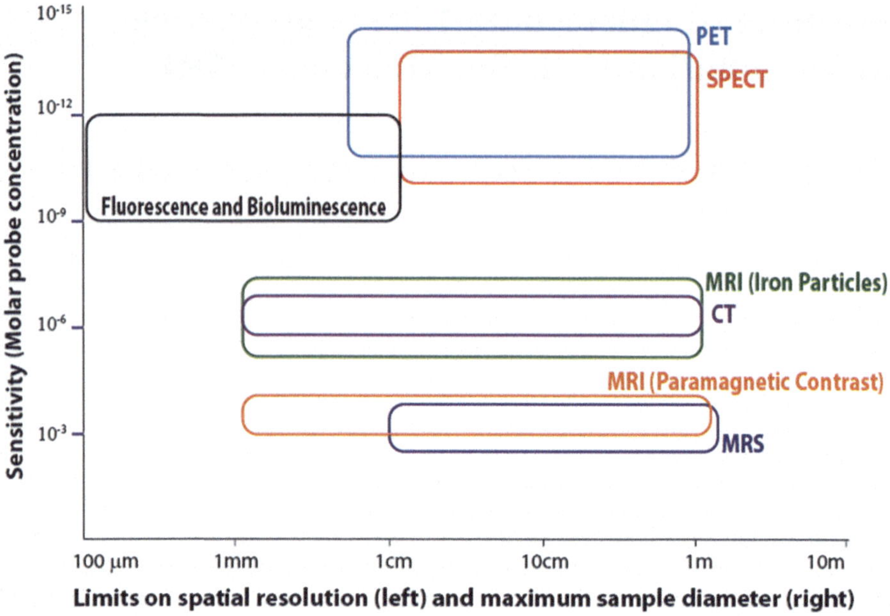

Fig. 1 Comparison of the capability of different large animal and human molecular imaging modalities. The matrix uses a log-log scale to indicate sensitivity on the vertical axis and spatial resolution on the horizontal axis. Maximum resolution corresponds to the left limit on the horizontal axis and maximum sample diameter corresponds to the right limit. The sensitivity range for PET is taken from Meikle et al. [71] and spatial resolution from Moses [6]. The sensitivity for SPECT is also taken from Meikle et al. [71] and spatial resolution is estimated at 1.2 cm; however, this is dependent on collimator choice and distance from collimator to object [72]. For fluorescence and bioluminescence imaging, resolution limit can be higher than the optical wavelength [73], i.e., much better than 100 μm, but decreases very quickly for an object of any thickness. A 1 cm limit for thickness is generous and corresponds to hybrid methods such as photoacoustic imaging [74]. For MRI with paramagnetic contrast agents, 1 mm^3 resolution at 3 T is easily achievable, with sample size being limited by available magnet bore [75]. For MRI with iron particles, refer to the section on estimated limits of sensitivity of a MRI reporter gene. For magnetic resonance spectroscopy (MRS), the spatial resolution for ^{31}P, ^{19}F, ^{23}Na, or ^{1}H (from non-water protons) is limited to approximately 1 cm due to gyromagnetic ratio and/or concentration of the isotope [76]. The values for CT were taken from Gore et al. [76]

resolution of small animal imaging does not match the scale of large animal/human systems [6]. There is also evidence that large animal studies (e.g., in dogs, pigs) are more indicative of the human condition [7] and will improve translational efficiency. To image on this clinical scale, only nuclear medicine reporter genes currently offer the needed sensitivity (comparable to optical); however, the spatial resolution is limited to about 64 mm^3 (Fig. 1). To capitalize on the superb spatial (approximately 1 mm^3 on clinical scanners) and temporal resolution of magnetic resonance imaging (MRI), further development of a magnetite-based MR reporter gene would substantially improve molecular imaging in both small and large animals, providing a route for seamless translation of medical imaging technology to human care.

The idea of using magnetotactic bacterial genes as noninvasive reporters of cellular activity for molecular MRI has recently been put forward [8–10]. Magnetotactic bacteria form magnetosomes [11, 12], membrane-enclosed iron biominerals that respond to the earth's magnetic field and enable magnetotaxis. With these attributes, the motile microaerophilic bacteria may navigate toward their preferred oxic-anoxic zones in aquatic sediments [13]. Magnetosomes are also similar in size and magnetic properties [14] to superparamagnetic iron oxide (SPIO) nanoparticles. While the latter have been used successfully to track cells in both research and clinical settings [15–18], within the genetic determinants of magnetosome synthesis is an opportunity to specify MR contrast as a direct response of select gene expression.

Approximately 20 years ago, a few reports were published about genes related to the magnetic properties of species of *Alphaproteobacteria* [19, 20]; although, it is only in the last decade that a clearer definition of the magnetosome and its constituent proteins has emerged [21]. The magnetosome is formed by a group of nonessential genes, suggesting that this bacterial structure is dispensable and confers an auxiliary function to the cell, i.e., magnetotaxis. When present, the magnetosome comprises an iron biomineral that is compartmentalized within a specialized lipid bilayer, protecting the cell from iron toxicity and confining the biomineral to a defined subcellular location. As detailed below, the magnetosome membrane contains a number of proteins that direct its location and crystal composition, size, and shape. In short, the magnetosome is an ideal structure by which cellular and molecular MRI may be refined.

Recent progress in defining the magnetosome in molecular terms provides an opportunity to further develop genetically engineered, MR contrast for effective molecular MRI [22, 23]. Such a tool would address the critical need to identify molecular activities that define the early stages of disease progression, ahead of the irreversible damage to tissue that leads to chronic illness. This is where the true strength of noninvasive reporter gene expression lies. The ability to detect transcription factor activity that prompts disease-related changes in gene expression is the key to understanding many, if not most medical conditions, including cancer [24, 25], inflammation [26, 27], and the fibrosis that leads to heart disease [28, 29]. Effective use of MRI reporter gene expression vectors, which create and strictly regulate magnetosome-like particles in mammalian cells, could provide the spatial and temporal information necessary to track disease processes and influence healthcare management and rate of cure.

This chapter describes recent progress in understanding how the magnetosome is formed in bacteria and how these mechanisms may be adapted to the formation of magnetosome-like particles in mammalian cells. From an MR imaging perspective, the expression of magnetotactic bacterial genes *magA* and *mms6* in mammalian cells provides the basis for a discussion on future development of MR detection methods, needed to optimize the use of gene-based MR contrast and its application in diagnostic medical imaging.

2 Design

2.1 Formation of Magnetosome-Like Nanoparticles in Mammalian Cells

To date, reports assessing the function of magnetotactic bacterial protein expression in mammalian cells have centered around MagA and to a lesser degree Mms6. To give perspective to this body of work, we first describe the bacterial magnetosome compartment that has inspired this approach to the development of gene-based MR contrast. Based on the current understanding of magnetosome formation, we then categorize the genes identified in terms of essential versus auxiliary function(s). Finally, we highlight useful features for the design of magnetosome-like nanoparticles in mammalian cells before discussing their applications in MRI.

Magnetosomes are subcellular structures encoded by approximately 30 genes, many of which reside on a conserved magnetosome genomic island (MAI), are not essential for survival, and are not expressed when bacterial cells are grown in nutrient-rich broth and have little need for magnetotaxis [30]. Deletion of either the MAI or one of its gene clusters, the *mamAB* operon [31], results in the loss of magnetosome formation [32] and underlines the important regulatory role of select magnetosome genes. As the structure and activity of individual magnetosome-associated proteins have been reported, models of magnetosome assembly have been proposed [12, 33] and refined. A recent model outlines four main stages: (a) vesicle formation, (b) magnetosome protein sorting, (c) cytoskeletal attachment, and (d) biomineralization [12]. Figure 2 depicts this process and identifies putative roles of select genes in membrane biogenesis and recruitment of proteins that facilitate vesicle formation, organization of magnetosomes into a chain, initiation of biomineralization, and definition of the mature crystal structure. Over the last 10 years, much evidence has accumulated indicating that magnetosome formation is an ordered process and relies on specific protein interactions. As the understanding of these protein activities becomes clearer, so does the means by which this technology can be adapted for medical imaging, among other applications. Here, we provide an MR imaging perspective and assess the genes involved in magnetosome synthesis in terms of essential versus auxiliary functions. With this categorization, we draw on one of the best cellular models of biomineralization to provide a context for continued development of the next stage of mammalian cell tracking and reporter gene expression for MRI.

2.2 Formation of a Magnetosome-Like Vesicle

All organisms require iron and carefully manage its redox chemistry through an elaborate set of regulatory mechanisms [22, 34]. Although iron is an essential cofactor for the function of many proteins, in general iron biominerals are not. Where they occur

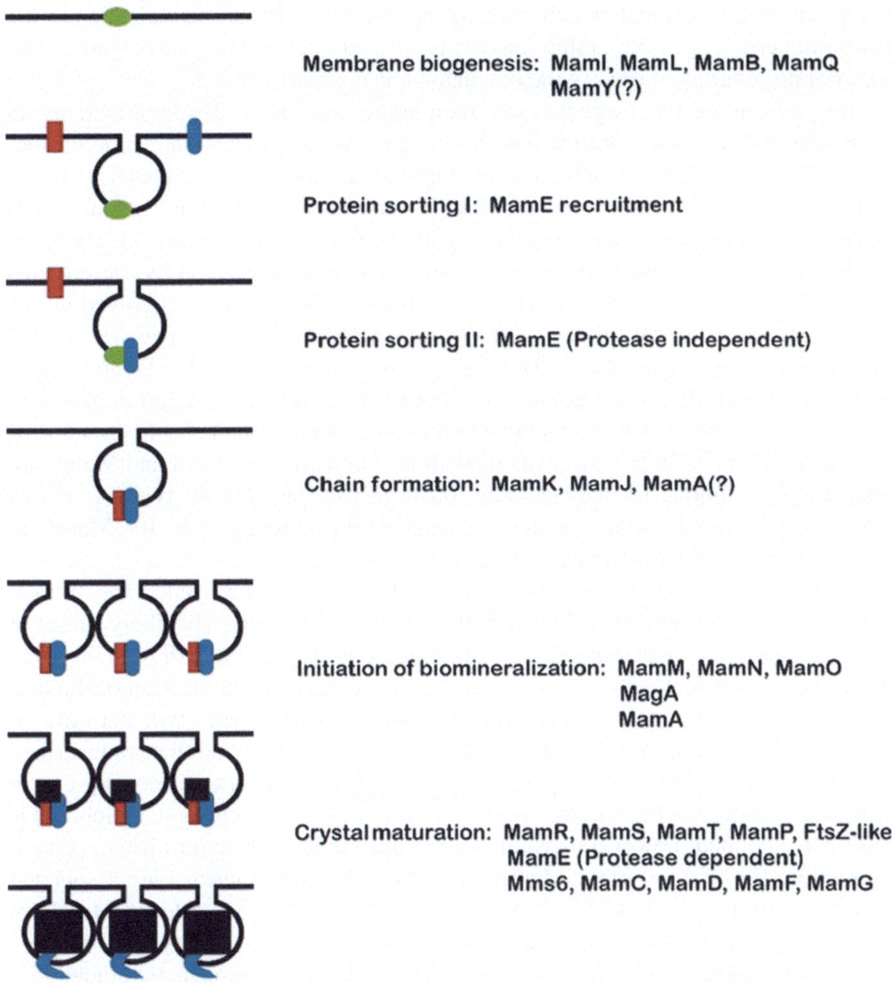

Fig. 2 Hypothetical model of magnetosome formation. Partial characterization of magnetotactic bacterial genes, particularly from the *Alphaproteobacteria*, provides further support for protein-directed assembly of the magnetosome. Largely based on genes located on the magnetosome genomic island found in multiple species of magnetotactic bacteria, the depicted stages of magnetosome synthesis include magnetosome membrane biogenesis through recruitment of the needed proteins for vesicle formation, arrangement of these magnetosome vesicles into a chain, followed by initiation and maturation of the iron biomineral. In the latter stage, factors that control crystal size and morphology vary among different classes of magnetotactic bacteria. Reproduced with permission from Ref. [12]

naturally, even as stored in ferritin, the iron biomineral is invariably sequestered to protect the cell from its potential toxicity. Accordingly, in magnetotactic bacteria the membrane-enclosed vesicle that will sequester an iron biomineral is recognized as the first step in magnetosome synthesis. Thus, in the design of magnetosome-like nanoparticles for mammalian cell tracking, appropriate compartmentalization of the iron biomineral is an essential step and may be refined by examining the magnetotactic bacterial protein(s) that specify the magnetosome compartment.

Biosynthesis of the magnetosome membrane was originally identified as an invagination of the inner plasma membrane in *Magnetospirillum magneticum* species AMB-1 [35]. However, a broader examination of magnetotactic bacteria reveals multiple arrangements of magnetosomes [13] and raises the possibility that not all magnetosomes are, or remain, associated with the plasma membrane [12]. The independent existence of magnetosome vesicles is most likely defined by the proteins that sort to this location and specify vesicle function. Key proteins involved in this process are encoded by genes located within a DNA cluster that is widespread among classes of magnetotactic bacteria [36], the *mamAB* operon [31], and whose expression is coordinately regulated. Of these genes, individual deletion of *mamI*, *mamL*, *mamQ*, and *mamB* results in no magnetosome membrane; however, none of these genes alone is sufficient for its formation. Interestingly, MamI and MamL are small proteins unique to magnetotactic bacteria [32] and MamL is not found in greigite-producing *Deltaproteobacteria*, suggesting a unique role for MamL in magnetite-producing *Proteobacteria* [37].

It remains to be seen what combination of genes may be needed for optimal expression of a magnetosome-like particle in mammalian cells. The likely subset of magnetosome genes will depend on the functionality desired but will probably possess a root structure that provides the scaffold for compartmentalization of the iron biomineral. With this scaffold, the recruitment of specific genes will not only be feasible but also programmable, delivering the type of MR signal that is prescribed by selective expression of magnetosome protein (see reporter gene expression below). The nature of the required protein sorting to the magnetosome membrane is still poorly defined; however, several studies substantiate a role for protein-protein interactions [38]. For example, MamB interacts with other magnetosome-associated membrane proteins, MamM and MamE [39]. The important role of MamE in recruiting additional magnetosome proteins to the membrane for crystal formation has recently emerged [40]. Deletion of *mamE* results in a nonmagnetic mutant that can nevertheless form empty magnetosome vesicles [12]. It appears that MamE provides a link to biomineralization partly through its interactions with MamI and MamB [32]. These interactions also likely contribute to the correct orientation of magnetosome proteins in the membrane so that crystallization is appropriately initiated within the vesicle. Separate functional domains of MamE have also been partially characterized [40]. Putative serine protease and heme-binding activity is associated with the N-terminal domain while protease-independent function in the C-terminal domain may be principally involved in recruiting magnetosome membrane protein(s). In some species of *Deltaproteobacteria*, the N- and C-terminal domains of MamE are encoded by separate genes [37], suggesting that strategies for streamlining mammalian expression of magnetosome genes may include the expression of functional gene fragments.

2.3 Formation of a Magnetosome-Like Biomineral

Critical magnetosome genes for iron biomineralization include *mamE, mamO, mamM,* and perhaps *mamN*. These genes from the *mamAB* operon appear to facilitate the initiation of iron biomineralization but are not sufficient for obtaining the final size and shape of the desired crystal structure [12]. The distinction between early and late events in biomineralization again reinforces the ordered nature of this process (Fig. 2). As a consequence, the optimal expression of magnetosome-like nanoparticles in mammalian cells will benefit from a clearer understanding of the temporal relationship between the required biomineralization activities. Once initiated, the controlled expression of magnetosome genes in the correct sequence will provide landmarks by which changes in MR contrast may be measured and correlated to discrete cellular activities. Eventually, we envision this strategy would include complementary expression systems that when activated create a more refined magnetosome-like particle than was possible when individually expressed.

Based on the study of magnetotactic bacteria harboring deletions of select magnetosome genes, attempts are being made to understand which genes are essential to the basic magnetosome structure and which genes have more auxiliary roles, for example in defining the location and configuration of magnetosomes within the cell or in specifying the nature of the biomineral. Many of the MAI genes involved in crystal maturation have such auxiliary functions; their absence mitigates but does not abrogate magnetosome formation. The entire *mamCGDF* operon encodes some of the most abundant magnetosome proteins that are nevertheless not essential for the formation of a more rudimentary particle [41]. Likewise, the *mms6* operon is directly involved in iron biomineralization and its absence diminishes the process but does not completely interrupt it [32]. In addition, several more proteins encoded within the *mamAB* operon appear to have a role in the final stage(s) of iron biomineralization and could be excluded without interrupting the synthesis of the core structure. The presence of all these MAI genes provides a powerful argument for the feasibility of producing a magnetosome-like particle in mammalian cells using a subset of magnetotactic bacterial genes. Furthermore, within this subset of gene products are protein domains that may function similarly to known mammalian proteins (or their functional domains) and could be used in combination with unique bacterial magnetosome proteins. Zeytuni et al. have described the similarity between MamM, a putative cation diffusion facilitator (CDF) protein, and a mammalian member of the CDF superfamily implicated in type II diabetes [42]. They generated mutant MamM to model CDF polymorphisms present in human disease and the manner in which select mutations influence cation transport, in this case using the change(s) in magnetosome biomineralization to monitor the change in CDF function. This work establishes the broad utility of magnetosome synthesis for biomedical research, with important ramifications for the development of gene-based MR contrast using magnetosome-like nanoparticles.

There are homologues to several of the genes on the *mamAB* operon in other classes of magnetotactic bacteria. Species of *Deltaproteobacteria* that produce bullet-shaped crystals of greigite and/or magnetite have homologues of *mamI*, *mamL*, and *mamM* [37]. By comparing genomes, Lefèvre et al. have suggested that *mamA*, *mamB*, *mamE*, *mamK*, *mamO*, *mamP*, and *mamQ* are involved in the synthesis of all types of magnetosomes, whether greigite or magnetite. In addition, *mamI*, *mamL*, and *mamM* may be specific to magnetite crystals while a distinct set of genes, termed *magnetosome-associated Deltaproteobacteria* (*mad*) genes, specifies greigite biomineralization. These interesting projections are largely based on nucleotide and amino acid sequence alignments, which provide a useful roadmap for understanding magnetosome synthesis but will require experimental validation.

The best studied magnetosomes typically contain magnetite (Fe_3O_4) in a cubo-octahedral crystal [43]. However, the dynamic nature of magnetosome synthesis includes different types of biominerals, varying in composition (e.g., greigite, Fe_3S_4) [13, 21], crystal structure, and size [37]. For MRI, the ideal size of a magnetosome-like particle in mammalian cells may be smaller than that needed to establish a single magnetic domain. While larger biominerals might be required for MR-guided movement or thermal ablation [44], many applications such as magnetic particle imaging (MPI, discussed below) place constraints on biomineral size and composition. Hence, not every aspect of the bacterial magnetosome should necessarily be reproduced for mammalian cell tracking. By regulating the formation of a magnetosome-like particle in mammalian cells, we could draw on select features of the nanoparticle and avoid functions that are not indicated for a given application, such as unwanted heating or movement that might disrupt tissue at higher field strengths. Both thermal and kinetic properties may depend on the arrangement of magnetosome(-like) particles within the cell. Since MamJ and MamK have principle roles in chain formation, their optional expression may provide added versatility and would certainly streamline the number of genes required to create magnetosome-like particles in mammalian cells.

Recently, Kolinko et al. described the stepwise expression of MAI gene clusters that closely recapitulated the magnetosome structure in a previously nonmagnetic bacterium, *Rhodospirillum rubrum* [45]. This report provides further evidence that a subset of magnetosome genes may be used to impart magnetic properties and that potentially all types of cells, from bacteria to mammals, may accommodate this nanoparticle without cytotoxic consequences.

3 Applications

3.1 MagA-Derived Iron-Labeling and MR Contrast

A putative iron transport protein, MagA, has been cloned from both MS-1 [10] and AMB-1 [8] species of magnetotactic bacteria and shown to increase MR contrast in stably transfected mammalian cells, in response to an iron supplement. Compared to overexpression of a modified form of ferritin, lacking iron response elements to enable continuous expression, MagA-derived MR contrast appears sooner in mouse tumor xenografts growing subcutaneously from transplanted cells and with greater contrast to noise ratio (CNR) [46]. An in vitro analysis of MR relaxation rates confirmed that iron-supplemented MagA-expressing cells provide significant increases in transverse relaxation rates (R_2*, R_2, and R_2'), with little or no change in longitudinal relaxation [47]. Elemental iron analysis in these cells also correlated an increase in iron content with the increase in transverse relaxation rate and the reversible R_2' component in particular [22].

To better understand the mechanisms of transverse relaxation in MagA-expressing iron-labeled cells, Lee et al. performed nuclear magnetic resonance experiments to study the relationship between R_2 and interecho time (2τ), as assessed with a Carr-Purcell-Meiboom-Gill (CPMG) sequence [48]. The R_2 versus 2τ curves were analyzed using a previously developed numerical model [49] that provided estimates of the so-called spatial correlation length, representing the distance scale of microscopic magnetic field variation. In a model where this magnetic field variation is caused by uniformly magnetized spheres within tissue, they showed that the spatial correlation length is approximately equal to the sphere radius. Using this method, the spatial correlation length estimated by Lee et al. in iron-supplemented, MagA-expressing MDA-MB-435 cells was on the order of 250–450 nm, reasonably consistent with transmission electron micrographs of MagA-expressing 293FT cells [10]. In these micrographs, it should be noted that the size of dense core clusters and the size of individual particles within these clusters (estimated at 3–5 nm [10]) are approximately 100-fold different. If the putative magnetosome-like particle size created in mammalian systems (i.e., dense core clusters) through the expression of a single magnetotactic bacterial gene is comparable to the bacterial magnetosome (~50 nm), then the biomineral structure and magnetic properties are still poorly developed. This is not surprising given the number of MAI genes used by magnetotactic bacteria, especially for growth of the biomineral (Fig. 2). Even the expression of Mms6 alone, a magnetosome protein involved in crystal maturation [50], provided an MR signal and particle size in mammalian cells that was no better than MagA-derived contrast [9]. Taken together, these results and the current understanding of magnetosome synthesis suggest that a combination of genes is likely needed to improve the biomineral structure and MR signal derived from a magnetosome-like particle in mammalian cells.

However rudimentary, MagA and Mms6 expression each provide a baseline MR signal upon which to build. These magnetotactic bacterial genes are compatible with mammalian cell culture models and/or tumor cell biology [46], and in a variety of mammalian systems MagA expression poses no apparent immune or cytotoxic responses [8, 51, 52]. In addition, the feasibility of inducible MagA expression in mouse embryonic stem cells was recently demonstrated using intracranial grafts and 7 T MRI [53]. Despite these proof-of-principle studies, relatively few magnetosome-associated genes have been tested in mammalian cell systems; however, a more thorough examination of these bacterial genes may further augment and refine gene-based MR contrast. The research in mammalian models should also help clarify which magnetotactic bacterial genes are essential for a properly functioning magnetosome and how best to modify this structure for different biotechnological applications.

3.2 Reporter Gene Expression

Collingwood and Davidson have recently reviewed methods of measuring iron biominerals, their localization, and quantification in the cell using synchrotron technology [54]. The authors conclude that the most important thing to understand about the role of iron in neuropathology is not the total concentration of iron or its localization in the cell but rather the interactions between iron and iron-handling protein(s). If correct, then the interactions of magnetosome and mammalian proteins should be invaluable for enhancing the influence of iron biomineral properties on the MRI signal and for identifying distinct MR signatures.

Virtually, all models of magnetosome assembly now incorporate the notion of a protein scaffold, in which sequential addition of proteins that interact generates the molecular structure required for optimal function. The scaffold is like what you find on a construction site when laborers need to work on the roof. If pieces are missing, then you cannot build a structure high enough to complete the job. Similarly, the framework upon which one builds a magnetosome-like nanoparticle in mammalian cells may entail several proteins that do not produce contrast but without which effective MR contrast cannot be achieved. This is an opportunity for reporter gene expression of genes that somehow complement the structure of the iron biomineral, be that proper formation of the magnetosome compartment, arrangement within the cell, composition of the crystal, its shape or size.

3.3 MRI Detection of Iron Metabolism

Perhaps an unexplored benefit of developing methods for detection of gene-based contrast is the potential use of MRI for measuring changes in iron metabolism [22]. In most cell culture models, MagA expression increases the amount of cellular iron, only in response to an extracellular iron supplement [47]. In this way, MagA acts as a beacon indicating a change in the extracellular environment, which may ultimately be a useful diagnostic tool.

For multipotent P19 cells, in which the parental line displays high iron import and export, similar to the iron recycling phenotype of M2 macrophages, the influence of MagA expression modulates iron export but has little effect on iron uptake [51]. Although further studies are needed to delineate the mechanism of MagA function in mammalian cells, there may be a therapeutic role for MagA expression in iron regulation or dysregulation. With the increasing awareness and understanding of inappropriate iron handling in neurological disorders [54], blood disorders [55], and inflammation [28, 56], future cell therapies might benefit from the magnetosome-like particle as a vehicle for removal or delivery of iron.

3.4 Estimated Limits of Sensitivity of MRI Reporter Gene Expression Based on the Magnetosome

Assuming that mammalian cells can eventually be engineered to express a magnetosome-like particle with the same capacity to produce iron biominerals as magnetotactic bacteria, per unit volume, we have estimated the limit of sensitivity of MRI reporter gene expression. We will express this limit as the number of mammalian cells required to produce a change in R_2 (ΔR_2) of 1 s^{-1}. For this calculation, we have used the results of Benoit et al. [57] in which magnetotactic bacteria were imaged within mouse tumors. In this work, the amount of iron per bacterial cell was measured by magnetometry and gave 2.2×10^{-15} g (3.9×10^{-11} μmol Fe) when the bacteria were cultured in the presence of 40 μM ferric malate and 0.64×10^{-15} g (1.15×10^{-11} μmol Fe) when the iron supplement was 40 μM $FeCl_3$. The corresponding R_2 relaxivities reported were 48 s^{-1}/mM and 337 s^{-1}/mM, respectively. Using the latter value, the concentration of iron required to produce a ΔR_2 of 1 s^{-1} is equal to 1 s^{-1}/(337 s^{-1}/mM) $\approx$ 3 μM. Assuming that mammalian cells could hold approximately 100 times more iron than magnetotactic bacteria, this iron concentration corresponds to a cellular concentration of (3 μmol/L)/($100\times1.15\times10^{-11}$ μmol/cell) = 2.6×10^{9} cells/L. For imaging of large animals and humans, one can assume a 1 μL voxel, and hence the number of cells per voxel required to produce a ΔR_2 of 1 s^{-1} would be 2600. In order to extend this estimate to small animal (e.g., mouse) imaging using the same criterion ($\Delta R_2 = 1\ s^{-1}$), we must assume similar signal to noise ratio (SNR) as in the large animal case. We note that SNR scales with the volume of tissue (V_T) within the radiofrequency coil as $1/(V_T)^{5/6}$ [58]. For a mouse (20 g) compared to a human (70 kg) this corresponds to a factor of $(3500)^{5/6}$ for an estimate of ~900–1000 cells. Therefore, assuming similar sequence and coil design for mouse and human imaging, the corresponding mouse voxel would be 10^{-3} μL. Hence, the number of cells in this small animal voxel required to produce a ΔR_2 of 1 s^{-1} would be approximately 2600/1000 ~ 3. Based on this calculation, sensitive imaging of relatively few cells could theoretically be achieved by developing MRI reporter gene expression modeled on the magnetosome.

To approach this projection in vivo, mammalian cells expressing a magnetosome-like particle would presumably draw on cellular stores of iron. Typically, this involves the uptake of transferrin-bound iron from the circulation, which is introduced into cells through transferrin receptor-mediated endocytosis [22]. Once internalized, iron homeostasis is tightly regulated by Iron Binding Proteins. Excess iron is stored in ferritin while the labile iron pool provides metal ion cofactor for the immediate needs of the cell. Consistent with this, MagA-derived MR contrast in P19 cells reduces apparent iron export activity, with little or no influence on iron uptake [59]. This finding suggests that intracellular iron may be rerouted for different purposes and in response to magnetotactic bacterial transgene expression.

3.5 Applications in Magnetic Particle Imaging

Particles of maghemite (γ-Fe_2O_3), an oxidized form of magnetite [60], are used exclusively in MPI. The size of these superparamagnetic iron oxide nanoparticles (SPION) is smaller than the iron core needed to hold one magnetic field domain. Consequently, SPION have predominantly fast relaxation rates as the internal magnetization of the iron core rotates in response to an external time-changing magnetic field (Néel relaxation). When the size of an iron biomineral exceeds approximately 25 nm, as is the case for most magnetosomes produced by bacteria, the particle then holds one magnetic field domain and responds to an altered magnetic field, such as used in MPI, primarily by Brownian relaxation. This motion requires the particle itself to rotate in its environment. Since both the iron core and the nonmagnetic outer shell must physically rotate, Brownian relaxation is much slower than Néel relaxation. Thus, in general MPI is optimal (with respect to spatial resolution and SNR) when the iron core of the MPI tracer is just below the transition from Néel to Brownian relaxation [61]. In addition, the dependence of imaging frequency on particle size permits MPI to discriminate SPION of different sizes [62].

Depending on the species, magnetotactic bacteria synthesize magnetosomes that range in size and shape of the magnetite biomineral [13]. Regardless, these nanoparticles are large enough to hold one magnetic field domain and arrangement of magnetosomes in a chain-like structure enables magnetotaxis as a response to the earth's geomagnetic field. Thus, magnetosomes (~25–120 nm) [13, 57] typically exceed the Néel/Brownian transition size (20–25 nm). However, it should be feasible to tailor the synthesis of a magnetosome-like particle [21] such that the biomineral meets the desired size and shape for multispectral MPI. Since current sources of SPION, such as Resovist, have such a spread in size that they reduce MPI sensitivity and spatial resolution, developing a more uniform preparation from a biological source like magnetotactic bacteria would be worthwhile. MPI would benefit significantly if a) SPION of well-defined sizes could be reliably produced and b) the technology could incorporate a reporter gene. Unlike MRI, in which there is potential for development of nonmagnetite-based reporter gene expression, MPI reporter genes must be maghemite- and/or magnetite-based. Hence, preclinical MPI would benefit enormously from the expression of magnetosome-like magnetite nanoparticles and would rival reporter gene applications in preclinical MRI due to the superior sensitivity of MPI, which might be comparable to PET in the future [63]. In addition, MPI, like PET and ^{19}F-MRI [64], is intrinsically quantitative [65]. This is an important advantage over the use of MRI, which becomes problematic when the concentration of SPIO reaches the level of pg/cell [66].

Interestingly, there is evidence in the fossil record that much larger iron biominerals, on the order of 4 μm, were created by biological systems, possibly including eukaryotes [67]. Ortega et al. [68] used PC-12 cells as a model for the dopaminergic neuron in Parkinson's disease (PD) research. These cells differentiate into neurons in response to nerve growth factor. Using an extracellular iron supplement of 300 mM $FeSO_4$ for 24 h and synchrotron technology, they demonstrated that iron was present in 200 nm structures in the cytosol and in neurite outgrowths.

This is consistent with postmortem analysis of the human brain from individuals afflicted with PD, which shows iron in neuromelanin granules [54]. All these reports provide evidence of subcellular iron compartments, quite apart from the traditional ferritin storage. Hence, the real potential of MRI reporter gene expression and generation of sizable iron biominerals for molecular imaging may approach the type of SPIO nanoparticles that have been successfully used to label and track cells by MRI [69].

4 Conclusion

To develop effective reporter gene expression for MRI, the genetic ability of magnetotactic bacteria may be exploited to impart magnetic characteristics to mammalian cells through expression of select magnetosome-related genes. The formation of magnetosome-like nanoparticles in mammalian systems does not necessarily require all the features of the bacterial magnetosome [23, 45]. Based on the success of MagA- and Mms6-derived MR contrast, an imperfect magnetosome-like compartment and/or biomineral may be sufficient, if not desirable, for MRI. Moreover, the stepwise addition of some of the essential magnetosome genes may reconstitute aspects of the magnetosome-like structure that could be associated with changes in cell/tissue contrast and provide unique signatures for MRI reporter gene expression. Employing this type of gene-based MR contrast will not only provide a method for regulating cellular iron biominerals and quantifying the response to transcription factor stimulation, but also enhance the capability of hybrid imaging platforms, such as PET/MRI, and the simultaneous detection of multiple in vivo activities (e.g., edema, hemorrhage, ischemia, inflammation) with anatomical precision. Further development of this medical, molecular imaging tool will enable early characterization of disease progression in both small and large animal models, and forge a path for translation of cell therapies to patient care [70].

Acknowledgments The authors are supported by a grant from the Ontario Research Fund in partnership with Multi-Magnetics Inc., in a grant, entitled Heart Failure: Prevention Through Early Detection Using New Imaging Methods, and by the Cancer Imaging Network of Ontario through Cancer Care Ontario. FSP is the recipient of a Discovery Grant from the Natural Sciences and Engineering Research Council of Canada.

References

1. Bhaumik S, Gambhir S. Optical imaging of Renilla luciferase reporter gene expression in living mice. Proc Natl Acad Sci U S A. 2002;99:377–82.
2. Josserand V, Texier-Nogues I, Huber P, Favrot M-C, Coll J-L. Non-invasive in vivo optical imaging of the lacZ and luc gene expression in mice. Gene Ther. 2007;14:1587–93.
3. Tangney M, Francis K. In vivo optical imaging in gene & cell therapy. Curr Gene Ther. 2012;12:2–11.
4. Couzin-Frankel J. When mice mislead. Science. 2013;342:922–5.

5. Dewald O, Ren G, Duerr G, Zoerlein M, Klemm C, Gersch C, et al. Of mice and dogs: species-specific differences in the inflammatory response following myocardial infarction. Am J Pathol. 2004;164:665–77.
6. Moses W. Fundamental limits of spatial resolution in PET. Nucl Instrum Methods Phys Res A. 2011;648 Suppl 1:S236–40.
7. Thompson K, Wisenberg G, Sykes J, Thompson R. MRI/MRS evaluation of cariporide in a canine long-term model of reperfused ischemic insults. Magnetic resonance imaging/magnetic resonance spectroscopy. J Magn Reson Imaging. 2003;17:136–41.
8. Goldhawk D, Lemaire C, McCreary C, McGirr R, Dhanvantari S, Thompson R, et al. Magnetic resonance imaging of cells overexpressing MagA, an endogenous contrast agent for live cell imaging. Mol Imaging. 2009;8:129–39.
9. Zhang X-Y, Robledo B, Harris S, Hu X. A bacterial gene, mms6, as a new reporter gene for magnetic resonance imaging of mammalian cells. Mol Imaging. 2014;13:1–12.
10. Zurkiya O, Chan AW, Hu X. MagA is sufficient for producing magnetic nanoparticles in mammalian cells, making it an MRI reporter. Magn Reson Med. 2008;59(6):1225–31.
11. Jogler C, Schuler D. Genomics, genetics, and cell biology of magnetosome formation. Annu Rev Microbiol. 2009;63:501–21.
12. Komeili A. Molecular mechanisms of compartmentalization and biomineralization in magnetotactic bacteria. FEMS Microbiol Rev. 2012;36:232–55.
13. Araujo A, Abreu F, Tavares Silva K, Bazylinski D, Lins U. Magnetotactic bacteria as potential sources of bioproducts. Mar Drugs. 2015;13:389–430.
14. Boucher M, Ginet N, Geffroy F, Preveral S, Adryanczyk-Perrier G, Pean M, et al. Genetically functionalized magnetosomes as MRI contrast agent suitable for molecular imaging. International Society for Magnetic Resonance in Medicine (2015) Abstract 0696; Toronto, Canada.
15. Bulte J. In vivo MRI cell tracking: clinical studies. AJR Am J Roentgenol. 2009;193:314–25.
16. de Chickera S, Willert C, Mallet C, Foley R, Foster P, Dekaban G. Cellular MRI as a suitable, sensitive non-invasive modality for correlating in vivo migratory efficiencies of different dendritic cell populations with subsequent immunological outcomes. Int Immunol. 2011;24:29–41.
17. Korchinski D, Taha M, Yang R, Nathoo N, Dunn J. Iron oxide as an MRI contrast agent for cell tracking. Magn Reson Insights. 2015;8(S1):15–29.
18. Nejadnik H, Ye D, Lenkov O, Donig J, Martin J, Castillo R, et al. Magnetic resonance imaging of stem cell apoptosis in arthritic joints with a caspase activatable contrast agent. ACS Nano. 2015;9:1150–60.
19. Matsunaga T, Nakamura C, Burgess J, Sode K. Gene transfer in magnetic bacteria: transposon mutagenesis and cloning of genomic DNA fragments required for magnetosome synthesis. J Bacteriol. 1992;174:2748–53.
20. Nakamura C, Burgess JG, Sode K, Matsunaga T. An iron-regulated gene, *magA*, encoding an iron transport protein of *Magnetospirillum* sp. strain AMB-1. J Biol Chem. 1995;270:28392–6.
21. Stanisland S. Nanoparticle biosynthesis, an accommodating host. Nat Nanotechnol. 2014;9:163–4.
22. Goldhawk D, Gelman N, Sengupta A, Prato F. The interface between iron metabolism and gene-based iron contrast for MRI. Magn Reson Insights. 2015;8(S1):9–14.
23. Goldhawk D, Rohani R, Sengupta A, Gelman N, Prato F. Using the magnetosome to model effective gene-based contrast for magnetic resonance imaging. WIRES Nanomed Nanobiotechnol. 2012;4:378–88.
24. Brennan P, Donev R, Hewamana S. Targeting transcription factors for therapeutic benefit. Mol BioSyst. 2008;4:909–19.
25. Mees C, Nemunaitis J, Senzer N. Transcription factors: their potential as targets for an individualized therapeutic approach to cancer. Cancer Gene Ther. 2009;16:103–12.
26. Corna G, Campana L, Pignatti E, Castiglioni A, Tagliafico E, Bosurgi L, et al. Polarization dictates iron handling by inflammatory and alternatively activated macrophages. Haematologica. 2010;95:1814–22.
27. Fleming R. Hepcidin activation during inflammation: make it STAT. Gastroenterology. 2007;132:447–9.

28. Frangogiannis N. The inflammatory response in myocardial injury, repair, and remodelling. Nat Rev Cardiol. 2014;5:255–65.
29. Nahrendorf M, Pittet M, Swirski F. Monocytes: protagonists of infarct inflammation and repair after myocardial infarction. Circulation. 2010;121:2437–45.
30. Faivre D, Schuler D. Magnetotactic bacteria and magnetosomes. Chem Rev. 2008;108:4875–98.
31. Lohße A, Ullrich S, Katzmann E, Borg S, Wanner G, Richter M, et al. Functional analysis of the magnetosome island in *Magnetospirillum gryphiswaldense*: the *mamAB* operon is sufficient for magnetite biomineralization. PLoS One. 2011;6, e25561.
32. Nudelman H, Zarivach R. Structure prediction of magnetosome-associated proteins. Front Microbiol. 2014;5:article 9.
33. Murat D, Quinlan A, Vali H, Komeili A. Comprehensive genetic dissection of the magnetosome gene island reveals the step-wise assembly of a prokaryotic organelle. Proc Natl Acad Sci USA. 2010;107:5593–8.
34. Hentze M, Muckenthaler M, Galy B, Camaschella C. Two to tango: regulation of mammalian iron metabolism. Cell. 2010;142:24–38.
35. Komeili A, Li Z, Newman DK, Jensen GJ. Magnetosomes are cell membrane invaginations organized by the actin-like protein MamK. Science. 2006;311(5758):242–5.
36. Richter M, Kube M, Bazylinski DA, Lombardot T, Glockner FO, Reinhardt R, et al. Comparative genome analysis of four magnetotactic bacteria reveals a complex set of group-specific genes implicated in magnetosome biomineralization and function. J Bacteriol. 2007;189(13):4899–910.
37. Lefevre C, Trubitsyn D, Abreu F, Kolinko S, Jogler C, Gonzaga Paula de Almeida L, et al. Comparative genomic analysis of magnetotactic bacteria from the Deltaproteobacteria provides new insights into magnetite and greigite magnetosome genes required for magnetotaxis. Environ Microbiol. 2013;15:2712–35.
38. Green S, Komeili A. Biogenesis and subcellular organization of the magnetosome organelles of magnetotactic bacteria. Curr Opin Cell Biol. 2012;24:490–5.
39. Uebe R, Junge K, Henn V, Poxleitner G, Katzmann E, Plitzko J, et al. The cation diffusion facilitator proteins, MamB and MamM of Magnetospirillum gryphiswaldense have distinct and complex functions, and are involved in magnetite biomineralization and magnetosome membrane assembly. Mol Microbiol. 2011;82:818–35.
40. Quinlan A, Murat D, Komeili A. The HtrA/DegP family protease MamE is a bifunctional protein with roles in magnetosome protein localization and magnetite biomineralization. Mol Microbiol. 2011;80:1075–87.
41. Scheffel A, Gardes A, Grunberg K, Wanner G, Schuler D. The major magnetosome proteins MamGFDC are not essential for magnetite biomineralization in Magnetospirillum gryphiswaldense but regulate the size of magnetosome crystals. J Bacteriol. 2008;190(1):377–86.
42. Zeytuni N, Uebe R, Maes M, Davidov G, Baram M, Raschdorf O, et al. Bacterial magnetosome biomineralization—a novel platform to study molecular mechanisms of human CDF-related type-II diabetes. PLoS One. 2014;9:e97154.
43. Rahn-Lee L, Komeili A. The magnetosome model: insights into the mechanisms of bacterial biomineralization. Front Microbiol. 2013;4:352.
44. Geffroy F, et al. In vitro characterization of AMB1 magnetosomes as biogenic functionalized contrast agents dedicated to molecular MRI. World Molecular Imaging Congress (2015) Abstract LBAP 033; Honolulu, USA.
45. Kolinko I, Lohße A, Borg S, Raschdorf O, Jogler C, Tu Q, et al. Biosynthesis of magnetic nanostructures in a foreign organism by transfer of bacterial magnetosome gene clusters. Nat Nanotechnol. 2014;9:193–7.
46. Rohani R, Figueredo R, Bureau Y, Koropatnick J, Foster P, Thompson R, et al. Imaging tumor growth non-invasively using expression of MagA or modified ferritin subunits to augment intracellular contrast for repetitive MRI. Mol Imaging Biol. 2014;16:63–73.

47. Sengupta A, Quiaoit K, Thompson R, Prato F, Gelman N, Goldhawk D. Biophysical features of MagA expression in mammalian cells: implications for MRI contrast. Front Microbiol. 2014;5:29.
48. Lee C, Thompson R, Prato F, Goldhawk D, Gelman N. Investigating the relationship between transverse relaxation rate (R2) and interecho time in MagA-expressing iron-labeled cells. Mol Imaging. 2015;14:551–60.
49. Jensen J, Chandra R. NMR relaxation in tissues with weak magnetic inhomogeneities. Magn Reson Med. 2000;44:144–56.
50. Tanaka M, Mazuyama E, Arakaki A, Matsunaga T. MMS6 protein regulates crystal morphology during nano-sized magnetite biomineralization in vivo. J Biol Chem. 2011;286(8):6386–92.
51. Liu L. Characterization of MagA expression and iron uptake in P19 cells: implications for use as a gene-based contrast agent for MRI. Scholarship@Western electronic thesis and dissertation repository. London, Canada: Western University; 2015.
52. Quiaoit K. Towards the Development of a MagA Reporter gene expression construct for magnetic resonance imaging. Scholarship@Western electronic thesis and dissertation repository. London, Canada: Western University; 2015.
53. Cho I, Moran S, Paudya R, Piotrowska-Nitsche K, Cheng P-H, Zhang X, et al. Longitudinal monitoring of stem cell grafts in vivo using magnetic resonance imaging with inducible *magA* as a genetic reporter. Theranostics. 2014;4:972–89.
54. Collingwood J, Davidson M. The role of iron in neurodegenerative disorders: insights and opportunities with synchrotron light. Front Pharm. 2014;5:191.
55. Andrews N. Anemia of inflammation: the cytokine-hepcidin link. J Clin Invest. 2004;113:1251–3.
56. Wu X-N, Su D, Wang L, Yu F-L. Roles of the hepcidin-ferroportin axis and iron in cancer. Eur J Cancer Prev. 2014;23:122–33.
57. Benoit M, Mayer D, Barak Y, Chen I, Hu W, Cheng Z, et al. Visualizing implanted tumors in mice with magnetic resonance imaging using magnetotactic bacteria. Clin Cancer Res. 2009;15:5170–7.
58. Chen C, Hoult D. Biomedical magnetic resonance technology. Bristol and New York: Adam Hilger; 1989.
59. Liu L, Sengupta A, McGirr R, Thompson R, Prato F, Hoffman L, et al. Expression of the MRI reporter gene *magA* overrides iron export activity in P19 cells. World Molecular Imaging Congress (2015) Abstract LBAP 080; Honolulu, USA.
60. Thorek D, Chen A, Czupryna J, Tsourkas A. Superparamagnetic iron oxide nanoparticle probes for molecular imaging. Ann Biomed Eng [Internet]. 2006.
61. Bauer L, Situ S, Griswold M, Samia A. Magnetic particle imaging tracers: state-of-the-art and future directions. J Phys Chem Lett. 2015;6:2509–17.
62. Rahmer J, Halkola A, Gleich B, Schmale I, Borgert J. First experimental evidence of the feasibility of multi-color magnetic particle imaging. Phys Med Biol. 2015;60:1775–91.
63. Goodwill P, Lu K, Zheng B, Conolly S. An x-space magnetic particle imaging scanner. Rev Sci Instrum. 2012;83:033708.
64. Gaudet J, Ribot E, Chen Y, Gilbert K, Foster P. Tracking the fate of stem cell implants with fluorine-19 MRI. PLoS One. 2015;10:e0118544.
65. Bulte J, Walczak P, Janowski M, Krishnan K, Arami H, Halkola A, et al. Quantitative "hot spot" imaging of transplanted stem cells using superparamagnetic tracers and magnetic particle imaging (MPI). Tomography. 2015;1:91–7.
66. Liu W, Dahnke H, Rahmer J, Jordan E, Frank J. Ultrashort T2* relaxometry for quantitation of highly concentrated superparamagnetic iron oxide (SPIO) nanoparticle labeled cells. Magn Reson Med. 2009;61:761–6.
67. Schumann D, Raub T, Kopp R, Guerquin-Kern J, Wu T, Rouiller I, et al. Gigantism in unique biogenic magnetite at the paleocene-eocene thermal maximum. Proc Natl Acad Sci U S A. 2008;105:17648–53.
68. Ortega R, Cloetens P, Deves G, Carmona A, Bohic S. Iron storage within dopamine neurovesicles revealed by chemical nano-imaging. PLoS One. 2007;2:e925.

69. Graham J, Foltz W, Vaags A, Ward M, Yang Y, Connelly K, et al. Long-term tracking of bone marrow progenitor cells following intracoronary injection post-myocardial infarction in swine using MRI. Am J Physiol Heart Circ Physiol. 2010;299:H125–33.
70. Xu H, Belkacemi L, Jog M, Parrent A, Hebb M. Neurotrophic factor expression in expandable cell populations from brain samples in living patients with Parkinson's disease. FASEB J. 2013;27:4157–68.
71. Meikle S, Kench P, Kassiou M, Banati R. Small animal SPECT and its place in the matrix of molecular imaging technologies. Phys Med Biol. 2005;50:R45–61.
72. Cherry S, Sorenson J, Phelps M. Single photon emission computed tomography. Physics in nuclear medicine. 3rd ed., 2003. p. 299–324.
73. Chen B, Legant W, Wang K. Lattice light-sheet microscopy: imaging molecules to embryos at high spatiotemporal resolution. Science. 2014;346:1257998.
74. Ephrat P, Albert G, Roumeliotis M, Belton M, Prato F, Carson J. Localization of spherical lesions in tumor-mimicking phantoms by 3D sparse array photoacoustic imaging. Med Phys. 2010;37:1619–28.
75. James M, Gambhir S. A molecular imaging primer: modalities, imaging agents, and applications. Physiol Rev. 2012;92:897–965.
76. Gore J, Yankeelov T, Peterson T, Avison M. Molecular imaging without radiopharmaceuticals? J Nucl Med. 2009;50:999–1007.

PET/SPECT/MRI Multimodal Nanoparticles

Tang Tang*, **Joel Garcia***, **and Angelique Y. Louie**

1 Introduction

Medical imaging techniques have dramatically improved in the past few decades, leading to positive impacts on early detection and diagnosis of disease; however, further improvements are necessary to truly optimize diagnostic accuracy. Multimodal imaging emerged as a means to merge the strengths of different modalities and provide complementary information to better meet clinical diagnostic challenges. While magnetic resonance imaging (MRI) and computed tomography (CT) offer high-resolution images (in the μm range) with anatomic information, the nuclear imaging techniques of positron emission tomography (PET) and single-photon emission computed tomography (SPECT) can provide very high sensitivity (in the pM range) [1]. Hybrid imaging instruments have been developed that combine these modalities to allow acquisition of high-sensitivity and high-resolution information simultaneously, and overcome limitations of having to move a patient between modalities.

*Author contributed equally with all other contributors.

T. Tang
Department of Chemistry, University of California, Davis, CA 95616, USA

Chemistry Graduate Group, University of California, Davis, CA 95616, USA

J. Garcia
Department of Biomedical Engineering, University of California, Davis, CA 95616, USA

A.Y. Louie (✉)
Chemistry Graduate Group, University of California, Davis, CA 95616, USA

Department of Biomedical Engineering, University of California, Davis, CA 95616, USA
e-mail: aylouie@ucdavis.edu

J.W.M. Bulte, M.M.J. Modo (eds.), *Design and Applications of Nanoparticles in Biomedical Imaging*, DOI 10.1007/978-3-319-42169-8_10

In general, a multimodal system contains an imaging technique that renders structural information (such as MRI, CT, or ultrasound) and an imaging modality that provides high-sensitivity, functional, or molecular information (PET, SPECT, or optical). Compared to SPECT, PET has higher sensitivity (by approximately two to three orders of magnitude), higher spatial and temporal resolution, and better attenuation correction, but it comes at a higher cost. On the other hand, SPECT offers the possibility to widen the observational time window, due to the longer half-life of single-photon emitters. Generally, biological considerations are the ultimate deciding factors when choosing between PET vs. SPECT [2]. Both PET and SPECT lack anatomic information and have relatively poor resolution. This imaging limitation has led to the development of hybrid systems to fuse PET or SPECT with CT to provide anatomic context. Multimodal imaging attracted wide attention with its introduction in the 1990s through the first PET/CT scanners [3]. Since then many other types of multimodal systems have been investigated and there has been promising progress. Today, the PET/CT system has been successfully accepted in clinics worldwide [4], and the number of SPECT/CT systems is also increasing at a brisk pace. However, CT has a rather narrow range of soft tissue contrast and the use of ionization radiation from CT restricts its application when repeated scans are needed. Alternative hybrid systems such as PET/MRI or SPECT/MRI, where excellent soft tissue contrast is rendered without ionizing radiation, are the topics of most recent attention, and have shown promise for various applications such as imaging brain tumors and cardiovascular diseases [5].

Along with progress in the imaging instruments and techniques, novel multimodal imaging probes have been developed to support these synergistic systems. In this chapter, we will focus on the development of PET/MRI and SPECT/MRI probes that have been reported in the past 5 years. Probe designs will be highlighted in Sect. 2, with interesting or promising applications discussed in Sect. 3. We limit our review to the most unique designs and applications; review articles on general probe design and synthesis as well as clinical applications of multimodal nanoparticles can be found elsewhere [6–9].

2 Designs of PET/MRI and SPECT/MRI Probes

Multimodal imaging probes have the potential advantage of temporal and spatial correlation of different imaging modalities. Compared with using a cocktail of probes for different modalities, a single platform avoids differential biodistribution and allows simultaneous imaging of the same biomarker by different modalities. Nanoparticles have physical and chemical traits that are beneficial for construction of multimodal imaging probes. For example, the large surface area of nanoparticles allows for introduction of different functional groups for imaging and targeting. A schematic representation of typically reported, nanoparticle-based, multimodal imaging probes is shown in Fig. 1. In general, reported probes contained: (1) an MRI component, which is normally a gadolinium-based T_1 contrast agent or an iron oxide-based T_2 contrast agent; (2) a radionuclide component, through direct

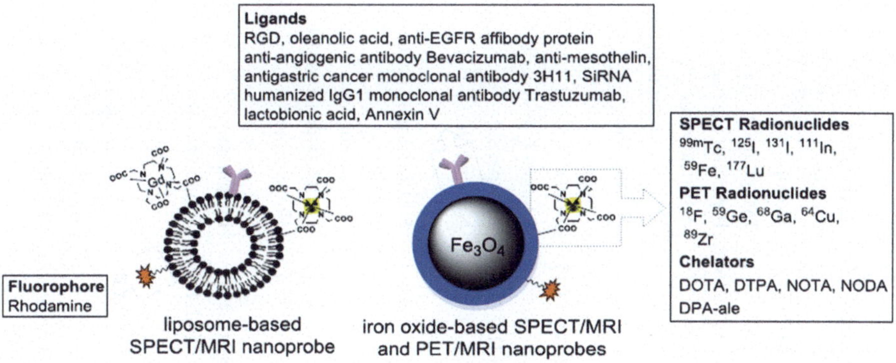

Fig. 1 Schematic representation of commonly used nanoparticle-based PET/MRI and SPECT/MRI probes

labeling with radiohalides (e.g., ^{18}F, ^{124}I) or attachment of radiometals for PET (e.g., ^{68}Ga, ^{64}Cu, ^{89}Zr) or SPECT (e.g., ^{99m}Tc, ^{111}In) through chelators; and (3) surface coating to improve stability and biocompatibility. For probes targeted to specific biomarkers, targeting moieties are conjugated on the outer surface to direct accumulation of probes at sites of interest. Summaries of the physical properties of PET/MRI and SPECT/MRI probes found in the recent literature are given in Tables 1 and 2, respectively. There have been two major classes of multimodal nanoparticle probes reported: iron oxide- and liposome/polymer-based nanoparticle probes. In this section, the unique designs for the nanoparticle core and surfaces will be summarized. Also, the chemical synthesis, radiolabeling, and stability assessment for multimodal nanoparticle-based probes will be briefly discussed.

2.1 Iron Oxide-Based Imaging Probes

Iron oxide nanoparticles (maghemite γ-Fe_2O_3 or magnetite Fe_3O_4) are commonly used platforms for constructing multimodal probes for PET/MR and SPECT/MR imaging because of their unique magnetic and biological properties. Iron oxide nanoparticles, in general, have an excellent biocompatibility profile at concentration ranges suitable for most diagnostic applications (0.1–10 μg/mL) because iron taken up by the liver and spleen is metabolized and joins the physiological iron pool. However, concerns for cytotoxicity could arise when concentrations of 100 μg/mL of iron oxide nanoparticles are exceeded [10] and adverse reactions such as hypotension and back pain have been reported for some formulations [11]. Besides biocompatibility, the large surface-to-volume ratio for iron oxide nanoparticles has been attractive for development of targeted agents. In addition, self-heating of these nanoparticles in the presence of alternating magnetic fields could facilitate controlled drug release or cytolysis of tumor cells [12], which has spurred interest in developing these magnetic nanoparticles as therapeutic agents for tumor treatments.

Table 1 Physical properties of nanoparticle-based PET/MRI probes

Nanoparticle	MRI component	PET component	Chelator	Core size (nm)	Hydrodynamic diameter (nm)	r_1 ($mM^{-1}s^{-1}$)	r_2 ($mM^{-1}s^{-1}$)	Application	Ref.
^{124}I-$Gd_3N@C_{80}$	Gd^{3+}	^{124}I	–	–	–	70	–	Tumor	[21]
^{64}Cu-NOTA-Au-IONP-Affibody	Iron oxide	^{64}Cu	NOTA	~10	24.4±2.0	–	143.2 (7T)	Tumor	[43]
*As-SPION@PAA	Iron oxide	$^{71,72,74}As$	–	~10	21	–	93.8 (4.7T)	Lymph node	[44]
Magh-1-PNPs-NODA	Iron oxide	^{68}Ga	NODA	8-12	44-55	0.5	182 (1.5T)	Pancreatic cancer	[45]
MDIO-^{64}Cu-DOTA	Iron oxide	^{64}Cu	DOTA	7-8	62.7	16.8	83.9 (1.4T)	Atherosclerotic plaques	[38]
SDIO-DO3A	Iron oxide	^{64}Cu	DO3A	5.9	~60	14	72 (1.4T)	Atherosclerotic plaques	[46]
$Gd^{3+}/Yb^{3+}/Er^{3+}$ doped $NaYF_4$	Gd^{3+}	^{18}F	–	22×19	28.2	0.40 (3T)	–	Nonspecific	[47]
NOTA-OA-IONP	Iron oxide	^{68}Ga	NOTA	<15	66.3±31.0	–	157 (4.7T)	Tumor	[48]
cRGD-SPIO	Iron oxide	^{64}Cu	NOTA	~10	68	–	101.9 (4.7T)	Tumor	[49]
^{68}Ga-SPIONs	Iron oxide	^{68}Ga	–	–	30	–	–	Sentinel lymph node	[50]
$Fe_3O_4@Al(OH)_3$	Iron oxide	^{18}F	–	–	21	4.9	121.9 (3T)	Nonspecific	[51]
$MnFe_2O_4@Al(OH)_3$	Mn doped iron oxide	^{18}F	–	–	21	8.2	20.1 (3T)	Nonspecific	[51]
$Co_{0.16}Fe_{2.84}O4@NaYF_4$	Iron oxide	^{18}F	–	10.3	44	5	102 (3T)	Sentinel lymph node (passive)	[13]
^{89}Zr-Feraheme	Iron oxide	^{89}Zr	–	5-10	18.1	–	–	Nonspecific	[39]
^{69}Ge-SPION@PEG	Iron oxide	^{69}Ge	–	10	23	–	93.8 (4.7T)	Lymph node (passive)	[36]
$Gd^{3+}/^{64}Cu^{2+}$-DTPA-MSNs	Gd^{3+}	^{64}Cu	DTPA	150	250±76	17.6 (1.4T)	–	Nonspecific	[22]
^{89}Zr-ferumoxytol	Iron oxide	^{89}Zr	desferrioxamine	–	17-35	–	89 (0.47T)	Lymph nodes	[52]
PEMs	Iron oxide	^{64}Cu	DOTA	100 (5 nm for single core)	140±7	2.15	265 (1.4T)	Tumor (passive)	[53]
^{68}GaNHFCNP	Iron oxide	^{68}Ga	NOTA	15.3	–	–	–	Target cancer cell though charge	[37]
AGuIX	Gd^{3+}	^{68}Ga	NODA (^{68}Ga), DOTA (Gd)	–	2.5±0.1	10.3	13.4 (1.4T)	Nonspecific	[20]

MSN mesoporous silica nanoparticles, *PEM* positron emitting magnetic nanoconstructs

Table 2 Physical properties of the reported nanoparticle-based SPECT/MRI probes

Nanoparticle	MRI component	SPECT component	Chelator	Core size (nm)	Hydrodynamic diameter (nm)	r_1 ($mM^{-1}s^{-1}$)	r_2 ($mM^{-1}s^{-1}$)	Application	Ref.
[^{59}Fe]-SPIONs	Iron oxide	^{59}Fe	–	5.0 ± 1.5	17 ± 6	–	97 ± 3	Nonspecific	[34]
^{111}In-doped iron oxide	Iron oxide	^{111}In	–	7.7	37.3	–	–	Nonspecific	[33]
^{99m}Tc-LBA-SPION	Iron oxide	^{99m}Tc	DTPA	12	–	–	–	Human hepatoblastoma	[17]
^{99m}Tc-NBRh1	Iron oxide	^{99m}Tc	–	–	160.8 ± 1.3	–	–	Tumor	[28]
^{99m}Tc-USPION-RGD	Iron oxide	^{99m}Tc	DTPA	3.5 ± 1.2	10	8.2	20.1	Tumor	[14]
^{99m}Tc-AF-SPIONs	Iron oxide	^{99m}Tc	–	11 ± 2	–	–	–	Lymph nodes (passive)	[27]
^{99m}Tc-DTPA-USPIO-Annexin V	Iron oxide	^{99m}Tc	DTPA	3-5	–	8.2	20.1	Vulnerable atherosclerotic plaques	[15]
cRGD-conjugated IO-NPs	Iron oxide	^{99m}Tc	–	10 ± 2	~200	–	–	Tumor	[12]
^{177}Lu-trastuzumab-iron oxide	Iron oxide	^{177}Lu	DOTA	9.0 ± 0.5	41 ± 15	–	–	Tumor	[35]
^{99m}Tc-Au–Fe_3O_4 core-shell NPs	Iron oxide	^{99m}Tc	–	7.9 ± 0.4	11	–	–	Nonspecific	[25]
^{99m}Tc-Au–Fe_3O_4 dumbbell-like NPs	Iron oxide	^{99m}Tc	–	26.1 ± 1.6	35	–	–	Nonspecific	[25]
Radiolabled PEGylated SPIO	Iron oxide	^{111}In, ^{14}C, ^{59}Fe	DTPA	15	40 ± 7	–	–	Nonspecific	[54]
Fe_3O_4–3H11- ^{125}I	Iron oxide	^{125}I	–	–	33.5	–	–	Tumor	[16]
^{99m}Tc-MNP	Iron oxide	^{99m}Tc	–	–	–	–	–	Nonspecific	[55]
^{125}I-cRGD-USPIO	Iron oxide	^{125}I	–	–	51.3	–	–	Breast cancer	[30]
^{131}I-hVEGF siRNA/SilenceMag	Iron oxide	^{131}I	–	–	–	–	–	Hepatocellular carcinoma	[31]
Fe_3O_4-Ag^{125}I heterodimers	Iron oxide	^{125}I	–	14 (Fe_3O_4) 9 (Ag^{125}I)	–		139.8	Nonspecific	[41]
RGD-liposomes	Gd^{3+}	^{111}In	–		182	2.7	16.2 (6.3 T)	Tumor	[19]
^{99m}Tc-DPA-ale-Endorem	Iron oxide	^{99m}Tc	DPA	5	106 ± 60	–	26 (9.4 T)	Nonspecific	[24]
^{99m}Tc-SPIONs	Iron oxide	^{99m}Tc	–	11	18	–	–	Sentinel lymph node (passive)	[23]
^{111}In-mAbMB-SPIONs	Iron oxide	^{111}In	DTPA	15.6	76.6	0.59	468.57	Tumor	[32]
^{111}In-liposomes	Gd^{3+}	^{111}In	–	–	110.4–131.3	–	–	Tumor	[18]
^{99m}Tc-USPIO-bevacizumab	Iron oxide	^{99m}Tc	DTPA	<20	–	–	–	Hepatocellular carcinoma	[26]
^{125}I-fSiO_4@SPIOs	Iron oxide	^{125}I	–	6	20	–	165 (1.41 T)	Stem cell tracking	[29]

DPA dipicolylamine, *LBA* lactobionic acid

The synthetic designs of recently reported PET/MRI and SPECT/MRI nanoparticle-based multimodal probes are heavily dependent on the proposed usage of these probes in vivo. Some properties can be engineered into the core of the nanoparticle itself, and preparation of the iron oxide nanoparticles under controlled conditions (concentration of precursors, temperature, and solvent system) can modulate core properties (composition, hydrodynamic diameter, polydispersity, and shape). These properties of the core have been tailored to influence the biodistribution, blood circulation half-life, and magnetic anisotropy. For example, incorporation of cation dopants into the core of iron oxide nanoparticles can alter magnetic anisotropy and can provide additional imaging capability such as fluorescent imaging. Cui et al. developed hybrid inorganic composites and their variants: Fe_3O_4@$NaYF_4$ core/shell nanoparticles, and bisphosphonate polyethylene glycol (BP-PEG)-coated $Co_{0.16}Fe_{2.84}O_4$@$NaYF_4$(Yb,Er)-BP-PEG and Fe_3O_4@$NaYF_4$(Yb,Tm)-BP-PEG nanoparticles. These nanocomposites contained different lanthanide cation dopants Yb (sensitizer), Er, or Tm in $NaYF_4$ shell for up-conversion fluorescence, and Co incorporated in the Fe_3O_4 core to adjust the magnetic properties of the nanoparticles [13]. Up-conversion fluorescence (emission of light at a shorter wavelength than the excitation wavelength) with emission peaks at 700 and 800 nm was observed for Fe_3O_4@$NaYF_4$(Yb,Tm)-BP-PEG nanocomposites while emission signals centered at 525, 550, and 650 nm were obtained for $Co_{0.16}Fe_{2.84}O_4$@$NaYF_4$(Yb,Er)-BP-PEG nanoparticles upon excitation with a 980-nm laser. The high affinity of $NaYF_4$ for [^{18}F-fluoride] and the ability of bisphosphonate moiety to chelate ^{99m}Tc and ^{64}Cu radiometals facilitated multimodal PET/MR/optical or SPECT/MR/optical imaging.

Surface modifications to introduce reactive moieties on the surface of the nanoparticles are often implemented. This includes attachment of chelators of MRI-active metals (e.g., Gd^{3+} and Mn^{2+}) and radiometals for multimodal imaging; addition of ligands that allow targeting; conjugation with fluorescent dyes enabling optical imaging; and loading with drugs for delivery. The surface functionalization of reported SPECT/MRI and PET/MRI probes typically involves modification of the surface with abundant $-NH_2$ or –COOH groups for conjugation using any number of conjugation chemistries. For example, the $-NH_2$ groups in the common coating material poly(ethylene glycol) 2-aminoethyl ether acetic acid (NH_2-PEG-COOH) can be activated by m-maleimidobenzoyl-*N*-hydroxysulfosuccinimidyl ester (sulfo-MBS) for subsequent functionalization with targeting groups through a thiol-maleimide linkage [14] (Scheme 1). A structural derivative of PEG, COOH-PEG-COOH, can be conjugated to amine-terminated ligands such as phospholipid-binding Annexin V [15], an anti-gastric cancer monoclonal antibody 3H11 [16], by activating the –COOH groups via NHS-based biolinkers such as 1-ethyl-3-(3-dimethylaminopropyl)carbodiimide hydrochloride (EDC)/sulfo-*N*-hydroxysuccinimide (sulfo-NHS). The –COOH of carboxymethyl dextran on the iron oxide nanoparticle surfaces can be conjugated to chelators or targeting ligands using EDC/NHS via carbodiimide coupling chemistry. Similarly, EDC/NHS chemistry can be used to conjugate asiologlycoprotein-targeted lactobionic acid to the free amine groups of dopamine [17].

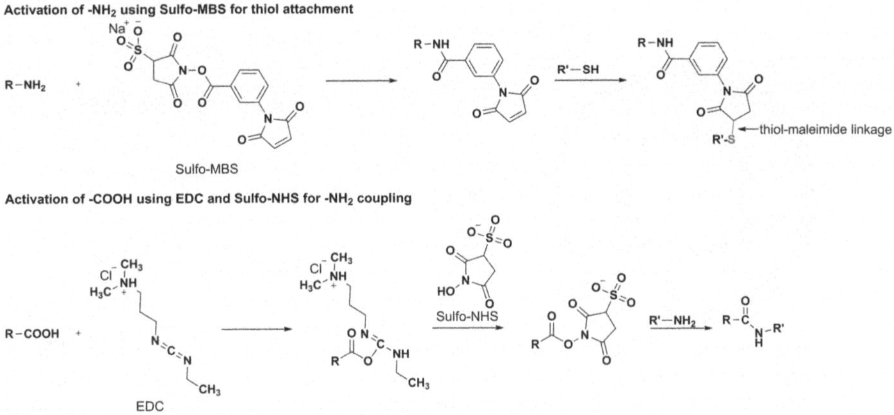

Scheme 1 Crosslinking strategies used in multimodal nanoparticle-based SPECT/MRI and PET/MRI probes for surface functionalization and subsequent ligand attachment

2.2 *Non Iron Oxide-Based Imaging Probes*

Liposomes have long been used as carriers for gadolinium-based MRI contrast agents. They are spherical vesicles composed of a bilayer of phospholipids with an aqueous interior [18]. By encapsulating large amounts of gadolinium ions, liposome-based nanoprobes have much higher longitudinal and transverse relaxivities compared with single molecule gadolinium chelates. The ability to accommodate different components (both lipophilic and hydrophilic compounds) makes the liposome an interesting vehicle for combined multimodal imaging and drug delivery systems. For example, de Vries et al. reported a multimodal liposome probe (182 nm in diameter) carrying gadolinium chelates, ^{111}In chelates, rhodamine dye, and tumor-targeted cyclic RGD peptides (Fig. 2) [19]. The liposomes were prepared by the widely used thin film hydration method, where a thin lipid film is hydrated using buffer or other aqueous solutions to form the liposome. A biocompatible polymer coating, polyethylene glycol (PEG), was added to offer improved in vivo stability and blood half-life. In addition to SPECT/MRI dual modality imaging and SPECT-based quantification, the liposome cargo allowed in vitro and in vivo determination of local/cellular concentrations of gadolinium (given that the ratio of gadolinium to radionuclide of the probe was known), and permitted calculation of relaxivities in biological environments. A similar liposomal nanoprobe targeted to tumors using the RGD peptide and the neuropeptide substance P (binds to neurokinin-1 receptor), however, revealed only moderate tumor uptake and no additive effect of dual-targeting approach [18]. These findings suggest that the liposome-based probes have potential for further targeting optimization.

Similar to liposome-based agents, polymer-based contrast agents, such as dendrimers or polysiloxane probes, can also increase the payload of gadolinium ions by permitting grafting of multiple gadolinium chelates to the same carrier.

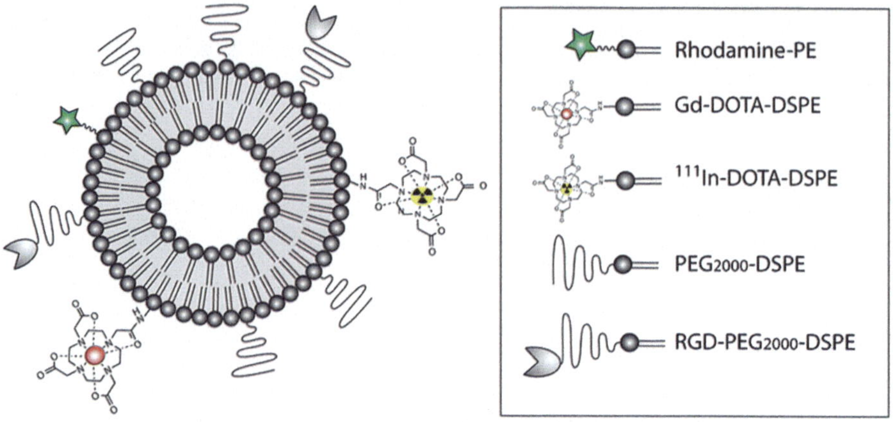

Fig. 2 Schematic representations of multimodal RGD-conjugated liposomal contrast agent. Figure adapted from Ref. [19]

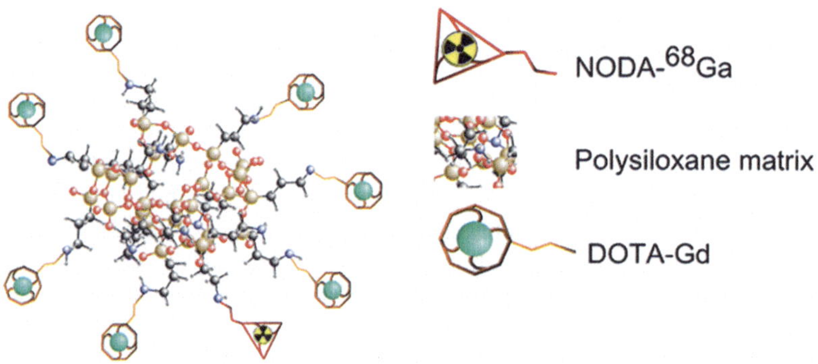

Fig. 3 Schematic representations of ultrasmall polysiloxane-based multimodal imaging probe. Figure adapted from Ref. [20]

A polysiloxane matrix was reported that carried Gd^{3+} and $^{68}Ga^{3+}$ macrocyclic chelates in a single particle, such as in a PET/MRI dual-modal imaging probe (Fig. 3) [20]. This interesting ultrasmall (2.5 nm in diameter), rigid platform was synthesized from a combination of bottom-up and top-down methods, where gadolinium oxide cores (1.7 nm) were first synthesized and induced hydrolysis-condensation of silane precursors on the surface to form monodispersed polysiloxane shells in diethylene glycol (bottom-up). Macrocyclic chelators, such as DOTAGA (1,4,7,10-tetraazacyclododecane-1-glutaric anhydride-4,7,10-triacetic acid), were then conjugated to the core-shell particles. When the particles were transferred in water, the macrocyclic chelators strongly accelerated the dissolution of Gd_2O_3 core in aqueous solutions. Since the polysiloxane layer was thin, the defects in this layer might have favored the migration of Gd^{3+} out of the core. The gadolinium ions were then chelated by the macrocyclic moieties and the hollow polysiloxane structure collapsed and fragmented into small and rigid scaffolds (top-down). The final particle has a slightly positive charge (+4.5 mV) at physiological pH and was demonstrated as a promising PET/MRI polymer-based imaging probe. Compared with traditional methods, where polymers/dendrimers are synthesized in complicated steps to become scaffolds large enough to conjugate gadolinium chelates, this method is simpler and more efficient. A uniform polysiloxane matrix was formed on the core surface and hosted about 8 chelates per particle.

Beside liposomes and polymers, researchers have also investigated inorganic nanomaterials as scaffolds for imaging probes. Fullerene cages were reported to provide large surface functionalization and biocompatibility [21]. Their ability to isolate the metals from the biological system by keeping them inside the carbon cages lowers systemic toxicity. Mesoporous silica nanoparticles, which also have very high surface area for metal chelates (Gd^{3+}/$^{64}Cu^{2+}$-DTPA) like other nanoparticles, demonstrated not only their ability as PET/MRI probes, but also their potential as therapeutic vectors. With their high pore volume (0.6–1 cm^3/g), drug loading of higher than 30 % weight can be achieved in these mesoporous nanoconstructs [22].

2.3 *Radiolabeling and Stability Studies*

Radiolabeling MRI probes adds a quantitative component and allows for the determination of the pharmacokinetics and biodistribution profile of nanoprobes through measuring the radioactivity in tissues. To this end, the radionuclide of choice depends on the imaging application of a particular probe. Radionuclides such as ^{99m}Tc [12, 14, 15, 23–28], ^{125}I [12, 16, 29, 30], ^{131}I [31], ^{111}In [18, 19, 32, 33], ^{59}Fe [34], and ^{177}Lu [35] have been explored for multimodal nanoparticle-based SPECT/MRI probes while ^{18}F [13], ^{59}Ge [36], ^{68}Ga [20, 37], ^{64}Cu [22, 38], and ^{89}Zr [39] have been reported for PET/MRI probes. ^{99m}Tc is the most commonly used γ-emitting SPECT radionuclide having a half-life of 6 h, useful γ-energy (140 keV), and widespread availability through ^{99}Mo-^{99m}Tc generators. The quantity of the reducing agent, $SnCl_2$ used in synthesis, is a critical parameter during direct labeling of a multimodal nanoparticle probe using $^{99m}TcO_4^-$ and needs to be optimized to circumvent any undesirable radiocolloids or free pertechnate resulting from incomplete reduction. The utility of ^{99m}Tc is limited by its relatively short half-life ($t_{1/2}$=6 h) and may not be applicable for use in tracking stem cells in vivo. Therefore, radiolabeling with radionuclides such as ^{125}I ($t_{1/2}$=59 days) and ^{69}Ge ($t_{1/2}$=39.05 days) with longer half-life for SPECT and PET imaging, respectively, has been proposed to monitor transplanted stem cells using SPECT imaging [29] or perform immuno-PET imaging using radiolabeled antibodies [36]. However, the damaging ionizing radiation of gamma rays at the cellular level may limit the use of gamma-emitting radionuclides for cell tracking. Gamma radiation is typically used more in therapeutic applications than diagnostic. Exposure to more than 1 Gy of gamma radiation may lead to acute radiation syndrome that may result in hematopoietic disorder, gastrointestinal disturbance, and cardiovascular instability; while absorption doses of 30–50 Gy can result in blistering and necrosis in 3 weeks [40].

In addition, radiolabeling of nanoparticles with ^{69}Ge using traditional chelator-based methods is challenging. Germanium species in aqueous solution exist in multiple forms $Ge(OH)_4$, $[Ge(OH)_3]^-$, $[GeO_2(OH)_2]^{2-}$, and $[[Ge(OH)_4]_8(OH)_3]^{3-}$ whose relative proportion is dependent upon the pH of the solvent [36]. The presence of different germanium species complicates the development of a chelator that can coordinate germanium in order for this radiometal to remain attached with the nanoparticle. To address the problem of the complex coordination chemistry of ^{69}Ge in aqueous solutions, Chakravarty et al. incorporated ^{69}Ge through specific ^{69}Ge sorption on the surface of the iron oxide nanoparticles, which are coated with poly(acrylic acid) (SPION@PAA). To label SPION@PAA, the pH of ^{69}Ge solution was adjusted to 7 using Na_2CO_3 solution and subsequently added to SPION@PAA (dispersed in 1 mM sodium hydroxide, pH 7–8). The radiolabeling yield of ^{69}Ge-SPION@PAA was about 90 % after 24 h of incubation at 37 °C. The specificity of ^{69}Ge adsorption toward the surface of the SPION was demonstrated using dense silica to block the SPION surface from ^{69}Ge and copper sulfide as nonmetal oxide nanoparticles for ^{69}Ge radiolabeling. Both cases resulted in low (<15 %) radiolabeling yield. On the other hand, the radiolabeling yield for ligand-free SPION was >95 % and free PAA did not exhibit any significant radioactivity indicating SPION-specific ^{69}Ge radiolabeling. The prepared ^{69}Ge-SPION was PEGylated to increase in vivo stability. In vitro incubation ^{69}Ge-SPION@PEG in mouse serum at 37 °C for 24 h was found to have ca. 75 % intact ^{69}Ge while non-PEGylated ^{69}Ge-SPION was less stable (<60 % radiolabeling yield) after 6 h of incubation. PET imaging studies of ^{69}Ge-SPION@PEG performed after intravenous injection in normal BALB/c mice revealed liver uptake indicating integrity of ^{69}Ge-SPION@PEG in vivo, while intravenous injection of free ^{69}Ge exhibited rapid renal clearance and nearly no accumulation in the liver [36].

The most commonly used radionuclides for clinical PET imaging such as ^{18}F ($t_{1/2}$=110 min) are limited by short half-lives. Research isotopes such as ^{64}Cu ($t_{1/2}$=12.7 h) and ^{68}Ga ($t_{1/2}$=68 min) have longer half-lives and can easily be chelated with 1,4,7,10-tetraazacyclododecane-1,4,7,10-tetraacetic acid (DOTA) and 1,4,7-triazacyclononane-1,4,7-triacetic acid (NOTA), respectively, forming highly stable complexes (pK_{CuDOTA}=22.3 [38], pK_{GaNOTA}=30.98 [20]) (Fig. 4). Integrity of radiolabeled nanoparticles and stability from dechelation of radiometals are typically assessed by incubating these nanoparticles in the presence of potent metal chelators including histidine, diethylenetriaminepentaacetic acid (DTPA), or serum (human and/or bovine serum albumin) at physiological temperature, and measuring radioactivity at different time points using instant thin layer chromatography using different mobile phases. In the work of Tsiapa et al. the evaluation of the ^{99m}Tc binding with iron oxide nanoparticles was performed by challenge stability studies in the presence of DTPA and His using instant thin layer chromatography with acetone and normal saline as mobile phases [12]. When acetone was used as mobile phase, $^{99m}TcO_4$ (R_f=0.7–1.0) can be separated from ^{99m}Tc-iron oxide nanoparticles (R_f=0). On the other hand, saline was used as mobile phase to separate ^{99m}Tc-His and ^{99m}Tc-DTPA (R_f= 1.0) from $^{99m}TcO_4$ and ^{99m}Tc-iron oxide nanoparticles (R_f=0).

Several research groups used doping into the lattice of iron oxide nanoparticles as an alternative approach to traditional radiolabeling by chelation, with the notion that containment in the nanoparticle core could prevent leaching due to transchelation of the radiometal or detachment of the radiolabeled ligand from the nanoparticles. Zeng

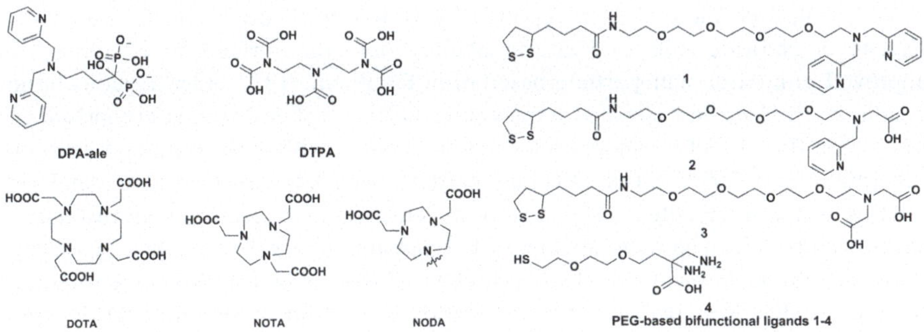

Fig. 4 Structures of chelators for SPECT and PET radiometals

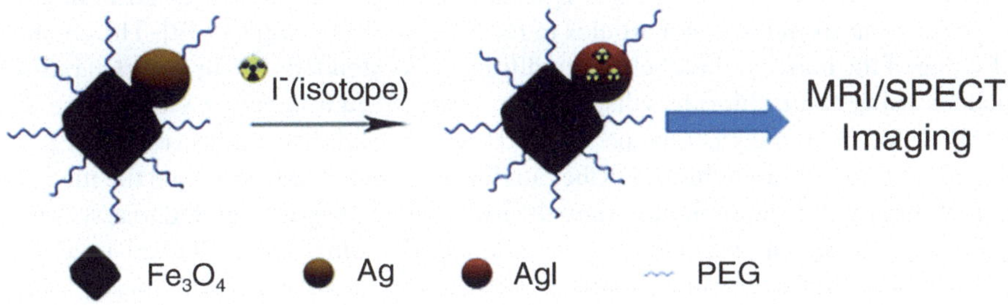

Fig. 5 PEGylated (*blue* squiggly lines) Fe_3O_4-$Ag^{125}I$ heterostructured nanoparticles for MR/SPECT imaging were prepared by introducing Ag component (*orange circle*) onto the Fe_3O_4 (*black square*) surface. The Ag component enabled ^{125}I attachment onto the Ag surface producing $Ag^{125}I$ (*red circle*) without changing the morphology of the Fe_3O_4-Ag heterodimer core. Reproduced with permission from Ref. [41]

et al. [33] incorporated ^{111}In into the Fe_3O_4 lattice, and the resulting ^{111}In-doped Fe_3O_4 nanoparticles were stable in saline and fetal bovine serum for at least 24 h of incubation. In a separate proof-of-concept study, Hoffman et al. [34] reported ^{59}Fe-labeled iron oxide nanoparticles ([^{59}Fe]-SPIONs) where ^{59}Fe was incorporated into the F_3O_4 nanoparticle lattice during thermal decomposition synthesis. No precipitation was observed when [^{59}Fe]-SPIONs was incubated in mouse plasma for at least 24 h demonstrating stability. Zhu et al. [41] introduced Ag component grown onto the Fe_3O_4 nanoparticles by adding silver acetate (Fig. 5). Introduction of the Ag component on the Fe_3O_4 allows $^{125}I^-$ to bind on the Ag surface to produce heterostructured Fe_3O_4-$Ag^{125}I$ nanoparticles with 80 % radiolabeling yield and low cell toxicity. While chemical doping provides an alternative approach for radiolabeling without the use of chelators, the doping process requires time-consuming full batch preparation and therefore may not be applicable for radionuclides with short half-lives [39]. Microwave-based synthesis techniques have been demonstrated by our research group for incorporating ^{64}Cu into dextran-coated iron oxide nanoparticles in a process that yielded 50 nm nanoparticles with heating times (5 min) significantly shorter than traditional methods (2 h) [42].

A chelate-free radiolabeling reported by Boros et al. involved the use of the oxalate or chloride salts of ^{89}Zr for labeling onto the surface of the clinically approved iron oxide nanoparticle-based drug Feraheme (FH) under aqueous basic conditions (pH 8) and elevated temperature (120 °C). Feraheme is comprised of iron oxide (Fe_2O_3:FeO) nanoparticles that are coated with carboxymethyl dextran. Incorporation of $^{89}Zr^{4+}$ into the FH nanoparticles was believed to involve metal ion binding to the magnetite crystal surface as supported by electron spin resonance spectroscopy data. The radiolabeling of FH nanoparticles was quenched by adding desferrioxamine B mesylate (DFO), DTPA, EDTA, or a DTPA/EDTA mixture. Stability studies were performed by incubating ^{89}Zr-FH in sterile saline, DTPA (pH 7.5), DTPA-EDTA mixture (pH 7), and human and rat plasma at 37 °C for 72 h and revealed >85 % intact ^{89}Zr-FH. However, lower (52 % remaining radioactivity) stability was obtained when challenged with desferrioxamine (DFO, pH 7). The relatively lower stability of ^{89}Zr-FH against challenge with DFO was attributed to Fe^{3+} extraction from the nanoparticles to form Fe^{3+}–DFO complex [39]. The chemical scope of the above radiolabeling conditions was extended to ^{111}In and ^{64}Cu radionuclides using their chloride salts, and the obtained radiolabeling yields were 91 and 66 % for ^{111}In and ^{64}Cu, respectively. However, stability studies were not performed for these radionuclides. Further studies are needed to understand the mechanism of binding for radionuclides onto the iron oxide nanoparticles and to assess the stability of these nanoparticles when radionuclides other than ^{89}Zr are used for radiolabeling. Improvements in stability will also be necessary before these probes are ready for the clinic.

3 In Vivo Imaging Applications

Nanoparticle-based imaging probes can be passively endocytosed by macrophages or accumulate in tumors or lymph nodes by the enhanced permeability retention (EPR) effect. This feature makes nanoparticles an attractive candidate for translational cancer diagnostic and treatment applications. In addition, the introduction of ligands to the nanoparticle surface enables specific targeting to disease biomarkers and generally results in a much higher labeling efficiency. In this section, we will briefly discuss PET/MRI and SPECT/MRI nanoprobes that have been used for imaging tumor, lymph node, and other targets.

3.1 PET/MRI: Tumor Imaging

3.1.1 Convection-Enhanced Delivery (CED)

As delivery of imaging probes and drugs to brain tumor is a challenging task, CED has been designed to overcome difficulties so that agents that would not normally cross the blood-brain barrier can be utilized for imaging or treatment. Probes can be delivered through catheters placed stereotactically directly within the tumor mass or around the tumor to improve deliver efficiency [56]. For example, an ^{124}I radiolabeled tri-gadolinium endohedral metallofullerene $Gd_3N@C_{80}$ nanoprobe platform was designed for simultaneous PET/MRI to examine brain tumors [21]. The surface of the fullerene cage was functionalized with carboxyl and hydroxyl groups and subsequently radiolabeled with ^{124}I to produce ^{124}I-f-$Gd_3N@C_{80}$ nanoprobe. Rats bearing T9 rat glioblastoma tumors on the right hemisphere of the brain were infused intratumorally with ^{124}I-f-$Gd_3N@C_{80}$ by CED using a microinjection pump. MRI and PET images are shown in Fig. 6. T_1- (panels b, e) and T_2- (panels c, f) weighted images showed bright and dark contrast, respectively, for the fullerene probe. This region was also highlighted by PET signals (panels j, k, l) from the probe. PET provides much stronger localization of the tumor, while MRI gives context for position of the tumor in the brain. This work has successfully shown the potential of metallofullerene-based multimodal probes and could be followed with further studies to optimize the surface functionality to target specific biomarkers. Substitution of ^{124}I with either ^{125}I or ^{131}I could serve for therapeutic purposes as well.

3.1.2 Passive Accumulation

Several nontargeted nanoparticles have been reported to detect xenografted tumors in animal models through passive accumulation by the EPR effect [37, 45, 53]. For example, positron emitting magnetic nanoconstructs (PEMs) of 140 nm in diameter comprised of three different compartments: a poly(lactic-co-glycolic acid) core, encapsulated USPIO; a phospholipid monolayer, labeled with DOTA-^{64}Cu or optical dye; and poly(ethylene glycol) chains on the outer compartment [53]. PEMs were intravenously injected via tail vein in nude mice bearing MDA-MB-231 human breast cancer xenografts and found to passively accumulate in tumors. Images taken at 1, 6, and 20 h postinjection showed highlighted vascular pool at first two time points, which washed out to allow highlighting of the tumor region at 20 h. PEMs have a circulation half-life estimated to be longer than 7 h, which was attributed to the higher structural rigidity and stability induced by the phospholipid and PEG coating. Biodistribution indicated 3.5 % ID/g accumulation in tumor, while the majority of the dose resided in the liver (22 % ID/g), spleen (~5 % ID/g), and kidney (~5 % ID/g). Although MRI only and PET/CT images were also acquired and both showed nonuniform distribution of PEMs surrounding the tumor, clear demonstration of the advantages of multimodal imaging was not explained in this work.

3.1.3 Active Accumulation

By adding targeting ligands to the outer surface of a particle, probes can be specifically targeted to tumor biomarkers. For example, integrin $\alpha_v\beta_3$ is a well-studied, cell-adhesion molecule that is expressed on both proliferating tumor vasculature and certain tumors cells but not on normal vasculature/cells, and thus has been widely targeted as a tumor biomarker. A common construct is to use the cyclic

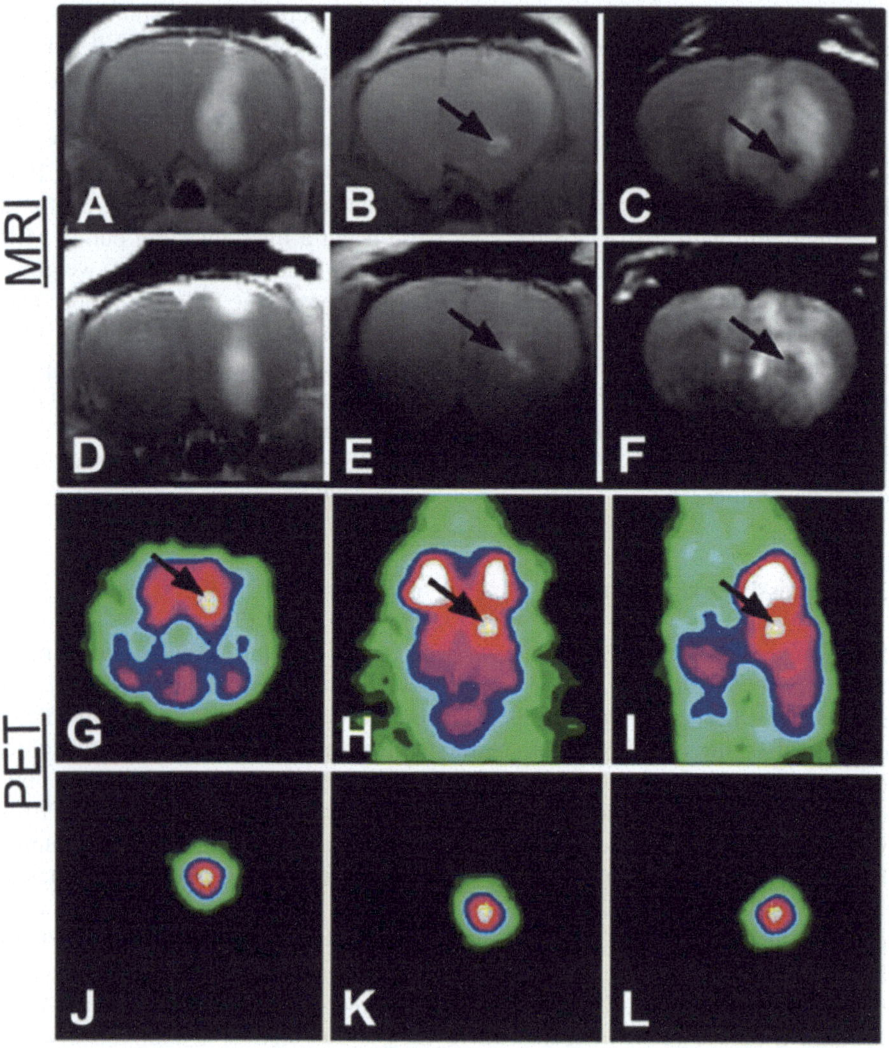

Fig. 6 MR images from two representative orthotopic tumor bearing rat brains and microPET images from one representative rat infused with the ^{124}I-f-$Gd_3N@C_{80}$ product. (**a**, **d**) T_1-weighted images with gadodiamide contrast agent. (**b**, **e**) T_1-weighted images showing bright contrast at the infusion site due to the ^{124}I-f-$Gd_3N@C_{80}$. (**c**, **f**) T_2-weighted images with dark contrast due to the ^{124}I-f-$Gd_3N@C_{80}$. (**j**) Coronal, (**k**) axial, and (**l**) sagittal microPET images showing signal from the ^{124}I-f-$Gd_3N@C_{80}$. (**g**) Coronal, (**h**) axial, and (**i**) sagittal microPET images following ^{18}F-FDG injection with the additive image signal allowing for localization of the ^{124}I-f-$Gd_3N@C_{80}$ within the right hemisphere of the rat brain (*arrows* point to infusion sites). Reproduced with permission from Ref. [21]

arginine-glycine-aspartic acid (cRGD) peptide, a potent integrin $\alpha_v\beta_3$ antagonist, to improve the specific targeting ability of nanoprobes [49]. Recent work described development of a ^{64}Cu-NOTA-labeled PEGylated SPIO nanocarrier of the anticancer drug doxorubicin, conjugated with the cRGD peptide. The cRGD-conjugation significantly improved cellular uptake of the nanocarrier and tumor accumulation. This pilot study did not investigate the in vivo therapeutic effect of released doxorubicin. One would expect to see a subsequent study where sufficient drug can be released from these nanocarriers and significantly reduce the tumor size after certain days.

Epidermal growth factor receptor (EGFR) is another well-established tumor biomarker, and anti-EGFR affibody, protein-modified, gold-iron oxide hetero-nanostructure was reported for specific diagnosis of various EGFR positive tumors [43]. Oleanolic acid, which has inhibitory effect on colon cancer cells, has also been utilized as a tumor-targeting ligand for PET/MRI nanoprobes [48]. More discussion on those tumor targeted probes can also be found on our previous review [6].

3.2 PET/MRI: Lymph Node Imaging

The sentinel lymph node (SLN) is considered the first regional lymph node receiving lymphatic drainage from tumor and probably the first node where metastatic cells anchor. Therefore, accurate localization and identification of lymph nodes is important for cancer staging and surgical planning. However, the current procedure for SLN identification and subsequent removal (along with the primary tumor) involves a surgical procedure guided by a mixture of a radiolabeled colloid (^{99m}Tc) and a blue dye; this colloid is nonspecific and labeled lymph nodes may or may not be cancerous. This SPECT method also suffers from limited spatial resolution.

A new approach takes advantage of the combined sensitivity and resolution of PET/MR imaging with the aid of multimodal, radiolabeled, iron oxide nanoparticles. The high resolution provided by PET/MR imaging enables detection of lymphatic micrometastases and better visualization of tumor invasion [8]. Particle size combined with surface properties plays an important role in enhancing lymphatic transport. Small particles with hydrodynamic diameter of <100 nm are transported and taken up more readily by the lymphatic ducts. For example, Cui et al. investigated the reaction of the popliteal lymph node in response to an acute inflammatory stimulus of lipopolysaccharide (LPS) injected into the foot [13]. After LPS induced inflammation, a solution of ^{18}F-labeled $Co_{0.16}Fe_{2.84}O_4$@$NaYF_4$(Yb,Er)-BP-PEG was injected into each of the two rear footpads of a female C57BL/6 mice. The MR image (Fig. 7) revealed a white spot with darkened background in the right popliteal lymph node, whereas a black spot with a white background was evident in the left counterpart. This observed contrast in the right popliteal lymph node was believed to be due to the LPS-induced response from macrophages migrating to draining lymph nodes, which tend to localize in the capsular layer of the lymph nodes. These macrophages may have consumed the radiolabeled nanoparticles resulting in lower PET signal intensity for the inflamed popliteal lymph node relative to the normal popliteal lymph node; hence, normal lymph node appears brighter than the inflamed lymph

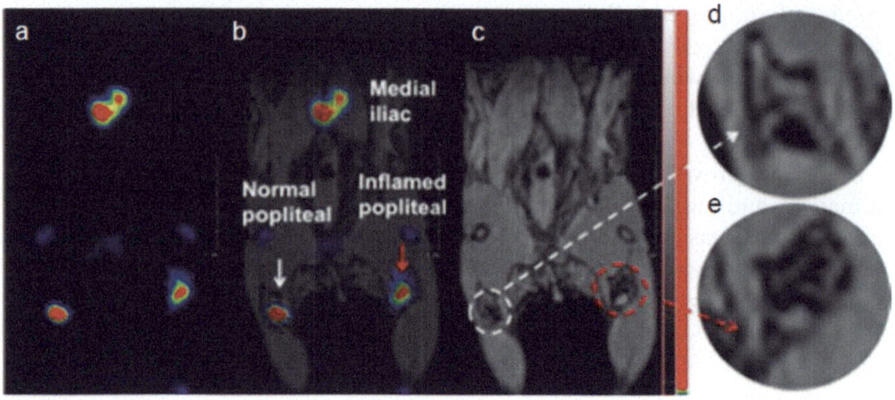

Fig. 7 Lymph node PET/MR imaging of a mouse with inflamed right leg using ^{18}F-labeled Fe_3O_4@$NaYF_4$(Yb,Tm)-BP-PEG nanoparticles: (**a**) PET image showing popliteal and iliac lymph nodes (coronal section); (**b**) PET/MR fused image (coronal section); (**c**) MR image (coronal section) with darkening contrast inside popliteal lymph node at left-rear (*white circle*) and "outside" lymph node at the inflamed right-rear (*red circle*) induced by injection of lipopolysaccharide 18 h prior to imaging, and at iliac lymph node; and (**d**, **e**) enlarged MR images of corresponding lymph nodes. The MR image revealed a white spot with darkened background in the right popliteal lymph node, whereas a black spot with a white background was evident in the left counterpart. In this case, the PET image was used for easier identification of the radiolabeled agents, while the MR images revealed distribution and uptake of the nanoparticles inside and outside the lymph nodes. Adapted with permission from Ref [13]

node as seen in Panels a and b of Fig. 7. While the PET images allowed easy detection of relevant lymph nodes, MRI provided resolution lacking in PET to delineate the distribution of contrast agents in and around the lymph node. The role for the $NaYF_4$ incorporated on the nanoparticles was not demonstrated in this work, but its fluorescent properties could be explored as a potential visual guide during surgery.

3.3 PET/MRI: Other Applications

3.3.1 Atherosclerosis

The rupture of vulnerable atherosclerotic plaques can lead to acute clinical events, such as stroke or myocardial infarction. Thus, noninvasive and sensitive detection of vulnerable plaques has gained wide interest and attention. Our group has studied a number of nanoparticle-based PET/MRI probes to visualize vulnerable atherosclerotic plaques. We have reported three types of agents, MDIO-^{64}Cu-DOTA, SDIO-DO3A, and mal-BSA-^{64}Cu/Gd-DOTA nanoparticles targeted to macrophage scavenger receptor A, to label macrophages accumulated in the vulnerable plaques [38, 46, 57]. We used sensitive, whole body, PET images to identify inflamed plaques, which then can guide the high-resolution MRI to acquire detailed local information (Fig. 8). The PET prescan (Fig. 8a) reveals sites of macrophage

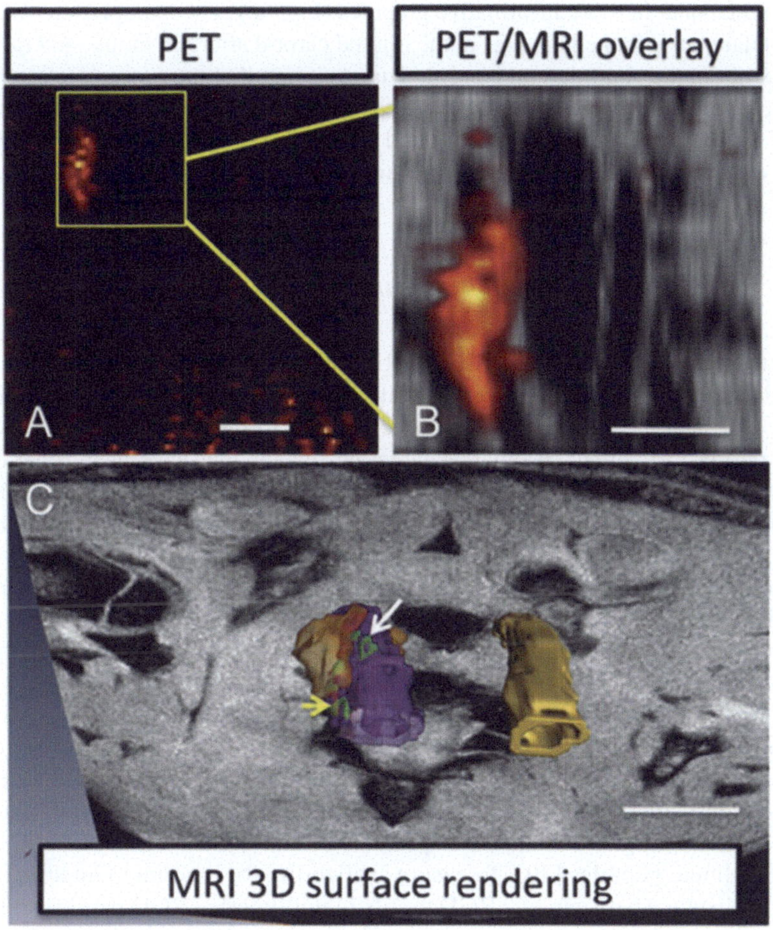

Fig. 8 Multimodal probe accumulate in the injured vessel of the rat carotid clamp injury model. (**a**) Coronal view PET image of the rat thorax shows a region of high signal intensity indicating probe accumulation in macrophages; however, it is difficult to interpret the tissue of origin for the signal without anatomical information. Scale bar = 20 mm. (**b**) Higher magnification MR images for the volume indicated by the boxed area in A reveal that the PET signal correlates to the carotid artery on the left. Scale bar = 5 mm. This vessel also shows elevated MR contrast and thicker vessel walls compared to the vessel on the right. (**c**) 3D reconstruction of MRI and PET data from the carotids is shown protruding from a plane in the MR image. This view illustrates that the probes are localized to the vessel wall of the injured carotid artery (*purple*), higher MR signal intensities relative to vessel background signal are rendered in green, and PET intensities are in orange. The contralateral uninjured vessel is *yellow*. Images rendered by segmentation. Scale bar = 5 mm. Reproduced with permission from Ref. [57]

accumulation that can direct the observer where to look more closely for detailed MRI signal (Fig. 8b). Rendering multimodal imaging sets is a challenge and we have been working with computer scientists to develop programs that can display multimodal data in more informative graphics (Fig. 8c). The nanoparticles actively accumulated in inflamed plaques in the injured carotid artery in mouse and rat models, but not in the contralateral control, demonstrating that the probes are promising for future clinical translations. We have recently characterized the binding properties of these nanoparticle agents and found that the binding affinities for the nanoparticles were similar to that of free ligands to the scavenger receptor [46].

3.4 SPECT/MRI: Tumor Imaging

Reported nanoparticle-based, SPECT/MRI probes have featured a variety of tumor-specific ligands. As mentioned earlier, attaching a ligand to a receptor found on tumor cells can increase tumor labeling compared to the passive approach, which relies on the EPR effect. Biomarker-targeting ligands that have been reported for nanoparticle-based, SPECT/MRI probes for tumor imaging include peptides (RGD) [19], monoclonal antibodies (antiangiogenic antibody Bevacizumab [26], antimesothelin [32], antigastric cancer monoclonal antibody 3H11 [16], humanized IgG1 monoclonal antibody Trastuzumab [35]), nucleic acids(siRNA [31]), and other molecules such as sugar acids (lactobionic acid) [17]. Peptides have received much attention as targeting ligands because of their small size, lower immunogenicity, and high stability.

As noted in the active accumulation section, iron oxide nanoparticles with tripeptide sequence Arg-Gly-Asp (RGD) target $\alpha_v\beta_3$ integrins, which are strongly expressed on the endothelium of angiogenic blood vessels in tumor angiogenesis [12]. A derivative of this peptide sequence, cyclic Arg-Gly-Asp-D-Phe-Lys (cRGDfK), has higher resistance to proteolysis and higher affinity to $\alpha_v\beta_3$ integrins than linear peptides [30]. To ensure availability for binding to integrin receptors, the distance between two RGD motifs in multimeric cRGD peptides must be long enough and spacers such as two tyrosine residues [30] installed between two cRGD motifs and tripeptide ornithine [12] have been used to facilitate binding of cRGD to $\alpha_v\beta_3$ integrins. These probes have demonstrated multimodal imaging of tumor angiogenesis. For example, ^{99m}Tc-USPION-RGD [14] administered to nude mice containing H1299 tumor cells showed hyper-intense (T_1-weighted) and hypo-intense (T_2-weighted) MR signals that were distributed mainly in the tumor peripheral regions and highlighted the tumor angiogenesis vessels. Enhanced contrast between tumor angiogenesis and the necrotic core was also observed as the MR signal intensity of the tumor necrotic core remained the same before and after probe injection. SPECT images revealed accumulation of the probe in the tumor peripheral regions. Overall, the use of MRI-active probes in this study highlighted tumor angiogenesis and provided contrast from the necrotic tumor core. The use of the radionuclide ^{99m}Tc enabled quantification of the probes accumulated in the tumor ($13.8 \pm 9.6\%$ ID/g) as well as determination of their biodistribution.

3.5 SPECT/MRI: Lymph Node Imaging

Similar to the PET/MRI examples, SPECT/MRI has employed ^{99m}Tc-SPIONs for SPECT/MRI sentinel lymph node (SLN) mapping in vivo. SPIONs were radiolabeled with high efficiency (99 %) and showed high accumulation in the lymph nodes [23]. In a subsequent study, a trimodal probe was developed to add an optical component as an intraoperative feature. As shown in Fig. 9, detection of SLN was shown for all three modalities: SPECT, MRI, and optical. The ^{99m}Tc-AF-SPIONs probe targeted SLN in rat models and was shown to be stable in human serum for up to 24 h, and have high accumulation in SLN [27]. Major accumulation of the probe was visualized

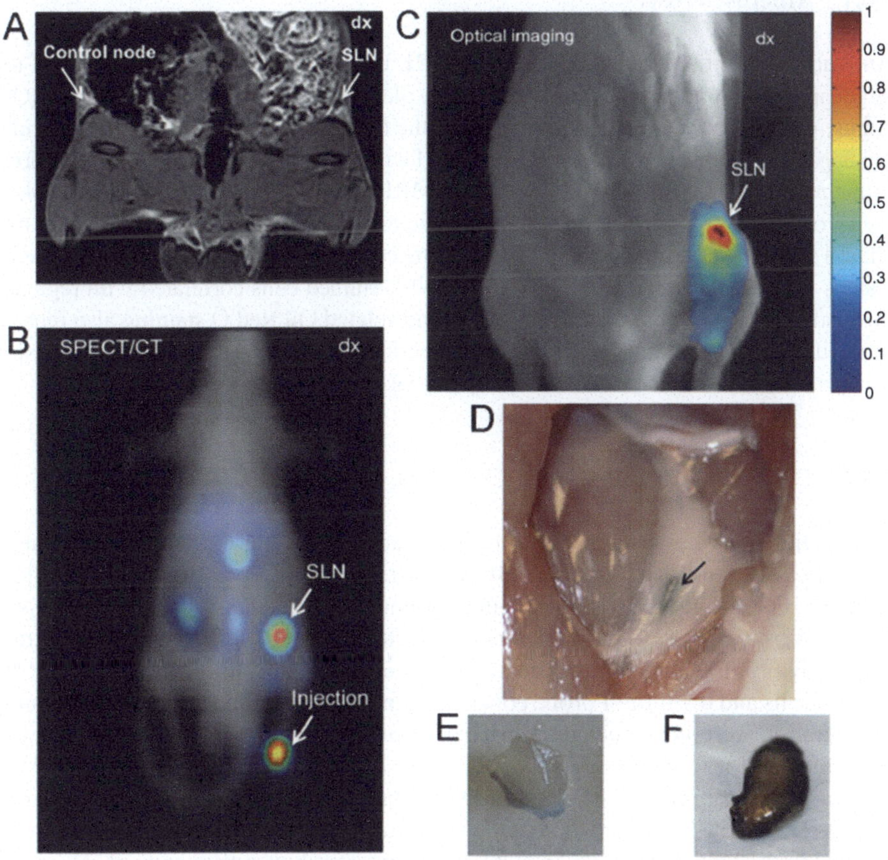

Fig. 9 (**a**) Representative coronal MR images of a Wistar rat injected subcutaneously with ^{99m}Tc-AF-SPIONs in the right hind paw. Accumulation of the ^{99m}Tc-AF-SPIONs in the SLN can be clearly visualized using GRE sequences. (**b**) SPECT/CT image of the same animal. SPECT is less affected by attenuation compared with optical imaging, and therefore is an invaluable tool to quantify and study the biodistribution of a newly developed agent. (**c**) Optical fluorescence image visualizing the SLN. (**d**) Intraoperative identification of the SLN using ^{99m}Tc-AF-SPIONs. Similar to the clinical procedure using blue dye, ^{99m}Tc-AF-SPIONs stain the SLN green, which makes it easy to be identified during surgery. (**e**) Reference node from the collateral side of the animal. (**f**) Resected SLN. Figure adapted from Ref. [27]

in the injection side (right hind paw) and right inguinal lymph node, while smaller amounts were also detected in the renal lymph node and left inguinal lymph node. This work demonstrated how whole-body, SPECT imaging sensitively delineates the biodistribution of the probe. Although multiple modalities were used here, MRI was not used to advantage over SPECT/CT in this example. The optical component (Alexa Fluor 647), however, added value to the probe. It played an intraoperative role by staining the SLN green, providing real-time, visual guidance during surgery.

3.6 SPECT/MRI: Other Applications

3.6.1 Atherosclerosis

An Annexin V-modified, hybrid, SPECT/MRI, nanoparticle probe was developed to target apoptosis of macrophages in vulnerable plaques [15]. The ^{99m}Tc-DTPA-USPIO-Annexin V probe accumulated in plaques in the thoracic and abdominal aorta providing negative contrast in MRI. SPECT signals facilitated recognition and quantification of the region of interest, while T_2-weighted MRI improved the localization and volumetry of plaques. The existence of macrophages in aortic lesions was confirmed by CD-68 staining and further verified as apoptotic cells through TUNEL staining. Severe lesions with higher ratios of CD-68 and TUNEL stained cells correlated with regions that had more intense radioactive signals. Lipid-related Oil Red O staining also reproduced the SPECT/MR images, proving the specificity of ^{99m}Tc-DTPA-USPIO-Annexin V as a targeted probe for vulnerable plaques that contain apoptotic macrophages.

3.6.2 Cell Tracking

Stem cell therapy has brought hope for the treatment of central nervous system and other diseases, and there is much effort toward methods to track stem cells after delivery. For example, ^{125}I-labeled fluorescent silica coated SPIOs (^{125}I-fSiO$_4$@SPIOs) were designed to track mesenchymal stem cells (MSCs) transplanted intracerebrally or intravenously in a rat stroke model. With low radioactivity of 20 μCi/million cells and 0.1 mM of probe concentration, MSC viability, differentiation, and proliferation were not affected after labeling in vitro. This concentration was thus implemented for in vivo imaging. SPECT and MR images demonstrated that 35 % of intracerebrally-injected MSCs migrated along the corpus callosum to the lesion area as shown in Fig. 10, while 90 % of intravenously injected MSCs remained trapped in the lung at 14 days. In spite of the apparent lower efficiency of IV injection, significantly improved neurobehavioral outcomes, such as increased levels of vascular endothelial growth factor, insulin growth factor, and tissue inhibitor of matrix metalloproteinase-3, were observed by both administration routes. ^{125}I-fSiO$_4$@SPIOs were stable for long-term tracking and appear promising for brain ischemia treatment. In these studies, MRI signal was still detected even after 14 days, and the anatomic information provided by MRI is critical to localizing where the trafficking of the cells in brain tissue. The radionuclear signal provides a beacon for where to focus more detailed study of the MR images.

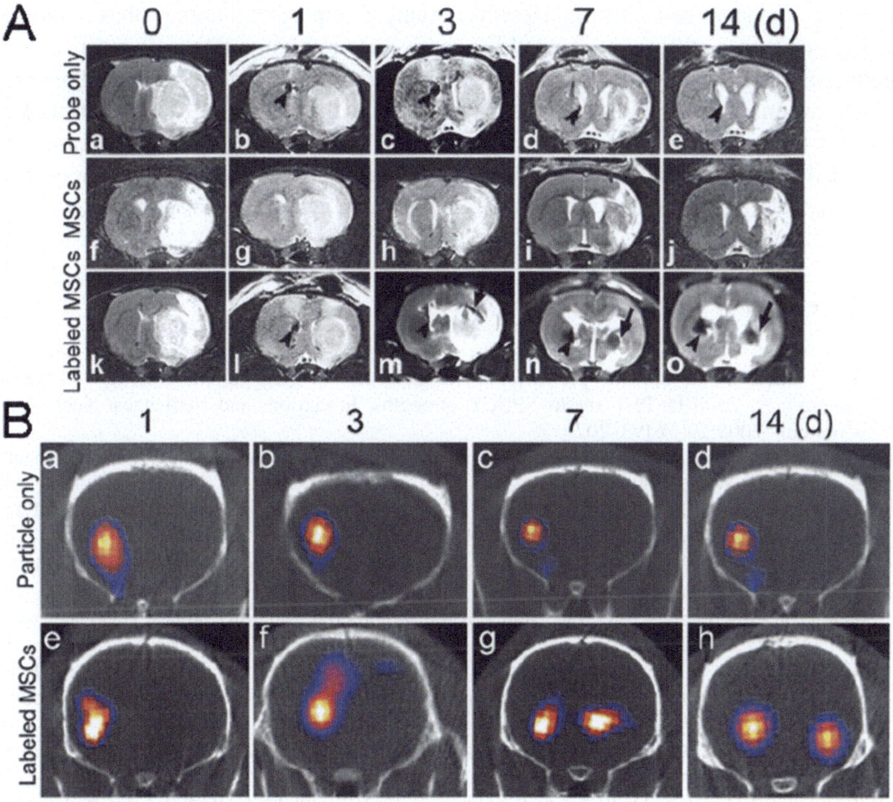

Fig. 10 MRI and SPECT/CT tracking of ^{125}I-fSiO4@SPIO-labeled MSCs in stroked rats after IC injection. (**a**) T_2-weighted MR images of ischemic rat brain before (a,f,k) and after IC injection of ^{125}I-fSiO4@SPIO particles, unlabeled MSCs, or ^{125}I-fSiO4@SPIO-labeled MSCs at day 1 (b,g,l), day 3 (c,h,m), day 7 (d,i,n), and day 14 (e,j,o). Arrows indicated hypointense signals from ^{125}I-fSiO4@SPIO particles and ^{125}I-fSiO4@ SPIO-labeled MSCs after IC injection. (**b**) SPECT/CT tracking of ^{125}I-fSiO4@SPIO-labeled MSCs in ischemic rat brains after IC injection. Figure adapted from Ref. [29]

4 Conclusions

Multimodality imaging is a promising method to achieve high-sensitivity and high-resolution imaging of targets deep in the body. We have shown in this chapter that nanoparticles are valuable platforms to build PET/MRI or SPECT/MRI multimodal imaging probes and have great potential to be translated into clinical applications. The PET or SPECT component offers high-sensitive signals for screening, while MRI can obtain high-resolution images with detailed anatomical and/or molecular information from regions of interest. In addition, PET and SPECT tracers can provide accurate quantification of the local probe concentration. Currently, PET/MRI and SPECT/MRI probes have been effectively used for a variety of preclinical applications such as tumor detection, lymph node labeling, atherosclerotic plaque imaging, and stem cell tracking. However, many nanoparticle-based probes reported to date are nonspecific and designed for passive accumulation rather than active targeting. Further development of multimodal imaging probes requires close collaboration between scientists and clinicians to truly harness the advantage of the multimodal species for a disease target that can benefit from the complementary information afforded by hybrid imaging. It is only through demonstration of improved diagnostic or therapeutic potential that multimodal imaging can achieve wide adoption.

References

1. Lee S, Chen X. Dual-modality probes for in vivo molecular imaging. Mol Imaging. 2009;8(2):87.
2. Rahmim A, Zaidi H. PET versus SPECT: strengths, limitations and challenges. Nucl Med Commun. 2008;29(3):193–207.
3. Cherry SR, editor. Multimodality imaging: beyond pet/ct and spect/ct. Seminars in nuclear medicine. Elsevier; 2009.
4. Beyer T, Townsend DW, Brun T, Kinahan PE, Charron M, Roddy R, et al. A combined PET/CT scanner for clinical oncology. J Nucl Med. 2000;41(8):1369–79.
5. Cherry SR, Louie AY, Jacobs RE. The integration of positron emission tomography with magnetic resonance imaging. Proc IEEE. 2008;96(3):416–38.
6. Garcia J, Tang T, Louie AY. Nanoparticle-based multimodal PET/MRI probes. Nanomedicine. 2015;10(8):1343–59.
7. Puttick S, Bell C, Dowson N, Rose S, Fay M. PET, MRI, and simultaneous PET/MRI in the development of diagnostic and therapeutic strategies for glioma. Drug Discov Today. 2015;20(3):306–17.
8. Rosales R. Potential clinical applications of bimodal PET-MRI or SPECT-MRI agents. J Label Compd Radiopharm. 2014;57(4):298–303.
9. Bouziotis P, Psimadas D, Tsotakos T, Stamopoulos D, Tsoukalas C. Radiolabeled iron oxide nanoparticles as dual-modality SPECT/MRI and PET/MRI agents. Curr Top Med Chem. 2012;12(23):2694–702.
10. Singh N, Jenkins GJ, Asadi R, Doak SH. Potential toxicity of superparamagnetic iron oxide nanoparticles (SPION). Nano Rev. 2010;1.
11. Wang Y-XJ. Superparamagnetic iron oxide based MRI contrast agents: current status of clinical application. Quant Imaging Med Surg. 2011;1(1):35.
12. Tsiapa I, Efthimiadou EK, Fragogeorgi E, Loudos G, Varvarigou AD, Bouziotis P, et al. ^{99m}Tc-labeled aminosilane-coated iron oxide nanoparticles for molecular imaging of $\alpha\nu\beta3$-mediated tumor expression and feasibility for hyperthermia treatment. J Colloid Interface Sci. 2014;433:163–75.
13. Cui X, Mathe D, Kovács N, Horvath I, Jauregui-Osoro M, Torres Martin de Rosales R, et al. Synthesis, characterization and application of core-shell Co0.16Fe2.84O4@NaYF4 (Yb, Er) and Fe_3O_4@NaYF4 (Yb, Tm) nanoparticle as tri-modal (MRI, PET/SPECT and optical) imaging agents. Bioconjug Chem. 2015.
14. Xue S, Zhang C, Yang Y, Zhang L, Cheng D, Zhang J, et al. ^{99m}Tc-labeled iron oxide nanoparticles for dual-contrast (T 1/T 2) magnetic resonance and dual-modality imaging of tumor angiogenesis. J Biomed Nanotechnol. 2015;11(6):1027–37.
15. Cheng D, Li X, Zhang C, Tan H, Wang C, Pang L, et al. Detection of vulnerable atherosclerosis plaques with a dual-modal single-photon-emission computed tomography/magnetic resonance imaging probe targeting apoptotic macrophages. ACS Appl Mater Interfaces. 2015;7(4):2847–55.
16. Liu S, Jia B, Qiao R, Yang Z, Yu Z, Liu Z, et al. A novel type of dual-modality molecular probe for MR and nuclear imaging of tumor: preparation, characterization and in vivo application. Mol Pharm. 2009;6(4):1074–82.

17. Lee CM, Jeong HJ, Kim EM, Kim DW, Lim ST, Kim HT, et al. Superparamagnetic iron oxide nanoparticles as a dual imaging probe for targeting hepatocytes in vivo. Magn Reson Med. 2009;62(6):1440–6.
18. Rangger C, Helbok A, Sosabowski J, Kremser C, Koehler G, Prassl R, et al. Tumor targeting and imaging with dual-peptide conjugated multifunctional liposomal nanoparticles. Int J Nanomedicine. 2013;8:4659.
19. de Vries A, Kok MB, Sanders HM, Nicolay K, Strijkers GJ, Grüll H. Multimodal liposomes for SPECT/MR imaging as a tool for in situ relaxivity measurements. Contrast Media Mol Imaging. 2012;7(1):68–75.
20. Truillet C, Bouziotis P, Tsoukalas C, Brugière J, Martini M, Sancey L et al. Ultrasmall particles for Gd-MRI and ^{68}Ga-PET dual imaging. Contrast Media Mol Imaging. 2014.
21. Luo J, Wilson JD, Zhang J, Hirsch JI, Dorn HC, Fatouros PP, et al. A dual PET/MR imaging nanoprobe: ^{124}I labeled Gd3N@ C80. Appl Sci. 2012;2(2):465–78.
22. Laprise-Pelletier M, Bouchoucha M, Lagueux J, Chevallier P, Lecomte R, Gossuin Y, et al. Metal chelate grafting at the surface of mesoporous silica nanoparticles (MSNs): physicochemical and biomedical imaging assessment. J Mater Chem B. 2015;3(5):748–58.
23. Madru R, Kjellman P, Olsson F, Wingårdh K, Ingvar C, Ståhlberg F, et al. ^{99m}Tc-labeled superparamagnetic iron oxide nanoparticles for multimodality SPECT/MRI of sentinel lymph nodes. J Nucl Med. 2012;53(3):459–63.
24. de Rosales RTM, Tavaré R, Glaria A, Varma G, Protti A, Blower PJ. ^{99m}Tc-bisphosphonate-iron oxide nanoparticle conjugates for dual-modality biomedical imaging. Bioconjug Chem. 2011;22:455–65.
25. Felber M, Alberto R. ^{99m}Tc radiolabelling of Fe_3O_4–Au core–shell and Au–Fe_3O_4 dumbbell-like nanoparticles. Nanoscale. 2015;7(15):6653–60.
26. Zhao Y, Yao Q, Tan H, Wu B, Hu P, Wu P, et al. Design and preliminary assessment of ^{99m}Tc-labeled ultrasmall superparamagnetic iron oxide-conjugated bevacizumab for single photon emission computed tomography/magnetic resonance imaging of hepatocellular carcinoma. J Radioanal Nucl Chem. 2014;299(3):1273–80.
27. Madru R, Svenmarker P, Ingvar C, Ståhlberg F, Engels S-A, Knutsson L, et al. Development of a hybrid nanoprobe for triple-modality MR/SPECT/optical fluorescence imaging. Diagnostics. 2014;4(1):13–26.
28. Psimadas D, Baldi G, Ravagli C, Bouziotis P, Xanthopoulos S, Franchini MC, et al. Preliminary evaluation of a ^{99m}Tc labeled hybrid nanoparticle bearing a cobalt ferrite core: in vivo biodistribution. J Biomed Nanotechnol. 2012;8(4):575–85.
29. Tang Y, Zhang C, Wang J, Lin X, Zhang L, Yang Y, et al. MRI/SPECT/fluorescent tri-modal probe for evaluating the homing and therapeutic efficacy of transplanted mesenchymal stem cells in a rat ischemic stroke model. Adv Funct Mater. 2015;25(7):1024–34.
30. Deng S, Zhang W, Zhang B, Hong R, Chen Q, Dong J, et al. Radiolabeled cyclic arginine-glycine-aspartic (RGD)-conjugated iron oxide nanoparticles as single-photon emission computed tomography (SPECT) and magnetic resonance imaging (MRI) dual-modality agents for imaging of breast cancer. J Nanopart Res. 2015;17(1):1–11.
31. Chen J, Zhu S, Tong L, Li J, Chen F, Han Y, et al. Superparamagnetic iron oxide nanoparticles mediated 131I-hVEGF siRNA inhibits hepatocellular carcinoma tumor growth in nude mice. BMC Cancer. 2014;14(1):114.
32. Misri R, Meier D, Yung AC, Kozlowski P, Häfeli UO. Development and evaluation of a dual-modality (MRI/SPECT) molecular imaging bioprobe. Nanomedicine. 2012;8(6):1007–16.
33. Zeng J, Jia B, Qiao R, Wang C, Jing L, Wang F, et al. In situ 111 in-doping for achieving biocompatible and non-leachable ^{111}In-labeled Fe_3O_4 nanoparticles. Chem Commun. 2014;50(17):2170–2.
34. Hoffman D, Sun M, Yang L, McDonagh PR, Corwin F, Sundaresan G, et al. Intrinsically radiolabelled [^{59}Fe]-SPIONs for dual MRI/radionuclide detection. Am J Nucl Med Mol Imaging. 2014;4(6):548.
35. Rasaneh S, Rajabi H, Daha FJ. Activity estimation in radioimmunotherapy using magnetic nanoparticles. Chin J Cancer Res. 2015;27(2):203.

36. Chakravarty R, Valdovinos HF, Chen F, Lewis CM, Ellison PA, Luo H, et al. Intrinsically germanium-69-labeled iron oxide nanoparticles: synthesis and in-vivo dual-modality PET/MR imaging. Adv Mater. 2014;26(30):5119–23.
37. Cho B-B, Park JH, Jung SJ, Lee J, Lee JH, Hur MG et al. Synthesis and characterization of 68Ga labeled Fe3O4 nanoparticles for positron emission tomography (PET) and magnetic resonance imaging (MRI). J Radioanal Nucl Chem. 2015:1–10.
38. Tu C, Ng TS, Jacobs RE, Louie AY. Multimodality PET/MRI agents targeted to activated macrophages. J Biol Inorg Chem. 2014;19(2):247–58.
39. Boros E, Bowen AM, Josephson L, Vasdev N, Holland JP. Chelate-free metal ion binding and heat-induced radiolabeling of iron oxide nanoparticles. Chem Sci. 2015;6(1):225–36.
40. Bhattacharya S. Radiation injury. Indian J Plastic Surg. 2010;43(Suppl):S91.
41. Zhu J, Zhang B, Tian J, Wang J, Chong Y, Wang X, et al. Synthesis of heterodimer radionuclide nanoparticles for magnetic resonance and single-photon emission computed tomography dual-modality imaging. Nanoscale. 2015;7(8):3392–5.
42. Wong RM, Gilbert DA, Liu K, Louie AY. Rapid size-controlled synthesis of dextran-coated, ^{64}Cu-doped iron oxide nanoparticles. ACS Nano. 2012;6(4):3461–7.
43. Yang M, Cheng K, Qi S, Liu H, Jiang Y, Jiang H, et al. Affibody modified and radiolabeled gold–iron oxide hetero-nanostructures for tumor PET, optical and MR imaging. Biomaterials. 2013;34(11):2796–806.
44. Chen F, Ellison PA, Lewis CM, Hong H, Zhang Y, Shi S, et al. Chelator-free synthesis of a dual-modality PET/MRI agent. Angew Chem Int Ed. 2013;52(50):13319–23.
45. Locatelli E, Gil L, Israel LL, Passoni L, Naddaka M, Pucci A, et al. Biocompatible nanocomposite for PET/MRI hybrid imaging. Int J Nanomed. 2012;7:6021.
46. Tang T, Tu C, Chow SY, Leung KH, Du S, Louie AY. Quantitative assessment of binding affinities for nanoparticles targeted to the vulnerable plaque. Bioconjug Chem. 2015;26(6):1086–94.
47. Zhou J, Yu M, Sun Y, Zhang X, Zhu X, Wu Z, et al. Fluorine-18-labeled Gd 3+/Yb 3+/Er 3+ co-doped NaYF 4 nanophosphors for multimodality PET/MR/UCL imaging. Biomaterials. 2011;32(4):1148–56.
48. S-m K, Chae MK, Yim MS, Jeong IH, Cho J, Lee C, et al. Hybrid PET/MR imaging of tumors using an oleanolic acid-conjugated nanoparticle. Biomaterials. 2013;34(33):8114–21.
49. Yang X, Hong H, Grailer JJ, Rowland IJ, Javadi A, Hurley SA, et al. cRGD-functionalized, DOX-conjugated, and ^{64}Cu-labeled superparamagnetic iron oxide nanoparticles for targeted anticancer drug delivery and PET/MR imaging. Biomaterials. 2011;32(17):4151–60.
50. Madru R, Tran TA, Axelsson J, Ingvar C, Bibic A, Ståhlberg F, et al. ^{68}Ga-labeled superparamagnetic iron oxide nanoparticles (SPIONs) for multi-modality PET/MR/Cherenkov luminescence imaging of sentinel lymph nodes. Am J Nucl Med Mol Imaging. 2014;4(1):60.
51. Cui X, Belo S, Krüger D, Yan Y, de Rosales RT, Jauregui-Osoro M, et al. Aluminium hydroxide stabilised $MnFe_2O_4$ and Fe_3O_4 nanoparticles as dual-modality contrasts agent for MRI and PET imaging. Biomaterials. 2014;35(22):5840–6.
52. Thorek DL, Ulmert D, Diop N-FM, Lupu ME, Doran MG, Huang, R et al. Non-invasive mapping of deep-tissue lymph nodes in live animals using a multimodal PET/MRI nanoparticle. Nat Commun. 2014;5(3097).
53. Aryal S, Key J, Stigliano C, Landis MD, Lee DY, Decuzzi P. Positron emitting magnetic nanoconstructs for PET/MR imaging. Small. 2014;10(13):2688–96.
54. Wang H, Kumar R, Nagesha D, Duclos RI, Sridhar S, Gatley SJ. Integrity of ^{111}In-radiolabeled superparamagnetic iron oxide nanoparticles in the mouse. Nucl Med Biol. 2015;42(1):65–70.
55. Tsai CS, Liu WC, Chen HY, Hsu WC, editors. Preparation and characterization of Fe_3O_4 Magnetic nanoparticles labeled with Technetium-99m pertectnetate. App Mech Mater. 2014. Trans Tech Publ.
56. Debinski W, Tatter SB. Convection-enhanced delivery for the treatment of brain tumors. 2009.
57. Jarrett BR, Correa C, Ma KL, Louie AY. In vivo mapping of vascular inflammation using multimodal imaging. PLoS One. 2010;5(10):e13254.

Nanoparticles for PET Imaging of Tumors and Cancer Metastasis

Marie-Caline Z. Abadjian, Jaeyeon Choi, and Carolyn J. Anderson

1 Overview of PET Radionuclides and Chelation Chemistry

Positron Emission Tomography (PET) is a nuclear imaging modality that provides biochemical information with exquisite sensitivity for monitoring a variety of molecular processes using nM to pM concentration of radiotracers. PET imaging occurs after the administration of molecules labeled with radionuclides (typically called PET tracers or radiopharmaceuticals). The mass of PET tracer that is injected into the subject is extremely small (at the level of nmol to pmol), and causes minimal pharmacological effect. In this regard, PET enables the imaging and monitoring of disease in a noninvasive manner. PET has become a widely used diagnostic imaging tool by clinicians throughout the world. Although thousands of PET tracers have been developed for potential use in a clinical imaging setting, at present, only

M.-C. Z. Abadjian
Department of Medicine, University of Pittsburgh, 3501 Fifth Avenue, Pittsburgh, PA 15260, USA

Department of Radiology, University of Pittsburgh, Pittsburgh, PA, USA

J. Choi
Department of Bioengineering, University of Pittsburgh, Pittsburgh, PA, USA

C.J. Anderson, Ph.D. (✉)
Department of Medicine, University of Pittsburgh, 3501 Fifth Avenue, Pittsburgh, PA 15260, USA

Department of Radiology, University of Pittsburgh, Pittsburgh, PA, USA

Department of Bioengineering, University of Pittsburgh, Pittsburgh, PA, USA

Departments of Pharmacology and Chemical Biology, University of Pittsburgh, Pittsburgh, PA 15260, USA
e-mail: andersoncj@upmc.edu

J.W.M. Bulte, M.M.J. Modo (eds.), *Design and Applications of Nanoparticles in Biomedical Imaging*, DOI 10.1007/978-3-319-42169-8_11

Table 1 Decay characteristics of PET radionuclides for radiolabeling nanoparticles

Isotope	$T_{1/2}$	β^- MeV (%)	β^+ MeV (%)	EC (%)	γ MeV (%)
^{18}F	110 min	–	0.634 (97 %)	3 %	0.511 (194 %) 1.66 (3.1 %)
^{64}Cu	12.7 h	0.573 (38.4 %)	0.655 (17.8 %)	43.8 %	0.511 (35.6 %) 1.35 (0.6 %)
^{68}Ga	68 min	–	1.90 (87.7 %) 0.82 (1.2 %)	11 %	0.511 (178 %)
^{89}Zr	78.4 h	–	0.902 (22.8 %)	77.2 %	0.511 (45.6 %) 0.909 (99 %)
^{124}I	100.2 h	–	1.54 (11.7 %) 2.14 (10.8 %)	77 %	0.511 (45 %) 0.603 (63 %) 1.69 (10.9 %) 0.723 (10.4 %)

a handful (including [^{18}F]-labeled 2-fluorodeoxyglucose (FDG) and three Alzheimer's disease agents) are approved for routine clinical evaluation. Small animal PET systems for rodent imaging with 1–2 mm resolution are used to evaluate PET tracers in mouse and rat models to evaluate their suitability for humans.

1.1 Positron-Emitting Radionuclides

Traditional positron-emitting radionuclides include short-lived isotopes such as fluorine-18 (^{18}F; $T_{1/2}=110$ min), carbon-11 (^{11}C; $T_{1/2}=20$ min), nitrogen-13 (^{13}N; $T_{1/2}=10$ min), and oxygen-15 (^{15}O; $T_{1/2}=2$ min). Nontraditional positron-emitting radionuclides, particularly those of the transition metals, have gained considerable interest for imaging with PET because of increased production and availability (Table 1). For example, radionuclides of copper (^{64}Cu; $T_{1/2}=12.7$ h) and zirconium (^{89}Zr; $T_{1/2}=78.4$ h) allow radiolabeling of agents that clear the blood circulation more slowly [1].

1.2 Chelation Chemistry

The most commonly used chelators for complexing ^{64}Cu to nanoparticles are tetraazamacrocyclic ligands with pendant arms. The chelators DOTA (1,4,7,10-tetraazacyclododecane-1,4,7,10-tetraacetic acid), NOTA (1,4,7-triazacyclononane-1,4,7-triacetic acid), and TETA (1,4,8,11-tetraazacyclotetradecane-1,4,8,11-tetraacetic acid) have been the most widely used in ^{64}Cu-labeled nanoparticles, with DOTA being the most popular (Fig. 1). Anderson and colleagues have shown that cross-bridged macrocycles form more kinetically stable $^{64}Cu(II)$ complexes in vivo [2]; however, it has been demonstrated that there are only

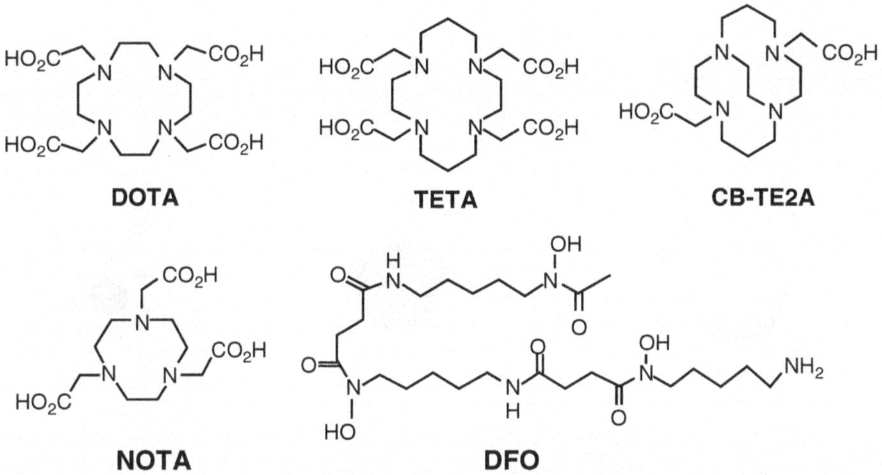

Fig. 1 Chelators that have been employed for complexing radiometals to nanoparticles

minimal differences in mouse biodistribution between DOTA and CB-TE2A conjugates of micellular nanoparticles [3]. Although the development of stable chelators for ^{89}Zr is an active area of research [4, 5], thus far, the chelator deferrioxamine (DFO) is the chelator of choice for attaching ^{89}Zr to nanoparticles. Examples of the nanoparticles labeled with radiometal chelates will be provided in the forthcoming sections.

2 Silica-Based Nanoparticle PET Tracers

Silica-based nanoparticles that have been employed in PET imaging are typically modified forms of mesoporous silica or coated dense silica nanoparticles ($dSiO_2$). Since the late 1990s, mesoporous silica nanoparticles (MSNs) have been studied extensively for a variety of applications due to their biocompatibility, large surface areas and ease of surface modification [6–9]. Various groups have coupled targeting moieties, drugs, and imaging agents to MSNs to investigate them as efficiency delivery vectors to tumors (Fig. 2) [10, 11].

MSNs also experience the EPR effect when their diameter is 100–130 nm making them suitable for imaging tumors [12]. For example, aza-dibenzocyclooctyne (DBCO) PEGylated MSNs have been coupled via a biorthogonal in vivo click reaction to ^{18}F-labeled azides to give ^{18}F-DBCO-PEG-MSNs (~150 nm) (Fig. 2c) [13, 14]. This study investigated these PET agents in female nude mice bearing subcutaneously (s.c.) U87MG tumors (Fig. 3). The DBCO-PEG-MSNs were injected in mice 24 h before the ^{18}F-labeled azide, allowing the MSNs to accumulate in the tumors prior to giving the ^{18}F-azide, which would allow localization of ^{18}F to the tumor, and clearance of all unreacted ^{18}F-azide.

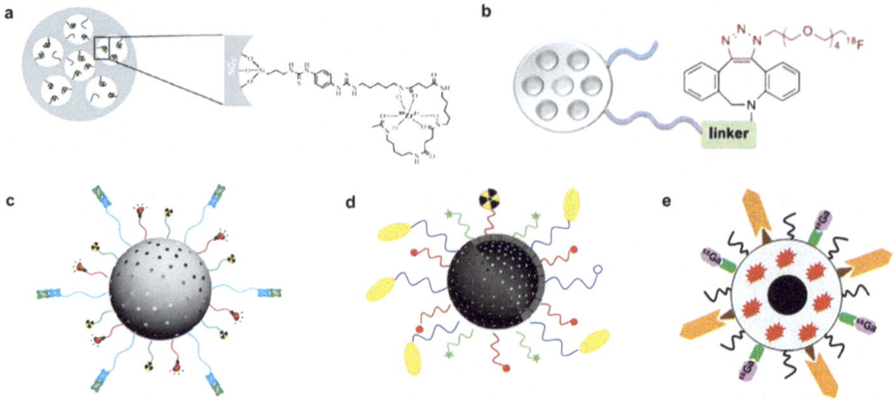

Fig. 2 Silica-based nanoparticles PET tracers: (**a**) representation of MSNs functionalized with APTMS, coupled with DFO-NCS and radiolabeled with radionuclide ^{89}Zr [15]; (**b**) ^{18}F-labeled peptide radiotracers [14]; (**c**) ^{64}Cu-MSN-800CW TRC105(Fab) [16]; (**d**) ^{64}Cu-NOTA-HMSN-fluorecein-PEG-cRGDyK nanoconjugate for drug delivery studies [18]; (**e**) ^{68}Ga was labeled with the MF-uMUC-1 [17]

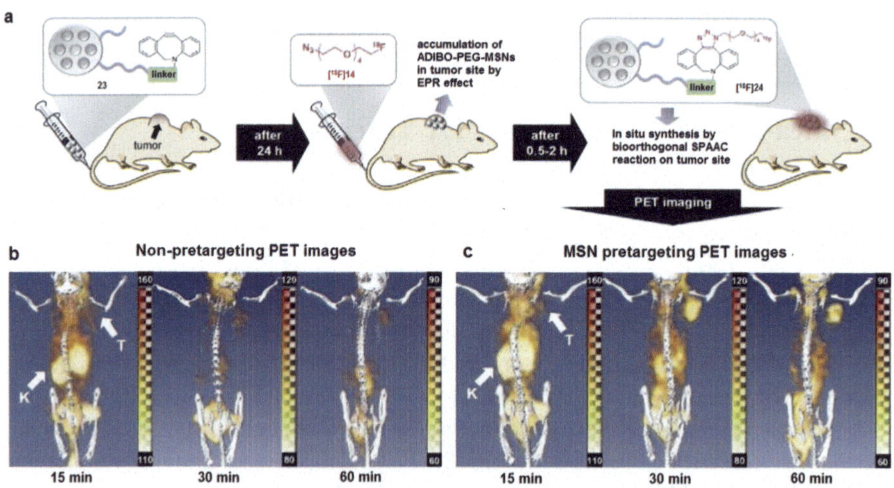

Fig. 3 Pretargeting PET imaging study by bioorthogonal covalent ^{18}F-labeling. (**a**) The procedure for the in situ synthesis of ^{18}F -DBCOT-PEG- MSNs in a living specimen by a bioorthogonal SPAAC reaction for the DBCO-PEG-MSN-pretargeting PET-imaging study. Three-dimensional reconstruction (*upper*) and transverse section (*lower*) combined PET-CT images of ^{18}F-labeled azide ([^{18}F]2 ; 2.6 MBq) in a U87 MG tumor-bearing mouse given only [^{18}F]2 alone (non-pretargeted ; (**b**) or a mouse given DBCO-PEG-MSNs 24 h earlier (pretargeted ; **c**) recorded at 15, 30, 60, and 120 min after injection of [^{18}F]2. *T* tumor, *K* kidneys [13]

PET images of the clicked MSNs in tumor were observed 2 h postinjection (p.i.) of the azide. Another group designed a new ^{89}Zr desferrioxamine (DFO) MSNs (~180 nm) to image SCID mice with s.c. prostate carcinoma cell lines (LNCaP and C4-2) finding minimal dissociation of $^{89}Zr^{4+}$ and typical biodistribution compared to other PET MSNs (Fig. 2a) [15].

MSNs can accommodate various surface modifications making them multimodality imaging agents not only for PET, but also near infrared fluorescence (NIRF) imaging and MRI. The surface of these nanoparticles can also be coupled to targeting agents. PET/ NIRF MSNs were developed with vasculature targeting capability to image 4 T1 murine breast cancer tumors in mice (Fig. 2d) [16]. Surface coupling of a human/murine chimeric IgG1 monoclonal antibody (TRC105(Fab)), NIR dye (800CW), and ^{64}Cu-NOTA labeling resulted in MSNs (175.3 ± 9.7 nm) imaging tumors with 5.4 ± 0.2 % ID/g at 4 h p.i.. Another example of multimodality silica NPs was comprised of PET, NIRF, MRI agents and a targeting moiety that successfully imaged BT-20 cells in a nude mouse model (Fig. 2f) [17]. The silica-based nanoparticles (81 nm) have a cobalt ferrite core with rhodamine B in the silica shell and surface coupled underglycosylated mucin-1 antigen (uMUC-1 aptamer), along with NOTA for Ga-68 labeling.

Targeting of tumors has also been shown in hollow mesoporous silica nanoparticles (HMSNs) in U87MG tumor-bearing female athymic nude mice (Fig. 2e) [18]. Dense silica nanoparticles coated with MSNs have surface coupled cyclic arginine-glycine-aspartic acid (cRGDyK) peptide, ^{64}Cu-NOTA as well as anticancer drug (Sunitinib) loaded in the core. PET images showed relatively high uptake (7.2 ± 0.6 % ID/g) at 0.5 h postinjection.

3 Gold Nanoparticle (AuNP) PET Tracers

Gold nanoparticles (AuNPs) are a promising platform for biomedical applications and have rapidly advanced toward multifunctional particles for imaging and treatment of cancer. AuNPs have been at the forefront of cancer research in recent years owing to the high biocompatibility via functionalization, their low toxicity of the gold core, and availability in a range of sizes and shapes [19, 20].

The readers are referred to a highly comprehensive review by Daniel and Astruc for information on the structure and properties of AuNPs [21]. Surface plasmon resonance (SPR), a unique plasmatic absorption band phenomenon of AuNPs, can be converted to strong infrared spectral ranges, thereby allowing vital optical imaging in tissues where light exhibits minimal absorption and deep penetration in tissue. These specialized properties have also been used for photothermal therapy, which is a noninvasive, accurately targeted hyperthermia cancer treatment based on the optical absorbance of AuNPs in the intrinsic near-infrared (NIR) (650–900 nm) [22]. AuNPs also provide attractive scaffolds for many biomedical imaging modalities, including surface-enhanced Raman scattering (SERS), two-photon photoluminescence (TPL), magnetic resonance imaging (MRI), positron emission tomography (PET), and X-ray computer tomography (CT) imaging [23, 24].

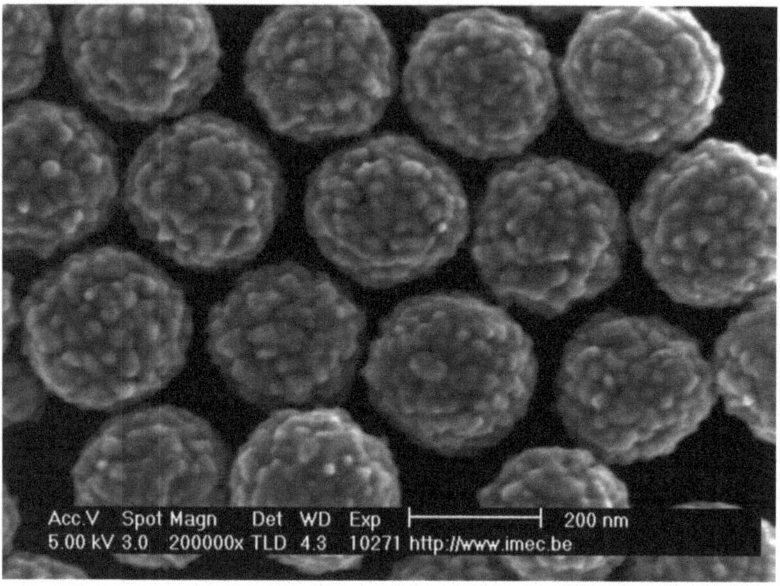

Fig. 4 SEM image of a monolayer portion showing self-assembled silica/gold nanoshells on a silane functionalized quartz substrate [25]

3.1 Types of AuNPs

Typically, the dual imaging AuNPs are synthesized from commercial HAu_3Cl_4 by the reduction of Au^{3+} ions to metallic Au atoms (Au^0) using citrate ions as a reducing and capping agent. There are different types of gold nanoparticles in the size range of 9–120 nm that have been developed with various shape, size, and physical properties (Fig. 4) [25].

Gold nanosheres (gold colloids) in the size range of 2–100 nm can be produced by chemical reduction of gold chloride, with the properties controlled by citrate/gold ratio. The absorption spectra of gold nanospheres (visible range of 510–550 nm) are related to the size distribution [26–28].

Gold nanorods are synthesized by a special template method based on the electrochemical deposition of gold within the cylindrical pores of rigid matrices, such as nanoporous polycarbonate or alumina template membranes. Gold nanorods have absorption and scattering peaks that can be converted to the visible and near IR spectra, generating heat when excited by IR light. The characteristic has been widely applied to selectively destroy cancer cells [29].

Nanoshells, silica coated nanoparticles with a thin film of gold, have been used for an optical imaging with SPR in the visible to the NIR region [30, 31].

Gold nanocages are synthesized by reacting silver nanoparticles with chloroauric acid in aqueous conditions, and range in size from 10 to 150 nm. As the optical resonance peaks shift to near-infrared light, the strong absorption of gold nanocages enhances the contrast and photothermal effect for cancer diagnostic and therapy [32, 33].

3.2 *Radiolabeling Chemistry of AuNPs*

The surface of AuNPs can be modified with ligands, surfactants, polymers, and dendrimers. Upon functionalizing with thiol groups that strongly bind with gold, AuNPs can be readily conjugated to chelators or targeting molecules [34], enabling the conjugation of biofunctional chelates such as 1,4,7-triazacyclononane-1,4,7 -triacetic acid (NOTA) or 1,4,7,10-tetraazacyclodedecane-1,4,7,10-tetraacetic acid (DOTA) via a thiol-maleimide coupling chemistry. Bifunctional chelators play an important role in conjugation of metal radionuclides on the AuNPs surface. For example, Gd^{3+} is attached on a chelate on the AuNPs surface for MRI imaging, whereas for PET imaging, radionuclides such as ^{64}Cu, ^{68}Ga, and ^{89}Zr are generally used for PET/MR tracers. Targeting molecules added to the surface of AuNPs can render them tumor-specific for targeting via receptors expressed exclusively on target cancer cells, improving tumor localization over the nonspecific enhanced permeability and retention effect (EPR) [35].

^{64}Cu-labeled RGD peptide-gold nanoshells (^{64}Cu-NS-RGDfK) were developed for targeting integrin $\alpha_v\beta_3$ on tumor cells. The NSs were conjugated with RGDfK and DOTA through bifunctional PEG, and the surface modified size was ~170 nm. PET/CT imaging of two rats bearing head and neck xenografts showed high tumor uptake beginning at 4 h postinjection and reaching a maximum at 20 h, with a decrease in tumor accumulation until 44 h postinjection [36].

Nonchelator conjugated [^{64}Cu]CuS AuNPs (11 nm) was developed for a tumor theranostic probe both for PET imaging and as photothermal ablation agents using a passive targeting strategy in a breast cancer mouse model. At 24 h, biodistribution in mice bearing subcutaneous U87 glioma xenografts showed that the PEG-[^{64}Cu]CuS NPs reduced liver/spleen uptake and enhanced tumor uptake ratio (7.6 ± 1.4 %ID/g) [37].

Other chelate-free ^{64}Cu-labeled alloyed AuNPs (27 nm hydrodynamic (HD) size) were developed for cancer imaging. In these AuNPs, ^{64}Cu is directly incorporated into the lattice of the gold nanoparticle structure, maintaining high stability in vivo. The PET/CT image using EMT-6 tumor-bearing mouse showed 4.93±0.32 % ID/g of tumor uptake ratio at 1 h of postinjection, increasing at 48 h pi to 16.8±0.98 %ID/g, with a tumor/muscle ratio of 16.2 ± 1.07 [38]. However, there was very high liver and spleen uptake of these particles at 48 h postinjection (~45 % ID/g liver and ~200 %ID/g spleen), (Fig. 5). The surface of these ^{64}CuAuNCs was then PEGylated with different-sized PEG chains (^{64}Cu-labeled AuNCs-PEG350, HD size of 4.3 nm vs. AuNCPEG1000, HD size of 6.9 nm) [39]. Biodistribution in PC3 tumor-bearing mice showed that both particles had dramatically decreased spleen uptake at 48 h (5 % ID/g or less), and the liver uptake was also <20 % ID/g. There was also significantly lower tumor uptake (3–5 % ID/g) for both sized AuNCs, with tumor:muscle ratios ~2.5 [38].

^{64}Cu-labeled hollow gold nanospheres (HAuNS) showed differential tumor uptake after hepatic intra-arterial (IA) and intravenous (IV) injection in VX2 liver tumor-bearing rabbits. RGD-PEG-HAuNS had an average diameter of ~40 nm. A DOTA analog chelator was attached to RGD-PEG-HAuNS for radiolabeling with Cu-64. PEG-HAuNS with lipiodol (IA-PEG-HAuNS-lipiodol) after hepatic

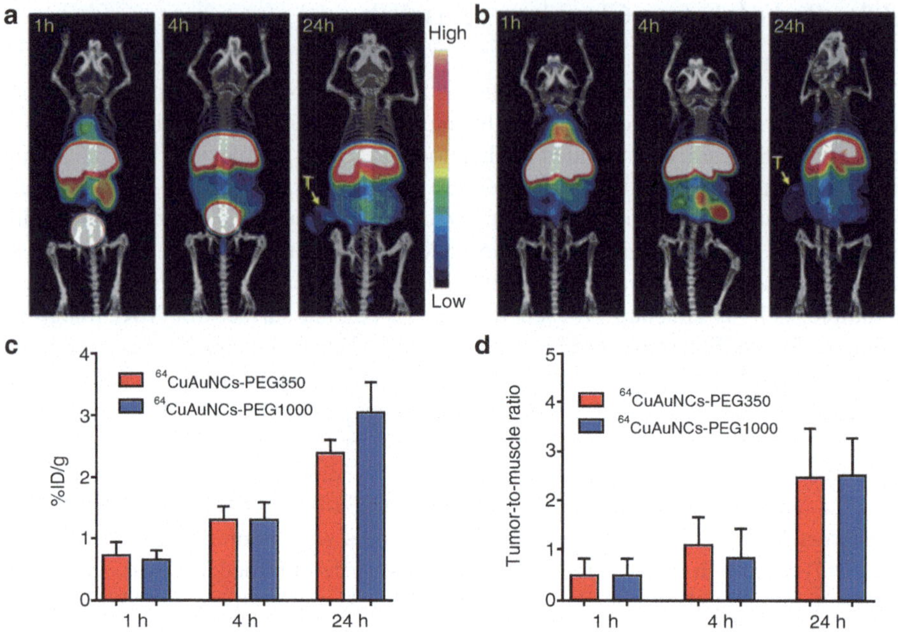

Fig. 5 PET/CT images at 1, 4, 24 h postinjection of ^{64}Cu-AuNCs-PEG350 (**a**) and ^{64}Cu-AuNCs-PEG1000 (**b**) in PC3 prostate-tumor-bearing mice. Quantitative tumor uptake (**c**) and tumor:muscle ratios (**d**) of the two agents (38)

intra-arterial injection had the highest tumor uptake (0.3319 ± 0.0711 ID%/g) compared to that of other ^{64}Cu-labeled PEG-HAuNS and ^{64}Cu radiolabeled RGD peptide-HAuNS after IA and IV injection, respectively. The result showed that adding iodized oil to PEG-HAuNS group can increase delivery of nanoparticles to hepatic tumors [40].

An RGD peptide was coupled with Gd-chelate coated gold nanoparticle (Au@DTDTPA-RGD) and labeled with ^{68}Ga for a dual PET/MRI imaging modality targeting integrin $\alpha_v\beta_3$ receptor-positive U87MG cancer cells. Biodistribution studies showed that the tumor to muscle ratio increased from 1 to 2 h postinjection (3.71 ± 0.22 and 4.69 ± 0.09 respectively) [41].

^{89}Zr-labeled anti-CD105 was coupled with gold nanoparticles (^{89}Zr-anti-CD105-AuNPs-PPAA), having a mean diameter of 102.6 ± 4.0 nm. Two groups of B16 melanoma-bearing mice were injected with ^{89}Zr-Df-Bz-NCS-anti-CD105 antibody or with ^{89}Zr-anti-CD105-AuNPs-PPAA, respectively. The similar PET imaging pattern was observed, but ^{89}Zr-anti-CD105-AuNPs-PPAA had higher uptake in liver, spleen, and lung compared to that of ^{89}Zr-Df-Bz-NCS-anti-CD105 antibody due to its large size of the nanoparticle. The biodistribution study between ^{89}Zr-anti-CD105-AuNPs-PPAA and ^{89}Zr-Df-Bz-NCS-anti-CD105 showed that no significant differences in tumor:background ratios were observed [42].

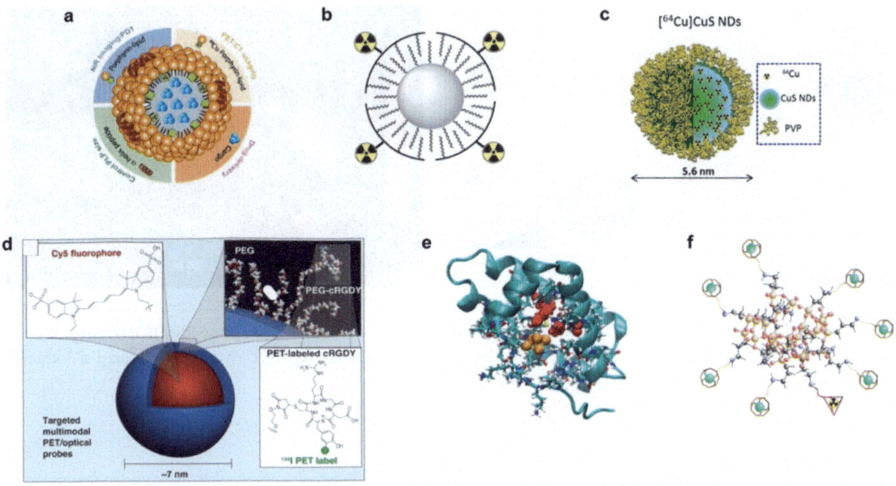

Fig. 6 Ultrasmall nanoparticles: (**a**) PLP structure [51]; (**b**) USPIO@OPA [52]; (**c**) [^{64}Cu]CuS ND [53]; (**d**) C dots [54]; (**e**) [^{64}Cu]CuNC@BSALHRH [56, 57]; (**f**) AGuIX nanoparticles [58]

4 Ultrasmall Nanoparticle PET Tracers

Ultrasmall nanoparticles are defined by their size (less than 10 nm), falling within the renal glomerular filtration size cutoff (~10 nm) [43–46]. Nanoparticles that fall between 10 and 200 nm are known to accumulate in the reticuloendothelial system (RES), which includes the liver, lymph nodes, bone marrow, and spleen, leading to longer retention in the body that could cause long-term adverse effects [47–50]. Small nanoparticles (~10–100 nm) and ultrasmall nanoparticles have unique properties, sparking interest in targeted therapies using PET imaging. Small and ultrasmall nanoparticles have provided biocompatible vectors compact enough to study metabolic pathways and therapies (Fig. 6). Some small nanoparticles mimic lipoprotein structures to sustain longer circulation as an alternative to PEGylation. A porphylipoprotein (PLP) nanoplatform (20.6 ± 5.2 nm) has been developed for fluorescence imaging and photodynamic light therapy of glioblastoma multiforme (Fig. 6a) [51]. PET imaging using ^{64}Cu-labeled versions of PLP and PEGylated PLP in SKOV3 orthotopic ovarian cancer model showed similar uptake in tumors, but striking differences in spleen uptake, 3.4 ± 0.2 % ID/g versus 19.8 ± 1.6 % ID/g, respectively. Other compositions of similar sized nanoparticles like ultrasmall superparamagnetic iron oxide nanoparticles (USPIOs) have also been studied for their biocompatibility and cellular uptake effi ciency (Fig. 6b) [52]. USPIOs were coated with octylamine-modified polyacrylic acid (OPA) (10 ± 2 nm) and conjugated with a ^{64}Cu chelator (DMPTACN) for PET imaging of mice bearing tumors of various human cell lines (epidermoid carcinoma A431, squamous cell carcinoma FaDu, ductal carcinoma MDA-MB 435S, umbilical vein endothelial HUVEC), finding generally low toxicity.

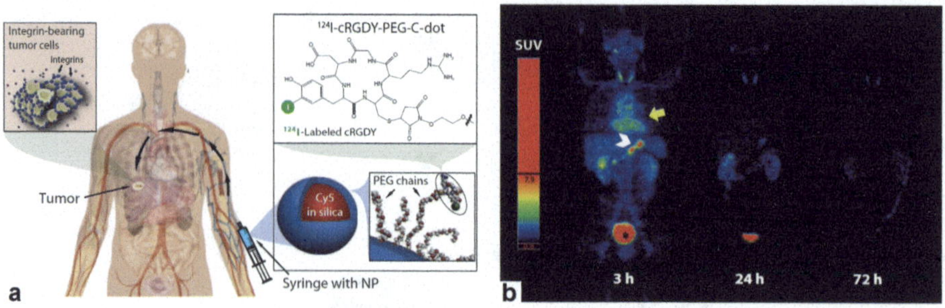

Fig. 7 (**a**) Cornell dots (C dots) conjugated with ^{124}I-labeled cRGDY have been investigated in humans with cancer for safety, biodistribution, and dosimetry; (**b**) PET image in patient #3 showing activity in bladder, heart, and bowel [55]

Nanodots also fall into the category of ultrasmall nanoparticles and have shown interesting applications as photothermal agents and in tumor imaging. Copper-64 labeled copper sulfi de nanodots ([^{64}Cu]CuS NDs) coated with polyvinylpyrrolidone (~5.6 nm) were found to have tumor uptake in 4 T1 tumors in Balb/c mice (3.62 ± 0.50 % ID/g, 2 h p.i.) and fast renal clearance (Fig. 6c) [39, 53]. Cornell dots (C dots) are ultrasmall inorganic optical-PET imaging nanoparticles probes. The silica-based core contains encapsulated Cy5 fluorophore and the surface is PEG-coated with conjugated 124 I-labeled cyclic arginine-glycine-aspartic acid (cRGD, for $\alpha v\beta 3$ integrin targeting) (Fig. 6d) [54]. The C dots were found to have high receptor-binding specificity, and biodistribution and targeting kinetics data, clearance and dehalogenation profiles, blood/tissue residence times, and bioavailability and radiation dosimetry were reported (Fig. 7). This ultrasmall nanoparticle has received FDA-approval for a first-in-human clinical trial and found to be safe in human (five patients) as a diagnostic for metastatic melanoma [55]. The preliminary trial found that the agent was well tolerated, cleared quickly from the body through the kidneys, and the 124 I-label was stable in vivo. Two patients showed tumor uptake in a pituitary and a liver lesion. To date, this is one of the few nanoparticle-based PET tracers to be evaluated in human cancer patients. Some ultrasmall nanoparticles are small enough to be considered clusters or matrices. One group has made chelator-free labeled [^{64}Cu]Cu nanoclusters coated with bovine serum albumin (BSA) for PET imaging for tumors (Fig. 6e) [56, 57]. These nanoclusters have been conjugated with a tumor target peptide, lutenizing hormone releasing hormone (LHRH), giving [^{64}Cu]Cu NC @BSA-LHRH (3.8 ± 0.5 nm) that were studied in mice bearing orthotopic A549 lung tumors (Fig. 8). The targeted nanoclusters were found to have much higher tumor uptake 12 % ID/g than nontargeted 3 % ID/g at 4 h p.i. Other interesting ultrasmall nanoparticles were made from a polysiloxane matrix coated with chelators for PET/MR dual imaging [58]. These ultrasmall nanoparticles (AGuIX) are only 2.5 ± 0.1 nm in diameter (Fig. 6f). They have been well-characterized and have shown no toxicity in U87MG (human primary glioblastoma) and HEKβ3 (human embyonic kidney) cell lines.

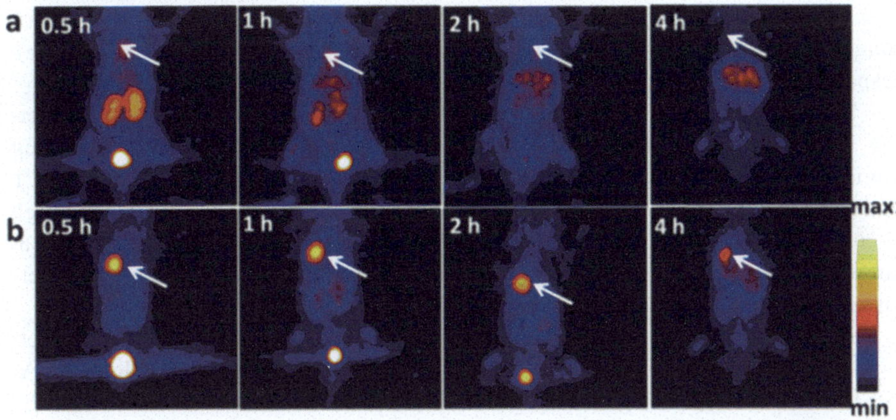

Fig. 8 Representative PET images of coronal single slices on orthotopic A549 lung tumor-bearing mice after intravenous injection of 6.7 MBq of [^{64}Cu]CuNC@BSA (**a**) and [^{64}Cu]CuNC@BSA-LHRH (**b**). Images were acquired at 0.5, 1, 2, and 4 h. White arrows indicate the lung tumor [57]

PET/CT imaging using ^{68}Ga-labeled AGuIX was found to have rapid clearance and coherence with MR images. The development of ultrasmall nanoparticles is becoming a new avenue for targeted imaging and therapy.

5 Radiolabeled Iron Oxide Nanoparticles (IONPs)

Iron oxide particles have been widely studied as an excellent MRI contrast agent in clinical trials due to its ideal paramagnetic and low toxicity [59, 60]. MR imaging is optimized with functional parameters such as spin-lattice relaxation time (T1) and spin–spin relaxation time (T2), which are a function of the local chemical structure of the molecules being imaged. The core of iron oxide particles is composed of iron and oxygen atoms, generating mostly magnetite (Fe_3O_4), maghemite, (γ-Fe_2O_3), and hematite (α-Fe_2O_3), which exhibit superparamagnetic physical properties at ambient temperature if the core diameter is relatively small (<20 nm) [61–63]. Iron oxide nanoparticles (IONPs) have a number of advantages for multimodality (e.g., PET/MRI) imaging: 1) they can be modified on the surface for conjugation with a large number of targeting ligands, thus enhancing the biological specificity and affinity to targeted molecules; 2) the nanoparticles can be coupled with chelators for labeling with metal radionuclides for PET or SPECT imaging; and 3) iron oxide nanoparticles can circulate in the blood vessels for a relatively long time, moving larger amount of radionuclides or other cargo to targeted organs [64]. *Please see Chap. 10 for more detailed information on structural properties of IONPs.*

^{64}Cu-DOTA-RGD-conjugated IONPs were developed for targeting tumor $\alpha_v\beta_3$ integrin. Polyaspartic acid (PASP)–coated iron oxide was coupled with RGD and DOTA via surface amino groups for ^{64}Cu-labeling. The hydrodynamic diameter is

45 ± 10 nm. In vivo PET studies in U87MG tumor-bearing mice showed tumor imaging at 1 h (7.9 ± 0.8 %ID/g), 4 h (10.1 ± 2.1 %ID/g), and 21 h (9.8 ± 3.2 %ID/g), respectively, after injection of ^{64}Cu-DOTA-iron oxide-RGD [65].

^{89}Zr-Deferoxamine-RGD-IONP, another RGD-iron oxide, was developed targeting $\alpha\nu\beta3$ and $\alpha\nu\beta5$ integrins overexpressed in nascent endothelial cells during angiogenesis in various tumors, but not in inactive endothelial cells. The PET/CT and MRI imaging after intravenously injection in tumor-bearing mice showed high accumulation in the liver and spleen at 1 h after injection, which remained high at subsequent time points. After 24 h, accumulation in tumor was observed with persistent and intense signal until 72 h, clearly delineating the tumor [66].

Protein-based ligands, such as affibodies, have been investigated as tumor targeting agents. An anti-EGFR affibody was conjugated to Au-Iron Oxide NPs (NOTA-Au-IONP-affibody) having an average size 24.4 ± 2.0 nm, and was labeled with ^{64}Cu. An in vitro study showed that NOTA-Au-IONP-Affibody probes had a higher cellular uptake in EGFR-positive tumor A431 cells at 1 and 2 h time points, respectively, compared with blocking samples (Fig. 9). The small animal PET images showed high tumor accumulation (4.6 % ID/g at 24 h p.i.), and high tumor-to-normal tissue contrast for ^{64}Cu-NOTA-Au-IONP-affibody. The biodistribution studies also indicated that the nanoprobe had higher tumor uptake value compared to the blocking group [67].

6 Liposome-Based PET Agents

Liposomes used in PET imaging are an attractive vector for imaging tumors and diseased tissues, as well as drug targeting. The study of liposomes as imaging agents started in the late 1980s [68], although it was not until the late 1990s that liposomes were radiolabeled for imaging purposes [69–72]. Generally, liposomes are lipid vesicles formed from thin lipid films or cakes. Upon hydration, layers of the bilayer sheets self-assemble into multilamellar vesicles (MLVs) where the hydrophilic portions face the water surrounding and within the vesicles. The size of these liposomes can then be reduced to unilamellar vesicles by input of either sonic energy (sonication) or mechanical energy (extrusion). Liposomes are typically characterized by their particle size (dynamic light scattering), zeta potential (laser doppler microelectrophoresis), and uniform size distribution. Like all nanoparticles used in medicine, the size and charge of the nanoparticle play a big role in their fate in vivo. Most liposomes range in size from 90 to 925 nm and are either negatively charged or neutral. This size range is meant to be above the renal threshold for longer circulation.

The success of liposomes in clinical and investigational research has led researchers to combine the physiological properties of liposomes with the quantitative imaging capability of PET. The longer-lived PET radionuclides (^{64}Cu, ^{62}Cu, ^{124}I, ^{89}Zr) have been typically investigated with liposomes to take advantage of the EPR effect; however, shorter-lived radionuclides (^{11}C, ^{18}F, ^{15}O) have also been studied for PET imaging [73]. Incorporating a radionuclide for PET imaging into a liposome can be achieved by (1) encapsulation; (2) remote loading (after-loading); (3) bilayer chelation; or (4) surface chelation.

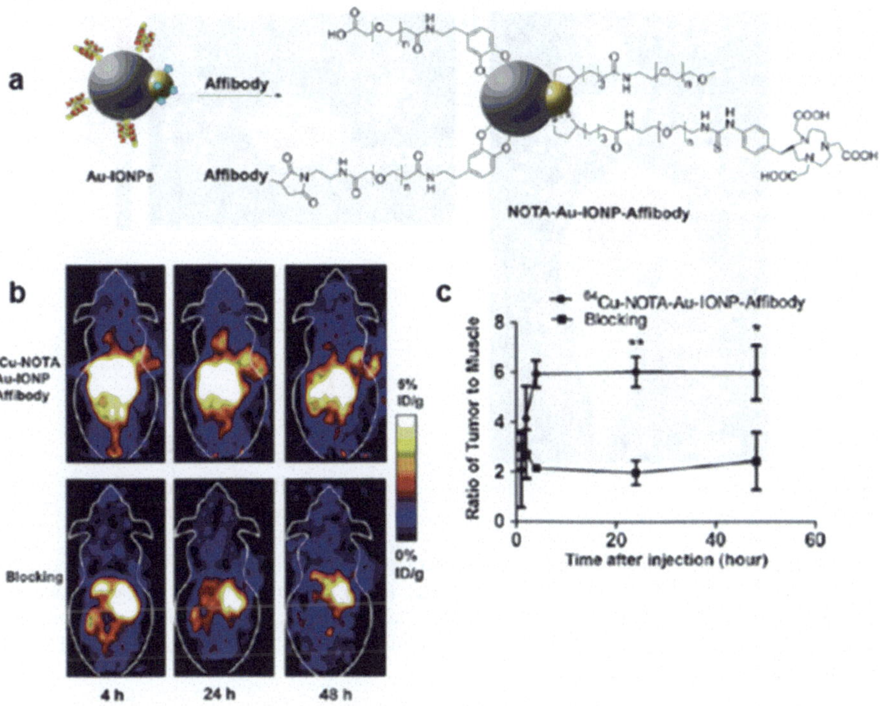

Fig. 9 (**a**) Schematic illustration of Affibody binding domain with Au-IONP and its functionalization and conjugation with affibodies and the NOTA chelator; (**b**) PET image of A431 tumor-bearing mice at 4, 24, and 48 h postinjection of ^{64}Cu-NOTA-Au-IONP-Affibody, with and without blocking; (**c**) tumor:muscle ratios with and without blocking [67]

A ^{18}F-labeled cholesteryl ether (^{18}FCE) PEGylated liposome employed bilayer chelation to incorporate radioactive lipophilic labels in the liposome membrane (Fig. 11a) [74]. The liposome ^{18}FCE was found to accumulate in NCI-H727 human lung carcinoid tumors of tumor-bearing nude mice after 8 h (2.25±0.23 %ID/g). Another ^{18}F-labeled liposome (^{18}F-TCO-liposome) incorporated the fast tetrazine (Tz)/*trans*-cyclooctene (TCO) inverse electron demand Diels-Alder cycloaddition (IEDDA) and pH (low) insertion pepide (pHLIP) to speed up tumor accumulation (Fig. 11b) [75]. Tumor-bearing athymic nude mice were injected with SKOV3 ovarian cancer cells and pHLIP-Tz (Fig. 10). The ^{18}F-TCO-liposome could be detected as early as 30 min after injection of the liposome. After 120 min, tumors marked with pHLIP-Tz had higher activity (3.5±1.2 %ID/g) compared to tumor without pHLIP-Tz (0.46±0.04 %ID/g).

For longer circulating PET imaging agents, Zr-89 is favored due to its half-life (78.4 h) and relatively lower fraction of gamma radiation than ^{124}I and ^{86}Y [76, 77]. Several groups have utilized the combination of the liposome's EPR effect over time and ^{89}Zr as a PET imaging agent. ^{89}Zr-PEGylated-liposomes radiolabeled by surface chelation have shown tumor accumulation peaks at 24–48 h postinjection

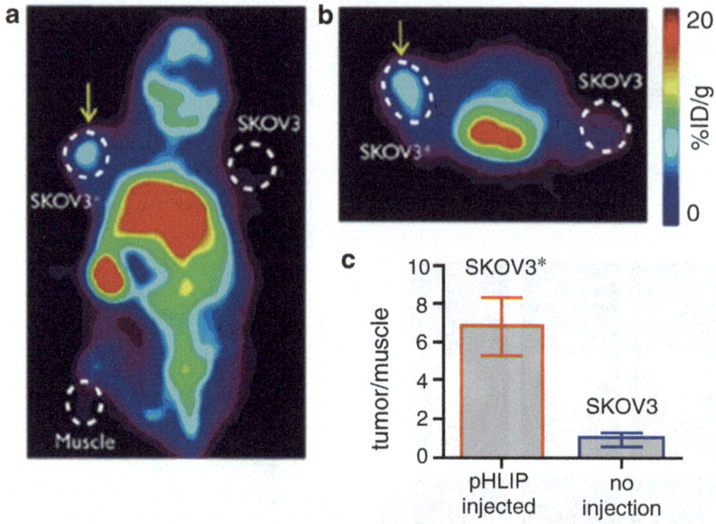

Fig. 10 Biodistribution and PET imaging of bioorthogonal ^{18}F-liposomes as a selective marker for tetrazine-labeled SKOV3 ovarian cancer xenografts. (**a, b**) PET imaging at 2 h postinjection in a mouse bearing both pHLIP-Tz-treated (SKOV3, left shoulder, *yellow arrow*) and untreated (SKOV3 right shoulder) tumors. Normalized tumor:muscle ratios for treated and untreated tumors at 2 h post-injection [75]

(Fig. 11c) [78, 79]. The PEG groups on the surface of the liposomes assist with longer circulation, giving the liposomes time to accumulate in/around tumors. Another ^{89}Zr-PEGylated-liposome also included a targeting agent, octreotide, on the surface of the liposome (OL) (Fig. 11d) [80]. Octreotide is a peptide targeting human somatostatin receptor subtype 2 (SSTr2) found on tumor cells. The ^{89}Zr-Gd-OL imaging agent was found to specifically accumulate at the tumor site at 50 h postinjection.

Other radionuclides, especially Cu-64, have gained attention as a viable radionuclide coupled to liposomes for PET imaging. Several groups have used Cu-64 with their liposomes to study a variety of tumor models. Remote loading and surface chelation are some of the more common methods for radiolabeling liposomes. Remote loading using 2-hydroxyquinoline has been shown to be highly efficient compared to other ionophores [81]. Unassisted radiolabeling or remote loading without the use of ionophores has shown evidence of similar radiolabeling efficient compared with ionophores (Fig. 11f) [82]. Copper-64 has been remotely loaded into liposomes containing surface targeting SSTr2 in human neuroendocrine carcinoma in mouse models (Fig. 11g) [83] and mannose-coated liposomes targeting macrophages in the tumor microenvironment (TME) [84]. Surface chelation and modification using ^{64}Cu has yielded informative findings in tumor targeting and organ uptake. DOTA and TETA macrocycle analogs have been used to coordinate the radionuclides to the liposome giving a stable radiolabeled liposome (Fig. 11e) [85, 86].

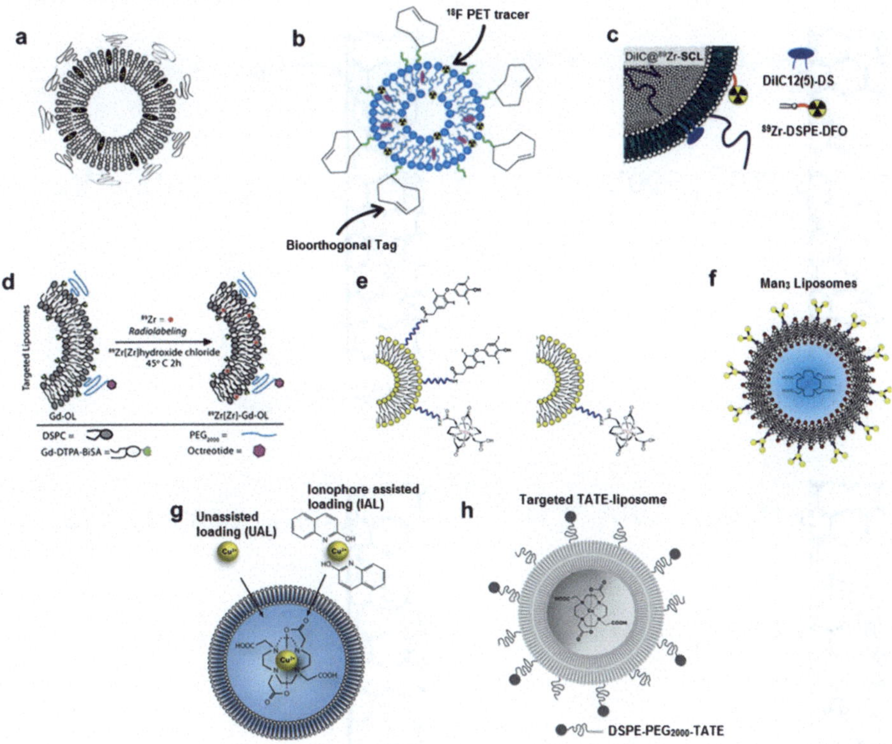

Fig. 11 Liposome-based PET agents: (**a**) PEGylated liposome incorporating radioactive lipophilic labels in the membrane [74]; (**b**) ^{18}F-TCO-liposomes and pHLIP-Tz coupling for pretargeting of acidosis [75]; (**c**) dual-labeled liposome DiIC@89Zr-SCL [78]; (**d**) OCT was conjugated to preformed Gd-Control LPs (CL) resulting in targeted OCT-LP (OL) [80]; (**e**) tetrac/^{64}Cu-DOTA-liposome (*left*) and ^{64}Cu-DOTA-liposome (*right*) [85]; (**f**) Mannose liposomes for remote loading of the ^{64}Cu-DOTA [84]; (**g**) Loading of $^{64}Cu^{2+}$ into liposomes [82]; (**h**) PEGylated (DSPE-PEG2000) targeted TATE-liposome [83]

7 Additional Types of PET Nanoparticle Agents

7.1 Carbon Nanotubes (CNTs)

Carbon nanotubes are allotropes of carbon having a cylindrical structure. CNTs have interesting properties for use in a diverse array of materials, owing to their extraordinary thermal conductivity, mechanical, and electrical properties. CNTs are categorized as single walled (SWNT) and multiwalled (MWNT) [87]. SWNTs consist of a one-atom thick layer of graphene, whereas MWNTs consist of multiple rolled layers of graphene. The normal organ biodistribution and pharmacokinetics of ^{86}Y-labeled SWNTs were investigated in non-tumor-bearing athymic nude mice

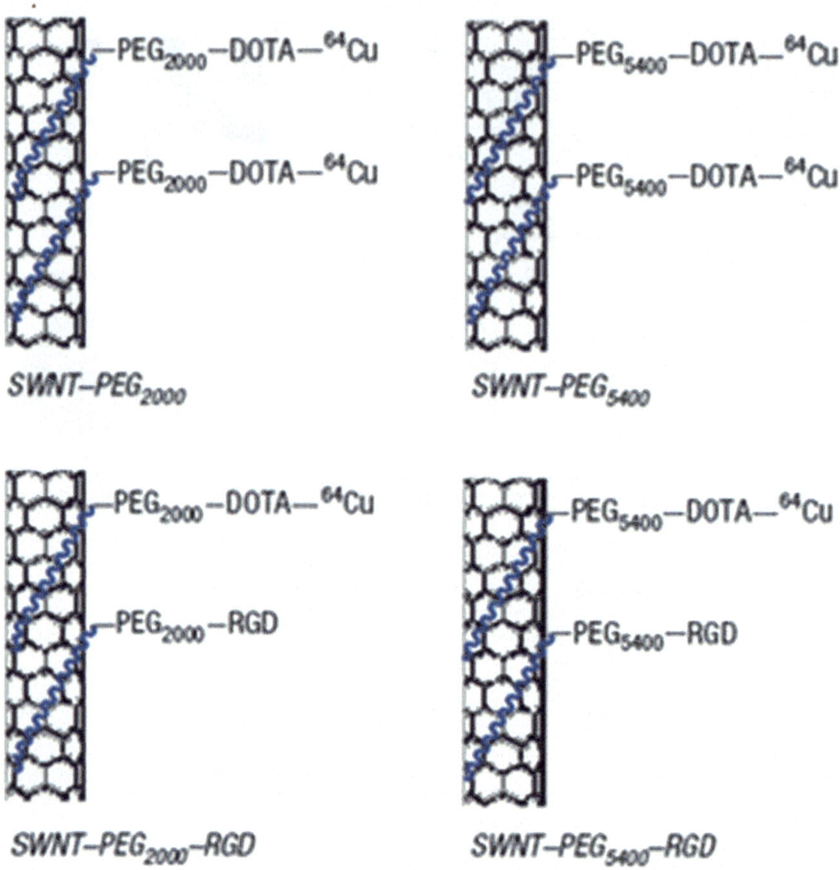

Fig. 12 Schematic drawing of noncovalently functionalized SWNTs with PEG2000 and PEG5400 with ^{64}Cu, either with or without RGD peptides for targeting integrin $\alpha_4\beta_1$ [50]

to determine the clearance patterns, which indicated blood clearance within 3 h and distribution predominantly to the kidneys, liver, spleen, and bone [88]. RGD-conjugated (targeting integrin $\alpha_v\beta_3$) and nontargeted SWNTs (diameter: 1–5 nm; length: 100–300 nm) with varying lengths of PEG chains were labeled with Cu-64 and investigated in U87MG human glioma tumor-bearing mice (Fig. 12) [50]. The authors showed that PEG_{5400} extended the circulation time of the nanoparticles, and RGD peptides improved tumor uptake compared to nontargeted agents. Although for PET imaging studies there was no observable toxicity of the radiolabeled SWNTs [50], the carbon nanotube class of nanoparticle has fallen out of favor due to health and safety issues [56].

7.2 *Micelle-Based Nanoparticles*

Micelle-based nanoparticles have hydrophilic regions that are in contact with the surrounding solvent, typically aqueous media, with hydrophobic regions in the center of the micelle that can sequester lipophilic drugs or other cargo. Micelles can be aggregates of small surfactant molecules (MW in the hundreds), or aggregates of larger amphiphilic block copolymers (MW in the thousands or tens of thousands).

Diblock co-polymer shell cross-linked nanoparticles (SCK-NPs) have been incorporated with DOTA (Fig. 13a) [89] and TETA [90] chelators for labeling with Cu-64. In one of the first studies of ^{64}Cu-labeled SCK-NPs (PEG_{1600} spacer; ~20 nm), Cu-64 was labeled using the TETA chelator to nontargeted and folate-targeted nanoparticles, and biodistribution was determined in folate-receptor positive KB tumor-bearing mice [90]. Targeted and nontargeted agents cleared through the liver, lung, and spleen, and there was no significant difference in tumor uptake. Zeng et al. achieved very high specific activity by incorporation of azide groups in the core of the nanoparticles, followed by reaction of ^{64}Cu-DOTA conjugated to the strained alkyne, dibenzylcyclooctyne (DBCO) under copper-free click chemistry conditions. They achieved 975 Ci/μmol, amplifying the amount of ^{64}Cu-labeling by a factor of 500 compared to direct labeling of chelator-SCK-NPs [89].

To address the question of whether incorporation of a more stable Cu(II) chelator positively impacts U87MG tumor uptake and nontarget tissue biodistribution of ^{64}Cu-labeled tri-block polymeric micelles (Fig. 13b) [3], DOTA was compared to CB-TE2A, which forms highly stable Cu(II) complexes in vivo [2, 91]. The ^{64}Cu-labeled CB-TE2A micelles showed somewhat slower blood clearance leading to higher tumor uptake. However, the tumor:blood ratios of the DOTA vs. CB-TE2A micelles were not significantly different. Nontarget tissue uptake of the two ^{64}Cu-labeled micelles was similar. The authors concluded that although DOTA is not an ideal ^{64}Cu-chelator for smaller molecules, it is adequate for labeling micellular nanoparticles.

Amphiphilic graft copolymers and associated comb nanoparticles (CNPs) were designed to be tunable with respect to their composition and to be modified with a controlled number of RGD peptide moieties (Fig. 13c) [92]. The 20–23 nm CNPs consisted of four building blocks: a) PEG units; b) a hydrophobic methyl mathacrylate backbone; c) DOTA chelator for Cu-64 labeling; and d) GRGDS linear integrin $\alpha_v\beta_3$ targeting peptides. The 20 % RGD CNPs demonstrated highest affinity for integrin $\alpha_v\beta_3$ in an in vitro assay, and the ^{64}Cu-labeled 20 % RGD CNPs showed the highest cell associated activity in cultured U87MG cells. A similar strategy was used to produce DOTA-CNPs modified with C-type natriuretic factor binding peptide (CANF), which binds to the cell surface natriuretic peptide clearance receptor (NPRC) [93]. The authors investigated imaging of prostate cancer in human CWR 22 tumor-bearing mice with ^{64}Cu-DOTA-CANF-CNPs, and found specific uptake in the tumor; however, histologically, it was shown that staining for NPRC is found in pools of inflammatory cells in the tumor, and not in the tumor itself. Targeting NPRC through the CANF-CNPs

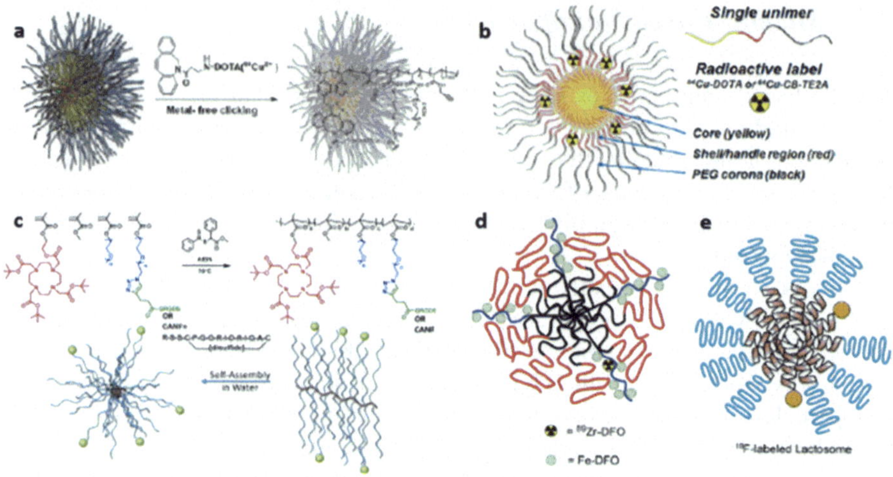

Fig. 13 Micelle-based nanoparticles for PET imaging. (**a**) strategy for metal-free click chemistry to label SCK-NPs with Cu-64, resulting in exceptionally high specific activity [89]; (**b**) tri-block polymeric micelles that are conjugated with the chelators DOTA or CB-TE2A for Cu-64 labeling [3]; (**c**) amphiphilic graft copolymers and associated comb nanoparticles (CNPs) that are tunable for adding the DOTA chelator and targeting peptides GRGDS or the CANF peptide [92, 93]; (**d**) amphiphilic diblock copolymer micelle that contains Fe-DFO for MRI and ^{89}Zr for PET [94]; (**e**) ^{18}F-labeled amphiphilic depsipeptide (Lactosome) composed of poly(L-lactic acid)-*block*-poly(sarcosine) [95]

may provide an interesting means of imaging and therapy of inflammatory cell types, although the specific cells targeted were not enumerated.

An amphiphilic diblock copolymer micelle was designed to incorporate Zr-89 and Fe(III) for PET and T_1-weighted MRI, through chelation of the two metals with deferoxamine (DFO) (Fig. 13d) [94]. The micelles were formed by self-assembly of a mixture of polybutadiene-b-polyethyleneoxide (PBD-b-PEO) and polybutadiene-b-polyacrylicacid (PBD-b-PAA) polymers, with the acetic acid residues functionalized with ^{89}Zr-DFO for PET and Fe-DFO for MRI. The dual PET/MRI-capable nontargeted particles were investigated in LS174R human colorectal tumor-bearing mice. Biodistribution based on gamma counting of Zr-89 showed high liver and spleen uptake (30–40 % ID/g), with tumor uptake being modest (4–6 % ID/g). Contrast of the subcutaneous tumors was observed with both imaging modalities, however.

Fluorine-18 has been labeled to an amphiphilic polydepsipeptide ("Lactosome") composed of poly(L-lactic acid)-*block*-poly(sarcosine) (Fig. 13e) [95], and was investigated for biodistribution and PET imaging in Hela-tumor-bearing mice. Due to the short half-life of F-18 ($T_{1/2} = 110$ min), the time limit for imaging was 6 h, and the blood activity remained high (25 % ID/g) with low tumor uptake (<5 % ID/g).

7.3 Nanogels

Almutairi and colleagues developed chelator crosslinking hydrogel nanoparticles (nanogels) that have applications for MRI and PET imaging. The initial nanogels were polyacrylamide (PAA)-based and incorporated acyclic (DTPA) and cyclic (DOTA) chelators for Gd^{3+} [96]. The DOTA macrocycle was *C*-substituted to allow coordination of all carboxylate moieties to the Gd^{3+}. These nontargeted PAA nanogels were then modified by incorporating NOTA chelators for labeling with Cu-64 for PET imaging, with comparison to the ^{64}Cu-DOTA-nanogels (Fig. 14a) [97]. In 4T1 mouse mammary tumor-bearing mice, the ^{64}Cu-NOTA-nanogels showed surprisingly low liver and spleen accumulation compared to the DOTA agent (~5–15 % vs. 15–20 % ID/g) with tumor:liver ratios >1 (Fig. 14b). Tumor uptake for the NOTA-nanogels increased to >15 % ID/g in subcutaneous 4T1 tumors at 48 h postinjection, and as high as 30 % ID/g in small tumor metastases, which is also much higher than reported for other ^{64}Cu-labeled nanoparticles in various tumor models [39, 93, 98, 99].

7.4 Porphysomes

^{64}Cu-porphysomes are self-assembled from a single porphyrin-lipid building block, and similar to the chelator-cross-linked nanogels by the Almutairi lab [97], they do not require conjugation of chelators to the nanoparticle itself (Fig. 15) [100]. Another advantage of the porphysome platform is that it is intrinsically fluorescent, allowing for optical imaging in the same study as PET imaging. The uptake of ^{64}Cu-porphysomes in orthotopic prostate tumors (PC3 and 22RV1) was relatively low (5–7 % ID/g) at 24 h postinjection, while liver was 15–20 % ID/g and spleen uptake was >100 % ID/g. Encouragingly, small prostate tumor bone metastases in the lower extremities were readily imaged with ^{64}Cu-porphysomes.

7.5 Zinc Oxide Nanoparticles

Zinc oxide nanoparticles (ZONs) have many applications and have been used as sensors, in electronics, cosmetics, food additives, and the medical industry [101, 102] and are attractive due to their low toxicity and biodegradability [103]. ZONs of varying sizes (20 and 100 nm) have been labeled with F-18 by click chemistry to trace their biodistribution after oral administration [104]. Based on the lack of bone uptake, it was surmised there was minimal defluorination, with the majority of ^{18}F-activity observed in the gut.

Hong et al. investigated ^{64}Cu-labeled ZONs (~80–100 nm) that were either nontargeted or conjugated to the TRC105 antibody against CD105, which is found

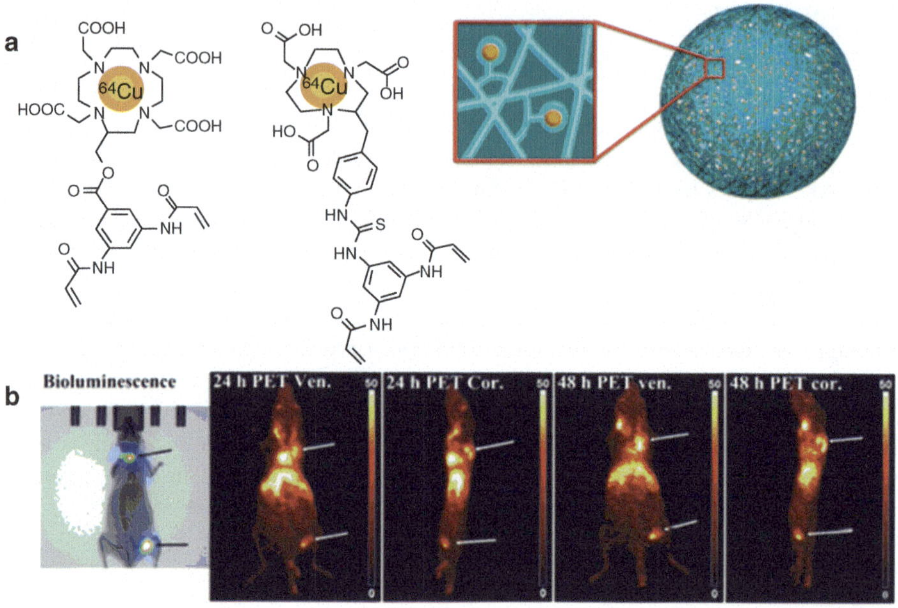

Fig. 14 (**a**) Schematic of ^{64}Cu-labeled DOTA and NOTA chelator cross-linked nanogels. (**b**) Bioluminescence and ^{64}Cu-NOTA-nanogel PET images of mice with 4 T1 tumors in their shoulder bone and lymph node in the hind leg. Uptake of ^{64}Cu-NOTA-nanogel in the metastases reached 20–30 % ID/g at 48 h postinjection [97]

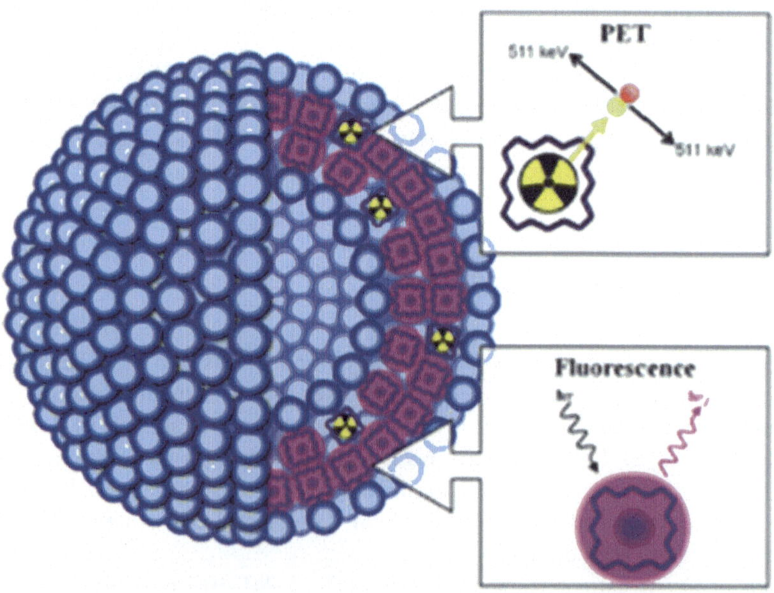

Fig. 15 Schematic diagram showing PET and fluorescence imaging properties of porphysome nanoparticles [101]

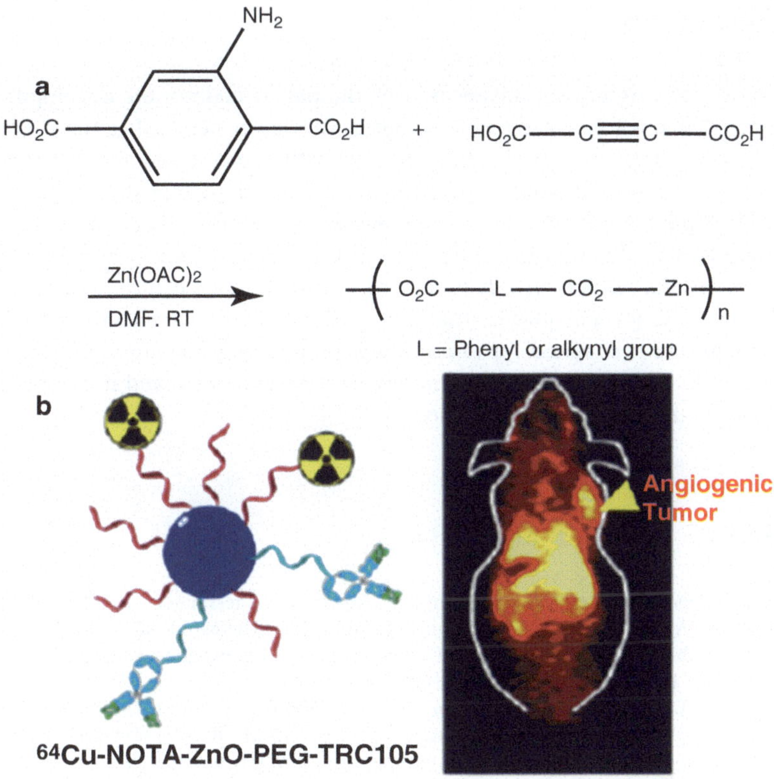

Fig. 16 (**a**) Preparation of coordination polymer precursor [107]; (**b**) red fluorescent ZnO nanoparticle conjugated with TRC105 for imaging CD105 as a marker of tumor vasculature. The nanoparticle is labeled with Cu-64 for PET imaging in 4T1 tumor-bearing mice [106]

on endothelial cells for imaging angiogenesis [105]. The ZONs used in this study incorporated a coordination polymer precursor consisting of a carboxylic acid functionalized organic molecule (Fig. 16a), and the particles were prepared by calcination at high temperatures (e.g., 550 °C) of the coordination polymers as described [106]. The authors found that the ^{64}Cu-NOTA-ZON-PEG-TRC105 particles were taken up in 4T1 tumors in significantly higher amounts than the nontargeted ZONs (~4–6 % ID/g vs. ~2–3 %ID/g) (Fig. 16b). Liver uptake was 15–20 % ID/g, and spleen was ~10 % ID/g. Overall, the tumor:non-tumor ratios were modest, even for the targeted particles.

8 Conclusions

There have been significant advances over the past decade in the development of radiolabeled nanoparticles for PET imaging of cancer. The advantages of PET, which include the ability to visualize disease with nano- to picomolar concentrations of tracer, are well suited toward nanoparticles as diagnostic or theranostic agents. The plethora of classes of nanoparticles that have been investigated is encouraging, and although there are currently few agents that have moved on to human studies, it is anticipated that many more agents will advance in this direction over the next 5 years. The jury is still out regarding whether targeted or nontargeted nanoparticles are best suited for tumor imaging; however, the answer to this question is most likely that both will ultimately be used clinically, and it will be highly dependent on the type of nanostructure.

References

1. Cook G. Oncological molecular imaging: nuclear medicine techniques. Br J Radiol. 2014.
2. Boswell CA, Sun X, Niu W, Weisman GR, Wong EH, Rheingold AL, et al. Comparative in vivo stability of copper-64-labeled cross-bridged and conventional tetraazamacrocyclic complexes. J Med Chem. 2004;47(6):1465–74.
3. Jensen AI, Binderup T, Kumar EP, Kjaer A, Rasmussen PH, Andresen TL. Positron emission tomography based analysis of long-circulating cross-linked triblock polymeric micelles in a U87MG mouse xenograft model and comparison of DOTA and CB-TE2A as chelators of copper-64. Biomacromolecules. 2014;15(5):1625–33. doi:10.1021/bm401871w.
4. Deri MA, Ponnala S, Zeglis BM, Pohl G, Dannenberg JJ, Lewis JS, et al. Alternative chelator for (8)(9)Zr radiopharmaceuticals: radiolabeling and evaluation of 3,4,3-(LI-1,2-HOPO). J Med Chem. 2014;57(11):4849–60. doi:10.1021/jm500389b.
5. Pandya DN, Pailloux S, Tatum D, Magda D, Wadas TJ. Di-macrocyclic terephthalamide ligands as chelators for the PET radionuclide zirconium-89. Chem Commun (Camb). 2015;51(12):2301–3. doi:10.1039/c4cc09256b.
6. Lee JE, Lee N, Kim T, Kim J, Hyeon T. Multifunctional mesoporous silica nanocomposite nanoparticles for theranostic applications. Acc Chem Res. 2011;44(10):893–902. doi:10.1021/ar2000259.
7. Rosenholm JM, Mamaeva V, Sahlgren C, Linden M. Nanoparticles in targeted cancer therapy: mesoporous silica nanoparticles entering preclinical development stage. Nanomedicine (Lond). 2012;7(1):111–20. doi:10.2217/nnm.11.166.
8. Meng H, Xue M, Xia T, Ji Z, Tarn DY, Zink JI, et al. Use of size and a copolymer design feature to improve the biodistribution and the enhanced permeability and retention effect of doxorubicin-loaded mesoporous silica nanoparticles in a murine xenograft tumor model. ACS Nano. 2011;5(5):4131–44. doi:10.1021/nn200809t.
9. Pan L, He Q, Liu J, Chen Y, Ma M, Zhang L, et al. Nuclear-targeted drug delivery of TAT peptide-conjugated monodisperse mesoporous silica nanoparticles. J Am Chem Soc. 2012;134(13):5722–5. doi:10.1021/ja211035w.
10. Caltagirone C, Bettoschi A, Garau A, Montis R. Silica-based nanoparticles: a versatile tool for the development of efficient imaging agents. Chem Soc Rev. 2015;44(14):4645–71. doi:10.1039/c4cs00270a.
11. Taylor KM, Kim JS, Rieter WJ, An H, Lin W, Lin W. Mesoporous silica nanospheres as highly efficient MRI contrast agents. J Am Chem Soc. 2008;130(7):2154–5. doi:10.1021/ja710193c.

12. Lu J, Liong M, Li Z, Zink JI, Tamanoi F. Biocompatibility, biodistribution, and drug-delivery efficiency of mesoporous silica nanoparticles for cancer therapy in animals. Small. 2010;6(16):1794–805. doi:10.1002/smll.201000538.
13. Lee SB, Kim HL, Jeong HJ, Lim ST, Sohn MH, Kim DW. Mesoporous silica nanoparticle pretargeting for PET imaging based on a rapid bioorthogonal reaction in a living body. Angew Chem Int Ed Engl. 2013;52(40):10549–52. doi:10.1002/anie.201304026.
14. Kim DW. Bioorthogonal click chemistry for fluorine-18 labeling protocols under physiological friend reaction condition. J Fluor Chem. 2015;174:142–7. doi:10.1016/j.jfluchem.2014.11.009.
15. Miller L, Winter G, Baur B, Witulla B, Solbach C, Reske S, et al. Synthesis, characterization, and biodistribution of multiple 89Zr-labeled pore-expanded mesoporous silica nanoparticles for PET. Nanoscale. 2014;6(9):4928–35. doi:10.1039/c3nr06800e.
16. Chen F, Nayak TR, Goel S, Valdovinos HF, Hong H, Theuer CP, et al. In vivo tumor vasculature targeted PET/NIRF imaging with TRC105(Fab)-conjugated, dual-labeled mesoporous silica nanoparticles. Mol Pharm. 2014;11(11):4007–14. doi:10.1021/mp500306k.
17. Kang WJ, Lee J, Lee YS, Cho S, Ali BA, Al-Khedhairy AA, et al. Multimodal imaging probe for targeting cancer cells using uMUC-1 aptamer. Colloids Surf B Biointerfaces. 2015;136:134–40. doi:10.1016/j.colsurfb.2015.09.004.
18. Chakravarty R, Goel S, Hong H, Chen F, Valdovinos HF, Hernandez R, et al. Hollow mesoporous silica nanoparticles for tumor vasculature targeting and PET image-guided drug delivery. Nanomedicine (Lond). 2015;10(8):1233–46. doi:10.2217/nnm.14.226.
19. Cobley CM, Chen J, Cho EC, Wang LV, Xia Y. Gold nanostructures: a class of multifunctional materials for biomedical applications. Chem Soc Rev. 2011;40(1):44–56. doi:10.1039/b821763g.
20. Sperling RA, Gil PR, Zhang F, Zanella M, Parak WJ. Biological applications of gold nanoparticles. Chem Soc Rev. 2008;37(9):1896–908. doi:10.1039/b712170a.
21. Daniel MC, Astruc D. Gold nanoparticles: assembly, supramolecular chemistry, quantum-size-related properties, and applications toward biology, catalysis, and nanotechnology. Chem Rev. 2004;104(1):293–346. doi:10.1021/cr030698+.
22. Chen J, Glaus C, Laforest R, Zhang Q, Yang M, Gidding M, et al. Gold nanocages as photothermal transducers for cancer treatment. Small. 2010;6(7):811–7. doi:10.1002/smll.200902216.
23. Liu Y, Yuan H, Kersey FR, Register JK, Parrott MC, Vo-Dinh T. Plasmonic gold nanostars for multi-modality sensing and diagnostics. Sensors. 2015;15(2):3706–20. doi:10.3390/s150203706.
24. Vo-Dinh T, Liu Y, Fales AM, Ngo H, Wang HN, Register JK, et al. SERS nanosensors and nanoreporters: golden opportunities in biomedical applications. Wiley Interdiscip Rev Nanomed Nanobiotechnol. 2015;7(1):17–33. doi:10.1002/wnan.1283.
25. Khlebtsov N, Bogatyrev V, Dykman L, Khlebtsov B, Staroverov S, Shirokov A, et al. Analytical and theranostic applications of gold nanoparticles and multifunctional nanocomposites. Theranostics. 2013;3(3):167–80. doi:10.7150/thno.5716.
26. Brust M, Fink J, Bethell D, Schiffrin D, Kiely C. Synthesis and reactions of functionalised gold nanoparticles. J Chem Soc Chem Commun. 1995;16:1655–6.
27. Leff DV, Brandt L, Heath JR. Synthesis and characterization of hydrophobic, organically-soluble gold nanocrystals functionalized with primary amines. Langmuir. 1996;12(20):4723–30.
28. Anshup A, Venkataraman JS, Subramaniam C, Kumar RR, Priya S, Kumar TS, et al. Growth of gold nanoparticles in human cells. Langmuir. 2005;21(25):11562–7.
29. Huang X, El-Sayed IH, Qian W, El-Sayed MA. Cancer cell imaging and photothermal therapy in the near-infrared region by using gold nanorods. J Am Chem Soc. 2006;128(6):2115–20.
30. Caruso F, Spasova M, Salgueiriño-Maceira V, Liz-Marzán L. Multilayer assemblies of silica-encapsulated gold nanoparticles on decomposable colloid templates. Adv Mater. 2001;13(14):1090–4.
31. Oldenburg SJ, Jackson JB, Westcott SL, Halas N. Infrared extinction properties of gold nanoshells. Appl Phys Lett. 1999;75(19):2897–9.
32. Chen J, McLellan JM, Siekkinen A, Xiong Y, Li Z-Y, Xia Y. Facile synthesis of gold-silver nanocages with controllable pores on the surface. J Am Chem Soc. 2006;128(46):14776–7.

33. Chen J, Saeki F, Wiley BJ, Cang H, Cobb MJ, Li Z-Y, et al. Gold nanocages: bioconjugation and their potential use as optical imaging contrast agents. Nano Lett. 2005;5(3):473–7.
34. Giersig M, Mulvaney P. Preparation of ordered colloid monolayers by electrophoretic deposition. Langmuir. 1993;9(12):3408–13.
35. Pirollo KF, Chang EH. Does a targeting ligand influence nanoparticle tumor localization or uptake? Trends Biotechnol. 2008;26(10):552–8. http://dx.doi.org/10.1016/j.tibtech.2008.06.007.
36. Xie H, Diagaradjane P, Deorukhkar AA, Goins B, Bao A, Phillips WT, et al. Integrin α(v) β(3)-targeted gold nanoshells augment tumor vasculature-specific imaging and therapy. Int J Nanomedicine. 2011;6:259–69. doi:10.2147/ijn.s15479.
37. Zhou M, Zhang R, Huang M, Lu W, Song S, Melancon MP, et al. A chelator-free multifunctional [64Cu] CuS nanoparticle platform for simultaneous micro-PET/CT imaging and photothermal ablation therapy. J Am Chem Soc. 2010;132(43):15351–8.
38. Zhao Y, Sultan D, Detering L, Cho S, Sun G, Pierce R, et al. Copper-64-alloyed gold nanoparticles for cancer imaging: improved radiolabel stability and diagnostic accuracy. Angew Chem Int Ed. 2014;53(1):156–9.
39. Zhao Y, Sultan D, Detering L, Luehmann H, Liu Y. Facile synthesis, pharmacokinetic and systemic clearance evaluation, and positron emission tomography cancer imaging of 64Cu-Au alloy nanoclusters. Nanoscale. 2014;6(22):13501–9. doi:10.1039/c4nr04569f.
40. Tian M, Lu W, Zhang R, Xiong C, Ensor J, Nazario J, et al. Tumor uptake of hollow gold nanospheres after intravenous and intra-arterial injection: PET/CT study in a rabbit VX2 liver cancer model. Mol Imaging Biol. 2013;15(5):614–24. doi:10.1007/s11307-013-0635-x.
41. Tsoukalas C, Laurent G, Sánchez GJ, Tsotakos T, Bazzi R, Stellas D, et al. Initial in vitro and in vivo assessment of Au@ DTDTPA-RGD nanoparticles for Gd-MRI and 68Ga-PET dual modality imaging. EJNMMI Phys. 2015;2 Suppl 1:A89.
42. Karmani L, Bouchat V, Bouzin C, Levêque P, Labar D, Bol A, et al. 89Zr-labeled anti-endoglin antibody-targeted gold nanoparticles for imaging cancer: implications for future cancer therapy. Nanomedicine. 2014;9(13):1923–37.
43. Choi CH, Zuckerman JE, Webster P, Davis ME. Targeting kidney mesangium by nanoparticles of defined size. Proc Natl Acad Sci U S A. 2011;108(16):6656–61. doi:10.1073/pnas.1103573108.
44. Burns AA, Vider J, Ow H, Herz E, Penate-Medina O, Baumgart M, et al. Fluorescent silica nanoparticles with efficient urinary excretion for nanomedicine. Nano Lett. 2009;9(1):442–8. doi:10.1021/nl803405h.
45. Longmire M, Choyke PL, Kobayashi H. Clearance properties of nano-sized particles and molecules as imaging agents: considerations and caveats. Nanomedicine. 2008;3(5):703–17. doi:10.2217/17435889.3.5.703.
46. Kharissova OV, Kharisov BI, Jiménez-Pérez VM, Flores BM, Méndez UO. Ultrasmall particles and nanocomposites: state of the art. RSC Adv. 2013;3(45):22648–82.
47. Schipper ML, Iyer G, Koh AL, Cheng Z, Ebenstein Y, Aharoni A, et al. Particle size, surface coating, and PEGylation influence the biodistribution of quantum dots in living mice. Small. 2009;5(1):126–34. doi:10.1002/smll.200800003.
48. Jokerst JV, Lobovkina T, Zare RN, Gambhir SS. Nanoparticle PEGylation for imaging and therapy. Nanomedicine. 2011;6(4):715–28. doi:10.2217/nnm.11.19.
49. You J, Zhou J, Zhou M, Liu Y, Robertson JD, Liang D, et al. Pharmacokinetics, clearance, and biosafety of polyethylene glycol-coated hollow gold nanospheres. Part Fibre Toxicol. 2014;11(1):1–14.
50. Liu Z, Cai W, He L, Nakayama N, Chen K, Sun X, et al. In vivo biodistribution and highly efficient tumour targeting of carbon nanotubes in mice. Nat Nanotechnol. 2007;2(1):47–52. doi:10.1038/nnano.2006.170.
51. Cui L, Lin Q, Jin CS, Jiang W, Huang H, Ding L, et al. A PEGylation-free biomimetic porphyrin nanoplatform for personalized cancer theranostics. ACS Nano. 2015;9(4):4484–95. doi:10.1021/acsnano.5b01077.
52. Pombo-García K, Zarschler K, Barreto JA, Hesse J, Spiccia L, Graham B, et al. Design, synthesis, characterisation and in vitro studies of hydrophilic, colloidally stable, 64Cu(ii)-

labelled, ultra-small iron oxide nanoparticles in a range of human cell lines. RSC Adv. 2013;3(44):22443. doi:10.1039/c3ra43726d.

53. Zhou M, Li J, Liang S, Sood AK, Liang D, Li C. CuS nanodots with ultrahigh efficient renal clearance for positron emission tomography imaging and image-guided photothermal therapy. ACS Nano. 2015;9(7):7085–96. doi:10.1021/acsnano.5b02635.
54. Benezra M, Penate-Medina O, Zanzonico PB, Schaer D, Ow H, Burns A, et al. Multimodal silica nanoparticles are effective cancer-targeted probes in a model of human melanoma. J Clin Invest. 2011;121(7):2768–80. doi:10.1172/JCI45600.
55. Phillips E, Penate-Medina O, Zanzonico PB, Carvajal RD, Mohan P, Ye Y, et al. Clinical translation of an ultrasmall inorganic optical-PET imaging nanoparticle probe. Sci Transl Med. 2014;6(260):260ra149. doi:10.1126/scitranslmed.3009524.
56. Liu Y, Zhao Y, Sun B, Chen C. Understanding the toxicity of carbon nanotubes. Acc Chem Res. 2013;46(3):702–13. doi:10.1021/ar300028m.
57. Gao F, Cai P, Yang W, Xue J, Gao L, Liu R, et al. Ultrasmall [(64)Cu]Cu nanoclusters for targeting orthotopic lung tumors using accurate positron emission tomography imaging. ACS Nano. 2015;9(5):4976–86. doi:10.1021/nn507130k.
58. Truillet C, Bouziotis P, Tsoukalas C, Brugiere J, Martini M, Sancey L, et al. Ultrasmall particles for Gd-MRI and (68) Ga-PET dual imaging. Contrast Media Mol Imaging. 2015;10(4):309–19. doi:10.1002/cmmi.1633.
59. Yang BY, Moon S-H, Seelam SR, Jeon MJ, Lee Y-S, Lee DS, et al. Development of a multimodal imaging probe by encapsulating iron oxide nanoparticles with functionalized amphiphiles for lymph node imaging. Nanomedicine. 2015;10(12):1899–910.
60. Lee S, Chen X. Dual-modality probes for in vivo molecular imaging. Mol Imaging. 2009;8(2):87.
61. Lee JH, Sherlock SP, Terashima M, Kosuge H, Suzuki Y, Goodwin A, et al. High-contrast in vivo visualization of microvessels using novel FeCo/GC magnetic nanocrystals. Magn Reson Med. 2009;62(6):1497–509.
62. Yang H, Zhang C, Shi X, Hu H, Du X, Fang Y, et al. Water-soluble superparamagnetic manganese ferrite nanoparticles for magnetic resonance imaging. Biomaterials. 2010;31(13):3667–73.
63. Tromsdorf UI, Bigall NC, Kaul MG, Bruns OT, Nikolic MS, Mollwitz B, et al. Size and surface effects on the MRI relaxivity of manganese ferrite nanoparticle contrast agents. Nano Lett. 2007;7(8):2422–7.
64. Xie J, Chen K, Huang J, Lee S, Wang J, Gao J, et al. PET/NIRF/MRI triple functional iron oxide nanoparticles. Biomaterials. 2010;31(11):3016–22.
65. Lee H-Y, Li Z, Chen K, Hsu AR, Xu C, Xie J, et al. PET/MRI dual-modality tumor imaging using arginine-glycine-aspartic (RGD)-conjugated radiolabeled iron oxide nanoparticles. J Nucl Med. 2008;49(8):1371–9.
66. Groult H, Ruiz-Cabello J, Pellico J, Lechuga-Vieco AV, Bhavesh R, Zamai M, et al. Parallel multifunctionalization of nanoparticles: a one-step modular approach for in vivo imaging. Bioconjug Chem. 2014;26(1):153–60.
67. Yang M, Cheng K, Qi S, Liu H, Jiang Y, Jiang H, et al. Affibody modified and radiolabeled gold–iron oxide hetero-nanostructures for tumor PET, optical and MR imaging. Biomaterials. 2013;34(11):2796–806.
68. Caride VJ. Liposomes as carriers of imaging agents. Crit Rev Ther Drug Carrier Syst. 1985;1(2):121–53.
69. Tilcock C, Ahkong QF, Fisher D. Polymer-derivatized technetium 99mTc-labeled liposomal blood pool agents for nuclear medicine applications. Biochim Biophys Acta Biomembr. 1993;1148(1):77–84.
70. Krause W, Klopp R, Leike J, Sachse A, Schuhmann-Giampieri G. Liposomes in diagnostic imaging-comparison of modalities-in-vivo visualization of liposomes. J Liposome Res. 1995;5(1):1–26.

71. Tilcock C, Philippot J, Schuber F. Imaging tools: liposomal agents for nuclear medicine, computed tomography, magnetic resonance, and ultrasound. Liposomes as tools in basic research and industry. 1995. p. 225–40.
72. Tilcock C. Delivery of contrast agents for magnetic resonance imaging, computed tomography, nuclear medicine and ultrasound. Adv Drug Deliv Rev. 1999;37(1):33–51.
73. Phillips WT, Goins BA, Bao A. Radioactive liposomes. Wiley Interdiscip Rev Nanomed Nanobiotechnol. 2009;1(1):69–83.
74. Jensen AT, Binderup T, Andresen TL, Kjaer A, Rasmussen PH. PET imaging of liposomes labeled with an [(1)(8)F]-fluorocholesteryl ether probe prepared by automated radiosynthesis. J Liposome Res. 2012;22(4):295–305. doi:10.3109/08982104.2012.698418.
75. Emmetiere F, Irwin C, Viola-Villegas NT, Longo V, Cheal SM, Zanzonico P, et al. (18) F-labeled-bioorthogonal liposomes for in vivo targeting. Bioconjug Chem. 2013;24(11):1784–9. doi:10.1021/bc400322h.
76. Perk LR, Visser GW, Vosjan MJ, Stigter-van Walsum M, Tijink BM, Leemans CR, et al. (89) Zr as a PET surrogate radioisotope for scouting biodistribution of the therapeutic radiometals (90)Y and (177)Lu in tumor-bearing nude mice after coupling to the internalizing antibody cetuximab. J Nucl Med. 2005;46(11):1898–906.
77. Lubberink M, Herzog H. Quantitative imaging of 124I and 86Y with PET. Eur J Nucl Med Mol Imaging. 2011;38 Suppl 1:S10–8. doi:10.1007/s00259-011-1768-2.
78. Perez-Medina C, Abdel-Atti D, Zhang Y, Longo VA, Irwin CP, Binderup T, et al. A modular labeling strategy for in vivo PET and near-infrared fluorescence imaging of nanoparticle tumor targeting. J Nucl Med. 2014;55(10):1706–11. doi:10.2967/jnumed.114.141861.
79. Seo JW, Mahakian LM, Tam S, Qin S, Ingham ES, Meares CF, et al. The pharmacokinetics of Zr-89 labeled liposomes over extended periods in a murine tumor model. Nucl Med Biol. 2015;42(2):155–63. doi:10.1016/j.nucmedbio.2014.09.001.
80. Abou DS, Thorek DL, Ramos NN, Pinkse MW, Wolterbeek HT, Carlin SD, et al. (89) Zr-labeled paramagnetic octreotide-liposomes for PET-MR imaging of cancer. Pharm Res. 2013;30(3):878–88. doi:10.1007/s11095-012-0929-8.
81. Petersen AL, Binderup T, Rasmussen P, Henriksen JR, Elema DR, Kjaer A, et al. 64Cu loaded liposomes as positron emission tomography imaging agents. Biomaterials. 2011;32(9):2334–41. doi:10.1016/j.biomaterials.2010.11.059.
82. Henriksen JR, Petersen AL, Hansen AE, Frankaer CG, Harris P, Elema DR, et al. Remote loading of (64)Cu(2+) into liposomes without the use of ion transport enhancers. ACS Appl Mater Interfaces. 2015;7(41):22796–806. doi:10.1021/acsami.5b04612.
83. Petersen AL, Binderup T, Jolck RI, Rasmussen P, Henriksen JR, Pfeifer AK, et al. Positron emission tomography evaluation of somatostatin receptor targeted 64Cu-TATE-liposomes in a human neuroendocrine carcinoma mouse model. J Contr Release. 2012;160(2):254–63. doi:10.1016/j.jconrel.2011.12.038.
84. Locke LW, Mayo MW, Yoo AD, Williams MB, Berr SS. PET imaging of tumor associated macrophages using mannose coated 64Cu liposomes. Biomaterials. 2012;33(31):7785–93. doi:10.1016/j.biomaterials.2012.07.022.
85. Kang CM, Koo HJ, Lee S, Lee KC, Oh YK, Choe YS. 64Cu-labeled tetraiodothyroacetic acid-conjugated liposomes for PET imaging of tumor angiogenesis. Nucl Med Biol. 2013;40(8):1018–24. doi:10.1016/j.nucmedbio.2013.08.003.
86. Wong AW, Ormsby E, Zhang H, Seo JW, Mahakian LM, Caskey CF, et al. A comparison of image contrast with (64)Cu-labeled long circulating liposomes and (18)F-FDG in a murine model of mammary carcinoma. Am J Nucl Med Mol Imaging. 2013;3(1):32–43.
87. Gong H, Peng R, Liu Z. Carbon nanotubes for biomedical imaging: the recent advances. Adv Drug Deliv Rev. 2013;65(15):1951–63. doi:10.1016/j.addr.2013.10.002.
88. McDevitt MR, Chattopadhyay D, Jaggi JS, Finn RD, Zanzonico PB, Villa C et al. PET imaging of soluble yttrium-86-labeled carbon nanotubes in mice. PLoS One. 2007;2(9):e907. 10.1371/journal.pone.0000907.
89. Zeng D, Lee NS, Liu Y, Zhou D, Dence CS, Wooley KL, et al. 64Cu Core-labeled nanoparticles with high specific activity via metal-free click chemistry. ACS Nano. 2012;6(6):5209–19. doi:10.1021/nn300974s.

90. Rossin R, Pan D, Qi K, Turner JL, Sun X, Wooley KL, et al. 64Cu-labeled folate-conjugated shell cross-linked nanoparticles for tumor imaging and radiotherapy: synthesis, radiolabeling, and biologic evaluation. J Nucl Med. 2005;46(7):1210–8.
91. Sun X, Wuest M, Weisman GR, Wong EH, Reed DP, Boswell CA, et al. Radiolabeling and in vivo behavior of copper-64-labeled cross-bridged cyclam ligands. J Med Chem. 2002;45:469–77.
92. Shokeen M, Pressly ED, Hagooly A, Zheleznyak A, Ramos N, Fiamengo AL, et al. Evaluation of multivalent, functional polymeric nanoparticles for imaging applications. ACS Nano. 2011;5(2):738–47. doi:10.1021/nn102278w.
93. Pressly ED, Pierce RA, Connal LA, Hawker CJ, Liu Y. Nanoparticle PET/CT imaging of natriuretic peptide clearance receptor in prostate cancer. Bioconjug Chem. 2013;24(2):196–204. doi:10.1021/bc300473x.
94. Starmans LW, Hummelink MA, Rossin R, Kneepkens EC, Lamerichs R, Donato K, et al. Zr- and Fe-labeled polymeric micelles for dual modality PET and T-weighted MR imaging. Adv Healthc Mater. 2015. doi:10.1002/adhm.201500414.
95. Yamamoto F, Yamahara R, Makino A, Kurihara K, Tsukada H, Hara E, et al. Radiosynthesis and initial evaluation of (18)F labeled nanocarrier composed of poly(L-lactic acid)-block-poly(sarcosine) amphiphilic polydepsipeptide. Nucl Med Biol. 2013;40(3):387–94. doi:10.1016/j.nucmedbio.2012.12.008.
96. Lux J, Chan M, Elst LV, Schopf E, Mahmoud E, Laurent S, et al. Metal chelating crosslinkers form nanogels with high chelation stability. J Mater Chem B Mater Biol Med. 2013;1(46):6359–64. doi:10.1039/C3TB21104E.
97. Lux J, White AG, Chan M, Anderson CJ, Almutairi A. Nanogels from metal-chelating crosslinkers as versatile platforms applied to copper-64 PET imaging of tumors and metastases. Theranostics. 2015;5(3):277–88. doi:10.7150/thno.10904.
98. Black KC, Wang Y, Luehmann HP, Cai X, Xing W, Pang B, et al. Radioactive 198Au-doped nanostructures with different shapes for in vivo analyses of their biodistribution, tumor uptake, and intratumoral distribution. ACS Nano. 2014;8(5):4385–94. doi:10.1021/nn406258m.
99. Zhao Y, Sultan D, Detering L, Cho S, Sun G, Pierce R, et al. Copper-64-alloyed gold nanoparticles for cancer imaging: improved radiolabel stability and diagnostic accuracy. Angew Chem Int Ed Engl. 2014;53(1):156–9. doi:10.1002/anie.201308494.
100. Liu TW, Macdonald TD, Jin CS, Gold JM, Bristow RG, Wilson BC, et al. Inherently multimodal nanoparticle-driven tracking and real-time delineation of orthotopic prostate tumors and micrometastases. ACS Nano. 2013;7(5):4221–32. doi:10.1021/nn400669r.
101. Heng BC, Zhao X, Xiong S, Ng KW, Boey FY, Loo JS. Cytotoxicity of zinc oxide (ZnO) nanoparticles is influenced by cell density and culture format. Arch Toxicol. 2011;85(6):695–704. doi:10.1007/s00204-010-0608-7.
102. Osmond MJ, McCall MJ. Zinc oxide nanoparticles in modern sunscreens: an analysis of potential exposure and hazard. Nanotoxicology. 2010;4(1):15–41. doi:10.3109/17435390903502028.
103. Zhou J, Xu NS, Wang ZL. Dissolving behavior and stability of ZnO wires in biofluids: a study on biodegradability and biocompatibility of ZnO nanostructures. Adv Mater. 2006;18:2432–5.
104. Lee CM, Jeong HJ, Kim DW, Sohn MH, Lim ST. The effect of fluorination of zinc oxide nanoparticles on evaluation of their biodistribution after oral administration. Nanotechnology. 2012;23(20):205102. doi:10.1088/0957-4484/23/20/205102.
105. Hong H, Wang F, Zhang Y, Graves SA, Eddine SB, Yang Y, et al. Red fluorescent zinc oxide nanoparticle: a novel platform for cancer targeting. ACS Appl Mater Interfaces. 2015;7(5):3373–81. doi:10.1021/am508440j.
106. Shi H-Y, Deng B, Zhong S-L, Wang L, Xu A-W. Synthesis of zinc oxide nanoparticles with strong, tunable and stable visible light emission by solid-state transformation of Zn (II)-organic coordination polymers. J Mater Chem. 2011;21(33):12309–15.

Theranostic Nanoplatforms for PET Image-Guided Drug Delivery

Rubel Chakravarty, Feng Chen, Ashutosh Dash, and Weibo Cai

1 Introduction

Nanomedicine is often heralded as one of the major leaps forward for twenty-first century clinical practice [1–3]. The use of nanomaterials for cancer diagnosis and therapy is arguably the most active area of nanomedicine research. Despite extensive research input and huge investments, cancer remains a major public health concern worldwide [4]. Basically, this disease encompasses a heterogeneous spectrum of conditions and is highly unpredictable in majority of cases [4]. As per the statistics provided by World Health Organization (WHO), 8.2 million people worldwide died from cancer in 2012, out of which at least 30 % of cancer deaths could have been prevented if necessary treatment was provided at an earlier time point [5]. Early diagnosis and treatment are essential to minimize the morbidity and mortality associated with the disease. In fact, early diagnosis of cancer is the crucial factor in majority of cases that directs the treatment regime and the choice of therapeutic intervention [4]. Successful cancer management relies on several factors that can be uniquely addressed via nanomedicine [1].

R. Chakravarty, Ph.D. (✉) • A. Dash
Isotope Production and Applications Division, Bhabha Atomic Research Centre, Mumbai 400 085, India
e-mail: rubelc@barc.gov.in

F. Chen
Department of Radiology, University of Wisconsin, Madison, WI 53792-3252, USA

W. Cai, Ph.D. (✉)
Department of Radiology, University of Wisconsin, Madison, WI 53792-3252, USA

Department of Medical Physics, University of Wisconsin, Madison, WI 53705-2275, USA

Carbone Cancer Center, University of Wisconsin, Madison, WI 53792-3252, USA
e-mail: wcai@uwhealth.org

J.W.M. Bulte, M.M.J. Modo (eds.), *Design and Applications of Nanoparticles in Biomedical Imaging*, DOI 10.1007/978-3-319-42169-8_12

Conventional cancer treatment approaches rely on systemic administration of chemotherapeutic drugs that indiscriminately affect tumor and healthy tissue alike, and therefore are toxic to both types of tissues [6, 7]. Such strategies are limited by a narrow therapeutic index (ratio of therapeutic to toxic effects) and demonstrate severe systemic side effects [8]. Additionally, several routinely used chemotherapeutic drugs suffer from poor pharmacokinetics and inappropriate biodistribution that greatly limits the maximum allowable dose of the drug [3, 6, 8]. Therefore, many conventional drugs that have been shown to be highly effective in vitro are often relatively ineffective when administered in vivo. From this perspective, the use of nanoplatforms for targeted drug delivery can increase the selectivity of the treatment, improve drug concentration at the tumor site, and maximize the therapeutic response while minimizing toxic side effects [3, 7, 9].

The past 10 years have witnessed significant advances in the development and deployment of nanoplatforms for targeted drug delivery, and innovative applications of cancer nanomedicine are now coming to fruition [3, 6–10]. Numerous nanoparticle-based products for drug delivery have been approved for clinical applications, and even more are currently in clinical trials [3, 7]. An important breakthrough in this direction is the development of multifunctional nanoplatforms—nanoparticles that are capable of accomplishing multiple objectives such as imaging and targeted therapy or performing a single advanced function through incorporation of multiple functional units [10–12]. The synergistic utilization of a single nanoplatform for both molecular imaging as well as targeted drug delivery is known as "image-guided drug delivery," which is a promising attribute toward personalized cancer management [9, 11, 12].

By adopting molecular imaging approaches it is possible to noninvasively visualize how well these nanoplatforms can accumulate at the target site and specifically deliver the drug molecules [11]. It is also possible to preselect patients who are likely to respond to such treatment procedures. This strategy also offers a way to monitor how well patients would respond to nanomedicine-based therapeutic interventions, based on which drug doses and treatment protocols can be individualized and optimized during follow-up. Molecular imaging information on the possible accumulation of nanomedicine formulations in endangered healthy tissues may be used to exclude patients from further treatment.

The field of image-guided drug delivery has witnessed rapid advancement over the last few years which could be attributed to the phenomenal growth of molecular imaging technologies [11–13]. Various imaging modalities [e.g., positron emission tomography (PET), single photon emission computed tomography (SPECT), magnetic resonance imaging (MRI), optical, ultrasound] are now routinely used to assess specific molecular targets in preclinical and clinical settings [13–15]. Among these molecular imaging modalities, PET imaging is becoming more prevalent in clinical practices all over the world particularly because of its high sensitivity and possibility for accurate quantification which helps in understanding biological processes at the molecular and metabolic levels in vivo [16]. Despite excellent attributes of PET technology in the field of clinical molecular imaging, it must be stressed here that no single imaging modality can provide information on all aspects of structure and function [17]. The choice of a particular imaging modality is primarily dependent on the specific question to be addressed through molecular imaging approach.

This chapter focuses on the inherent feasibility and practicality of the concept of PET image-guided drug delivery using nanoplatforms. The development of theranostic nanoplatforms for site- and event-specific targeting and controlled drug release are summarized, and the great potential and fascinating prospects for forthcoming developments that might help in translating the research results from "bench-to-bedside" are discussed.

2 PET Imaging as a Tool to Guide Drug Delivery

PET is a sensitive and specific noninvasive imaging modality that employs external detectors to measure the three dimensional distribution and pharmacokinetics of injected drug-loaded nanoplatforms that have been radiolabeled with suitable positron emitting radioisotopes [16, 18]. After being emitted from the nucleus, the positron travels a short distance in the surrounding matter or tissue before it annihilates with an electron to produce two 511 keV γ-rays, which correspond to the rest masses of the positron and electron [16]. These γ-rays are emitted simultaneously in opposite directions and are then detected by an array of surrounding detectors connected in coincidence mode. Generally, a large number of coincidence events are acquired within a very narrow time interval (nanoseconds). The data is stored in the form two-dimensional matrices called sonograms that are then corrected for detector nonuniformity of response, attenuation of photons by body and scatter, and reconstructed into an image with information on the spatial distribution of radioactivity as a function of time.

The ability to measure pharmacokinetics of drug-loaded nanoplatforms in tissues with PET needs to be underpinned by strong radiochemistry input [16, 18]. Several positron-emitting radioisotopes can be used to radiolabel drug-loaded nanoplatforms for research and clinical use, summary of which along with their nuclear decay properties is provided in Table 1. The availability of a wide variety of radioisotopes makes it possible to carefully pick the specific nuclear properties that are needed for a particular application in PET image-guided drug delivery [16]. From the perspective of PET imaging, ^{18}F ($t_{½}$=110 min) and ^{68}Ga ($t_{½}$=68 min) appear to be ideal choices because of their almost perfect chemical and nuclear decay characteristics. Indeed, these two radioisotopes are most widely used for preparing conventional PET radiopharmaceuticals for routine clinical use all over the world [19, 20]. However, ^{18}F and ^{68}Ga may not be suitable for use in PET image-guided drug delivery as their short decay half-lives do not match the pharmacokinetics of most drug-loaded nanoplatforms. Moreover, the radiolabeling step might be time consuming, especially, in case of chelator-free radiolabeling of nanoplatforms [21–23], wherein use of these short-lived radioisotopes might not be economically viable. From this perspective, the use of nonconventional radioisotopes such as ^{44}Sc, ^{64}Cu, ^{69}Ge, ^{86}Y, ^{89}Zr, etc. (Table 1), with relatively longer half-lives, might be beneficial [24, 25]. However, it is pertinent to point out that most of these radioisotopes emit high energy γ-photons (with relatively high abundance) caused by a complex decay scheme. These γ-photons can be detected in addition to the coincidence photons resulting from the positron emission, thereby increasing the spurious event rate, which leads to inferior imaging quality and

Table 1 Nuclear decay characteristics of some positron emitting radioisotopes that can be used for PET image-guided drug delivery

Radionuclide	Half-life (h)	Mode of decay	β^+ Particle energy (MeV)[a,b]	β^+-Branching ratio (%)[b]	Major γ-photons emitted other than annihilation photons in MeV (% abundance)
^{18}F	1.8	β^+/ECD	0.633	97.0	None
^{44}Sc	3.9	β^+/ECD	1.474	94.3	1.157 (99.9)
^{64}Cu	12.7	β^+/β^-/ECD	0.653	17.4	1.346 (0.47)
^{68}Ga	1.1	β^+/ECD	1.889	88.0	1.077 (3.3)
^{69}Ge	39.1	β^+/ECD	1.205	21.0	0.574 (13.3), 0.872 (11.9)
^{72}As	26.0	β^+/ECD	2.499	64.2	0.630 (7.9), 0.834 (80.0),
^{86}Y	14.7	β^+/ECD	1.220	11.9	0.443 (16.9), 0.646 (9.2), 0.777 (22.4), 1.854 (17.4), 1.920 (20.8)
^{89}Zr	78.4	β^+/ECD	0.897	23.0	0.909 (100)

ECD electron capture decay
[a]Maximum β^+ energy is mentioned
[b]Only principal β^+ emission is indicated

degradation of the quantitative accuracy. Therefore, depending on the specific application, a balance between half-life of the radioisotope and the image quality which is dependent on nuclear decay characteristics of the radioisotope needs to be considered for choosing the appropriate radioisotope for labeling the desired nanoplatform.

PET images acquired after in vivo administration of radiolabeled nanoplatforms are generally analyzed by defining regions of interest and extracting radioactivity versus time-curves for the region [16, 18]. In more realistic circumstances, functional parametric images are generated on a voxel-to-voxel basis using generic kinetic analysis [18]. If effect of drug-loaded nanoplatforms needs to be quantified, mathematical kinetic modeling approach can be employed [16, 18]. This strategy helps to enhance data interpretation within a framework of important kinetic behavior in the region of interest to obtain quantitative parameters of relevance and universal comprehension. Mathematical modeling of PET data also enables pharmacological and physiological parameters that can be used to quantitatively assess the in vivo behavior of drug-loaded nanoplatforms [16, 18].

3 Radiolabeled Nanoplatforms for PET Image-Guided Drug Delivery

In order to achieve image-guided drug delivery, researchers have developed several nanoplatforms with diverse sizes, architectures, and surface properties for selective administration of the chemotherapeutic drug to a specific target location [11, 12]. Some typical examples of nanoplatforms that have been particularly found to be useful for PET-image-guided drug delivery include organic nanoparticles such as liposomes, micelles, endogenous nanostructures, and inorganic nanoparticles such as colloidal metals and oxide nanoparticles [11, 26]. For a particular application, the choice of the nanoplatform is influenced by the bioavailability, biodistribution, types of drugs that can be delivered, and the specificity and pharmacokinetics of delivery. The in vivo stability and fate of the drug-loaded nanoplatforms is decided by numerous factors such as size, rigidity, charge, solubility, and surface modifications of the nanoplatform. With the advances in material science and nanotechnology, it is now possible to specifically tailor these properties during the synthesis of nanoplatforms for targeted delivery of requisite doses of chemotherapeutic drugs and imaging contrast agents into cancerous lesions while sparing the healthy tissues. Such "smart" theranostic nanoplatforms hold out the possibility of radically changing the practice of cancer management, allowing easy diagnosis followed by effective targeted therapy at the early stages of the disease.

Generally, two strategies are used for loading drugs onto targeting nanoplatforms [11, 12]. In the first approach, drugs are directly conjugated to the nanoplatforms surface using suitable linkers [11, 12]. The major limitations of this approach include (a) potential alteration in property of the drug due to conjugation, (b) inevitable heterogeneity of the final product, and (c) the need to develop customized conjugation procedure for each particular drug that needs to be delivered using the nanoplatform [11, 12]. Many of these limitations could be circumvented to a considerable extent in the other approach, which involves loading of chemotherapeutic drugs onto high capacity nanoplatforms (e.g., liposomes, inorganic oxides, etc.) [11, 12]. Such drug-loaded nanoplatforms protect entrapped drugs from degradation during their delivery to the target and do not alter the biological efficacy of the drug [11, 12]. Despite excellent attributes of this strategy for PET image-guided drug delivery, drug release from the nanoplatforms cannot always be properly triggered to take place with the desired selectivity [27]. Moreover, homogeneous distribution and effective internalization of the drugs by the whole population of targeted cancerous cells are not always achievable [27]. Nevertheless, it would be possible to achieve sustained drug release over a prolonged period of time by modulating the porosity, surface charge, and biodegradability of the nanoplatforms.

The delivery of the nanoplatform to the target tissue can be achieved primarily in two ways—passive and active targeting (Fig. 1) [28]. Passive targeting takes advantage of the permeability of tumor tissue [28]. Due to rapid vascularization to serve fast-growing cancerous tissues, the capillary endothelium in cancerous tissue is more disorderly and thus more permeable toward nanoplatforms than the capillary endothelium in normal tissues. If the drug-loaded nanoplatform can stay in blood circulation for a reasonably long time, there will be enrichment of nanoplatforms into the tumor tissues. Furthermore, since the lymphatic system is not developed in tumor tissue, extravasated nanoparticles tend to stay inside the interstitial space in tumor tissues. This overall phenomenon of accumulation of nanoplatforms in tumor tissues is known as the enhanced permeability and retention (EPR) effect [28].

The EPR phenomenon is dependent on several factors such as particle size, particle surface charge and hydrophobicity, immunogenicity, tumor characteristics, etc., which results in many challenges in the optimization of passive targeting [28].

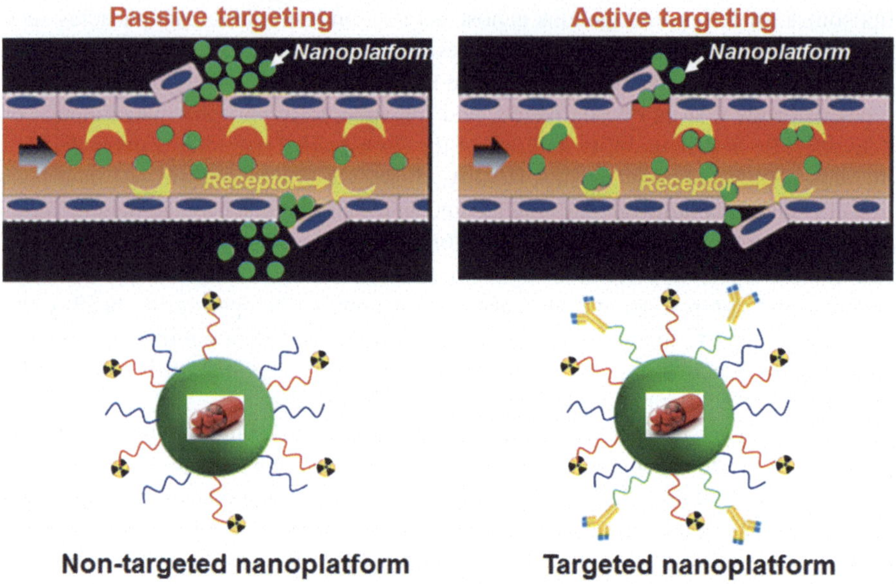

Fig. 1 Schematic depiction of EPR-mediated passive and active targeting using theranostic nanoplatforms. Nanoparticles can passively target tumors through preferential passage through larger interendothelial junctions compared to those of healthy tissues. Nanoparticles can also be conjugated with suitable targeting agents, such as antibodies that are specific to proteins (receptors) more highly expressed in tumors than healthy tissue, to actively target tumors. Adapted with permission from Ref. [28]

Further, specificity toward the tumor would be low in passive targeting and therefore therapeutic concentrations of the chemotherapeutic drug can be suboptimal at the tumor site resulting in poor therapeutic efficacy [11]. To overcome these limitations, active targeting would be a more viable approach that can be achieved by conjugating the functionalized nanoplatform to a suitable targeting moiety such as aptamers, peptides, or proteins, thereby allowing preferential accumulation of the drug in the tumor tissue [28]. The best results would be expected by combining the effects of both passive and active targeting to achieve maximal therapeutic efficacy [11].

A major hurdle in using nanoplatforms for image-guided drug delivery is their tendency to get trapped and cleared from the circulation by the reticuloendothelial system (RES) [27]. Additionally, nanoparticles can interact with plasma proteins effectively altering their surface properties [27]. The introduction of biocompatible hydrophilic polymer chains, such as polyethylene glycol (PEG), creates a hydrated brush-like coating on the nanoparticle surface that enhances nanoparticle solubility, prolongs blood circulation times, and delays RES clearance [29, 30]. It has been demonstrated that PEG-coated nanoplatforms have circulation times several orders of magnitude longer than uncoated nanoplatforms [29, 30]. Various nanoplatforms have been radiolabeled with different positron emitting radioisotopes for PET image-guided drug delivery, most of which are summarized in Table 2 and discussed in the following text.

Table 2 Representative examples of theranostic nanoplatforms utilized for PET image-guided drug delivery

Drug carrier	Targeting ligand	Target	Therapeutic agent	PET isotope	Disease model	Tumor uptake	References
Liposome	None (passive targeting)	None (passive targeting)	Model hydrophilic drug	^{18}F and ^{64}Cu	Met-1 tumors	(Not reported)	[33]
Micelles	cRGD peptide	Integrin $\alpha_v\beta_3$	Doxorubicin	^{64}Cu	Human glioblastoma	~7 % ID/g	[37]
Melanin nanoparticles	None (passive targeting)	None (passive targeting)	Sorafenib	^{64}Cu	Hepatocellular carcinoma	~5.5 % ID/g	[26]
Gold nanorods	cRGD peptide	Integrin $\alpha_v\beta_3$	Doxorubicin	^{64}Cu	Human glioblastoma	~6 % ID/g	[44]
Mesoporous silica nanoparticles	TRC105 antibody	CD105	Doxorubicin	^{64}Cu	Murine breast cancer	~6 % ID/g	[49]

3.1 Liposomes for PET Image-Guided Drug Delivery

Liposomes are good candidates as drug carriers and have been widely investigated in drug delivery systems [31, 32]. Basically, liposomes are self-assembled vesicles composed of a lipid bilayer, which forms a closed shell surrounding an internal aqueous phase. The major advantages of liposomal carriers for drug delivery are that they are biodegradable and nontoxic [31, 32]. Moreover, size, charge, and surface functionalization of liposomes are easily controllable and such systems are suitable for carrying both hydrophobic and hydrophilic drug molecules [31, 32]. Owing to these favorable characteristics, liposomes were the first nanoplatforms to make the transition from conceptual stage to clinical application, and are now an established technology platform with considerable clinical acceptance [7, 31, 32].

Paoli et al. reported the synthesis of liposomal formulations with particle size in the range 80–113 nm [33]. The liposomes were preconjugated with suitable fluorophores (calcein or AF-750) and radiolabeled with ^{18}F or ^{64}Cu for dual-modality PET/optical imaging. A model hydrophilic drug was encapsulated in the liposomal system and administered in mice bearing bilateral Met-1 tumors, to assess the relative stability and circulation kinetics of the drug-loaded liposomes, while maintaining temperature sensitivity. Using in vivo PET imaging and ex vivo fluorescent imaging of tumors, the authors could demonstrate that the accumulation of the drug was increased by up to 177-fold by liposomal encapsulation.

In a recent study, Lee et al. reported the synthesis of a chelator compound, 4-DEAP-ATSC, which serves as the ^{64}Cu loading and entrapment agent in liposomal formulations [34]. The authors demonstrated that the ^{64}Cu-DEAP-ATSC complex could be loaded into PEGylated liposomal doxorubicin (PLD) and HER2-targeted PLD (MM-302) with >90 % efficiency and that ^{64}Cu-loaded liposomal formulations were stable in human plasma up to 24 h. In vivo PET imaging studies in BT474-M3 (HER2-overexpressing breast carcinoma) xenografts after administration of ^{64}Cu-MM-302 showed heterogeneous distribution within tumors. The biodistribution profiles were quantitatively consistent with tissue-based analysis, and radioactivity uptake in the tumor (4.8 ± 0.7 %ID/g at 24 h postinjection) correlated with liposomal drug deposition. The promising results obtained in this study suggest that clinical translation of this strategy might aid in the identification of cancer patients who are most suited for undergoing liposomal therapy.

3.2 Micelles for PET Image-Guided Drug Delivery

Micelles are colloidal particles with a size usually within a range of 5–100 nm, and are currently under investigation for targeted delivery of hydrophobic anticancer drugs [35]. When tagged with suitable positron emitting radioisotopes, these systems can be used for image-guided drug delivery. Among the various micellar structures, the polymeric micelles are the most extensively used for drug delivery applications [35, 36]. The polymeric micelles generally consist of a unique core—shell structure.

The inner core is the hydrophobic part of the block copolymer, which encapsulates the hydrophobic drug. The outer shell or corona of the hydrophilic block of the copolymer is often composed of PEG, and it protects the drug from the aqueous environment and also imparts particle stability and excellent dispersibility in an aqueous solution. Because of these characteristics, polymeric micelles have several advantages as drug carriers such as enhancing the aqueous solubility of hydrophobic drugs, prolonging the circulation time of the drug in the blood, improving the in vivo stability of the drug, providing both passive and active tumor targeting abilities, and reducing nonspecific uptake by the RES [35, 36].

In the first use of micellar systems for PET image-guided drug delivery in tumor-bearing mice, Xiao et al. reported the synthesis of multifunctional unimolecular micelles made of a hyperbranched amphiphilic block copolymer [37]. The hyperbranched block copolymer of the micellar system was conjugated with cyclo(Arg-Gly-Asp-D-Phe-Cys) peptides (cRGD, for integrins $\alpha_v\beta_3$ targeting) and macrocyclic chelators [1,4,7-triazacyclononane-*N,N′,N″*-triacetic acid (NOTA)], for ^{64}Cu-labeling. An anticancer drug, doxorubicin (DOX), was also covalently conjugated onto the hydrophobic segments of the amphiphilic block copolymer arm to enable targeted drug delivery and pH-controlled drug release. In vitro studies showed that cRGD-conjugated unimolecular micelles exhibited a much higher cellular uptake in human glioblastoma (U87MG) cells due to integrin $\alpha_v\beta_3$-mediated endocytosis compared to nontargeted unimolecular micelles, thereby leading to a significantly higher cytotoxicity. In vivo PET imaging and biodistribution studies in U87MG (human glioblastoma) xenografts revealed that targeted unimolecular micelles (conjugated with cRGD peptide) exhibited a much higher level of tumor accumulation (~5 %ID/g) than nontargeted unimolecular micelles (~2.5 % ID/g) at 4 h postinjection (Fig. 2a). Administration of a blocking dose of cRGD peptide (10 mg/kg of mouse body weight) followed by administration of radiolabeled unimolecular micelle conjugated with cRGD peptide reduced the tumor uptake significantly (~2 % ID/g), which confirmed integrin $\alpha_v\beta_3$ specificity of the targeted micelle in vivo (Fig. 2a). These results were further confirmed by ex vivo optical imaging using the fluorescence signal of DOX (Fig. 2b).

The same group of authors further extended the work by conjugation of anti-CD105 monoclonal antibody (TRC105) with the unimolecular micelles (instead of cRGD peptide as done in the previous study) [38]. TRC105-conjugated unimolecular micelles showed a higher CD105-associated cellular uptake in human umbilical vein endothelial cells (HUVEC) compared with nontargeted unimolecular micelles. Similar to the previous study, in vivo PET imaging and biodistribution studies in 4T1 murine breast tumor-bearing mice showed that a tumor accumulation of ~6 % ID/g of targeted micelles (i.e., conjugated with TRC105) was higher than that of nontargeted micelles (~3 % ID/g) at 5 h postinjection. In a recent study, the same group of authors reported the development of a new type of unimolecular micelle formed by brush-shaped amphiphilic block copolymers [39]. As in the previous study, the unimolecular micelle was conjugated with TRC105, loaded with DOX and radiolabeled with ^{64}Cu for PET image-guided drug delivery in 4T1 tumor-bearing mice and similar results were obtained. The encouraging results obtained in all these studies clearly indicate that unimolecular micelles are promising form of nanomedicine for targeted cancer theranostics.

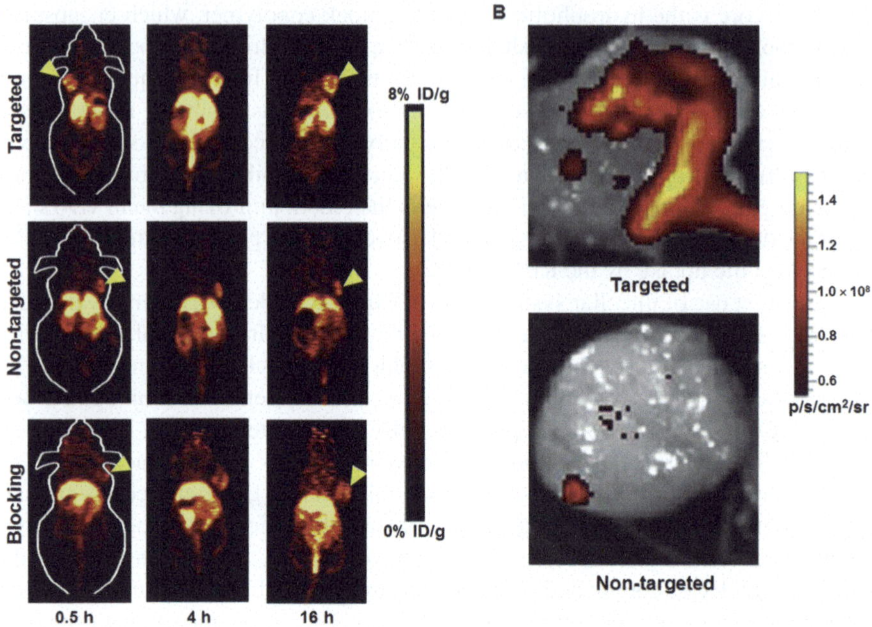

Fig. 2 PET image-guided drug delivery using unimolecular micelle. (**a**) PET imaging of U87MG tumor-bearing mice at different time points postinjection of ^{64}Cu-labeled unimolecular micelle loaded with DOX (nontargeted), ^{64}Cu-labeled unimolecular micelle conjugated to cRGD and loaded with DOX (targeted), and ^{64}Cu-labeled unimolecular micelle conjugated to cRGD and loaded with DOX with a blocking dose of cRGD (blocking). (**b**) Ex vivo fluorescence imaging of U87MG tumor, with the excitation and emission set for detecting DOX fluorescence, harvested from mice injected with targeted and nontargeted unimolecular micelles. Adapted with permission from Ref. [37]

3.3 Endogenous Nanosystems for PET Image-guided Drug Delivery

Over the last few years, there is growing interest toward the use of endogenous organic nanostructures, such as ferritins, melanin, etc. as drug delivery platforms due to their native biocompatibility and biodegradability [26, 40, 41]. In this direction, it is desirable to develop endogenous systems that intrinsically possess both contrast and drug delivery properties. Recently, Zhang et al. reported the synthesis of melanin nanoparticles as an efficient endogenous system for multimodality image-guided drug delivery [26]. Melanin is a biopolymer with good biocompatibility and biodegradability, intrinsic photoacoustic properties, and binding ability to various types of chemotherapeutic drugs [41]. The synthesized nanoparticles were PEGylated, loaded with an anticancer drug (Sorafenib), radiolabeled with a ^{64}Cu adopting chelator-free approach, and then used for dual modality PET and photoacoustic image-guided drug delivery. In vivo PET imaging and biodistribution studies after intravenous administration of the nanoplatforms in hepatocellular carcinoma

(HepG2) tumor-bearing mice revealed rapid tumor uptake by passive targeting (5.5 ± 0.3 % ID/g at 4 h postinjection) with a good tumor to background contrast. The results of PET imaging were further corroborated by photoacoustic imaging. Furthermore, the authors could successfully demonstrate the antitumor effects of drug-loaded melanin nanoparticles by studying the inhibition of tumor growth in vivo. The promising results obtained in this study prove that melanin nanoparticles are an efficient biosystem for multimodality image-guided drug delivery and hold potential for clinical translation in the foreseeable future.

3.4 Metallic Nanoparticles for PET Image-Guided Drug Delivery

Among the various metallic nanoparticles reported to date, gold nanostructures possess unique characteristics that enable their use as contrast agents, therapeutic entities, and frameworks to attach functional molecules, therapeutic cargo, or targeting ligands [42, 43]. Owing to their ease of synthesis, easy surface functionalization, and nontoxicity, gold nanostructures have emerged as powerful nanoplatforms for cancer theranostics [42]. The development of a multifunctional gold nanorod-based nanoplatform for PET image-guided drug delivery was reported by Xiao et al. [44]. The bare gold nanorods had a length and diameter of approximately 45 and 10 nm, respectively. The gold nanorods were PEGylated and an anticancer drug (DOX) and tumor targeting agent (cRGD) were conjugated to it. A chelator, NOTA, was attached onto the distal ends of the PEG arms for ^{64}Cu labeling. Based on flow cytometry analysis, cRGD-conjugated gold nanorods loaded with DOX exhibited a higher uptake and cytotoxicity in U87MG cells compared to nontargeted gold nanorods in vitro. However, in vivo PET imaging and biodistribution studies showed that targeted and nontargeted gold nanorods had a similar distribution pattern, in particular in respect to tumor uptake (~5 % ID/g). Despite this limitation, this study demonstrated the potential of gold nanorods as an efficient nanoplatform for cancer theranostics.

3.5 Inorganic Oxide Nanoparticles for PET Image-Guided Drug Delivery

Over the last few years, inorganic oxide nanoparticles such as silica and iron oxide nanoparticles have gained significant attention for image-guided drug delivery due to their easy synthesis, uniform morphology, adjustable pore volume, controllable diameter, modifiable surface potential, possibility for easy functionalization, and significant biocompatibility [45–47]. Owing to these favorable features, dye-doped ultrasmall and renal clearable silica nanoparticles (also known as C dots) have recently received the United States Food and Drug Administration (FDA) investigational new drug approval for a first-in-man clinical trial [48].

The utilization of silica nanoparticles for PET image-guided drug delivery in small animal models was reported by Chen et al. [49]. The authors synthesized uniform 80 nm sized mesoporous silica nanoparticles (MSNs), which were then surface functionalized with thiol groups, PEGylated, conjugated with NOTA chelator and TRC105 antibody (specific for CD105/endoglin). The nanoplatform was loaded with an anticancer drug, DOX, and radiolabeled with ^{64}Cu for PET image-guided drug delivery in 4T1 murine breast tumor-bearing mice. In vivo PET imaging and biodistribution studies showed high tumor uptake (~6 % ID/g) at 5 h postinjection. The tumor uptake (~3 % ID/g at 5 h postinjection) of nontargeted nanoparticles (not conjugated with TRC105) was lower than the tumor uptake observed with targeted nanoparticles, indicating that both active targeting and EPR effect were responsible for the enhanced tumor uptake. The CD105 specificity of the TRC105-conjugated nanoplatform was further confirmed by blocking studies, wherein administration of a blocking dose (1 mg/mouse) of TRC105 at 1 h before injection of radiolabeled nanoplatform could significantly reduce the tumor uptake to ~3 % ID/g at 5 h postinjection. The loading capacity of DOX in MSNs was estimated to range from 76.6 to 481.6 mg/g. Using ex vivo optical imaging, the authors could successfully demonstrate the feasibility of enhanced tumor targeted delivery of DOX using TRC105-conjugated MSNs. In another similar study from the same group, MSNs were PEGylated, conjugated with antivascular endothelial growth factor receptor (VEGFR) ligand, $VEGF_{121}$, loaded with anti-VEGFR drug (sunitinib), and radiolabeled with ^{64}Cu for PET image-guided drug delivery in human glioblastoma (U87MG) xenografts [50]. As observed in the previous study, significantly higher amount of drug could be delivered to the tumor by targeting VEGFR when compared with the nontargeted counterparts.

In order to achieve a higher drug loading capacity and improved tumor uptake than MSNs, Chen et al. reported the synthesis of functionalized hollow mesoporous silica nanoparticles (HMSNs) of 150 nm particle size [51]. The drug (DOX) loading capacity of HMSNs was found to be 3–15 times higher than previously reported MSNs [49, 51]. The hollow space inside the nanoplatform could be loaded with DOX, followed by conjugation with near-infrared (NIR) dye (ZW800) and the ^{64}Cu-chelator NOTA. The drug-loaded nanoplatform was subsequently PEGylated, conjugated with TRC105 antibody, and radiolabeled with ^{64}Cu for dual modality (PET/optical) image-guided drug delivery in 4T1 tumor-bearing mice. In vivo PET imaging and biodistribution studies revealed that tumor uptake of TRC105-conjugated HMSNs was ~10 % ID/g at 4 h postinjection, which was ~3 times higher than that of the nontargeted group (Fig. 3a). Administration of a blocking dose (1 mg/mouse) of free TRC105 at 1 h before injection of radiolabeled nanoplatform could significantly reduce the tumor uptake to ~5 % ID/g at 4 h postinjection, clearly demonstrating CD105 specificity of the targeted nanoplatform in vivo (Fig. 3a). These results were further corroborated by optical imaging studies. Enhanced DOX delivery was also demonstrated in 4T1 tumor-bearing mice by ex vivo optical imaging using fluorescence signal of DOX (Fig. 3b). This work was further extended by the same group, wherein ^{64}Cu-labeled HMSNs were loaded with anticancer drug (sunitinib), and conjugated with cRGD peptide for targeting integrin $\alpha_v\beta_3$ expression in U87MG tumor-bearing mice [52]. In vivo PET imaging studies indicated ~8 % ID/g tumor uptake of targeted nanoconju-

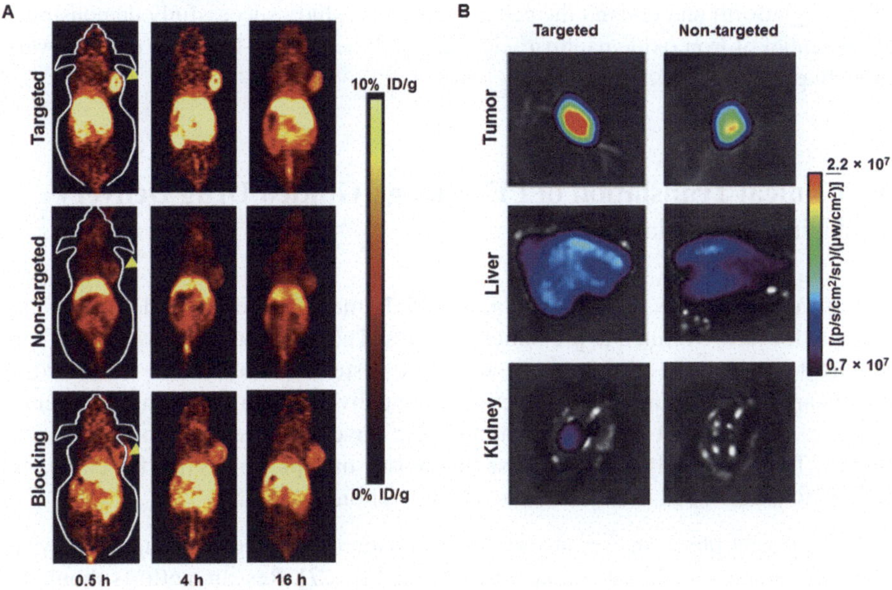

Fig. 3 PET image-guided drug delivery using HMSNs. (**a**) PET images of 4T1 tumor-bearing mice at different time points postinjection of ^{64}Cu-labeled HMSNs loaded with DOX (nontargeted), ^{64}Cu-labeled HMSNs conjugated to TRC105 and loaded with DOX (targeted), and ^{64}Cu-labeled HMSNs conjugated to TRC105 and loaded with DOX with a blocking dose of TRC105 (blocking). (**b**) Ex vivo optical imaging of major organs in 4T1 tumor-bearing mice with the excitation and emission set for detecting DOX fluorescence after intravenous injection of targeted and nontargeted HMSNs. Adapted with permission from Ref. [51]

gates, which correlated well with ex vivo biodistribution analyses. As in the previous study, enhanced tumor-targeted delivery of the anticancer drug was observed, thereby demonstrating the potential of HMSNs for targeted cancer imaging and drug delivery. Since silica-based nanoparticles have already received FDA approval for clinical use [48], the results obtained in these studies set the stage toward potential clinical translational of this promising nanoplatform for cancer theranostics.

Iron oxide nanoparticles are used as contrast agents in magnetic resonance imaging (MRI) [46, 47]. Additionally, such nanoparticles offer the scope of attaching probes containing multiple imaging motifs on its surface for multimodality molecular image-guided drug delivery [46, 47]. In an interesting study, Yang et al. reported the synthesis of cRGD-functionalized, DOX-conjugated, and ^{64}Cu-labeled superparamagnetic iron oxide nanoparticles (SPIOs) for targeted anticancer drug delivery and PET/MR imaging in human glioblastoma (U87MG) xenografts [53]. In vivo PET imaging and biodistribution studies showed that cRGD-conjugated SPIOs showed a much higher level of tumor accumulation (~5 % ID/g) than nontargeted, i.e., cRGD-free ones (<2 % ID/g). From relaxivity measurements, it was demonstrated that drug-loaded and cRGD-conjugated SPION could effectively serve as MR contrast agents. The authors also observed that cRGD-conjugated SPIOs induced a significant amount of cytotoxicity in the U87MG tumor cells, suggesting that DOX was released from the nanoplatform and entered the cell nucleus. This study successfully demonstrates the potential of iron oxide nanoparticles for combined tumor-targeting drug delivery as well as multimodality imaging.

4 Clinical Translation of PET Image-Guided Drug Delivery Using Nanoplatforms

Almost all the studies reported to date on PET image-guided drug delivery using nanoplatforms are limited to preclinical settings. Therefore, it must be admitted that this field is still in its nascent stage and more systematic studies are warranted to understand the mechanisms for targeting, drug delivery, and monitoring therapeutic efficacy, which might aid in translating these novel discoveries into real clinical impact. To successfully translate these theranostic nanoplatforms into the clinic, the following points need to be taken into consideration [54, 55].

- Several nanoplatforms are already used in patients for targeted drug delivery, as well as many are close to clinical translation [2, 3, 7]. Research efforts should be directed toward radiolabeling these nanoplatforms with suitable positron emitting radioisotopes to realize the scope PET image-guided drug delivery in near future. Given their established physicochemical versatility, and the large amount of preclinical studies already carried out, it can be expected that clinical translation of these radiolabeled nanomedicines will not be too problematic.
- Thorough knowledge needs to be gained regarding the pharmacokinetics and pharmacodynamics of the radiolabeled nanoplatforms loaded with chemotherapeutic drugs [54].
- It will be of paramount importance to establish that the degree of tumor accumulation of a radiolabeled nanoplatform loaded with drug as demonstrated by PET imaging should correspond, at least to some extent, to its therapeutic efficacy [54].
- Even if clear correlation between tumor accumulation of drug-loaded radiolabeled nanoplatform and its therapeutic efficacy is established, it will be essential to appropriately distinguish between low and high levels of target site accumulation. With the present knowledge, it is difficult to decide from what (relative) percentages of the injected dose onward, the patients can be expected to demonstrate adequately high levels of target site accumulation [54]. Obviously, these values will vary not only from one nanoplatform to the other but also from one type of malignancy to the other. Therefore, systematic and meticulously planned preclinical studies would be essential to provide answers to such questions.
- It would be important to understand how the level of target site accumulation of drug-loaded and radiolabeled nanoplatforms changes during the course of therapy [54]. In case the size, stage, perfusion, and/or permeability of tumors decrease significantly during the initial cycles of PET image-guided therapy, and if this also substantially lowers the degree of accumulation of the radiolabeled nanoplatform, then it will be essential to establish parameters and protocols to decide whether this treatment should be continued or not.

- Though not an essential criterion, it would be desirable to investigate how targeting and treating primary tumors with drug-loaded nanoplatforms correlates with targeting and treatment of metastases.

Besides these biological aspects, there are several technical and regulatory issues that are also of paramount importance for clinical translation of the radiolabeled nanoplatforms for PET image-guided drug delivery, as briefly enlisted later.

- As the theranostic nanoplatforms are intended for human use, their preparation must adhere to current good manufacturing process (cGMP) compliance to ensure that the quality of the final product meet the acceptance criteria. The United States FDA has approved a set of regulations describing production of molecular imaging agents according to cGMP, outlined in the Code of Federal Regulations [15, 56–59]. Enforcement of cGMP is intended to preclude patients at risk due to inadequate safety and quality, and to enhance consistency in the application of the regulatory requirements [15, 56–59]. Any deviation from the approved method of preparation would require considerable validation before patient use.
- While ensuring cGMP compliance is an appealing vision, it is a demanding task as it includes requirement of well-qualified personnel, use of controlled materials and procedures, accessibility of qualified equipment, preparation of radiolabeled and drug-loaded nanoplatforms in designated clean areas, applying only validated processes and analytical methods for each step, full documentation of the process, registration of the theranostic agent and clinical procedure to be adopted for PET image-guided drug delivery with national/regional health authorities, and release of the same for human use by an authorized personnel.
- While manual synthesis approach is generally adopted for synthesis of theranostics nanoplatforms for PET image-guided drug delivery in a preclinical setting due to the requirement of the theranostic agents in small quantities, use of manual synthesis procedure for large-scale clinical applications might be challenging. Therefore, it might be essential to consider the use of automated synthesis apparatus owing to the following advantages [60, 61]:
- Offer robust, repeatable synthesis of the theranostic nanoplatforms.
- Reduced operator intervention minimizes operational errors.
- Ensure radiation safety during the radiolabeling step through reduction or elimination of manual operations.
- The use of fully automated synthesis modules not only facilitates cGMP compliance, but also offers complete traceability of the process, an aspect of utmost importance because of the extensive regulatory burden.
- Precludes the risk of bacterial contamination of the drug-loaded and radiolabeled nanoplatforms that will be administered in human subjects.

Although automation strategy holds significant promise toward clinical translation of theranostics nanoplatforms, it is associated with the challenge of reconfiguring the synthesis module for new procedures requiring nonconventional chemistry while maintaining full automation and compliance with cGMP regulations. Nevertheless, in order to be effective in addressing the particular regulatory barriers, automated synthesis modules must be customized to local legislative, regulatory, and institutional conditions.

It is also pertinent to point out that besides scientific and technical issues, several socioeconomic and political factors might also affect clinical translation of theranostic nanoplatforms for PET image-guided drug delivery. The acceptance of a new approach in cancer management is generally an iterative process that requires substantial effort and sometimes luck. Extensive regulatory requirements and bureaucratic procedures in certain countries, limited potential market initially for such novel approaches, lobbying by the manufacturers of other approved cancer drugs, lack of reimbursement strategies by the insurance companies for these novel approaches, etc. might also impede the process of clinical translation of theranostic nanoplatforms for PET image-guided drug delivery. Despite these hurdles, the exciting results obtained to date indicate that theranostic nanoplatforms likely will have multifaceted applications in future clinical practice.

5 Conclusions and Future Perspectives

The versatility of theranostic nanoplatforms for PET image-guided drug delivery brings forth unique perspectives in cancer management that are evident in their discovery and rapid development. A single nanoplatform can now be used to detect tumors, treat them, monitor treatment response, and also guide therapeutic regimes. If the growth of this field continues at its present pace, the day may not be far out when PET image-guided drug delivery would become the norm rather than the exception in clinical oncology. However, before this optimism becomes a reality, several issues related to safety and complexity of theranostic nanoplatforms must be adequately addressed. Generally, vascular-targeted delivery is only achievable using theranostic nanoplatforms and penetration of cancerous drugs inside cancerous lesions has not yet fully succeeded [27]. Furthermore, cellular toxicity of several nanoplatforms proposed for cancer theranostics is an issue of serious concern [27]. In general, biocompatible and biodegradable nanomaterials would be the preferred choice for clinical use in PET image-guided drug delivery.

Despite excellent attributes of PET, there is now an overwhelming scientific consensus that no single molecular imaging modality is perfect and sufficient to gain all the necessary information [17]. The multifunctionality of the nanoplatforms would offer the scope for multimodality molecular imaging to provide synergistic advantages over any single modality alone. Similarly, it might be worth investigating the advantages of multimodality therapy wherein different types of therapeutic agents (such as chemotherapy, radiotherapy, or gene therapy agents) can be simultaneously delivered by the nanoplatforms at cancerous lesions for enhanced therapeutic effectiveness. While numerous challenges face all new technologies, bringing theranostic nanoplatforms for PET image-guided drug delivery to clinical market would require multidisciplinary effort at the intersection of manufacturing, regulation, and funding in order to turn these challenges into opportunities. Such a concerted effort may contribute toward developing more effective and less toxic treatment regimens for individual patients which would be a significant advancement toward achieving the ultimate goal of "personalized medicine" for cancer management.

Acknowledgments This work is supported, in part, by the University of Wisconsin—Madison, the Bhabha Atomic Research Centre (XII-N-R&D-004.01), the National Institutes of Health (NIBIB/NCI 1R01CA169365, P30CA014520, and 5T32GM08349), the Department of Defense (W81XWH-11-1-0644), the American Cancer Society (125246-RSG-13-099-01-CCE), and the National Science Foundation (DGE-1256259).

References

1. Bregoli L, Movia D, Gavigan-Imedio JD, Lysaght J, Reynolds J, Prina-Mello A. Nanomedicine applied to translational oncology: a future perspective on cancer treatment. Nanomedicine. 2016;12:81–103.
2. Lytton-Jean AK, Kauffman KJ, Kaczmarek JC, Langer R. Cancer nanotherapeutics in clinical trials. Cancer Treat Res. 2015;166:293–322.
3. Perez-Herrero E, Fernandez-Medarde A. Advanced targeted therapies in cancer: drug nanocarriers, the future of chemotherapy. Eur J Pharm Biopharm. 2015;93:52–79.
4. Jemal A, Center MM, DeSantis C, Ward EM. Global patterns of cancer incidence and mortality rates and trends. Cancer Epidemiol Biomarkers Prev. 2010;19(8):1893–907.
5. World Health Organization Cancer Factsheet. (http://www.who.int/mediacentre/factsheets/fs297/en/). Accessed 20 Sept 2015.
6. Zhang Y, Chan HF, Leong KW. Advanced materials and processing for drug delivery: the past and the future. Adv Drug Deliv Rev. 2013;65(1):104–20.
7. Marchal S, Hor AE, Millard M, Gillon V, Bezdetnaya L. Anticancer drug delivery: an update on clinically applied nanotherapeutics. Drugs. 2015;75:1601–11.
8. Terreno E, Uggeri F, Aime S. Image guided therapy: the advent of theranostic agents. J Control Release. 2012;161(2):328–37.
9. Chow EK, Ho D. Cancer nanomedicine: from drug delivery to imaging. Sci Transl Med. 2013;5(216):216rv4.
10. Chen F, Ehlerding EB, Cai W. Theranostic nanoparticles. J Nucl Med. 2014;55(12):1919–22.
11. Chakravarty R, Hong H, Cai W. Positron emission tomography image-guided drug delivery: current status and future perspectives. Mol Pharm. 2014;11(11):3777–97.
12. Chakravarty R, Hong H, Cai W. Image-guided drug delivery with single-photon emission computed tomography: a review of literature. Curr Drug Targets. 2015;16(6):592–609.
13. Jokerst JV, Gambhir SS. Molecular imaging with theranostic nanoparticles. Acc Chem Res. 2011;44(10):1050–60.
14. Pancholi K. A review of imaging methods for measuring drug release at nanometre scale: a case for drug delivery systems. Expert Opin Drug Deliv. 2012;9(2):203–18.
15. Chakravarty R, Chakraborty S, Dash A. Molecular imaging of breast cancer: role of RGD peptides. Mini Rev Med Chem. 2015.
16. Ametamey SM, Honer M, Schubiger PA. Molecular imaging with PET. Chem Rev. 2008;108(5):1501–16.
17. Jennings LE, Long NJ. "Two is better than one"—probes for dual-modality molecular imaging. Chem Commun (Camb). 2009;(24):3511–24
18. Aboagye EO, Price PM. Use of positron emission tomography in anticancer drug development. Invest New Drugs. 2003;21(2):169–81.
19. Alauddin MM. Positron emission tomography (PET) imaging with ^{18}F-based radiotracers. Am J Nucl Med Mol Imaging. 2012;2(1):55–76.
20. Rosch F. Past, present and future of ^{68}Ge/^{68}Ga generators. Appl Radiat Isot. 2013;76:24–30.
21. Chakravarty R, Valdovinos HF, Chen F, Lewis CM, Ellison PA, Luo H, et al. Intrinsically germanium-69-labeled iron oxide nanoparticles: synthesis and in-vivo dual-modality PET/MR imaging. Adv Mater. 2014;26(30):5119–23.
22. Chen F, Ellison PA, Lewis CM, Hong H, Zhang Y, Shi S, et al. Chelator-free synthesis of a dual-modality PET/MRI agent. Angew Chem Int Ed Engl. 2013;52(50):13319–23.

23. Chen F, Goel S, Valdovinos HF, Luo H, Hernandez R, Barnhart TE, et al. In vivo integrity and biological fate of chelator-free zirconium-89-labeled mesoporous silica nanoparticles. ACS Nano. 2015;9(8):7950–9.
24. Pagani M, Stone-Elander S, Larsson SA. Alternative positron emission tomography with non-conventional positron emitters: effects of their physical properties on image quality and potential clinical applications. Eur J Nucl Med. 1997;24(10):1301–27.
25. Kornyei J, Mikecz P, Toth G. PET radiopharmaceuticals: novelties and new possibilities. Magy Onkol. 2014;58(4):245–50.
26. Zhang R, Fan Q, Yang M, Cheng K, Lu X, Zhang L, et al. Engineering melanin nanoparticles as an efficient drug-delivery system for imaging-guided chemotherapy. Adv Mater. 2015;27(34):5063–9.
27. Lanza GM, Moonen C, Baker Jr JR, Chang E, Cheng Z, Grodzinski P, et al. Assessing the barriers to image-guided drug delivery. Wiley Interdiscip Rev Nanomed Nanobiotechnol. 2014;6(1):1–14.
28. Kunjachan S, Pola R, Gremse F, Theek B, Ehling J, Moeckel D, et al. Passive versus active tumor targeting using RGD- and NGR-modified polymeric nanomedicines. Nano Lett. 2014;14(2):972–81.
29. Ernsting MJ, Murakami M, Roy A, Li SD. Factors controlling the pharmacokinetics, biodistribution and intratumoral penetration of nanoparticles. J Control Release. 2013;172(3):782–94.
30. van Vlerken LE, Vyas TK, Amiji MM. Poly(ethylene glycol)-modified nanocarriers for tumor-targeted and intracellular delivery. Pharm Res. 2007;24(8):1405–14.
31. Goyal P, Goyal K, Vijaya Kumar SG, Singh A, Katare OP, Mishra DN. Liposomal drug delivery systems—clinical applications. Acta Pharm. 2005;55(1):1–25.
32. Medina OP, Zhu Y, Kairemo K. Targeted liposomal drug delivery in cancer. Curr Pharm Des. 2004;10(24):2981–9.
33. Paoli EE, Kruse DE, Seo JW, Zhang H, Kheirolomoom A, Watson KD, et al. An optical and microPET assessment of thermally-sensitive liposome biodistribution in the Met-1 tumor model: importance of formulation. J Control Release. 2010;143(1):13–22.
34. Lee H, Zheng J, Gaddy D, Orcutt KD, Leonard S, Geretti E, et al. A gradient-loadable ^{64}Cu-chelator for quantifying tumor deposition kinetics of nanoliposomal therapeutics by positron emission tomography. Nanomedicine. 2015;11(1):155–65.
35. Kedar U, Phutane P, Shidhaye S, Kadam V. Advances in polymeric micelles for drug delivery and tumor targeting. Nanomedicine. 2010;6(6):714–29.
36. Gothwal A, Khan I, Gupta U. Polymeric micelles: recent advancements in the delivery of anticancer drugs. Pharm Res. 2016;33:18–39.
37. Xiao Y, Hong H, Javadi A, Engle JW, Xu W, Yang Y, et al. Multifunctional unimolecular micelles for cancer-targeted drug delivery and positron emission tomography imaging. Biomaterials. 2012;33(11):3071–82.
38. Guo J, Hong H, Chen G, Shi S, Zheng Q, Zhang Y, et al. Image-guided and tumor-targeted drug delivery with radiolabeled unimolecular micelles. Biomaterials. 2013;34(33):8323–32.
39. Guo J, Hong H, Chen G, Shi S, Nayak TR, Theuer CP, et al. Theranostic unimolecular micelles based on brush-shaped amphiphilic block copolymers for tumor-targeted drug delivery and positron emission tomography imaging. ACS Appl Mater Interfaces. 2014;6(24):21769–79.
40. Huard DJ, Kane KM, Tezcan FA. Re-engineering protein interfaces yields copper-inducible ferritin cage assembly. Nat Chem Biol. 2013;9(3):169–76.
41. Ren G, Miao Z, Liu H, Jiang L, Limpa-Amara N, Mahmood A, et al. Melanin-targeted preclinical PET imaging of melanoma metastasis. J Nucl Med. 2009;50(10):1692–9.
42. Webb JA, Bardhan R. Emerging advances in nanomedicine with engineered gold nanostructures. Nanoscale. 2014;6(5):2502–30.
43. Mody VV, Siwale R, Singh A, Mody HR. Introduction to metallic nanoparticles. J Pharm Bioallied Sci. 2010;2(4):282–9.
44. Xiao Y, Hong H, Matson VZ, Javadi A, Xu W, Yang Y, et al. Gold nanorods conjugated with doxorubicin and cRGD for combined anticancer drug delivery and PET imaging. Theranostics. 2012;2(8):757–68.

45. Shi S, Chen F, Cai W. Biomedical applications of functionalized hollow mesoporous silica nanoparticles: focusing on molecular imaging. Nanomedicine. 2013;8(12):2027–39.
46. El-Hammadi MM, Arias JL. Iron oxide-based multifunctional nanoparticulate systems for biomedical applications: a patent review (2008—present). Expert Opin Ther Pat. 2015;25(6):691–709.
47. Sharifi S, Seyednejad H, Laurent S, Atyabi F, Saei AA, Mahmoudi M. Superparamagnetic iron oxide nanoparticles for in vivo molecular and cellular imaging. Contrast Media Mol Imaging. 2015.
48. US FDA Investigational New Drug approval for first-in-human trial of novel cancer-targeting nanoparticle. News Anal Ther Deliv. 2011;2(3):287.
49. Chen F, Hong H, Zhang Y, Valdovinos HF, Shi S, Kwon GS, et al. In vivo tumor targeting and image-guided drug delivery with antibody-conjugated, radiolabeled mesoporous silica nanoparticles. ACS Nano. 2013;7(10):9027–39.
50. Goel S, Chen F, Hong H, Valdovinos HF, Hernandez R, Shi S, et al. $VEGF_{121}$-conjugated mesoporous silica nanoparticle: a tumor targeted drug delivery system. ACS Appl Mater Interfaces. 2014;6(23):21677–85.
51. Chen F, Hong H, Shi S, Goel S, Valdovinos HF, Hernandez R, et al. Engineering of hollow mesoporous silica nanoparticles for remarkably enhanced tumor active targeting efficacy. Sci Rep. 2014;4:5080.
52. Chakravarty R, Goel S, Hong H, Chen F, Valdovinos HF, Hernandez R, et al. Hollow mesoporous silica nanoparticles for tumor vasculature targeting and PET image-guided drug delivery. Nanomedicine. 2015;10(8):1233–46.
53. Yang X, Hong H, Grailer JJ, Rowland IJ, Javadi A, Hurley SA, et al. cRGD-functionalized, DOX-conjugated, and ^{64}Cu-labeled superparamagnetic iron oxide nanoparticles for targeted anticancer drug delivery and PET/MR imaging. Biomaterials. 2011;32(17):4151–60.
54. Lammers T, Rizzo LY, Storm G, Kiessling F. Personalized nanomedicine. Clin Cancer Res. 2012;18(18):4889–94.
55. Theek B, Rizzo LY, Ehling J, Kiessling F, Lammers T. The theranostic path to personalized nanomedicine. Clin Transl Imaging. 2014;2(1):66–76.
56. Regulations for in vivo radiopharmaceuticals used for diagnosis and monitoring. Food and Drug Administration, HHS. Final rule. Fed Regist. 1999;64(94):26657–70.
57. Harolds J. What is now current good manufacturing practice for PET drugs? Clin Nucl Med. 2010;35(5):329.
58. Hung JC. USP general chapter <797> pharmaceutical compounding-sterile preparations. J Nucl Med. 2004;45(6):20N. 8N.
59. Hung JC. The potential impact of usp general chapter <797> on procedures and requirements for the preparation of sterile radiopharmaceuticals. J Nucl Med. 2004;45(6):21N–6.
60. Boschi S, Malizia C, Lodi F. Overview and perspectives on automation strategies in ^{68}Ga radiopharmaceutical preparations. Recent Results Cancer Res. 2013;194:17–31.
61. Chi YT, Chu PC, Chao HY, Shieh WC, Chen CC. Design of CGMP production of ^{18}F- and ^{68}Ga-radiopharmaceuticals. Biomed Res Int. 2014;2014:680195.

Triggered Drug Release and Enhanced Drug Transport from Ultrasound-Responsive Nanoparticles

James J. Kwan and Constantin C. Coussios

1 Introduction

There have been major advances in the ability to discover and develop novel drugs across nearly all diseases and drug classes [1–3]. However, for most fatal diseases, such as cancers, cardiovascular diseases, and neurological disorders, the therapeutic agents typically used are effective in treating the diseased tissue, but are either excessively toxic [4–7] or poorly distributed within the diseased tissue [8–11]. These limitations in drug delivery have impacted all routes of transport, such as: oral, nasal, aerosol, transdermal, and systemic. Each of the aforementioned drug delivery routes has their own associated set of challenges and opportunities. However, the scope of this chapter is focused on systemic drug delivery because it is one of the most widely used means of delivering a drug.

Efficacious yet highly toxic and nonspecific drugs often have limited bioavailability and distribution within diseased tissue [12]. These physiological challenges are not unique to specific diseases, but are present across cancers, cardiovascular lesions and occlusions, and the brain [13], and are therefore drug-class-agnostic. The need to overcome these challenges has resulted in a substantial increase in research devoted to techniques that promote site-targeted delivery and enhanced distribution of therapeutics. Many research groups have approached this need through a variety of "passive" and "active" drug delivery techniques. In this chapter, we focus on active processes that promote drug delivery in response to an ultrasound field.

J.J. Kwan • C.C. Coussios (✉)
Institute of Biomedical Engineering, Department of Engineering Science,
University of Oxford, Oxford OX3 7DQ, UK
e-mail: constantin.coussios@eng.ox.ac.uk

J.W.M. Bulte, M.M.J. Modo (eds.), *Design and Applications of Nanoparticles in Biomedical Imaging*, DOI 10.1007/978-3-319-42169-8_13

2 Applications of Drug Therapies

2.1 Cancer

Solid tumors represent a highly challenging environment for drug delivery, because of the chaotic vasculature, enhanced intratumoral pressure, dense extracellular matrix, and increased distance between a cancerous cell and the nearest blood vessel [14]. Tumors also present an unusual drug delivery opportunity by virtue of the leaky endothelial gaps that are typically present: this implies preferential accumulation or passage of therapeutics in the range 100–300 nm. Active delivery mechanisms typically have three roles to play in this context: enable increased extravasation of the therapeutic from the blood stream into the tumor, permit triggered release of the therapeutic at the tumor site only, and mediate improved transport and distribution of the therapeutic throughout the tumor mass.

Conventional chemotherapeutics include small molecular drugs, such as taxanes (e.g., paclitaxel [15]), anthracyclines (e.g., doxorubicin [16, 17]), and cytosines (e.g., arabinoside [18, 19]), which typically circulate well and have considerable diffusivity in tumors. However, all of these drugs are nonspecific, and are therefore highly cytotoxic to both healthy and cancerous tissues. In this context, ultrasound-mediated delivery could enable site-specific triggered release, as well as potentially enhance the penetration distance of the therapeutic from the perivascular space into the tumor mass.

Beside small molecular drugs, there is now an increasing trend towards using biologics, such as oncolytic viruses, peptides, and antibodies, to achieve more targeted cancer therapy. Oncolytic viruses selectively infect and kill cancer cells: although there are relatively few in the clinic, the first candidate for melanoma was recently approved by the FDA and EMA (T-Vec, Amgen) [20]. Peptides are typically used to block the production of vascular endothelial growth factor (VEGF) and prevent the proliferation of blood vessels [21]. Antibodies act by a wide range of mechanisms, with the most recent developments focusing on targeting cytotoxic T lymphocyte-associated antigen 4 (CTLA-4) [22] and the programmed cell death protein 1 pathway (PD-1/PD-L1) [23, 24].

Increased specificity does, however, come at the cost of greatly increased size and possibly much faster clearance in the systemic circulation. Peptides are typically on the order of 5 nm, antibodies are on the order of 10 nm and viruses range in size from 100 to 200 nm. The challenge in delivering these agents is therefore twofold. Their increased size implies greatly decreased extravasation and penetration into the tumor mass. Secondly, the very short half-life of agents, such as viruses, means that there is a very short time (on the order of 10 min) over which to convert a systemically administered dose into a therapeutic dose in the tumor. Once again, active delivery by ultrasound could facilitate both of those aspects.

2.2 Cardiovascular Diseases

Currently, there are several means to treat cardiovascular diseases. Treatments typically rely on regular doses of statins. However, for more acute instances of cardiovascular disease, treatments rely on the protein tissue plasminogen activators (tPA) to dissolve the occlusion and allow increased blood flow [25]. As a result, tPA therapies are used to treat embolisms, myocardial infarctions, and stroke that result from clot formations [26]. The drug catalyzes the enzymatic degradation of fibrin, a primary protein within clots. This drug, however, is very potent and nonspecific, and often needs to be delivered to the target under conditions of low or no blood flow. As a result, off-site bleeding is a substantial problem [27] and often prohibits some patients from this therapy. Active delivery mechanisms thus have a significant role to play in terms of enhanced transport, and improved specificity through triggered release.

2.3 Neurological Disorders

There is a wide range of therapeutics currently under investigation for the treatment of neurological disorders, such as Alzheimer's disease (AD) and Parkinson's disease (PD). Viruses, peptides, and other biologics have been utilized to help break down damaging plaque formations, influence neurotransmission, or enable the innate immune system to play a role in AD therapy [28]. Likewise, small drug molecules [29–31] and adenoviruses [32] have also been applied to PD. These therapeutic agents have shown great promise in treating these neurological disorders, as indicated by many of them entering into phase 1 and 2 clinical trials. However, many of these strategies rely on the passage beyond the blood–brain barrier (BBB) [33]. This is the key challenge for delivery of these classes of agent, and ultrasound has a key role to play both in reversible opening the BBB and helping transport agents across it.

3 Ultrasound for Drug Delivery and Transport

Ultrasound is a non-ionizing, non-destructive sound wave operating at frequencies above 20 kHz. These mechanical waves easily propagate through the human body, but are obstructed by bones or large gas cavities. Furthermore, the acoustic wave can be focused similarly to lens focusing of light [34], which allows for diseased tissue-specific targeting without harming nearby healthy tissue. It is therefore ideal for many diagnostic and therapeutic applications [35]. In the context for ultrasound-mediated drug delivery with nanoparticles, we broadly characterize the mechanisms of action from ultrasound into thermal or mechanical effects (Fig. 1).

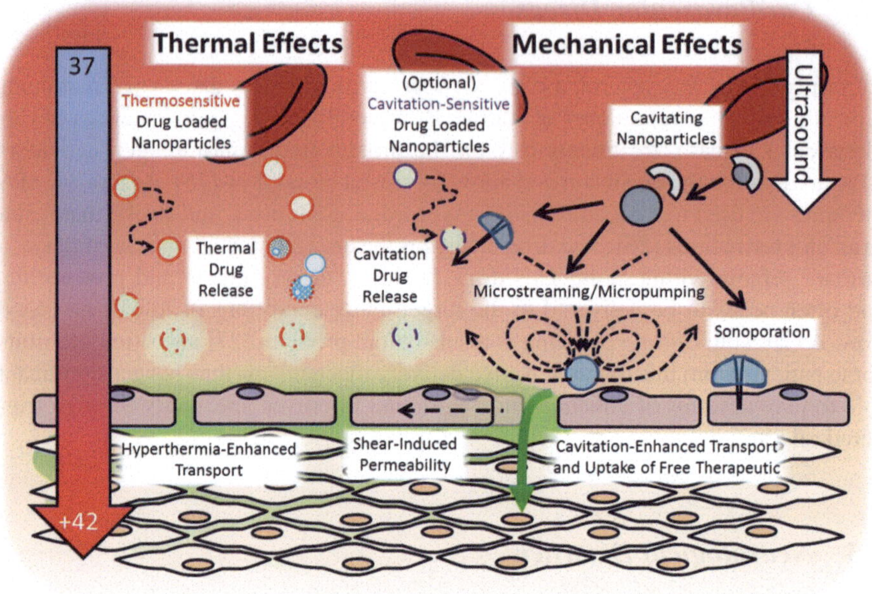

Fig. 1 Schematic of different mechanisms of ultrasound-mediated drug release, drug transport, and sonoportation from ultrasound-responsive nanoparticles

As an acoustic wave travels through a medium it is attenuated through reflection, scattering, and absorption [36]. Acoustic energy that is absorbed is converted to heat [35]. Ultrasound is thus one of the only modalities capable of generating highly localized mild hyperthermia (39–43 °C) at depth within the body. The resulting temperature rise can be monitored noninvasively using either ultrasound or MRI-based techniques, with MR-thermometry being most commonly used clinically in spite of its significant cost and limited spatiotemporal accuracy [37–40].

Ultrasonic waves are also capable of imposing mechanical effects, such as acoustic radiation force or cavitation. Acoustic radiation forces are the time averaged net force in the direction away from the ultrasound source [41]. Acoustic cavitation is the dynamic response of a gas and/or vapor cavity (i.e., a bubble) to an oscillating acoustic pressure amplitude [42]. However, bubble nucleation using ultrasound alone requires large pressure amplitudes [43]. In order to reduce the pressure amplitudes necessary for cavitation, cavitation nuclei are typically introduced via intravenous injection, and take the form of either microbubbles, also known as ultrasound contrast agents, or nanoscale cavitation nucleation agents [44, 45]. Occurrence of cavitation can be detected and monitored remotely through a technique known as passive cavitation detection, or passive acoustic mapping, whereby narrowband or broadband acoustic emissions arising from cavitating bubbles are remotely sensed [46, 47]. Acoustic cavitation has had significant impact on ultrasound based therapies [48–50] such as ultrasound-enhanced drug delivery, which is the focus of this chapter.

Non-inertial cavitation is the periodic oscillatory radial motion of a bubble, and is dependent on the size [51] and composition of the bubble [52]. This periodic motion perturbs the surrounding fluid over microsecond time scales, generating micro-streaming that results in convective transport of particles trapped in the currents [53–55], open up tight junctions between endothelial cells [56], disrupt cell membranes [57], and induce intercellular and intracellular bioeffects [58, 59]. Inertial cavitation occurs when the peak negative pressure amplitude becomes large enough to cause the bubble to unstably grow, and subsequently collapse during the positive pressure phase due to the inertia of the surrounding liquid. During the collapse phase of the bubble, jets, bubble fragments, and other asymmetric bubble shapes are often formed. The collapses emit shock waves that are detectable as broadband signals [60], which are useful for imaging techniques, such as passive acoustic mapping [61, 62]. As the collapses themselves can be periodic [63], inertial cavitation is also capable of generating micostreaming, along with the associated convective mass transport and bioeffects. As a result, inertial cavitation is a key enabler and facilitator in drug delivery.

We now describe nanoparticulate strategies that exploit the thermal and cavitational effects of ultrasound for ultrasound-triggered release, enhanced transport into biological targets, or improved delivery of therapeutics across the cellular membrane.

4 Ultrasound Triggered Drug Release from Nanoparticles

Drugs contained within biocompatible materials (e.g., lipids and polymers) to make drug-loaded nanoparticles increase disease specificity and reduce systemic toxicity of the encapsulated drug [64–67]. These drug delivery strategies are dependent on the size and composition of the drug-loaded nanoparticle [68]. However, it has also been shown that despite increased accumulation of drug-loaded nanoparticles, there is no associated increase in drug delivery within the tumor even when attempts to normalize the tumor environment are made [12, 69]. Without an external trigger to release the contents of the drug-loaded liposome, there remains a diffusive barrier of the encapsulation that prevents the bioavailability of the drug. In order to overcome these barriers, there has been a quest to develop external triggers to spatially and temporally release drugs from these nanoparticles, in order to achieve reduced systemic toxicity, enhanced drug accumulation, and improved bioavailability. Here we look at several designs of nanoparticles that release their payload in the presence of ultrasound.

4.1 Hyperthermia-Triggered Drug Release with Ultrasound

One of the key attributes for a well-designed drug delivery vehicle is incorporating a means to trigger the release of its contents. Encapsulating strategies that degrade in the presence of elevated temperatures have thus attracted considerable effort

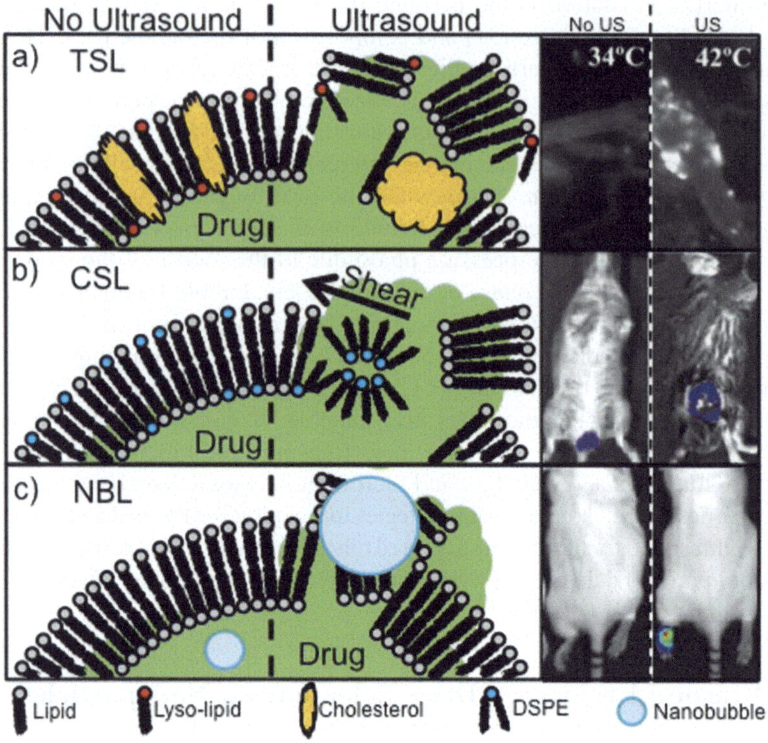

Fig. 2 Schematics and application of different ultrasound-mediated mechanisms for drug release from liposomal nanoparticles. (**a**) A cartoon of thermally sensitive liposomes (TSL) is shown before and after ultrasound exposure. Snapshots of blood vessels that indicate extravasation of a chemotherapeutic from thermosensitive liposomes before and after exposure to ultrasound-induced hyperthermia indicate presence of drug [74]. Image adapted with permission. (**b**) The mechanism of drug release from cavitation sensitive liposomes (CSL) upon exposure to ultrasound. In the presence of cavitation localized in a tumor demonstrates enhanced luciferin expression [100]. Image adapted with permission. (**c**) Drug release from nanobubble liposomes (NBL) is shown. Xenografted tumors in the foot of a mouse treated with NBL is shown to have increased luciferin expressions upon exposure to ultrasound [134]. Image adapted with permission

and interest. These thermosensitive drug delivery vehicles utilize heat-sensitive lyso-lipids, polymers, peptides, reactions, or a combination thereof in the shell of the vesicle (Fig. 2a). Here, we give an overview of the recent advances in thermosensitive drug delivery strategies that utilize ultrasound as the heat source.

Early work on thermosensitive liposomes primarily relied on the phase transition of the encapsulating material. For example, liposomes comprised of a combination of lipids with acyl chain lengths between 16 and 18 carbons will result in a leaky membrane between 41 and 54 °C, the phase transition temperatures of the respective lipid [70, 71]. Such mild hyperthermia is easily achievable with an extracorporeal ultrasound device. However, the effectiveness of such liposomes is not evident until more severe degrees of hyperthermia (>45 °C).

A key innovation towards liposomal drug delivery was the incorporation of temperature-sensitive lyso-lipids into the lipid bilayer shell [72]. These lyso-lipids begin to break down at temperatures around 39–40 °C, achieving maximum release at 41 °C [73]. A lyso-thermosensitive liposome (LTSL) formulation has been used to encapsulate doxorubicin [74] and is now in human trials under the trade name Thermodox (Celsion, USA) [75]. Subsequently, there has been continued work to develop other formulations [76–80] as well as demonstrate the capacity to trigger release from ultrasound-induced hyperthermia [81].

Other researchers have successfully encapsulated other chemotherapeutics into a thermally sensitive liposome. For example, cisplatin (a potent therapeutic for solid tumor present in head and neck, genitourinary, and lung cancers) was encapsulated in a formulation comprised predominately of hydrogenated soybean phosphatidylcholine and cholesterol [82]. The addition of a polyethylene glycol lipid enabled the liposome to be shielded from the innate immune response [83]. Unfortunately, this particular formulation has shown poor therapeutic efficacy despite the long circulation time, which has been shown to preferentially accumulate in tumors that have leaky blood vessels [84, 85]. Schroeder et al. [86] showed that when exposed to low frequency ultrasound (20 kHz), the STEALTH liposomal cisplatin released their contents in an in vivo murine model. Though no temperature or cavitation measurements were taken, Schroeder attributed their results to the large quantities of cholesterol in the shell of the liposome. Researchers have also utilized this shell composition to encapsulate other chemotherapeutics, such as 5-fluorouracil [87].

Instead of using lyso-lipids for temperature sensitivity, other researchers have added thermosensitive polymers to encapsulate drugs [88–90]. One polymer in particular has garnered much attention owing to its sensitivity to temperatures around 40 °C. *N*-isopropylacrylamide (NIPAM) is a polymer that undergoes a reversible phase transition between 32 and 40 °C, depending on the polymer chain length. This phase transition changes the hydrogel structure of NIPAM to a collapsed dehydrated state, losing up to 90 % of its initial volume. As a result, this polymer allows a change in the shell morphology, opening up pores, and allowing for drug release.

An alternative to changing the properties of the liposome shell for drug release is to instead generate bubbles from within the liposome. To do so, researchers have added ammonium bicarbonate to the liposomal core [91, 92]. This chemical decomposes to form ammonia, water, and carbon dioxide bubbles at temperatures above 36 °C. However, within the liposome, researchers have shown that the temperature sensitivity of the bubble generating liposome does not occur until temperatures above 40 °C. Once a carbon dioxide bubble is generated, the sudden expansion in volume disrupts the lipid membrane. This disruption opens pores within the membrane or ruptures it entirely, releasing the payload. Furthermore, these bubbles can be imaged using ultrasound, giving a clear indication of delivery.

Another means of bubble generation from a heat source is the use of gases, such as perfluorobutane and perfluorpentane, which have relatively low boiling points. Researchers have shown that this bubble may be stored initially as a meta-stable liquid. Once a liquid droplet is formed, interfacial forces enable these chemicals to remain as a liquid in elevated temperatures (such as those inside the body) [93, 94]. Heat from an ultrasound source is capable of temporarily disrupting this equilibrium, forcing the liquid to phase change into a gas. Because many of these gases are hydrophobic, researchers have dissolved hydrophobic drugs (such as many taxane-based chemotherapeutics [95–97]) into the nanodroplets. Upon ultrasound triggered phase-change, these droplets instantaneously release the therapeutic agents into the surrounding medium.

Though there are substantial advantages of triggered drug release from heat-sensitive nanoparticles, this technology suffers from the inability to monitor heat deposition. There are currently a limited number of techniques to noninvasively monitor temperature during treatment. The methods that are currently in use (MRI guided thermometry) are slow (i.e., not in real time) and expensive. Furthermore, current thermometry techniques are fairly inaccurate. Such inaccuracies may mean the difference between no-treatment and complete treatment. Thus the key challenge in heat-sensitive technologies lies not in the nanoparticle development, but instead with techniques to monitor the success of therapy in a safe and cost-effective way.

4.2 Mechanically Triggered Drug Release with Ultrasound

In order to avoid the imaging challenges presented by heat-based drug release therapies, there has been a surge in the utilization of mechanical means to disrupt drugs encapsulated by lipids, peptides, or polymers because these can potentially be more readily monitored by ultrasound. The goal of stimulus-responsive drug carriers is to release a drug in the presence of an externally triggered event. Similar to the heat-sensitive liposomes discussed earlier, researchers have also developed a mechanical energy analogue that exploits inertial cavitation to rupture the lipid bilayer.

In order to create a lipid shell that is sensitive to cavitation shockwaves, researchers used a lipid that has the propensity to change its solid phase structure in the presence of shear forces. Distearoyl-sn-glycero-3-phosphatidylethanolamine (DSPE) forms a lamellar gel structure in ambient conditions [98, 99]. In the presence of a shockwave, the structure of the solid phase transitions from a gel to an inverted hexagon. This change in structure destabilized the lipid bilayer, allowing for an abrupt release of the encapsulated payload. Cavitation-sensitive liposomes have recently been developed to break apart in the presence of a shockwave induced by inertially cavitating bubbles [100]. In the presence of artificial cavitation nuclei (in this case SonoVue microbubbles), these liposomes achieved close to 100 % release at peak rarefactional pressures on the order 1.5 MPa at 0.5 MHz. This represents a fraction of the pressure amplitude typically required to release thermo-sensitive liposomes (>4 MPa at 1 MHz) and is an operating regime that is potentially achievable by conventional diagnostic ultrasound scanners rather than highly specialized and expensive high intensity focused ultrasound systems.

Cavitation-sensitive liposomes require the proximity of artificial cavitation nuclei to generate the shockwave. These cavitation nuclei can range in size from as large as 5 μm to as small as 200 nm [101], with both types of nuclei having successfully demonstrated release As a result, the source of cavitation may have different pharmacokinetics depending on the size of the cavitation nuclei. To avoid the need for secondary nanoparticles, researchers have developed echogenic liposomes.

Unlike cavitation sensitive liposomes, echogenic liposomes are hypothesized to house nanoscopic gas pockets in the hydrophobic layer, either in the lipid bilayer shell or a micelle within the liposome. Upon exposure the ultrasound, the nanobubbles reportedly cavitate and destroy the integrity of the shell. Once broken, the contents of the liposome are released. Suzuki et al. [102, 103] successfully demonstrated improved luciferase coding plasmid DNA transfection to tumors from echogenic liposomes only in the presence of ultrasound. In addition to gene transfection, echogenic liposomes have also encapsulated tissue plasminogen activator in order to improve thrombolysis therapies [104–107]. Other hydrophilic and lipophilic therapeutics have also been encapsulated by echogenic liposomes [108].

Others have reported a polymeric nanobubble with a coating of a gene-loaded micelle [109]. These nanobubbles are capable of scattering ultrasound similar to microbubbles. These nanobubbles also demonstrated sustained acoustic response greater to that of gas-core liposomes. Moreover, in vivo survival studies with murine tumor models indicated that tumor volumes treated with ultrasound and gene-loaded polymeric nanobubbles were significantly more effective at controlling or reversing tumor growth.

5 Ultrasound-Enhanced Drug Transport from Nanoparticles

One of the key limitations of many drug therapies is not only their nonspecificity (which was addressed earlier), but also their inability to access tissue far beyond blood vessels. Such a challenge exists across all drug classes and has hindered the capacity to treat diseases, such as neurological disorders and cancer. To combat this challenge, many treatments rely on elevated drug doses that often come with severe side effects. This is perhaps best reflected in cancer treatments whereby the tumor itself hinders the passage of drug beyond 20–50 μm from a blood vessel [12]. The stunted travel distance of even small drug molecules often results from the tumor physiology, despite the leaky vasculature of a tumor. Moreover, physiological barriers, such as the tight junctions of the blood–brain barrier, preclude the use of nearly any drug administered simply by intravenous injection. It is therefore crucial that new generations of nanomedicines address this concern, and focus on methods to promote not only drug specificity but also its distribution within the diseased tissue.

To address this challenge without surgical intervention, elevated drug doses, or resorting to palliative care, researchers have looked towards ultrasound as a modality to interact with deep tissue. As mentioned in the previous sections, we have shown that nanoparticles that respond to ultrasound have been implemented for site-specific drug release. Note that the methodologies to be discussed are applicable to freely circulating drugs. This is an important distinction because the encapsulation of a drug presents a substantial regulatory and financial challenge.

Moreover, we want to distinguish between cellular transport (i.e., sonoporation), which is discussed in another section, and transport to extravascular tissue. Below, we show the capacity and advantages for nanoparticles to enhance the transport of therapeutics into a targeted tissue beyond blood vessels.

5.1 Improved Transport from Ultrasound-Induced Hyperthermia

There has been considerable effort in using ultrasound-induced hyperthermia to remotely trigger drug release from nanoparticles (as seen in our earlier section). Furthermore, there is evidence that hyperthermia itself enhances cell permeability, improving the efficacy of cellular drug transport [110–112]. It has been suggested that heat improves circulation in tissue with a dense microvasculature such as tumors, thereby improving local drug concentrations. However, recent work has demonstrated that hyperthermia does the opposite; blood-flow decreases due to the arteriolar–venular pressure gradient [113]. Thus, the exact mechanisms for enhanced transport of therapeutics from ultrasound-induced hyperthermia are still unclear.

Because the source of the heat is irrelevant, ultrasound provides a noninvasive means to locally increase temperatures in deep tissue, allowing for increased uptake of drugs such as monoclonal antibodies [112]. Beyond enhanced cellular uptake, it has been shown that hyperthermia from high-intensity focused ultrasound will disrupt the BBB, allowing easier passage of drugs into the brain [114, 115]. However, in these studies it was difficult to delineate the extents to which cavitation or hyperthermia disrupted the endothelial tight junctions. More recently, others [116] have shown that mild hyperthermia from ultrasound opens up the tight junctions of the BBB, allowing for increased drug uptake into the brain.

As mentioned earlier, phase-change nanodroplets have shown great promise in localized drug delivery activated by ultrasound-induced hyperthermia. Chen et al. have shown that these nanodroplets are capable of separating the tight junctions between the endothelial cells in the brain [117]. They demonstrated that there was preferential uptake of contrast agent into the brain following ultrasound exposure. Considering the beneficial effect of hyperthermia alone, it is difficult to distinguish drug transport induced by nanodroplets from hyperthermia alone.

5.2 Cavitation-Enhanced Transport of Small Molecular Drugs and Biologics with Ultrasound

Perhaps one of the most established means to promote transport of a drug beyond blood vessels is acoustic cavitation. As mentioned earlier, cavitation is typically generated by cavitation nuclei co-injected with the therapeutic. Upon exposure to

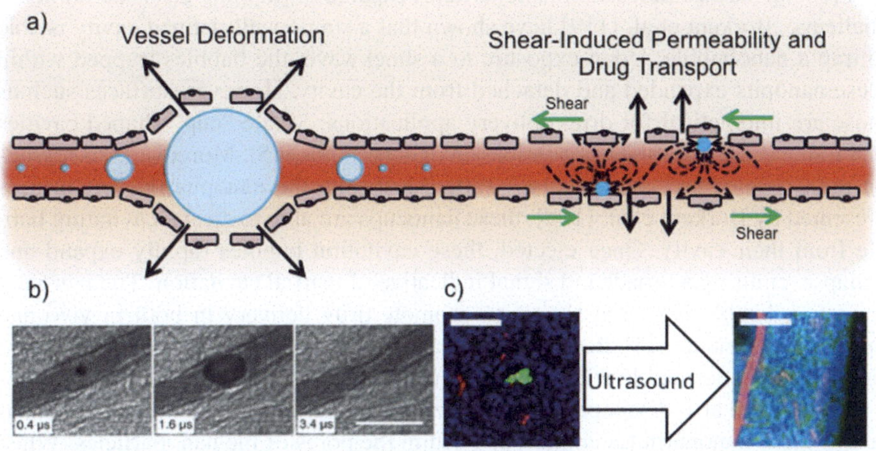

Fig. 3 (**a**) A cartoon of mechanical deformation and transport of a therapeutic beyond a microvessel from a cavitating bubble. Both vessel deformation and shear-induced permeability are shown (**b**) A microbubble inducing microbubble expansion and invagination as a direct result of cavitation in an ex vivo microvessel [118]. Image adapted with permission. The scale bar represents 50 μm. (**c**) Fluorescent microscopy images of a tumor treated with a co-injection of gas-stabilizing solid nanoparticles and a fluorescent antibody demonstrates the effect of ultrasound on drug-extravasation. Without ultrasound, antibody is co-localized with the blood vessel [45]. Image adapted with permission. *Blue, red*, and *green* represents the cancer cells, blood vessels, and antibody respectively

ultrasound, cavitation nuclei experience large volume changes that generate microstreams and pump drugs into the tissue, as well as open up endothelial junctions (Fig. 3) [118]. To date, these cavitation nuclei are typically of the micron size range and are primarily comprised of gas (i.e., a microbubble).

Microbubbles have long been established as the key cavitation agent used in biomedical technologies. However, many biomedical applications require submicron sizes and sustained cavitation response times. Microbubbles are suboptimal for these applications. As a result, there has been a surge in the development of submicron cavitation agents, such as nanobubbles and gas-stabilizing solid nanoparticles.

5.2.1 Gas-Stabilizing Solid Nanoparticles

Another means to generate bubbles from nanoscale nuclei is to partially stabilize a bubble on the surface of a solid nanoparticle. Much like bubbles in a champagne flute, nucleation of bubbles from solid surfaces requires defects (such as cracks and crevices on glass) that entrap gas. For bubbles on microparticles, these surface stabilized bubbles rapidly expand and detach from the surface when exposed to shockwaves. The expulsed bubble pushes the microparticle away, actively propelling it away from the cavitation site. This phenomenon has been well established on micron size particles, but there has been limited number of studies for gas-stabilizing nanoparticles, especially in the context of drug delivery.

Creating surface defects on nanoparticles capable of trapping gas is an immense challenge. Borkent et al. [119] have shown that a single well-defined cavity is able to trap a nanobubble. Upon exposure to a shockwave, the bubbles trapped within these nanopits expanded and detached from the cavity. However, surfaces such as those are impractical for drug delivery applications. Single "cup" shaped cavities that trap gas are nevertheless possible on nanoparticles [45]. Moreover, the cavities on these "nanocups" are tuneable [120]. Much like the nanopits on the surfaces presented by Borkent et al. [119], these nanocups are able to eject a cavitating bubble from their cavity. Once ejected, these cavitation bubbles rapidly expand and collapse, emitting a broadband signal indicative of inertial cavitation. The inertially cavitating bubble has been shown to promote drug delivery in both in vivo and in vitro experiments [45]. Because these particles exclusively emit broadband emissions, they are detectable with diagnostic ultrasound probes.

Others have also developed gas-stabilizing nanoparticles. In contrast to the nanocups, these nanoparticles contain gas within the pores of the nanoparticles. When exposed to ultrasound, gas from within the pores extends out, nucleating a cavitating bubble. Studies have shown that such cavitating bubbles are capable for diagnostic ultrasound [121]. However, there has not been any study that has evaluated the ability for these nanoparticles to promote drug transport.

In principle, these nanoparticles simply provide a source of nanobubbles. It is these bubbles that provide the means by which a circulating therapeutic extravasates beyond the blood vessel. The mechanism of action for enhanced extravasation is mechanical in nature, and as such is drug-class-agnostic. As a result, the key advantage of cavitation, inducing solid nanoparticles, is their capacity to promote the effectiveness of therapies across several clinical indications without the need to modify existing drugs.

6 Sonoporation

Gene therapies require the delivery of genes to the nucleus of the cell. However, there are substantial challenges in promoting cellular uptake of these genes. Sonoporation, therefore, is the use of ultrasound to temporarily permeate the cell membrane wall, allowing nucleic acid polymers (DNA, RNA, siRNA, etc.) to enter the cytoplasm.

Mechanistically speaking, sonoporation occurs from cavitation [122, 123]. Cavitating bubbles, as we discussed earlier, enable highly localized shear forces with shear rates on the $10^7\ s^{-1}$ [124]. Such shear rates near the cell wall force the cell membrane to temporarily open. Alternatively, shock waves generated by collapsing bubbles are also capable of disrupting the cell membrane wall. In addition to shock waves, jets formed by an inertial cavitation bubble [125] have also been shown to temporarily form pores on the surfaces of cells [126, 127]. These temporary openings in conjunction with the enhanced transport associated with cavitation results in an effective means to transfect diseased cells.

6.1 *Sonoporation from Ultrasound Induced Hyperthermia*

Earlier in sections 4 and 5, we discussed the use of phase-change nanodroplets for both drug release and drug transport. It is therefore not surprising that these nanodroplets have also been used for gene delivery. For example, Burgess and Porter [128] demonstrated successful transfection of cancer cells. To do so, they utilized a green fluorescent protein expressing siRNA freely suspended with phase-change nanodroplets. Expression of GFP only occurred in the presence of high intensity focused ultrasound (5 MHz center frequency at 6.2 MPa). Others were able to bind DNA to phase-change droplets. Upon exposure to ultrasound and DNA bound to phase-change droplets, Gao et al. [129] demonstrated a substantial increase in transfection of HepG2 cells.

Other temperature-sensitive materials (such as poly-NIPAM) have also shown promise for gene therapy [130]. However, there have been few (if any) studies that have manufactured nanoparticles with these materials and applied ultrasound induced hyperthermia for cell transfection.

6.2 *Mechanically Induced Sonoporation by Ultrasound-Mediated Cavitation*

A more direct route to permeate the cell membrane is to utilize the innate mechanical responses of pre-formed bubbles (Fig. 4). As of late there has been a surge of interest in the development of nanobubbles. These nanobubbles either exist as a stand-alone gas bubble with various coatings [131, 132] or are encapsulated within a hydrophobic shell or are contained within a liposome [133]. In our earlier

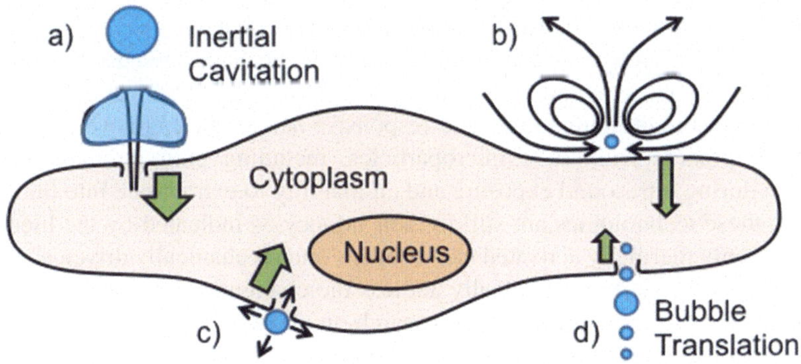

Fig. 4 A diagram of different mechanisms by which a cavitation bubble from an ultrasound-responsive nanoparticle can enhance permeation of a cell membrane and promote gene transfection is shown through (**a**) inertial cavitation, (**b**) microstreaming and micropumping, (**c**) membrane deformations, and (**d**) bubble translation as a result of acoustic radiation force

sections, we discussed the benefit of several of these nanobubble constructs for ultrasound-mediated drug release. Here we focus on their capacity to also promote cell transfection.

As already discussed, hydrophobic gases have been shown to preferentially reside in the hydrophobic regions of a lipid bilayer encapsulation or within a liposome encapsulation. Upon exposure to ultrasound, researchers have shown that the radial oscillations of the nanobubble will disrupt the lipid membrane, thereby releasing its payload. Thus, these echogenic liposomes have shown great promise in drug release. Beyond their capacity to release drugs, the cavitating nanobubbles also generate mechanical forces that affect the surrounding medium. For example, echogenic liposomes have been shown to improve gene transfection into cancer cells in a tumor, which requires the permeation of the cell membrane [102, 134–136].

7 Conclusion

In this chapter, we review the use of ultrasound-responsive nanoparticles, which cause either thermal or mechanical effects, to address the key clinical challenges of current drug therapies, and in particular (1) the triggered release of a drug from a nanoparticle, (2) the extravasation and enhanced transport of a free or recently released drug to diseased tissue, and (3) the enhanced cellular uptake of a gene therapy or other pharmacological agent.

There has been a substantial amount of work to demonstrate the effectiveness of thermally sensitive liposomes, polymer nanocapsules, and phase-change nanodroplets. Thus, we concluded that this technology has had and will continue to have a major impact in drug therapies, as further indicated by current clinical trials. In contrast, mechanical release via cavitation of therapeutics from nanoparticles remains a relatively young technology. However, there is much interest in further developing it due to its cost-effectiveness and ability to deliver and monitor the treatment using low-cost, quasi-diagnostic ultrasound systems.

Historically, micron-sized ultrasound contrast agents have been used to transport therapeutics beyond circulation and into tissue. However, there has been a recent surge of interest in utilizing ultrasound-responsive nanoparticles to overcome many limitations present with these microparticles, including poor circulation, rapid depletion during ultrasound exposure and an inability to extravasate into tumor tissue. Yet, these technologies are still in their infancy as indicated by the literature. Currently, only thermally activated nanodroplets and mechanically driven nanocups have been put forward to specifically address the challenges related to drug transport. That being said, we believe that research on ultrasound-responsive nanoparticles for drug transport will increase substantially.

Much like drug transport, there are few instances of using ultrasound-responsive nanoparticles for sonoporation. Thermally activated phase-change nanodroplets have demonstrated promise in transfecting cells with siRNA, but this work has not been conducted with in vivo models to date. Gene-carrying nanobubbles have shown that the cavitation generated from submicron gas particles is capable of temporarily perforating the cell membrane in both in vitro and in vivo models. It is clear that the use of ultrasound-responsive nanoparticles for sonoporation is still in its early stages, and thus there are many opportunities that still remain.

Unlike triggered release strategies from encapsulating nanoparticles, the ultrasound-enhanced delivery strategies presented in this chapter are ultimately drug-class-agnostic and rely on broadly applicable thermal and mechanical effects, which can be applied to free therapeutic macromolecules without modification. This portability across drug classes and clinical indications represents one of the major advantages of nanoparticle-enhanced ultrasound-mediated delivery, and could in future achieve the greatest impact in increasing the efficacy of current and emerging drug therapies.

References

1. Krall N, Pretto F, Decurtins W, Bernardes GJL, Supuran CT, Neri D. A small-molecule drug conjugate for the treatment of carbonic anhydrase IX expressing tumors. Angew Chem Int Ed. 2014;53(16):4231–5.
2. Hay M, Thomas DW, Craighead JL, Economides C, Rosenthal J. Clinical development success rates for investigational drugs. Nat Biotechnol. 2014;32(1):40–51.
3. O'Reilly LP, Luke CJ, Perlmutter DH, Silverman GA, Pak SC. C. elegans in high-throughput drug discovery. Adv Drug Deliver Rev. 2014;69:247–53.
4. Lipshultz. Long-term cardiovascular toxicity in children, adolescents, and young adults who receive cancer therapy: pathophysiology, course, monitoring, management, prevention, and research directions: a scientific statement from the American Heart Association (vol 128, p. 1927, 2013). Circulation. 2013;128(19):E394–E.
5. Kramer AH, Jenne CN, Zygun DA, Roberts DJ, Hill MD, Holodinsky JK, et al. Intraventricular fibrinolysis with tissue plasminogen activator is associated with transient cerebrospinal fluid inflammation: a randomized controlled trial. J Cerebr Blood F Met. 2015;35(8):1241–8.
6. Brott T, Broderick J, Kothari R, ODonoghue M, Barsan W, Tomsick T, et al. Intracerebral hemorrhage after intravenous t-PA therapy for ischemic stroke. Stroke. 1997;28(11):2109–18.
7. Goldstein LB. Acute ischemic stroke treatment in 2007. Circulation. 2007;116(13):1504–14.
8. Mitragotri S, Burke PA, Langer R. Overcoming the challenges in administering biopharmaceuticals: formulation and delivery strategies. Nat Rev Drug Discov. 2014;13(9):655–72.
9. Minchinton AI, Tannock IF. Drug penetration in solid tumours. Nat Rev Cancer. 2006;6(8):583–92.
10. Primeau AJ, Rendon A, Hedley D, Lilge L, Tannock IF. The distribution of the anticancer drug doxorubicin in relation to blood vessels in solid tumors. Clin Cancer Res. 2005;11(24):8782–8.
11. Baker JHE, Lindquist KE, Huxham L, Kyle AH, Sy JT, Minchinton AI. Direct visualization of heterogeneous extravascular distribution of trastuzumab in human epidermal growth factor receptor type 2 overexpressing xenografts. Clin Cancer Res. 2008;14(7):2171–9.
12. Tailor TD, Hanna G, Yarmolenko PS, Dreher MR, Betof AS, Nixon AB, et al. Effect of pazopanib on tumor microenvironment and liposome delivery. Mol Cancer Ther. 2010;9(6):1798–808.
13. Abbott NJ, Ronnback L, Hansson E. Astrocyte-endothelial interactions at the blood-brain barrier. Nat Rev Neurosci. 2006;7(1):41–53.
14. Carlisle R, Coussios CC. Mechanical approaches to oncological drug delivery. Ther Deliv. 2013;4(10):1213–5. Epub 2013/10/15.

15. Crown J, O'Leary M. The taxanes: an update. Lancet. 2000;355(9210):1176–8.
16. Hurley LH. DNA and its associated processes as targets for cancer therapy. Nat Rev Cancer. 2002;2(3):188–200.
17. Lee K, Qian DZ, Rey S, Wei H, Liu JO, Semenza GL. Anthracycline chemotherapy inhibits HIF-1 transcriptional activity and tumor-induced mobilization of circulating angiogenic cells. Proc Natl Acad Sci U S A. 2009;106(7):2353–8.
18. Karahoca M, Momparler RL. Pharmacokinetic and pharmacodynamic analysis of 5-aza-2'-deoxycytidine (decitabine) in the design of its dose-schedule for cancer therapy. Clin Epigenetics. 2013;5.
19. Chu MY, Fischer GA. A proposed mechanism of action of 1-beta-D-arabinofuranosyl-cytosine as an inhibitor of growth of leukemic cells. Biochem Pharmacol. 1962;11:423.
20. Hans R, Andtbacka I, Collichio FA, Amatruda T, Senzer NN, Chesney J, et al. OPTiM: a randomized phase III trial of talimogene laherparepvec (T-VEC) versus subcutaneous (SC) granulocyte-macrophage colony-stimulating factor (GM-CSF) for the treatment (tx) of unresected stage IIIB/C and IV melanoma. J Clin Oncol. 2013;31(18).
21. Folkman J. Antiangiogenesis in cancer therapy—endostatin and its mechanisms of action. Exp Cell Res. 2006;312(5):594–607.
22. Salama AKS, Hodi FS. Cytotoxic T-lymphocyte-associated antigen-4. Clin Cancer Res. 2011;17(14):4622–8.
23. Brahmer JR, Tykodi SS, Chow LQM, Hwu WJ, Topalian SL, Hwu P, et al. Safety and activity of anti-PD-L1 antibody in patients with advanced cancer. N Engl J Med. 2012;366(26):2455–65.
24. Topalian SL, Drake CG, Pardoll DM. Targeting the PD-1/B7-H1(PD-L1) pathway to activate anti-tumor immunity. Curr Opin Immunol. 2012;24(2):207–12.
25. Wardlaw JM, Murray V, Berge E, del Zoppo G, Sandercock P, Lindley RL, et al. Recombinant tissue plasminogen activator for acute ischaemic stroke: an updated systematic review and meta-analysis. Lancet. 2012;379(9834):2364–72.
26. Williams JM, Navin TJ, Levi CR, Jude M. Recombinant tissue plasminogen activator (rt-PA) utilisation by rural clinicians in acute ischaemic stroke: an audit of current practice and clinical outcomes. Int J Stroke. 2012;7:42–3.
27. Lapchak PA, Chapman DF, Zivin JA. Metalloproteinase inhibition reduces thrombolytic (tissue plasminogen activator)-induced hemorrhage after thromboembolic stroke. Stroke. 2000;31(12):3034–9.
28. Mangialasche F, Solomon A, Winblad B, Mecocci P, Kivipelto M. Alzheimer's disease: clinical trials and drug development. Lancet Neurol. 2010;9(7):702–16.
29. Olanow CW, Rascol O, Hauser R, Feigin PD, Jankovic J, Lang A, et al. A double-blind, delayed-start trial of rasagiline in Parkinson's disease. N Engl J Med. 2009;361(13):1268–78.
30. Schapira AHV, Bezard E, Brotchie J, Calon F, Collingridge GL, Ferger B, et al. Novel pharmacological targets for the treatment of Parkinson's disease. Nat Rev Drug Discov. 2006;5(10):845–54.
31. Goetz CG, Poewe W, Rascol O, Sampaio C. Evidence-based medical review update: pharmacological and surgical treatments of Parkinson's disease: 2001 to 2004. Mov Disord. 2005;20(5):523–39.
32. Kaplitt MG, Feigin A, Tang C, Fitzsimons HL, Mattis P, Lawlor PA, et al. Safety and tolerability of gene therapy with an adeno-associated virus (AAV) borne GAD gene for Parkinson's disease: an open label, phase I trial. Lancet. 2007;369(9579):2097–105.
33. Landreth G, Jiang QG, Mandrekar S, Heneka M. PPAR gamma agonists as therapeutics for the treatment of Alzheimer's disease. Neurotherapeutics. 2008;5(3):481–9.
34. Cobbold RS. Foundations of biomedical ultrasound. Oxford University Press on Demand; 2007.
35. Ter Haar G, Coussios C. High intensity focused ultrasound: physical principles and devices. Int J Hyperther. 2007;23(2):89–104.

36. Szabo TL. Diagnostic ultrasound imaging: inside out. Boston: Academic Press; 2004.
37. McDannold N, Clement G, Black P, Jolesz F, Hynynen K. Transcranial MRI-guided focused ultrasound surgery of brain tumors: initial findings in three patients. Neurosurgery. 2010;66(2):323.
38. Salomir R, Vimeux FC, de Zwart JA, Grenier N, Moonen CTW. Hyperthermia by MR-guided focused ultrasound: accurate temperature control based on fast MRI and a physical model of local energy deposition and heat conduction. Magnet Reson Med. 2000;43(3):342–7.
39. Rieke V, Pauly KB. MR thermometry. J Magn Reson Imaging. 2008;27(2):376–90.
40. Bradley Jr WG. MR-guided focused ultrasound: a potentially disruptive technology. J Am Coll Radiol. 2009;6(7):510–3. Epub 2009/06/30.
41. Sarvazyan AP, Rudenko OV, Nyborg WL. Biomedical applications of radiation force of ultrasound: historical roots and physical basis. Ultrasound Med Biol. 2010;36(9):1379–94. Epub 2010/08/31.
42. Leighton T. The acoustic bubble. London: Academic Press; 2012.
43. Maxwell AD, Cain CA, Hall TL, Fowlkes JB, Xu Z. Probability of cavitation for single ultrasound pulses applied to tissues and tissue-mimicking materials. Ultrasound Med Biol. 2013;39(3):449–65. Epub 2013/02/06.
44. Stride EP, Coussios CC. Cavitation and contrast: the use of bubbles in ultrasound imaging and therapy. Proc Inst Mech Eng H. 2010;224(H2):171–91.
45. Kwan JJ, Myers R, Coviello CM, Graham SM, Shah AR, Stride E, et al. Ultrasound-propelled nanocups for drug delivery. Small. 2015. Epub 2015/08/25.
46. Arvanitis CD, Bazan-Peregrino M, Rifai B, Seymour LW, Coussios CC. Cavitation-enhanced extravasation for drug delivery. Ultrasound Med Biol. 2011;37(11):1838–52. Epub 2011/10/04.
47. Ammi AY, Cleveland RO, Mamou J, Wang GI, Bridal SL, O'Brien Jr WD. Ultrasonic contrast agent shell rupture detected by inertial cavitation and rebound signals. IEEE Trans Ultrason Ferroelectr Freq Control. 2006;53(1):126–36. Epub 2006/02/14.
48. Roberts WW, Hall TL, Ives K, Wolf Jr JS, Fowlkes JB, Cain CA. Pulsed cavitational ultrasound: a noninvasive technology for controlled tissue ablation (histotripsy) in the rabbit kidney. J Urol. 2006;175(2):734–8. Epub 2006/01/13.
49. Wang YN, Khokhlova T, Bailey M, Hwang JH, Khokhlova V. Histological and biochemical analysis of mechanical and thermal bioeffects in boiling histotripsy lesions induced by high intensity focused ultrasound. Ultrasound Med Biol. 2013;39(3):424–38. Epub 2013/01/15.
50. Schade GR, Keller J, Ives K, Cheng X, Rosol TJ, Keller E, et al. Histotripsy focal ablation of implanted prostate tumor in an ACE-1 canine cancer model. J Urol. 2012;188(5):1957–64. Epub 2012/09/25.
51. Bloch SH, Short RE, Ferrara KW, Wisner ER. The effect of size on the acoustic response of polymer-shelled contrast agents. Ultrasound Med Biol. 2005;31(3):439–44. Epub 2005/03/08.
52. Dicker S, Mleczko M, Siepmann M, Wallace N, Sunny Y, Bawiec CR, et al. Influence of shell composition on the resonance frequency of microbubble contrast agents. Ultrasound Med Biol. 2013;39(7):1292–302. Epub 2013/05/21.
53. Collis J, Manasseh R, Liovic P, Tho P, Ooi A, Petkovic-Duran K, et al. Cavitation microstreaming and stress fields created by microbubbles. Ultrasonics. 2010;50(2):273–9. Epub 2009/11/10.
54. Liu X, Wu J. Acoustic microstreaming around an isolated encapsulated microbubble. J Acoust Soc Am. 2009;125(3):1319–30. Epub 2009/03/12.
55. Won JM, Lee JH, Lee KH, Rhee K, Chung SK. Propulsion of water-floating objects by acoustically oscillating microbubbles. Int J Precis Eng Man. 2011;12(3):577–80.
56. Samiotaki G, Vlachos F, Tung YS, Konofagou EE. A quantitative pressure and microbubble-size dependence study of focused ultrasound-induced blood-brain barrier opening reversibility in vivo using MRI. Magnet Reson Med. 2012;67(3):769–77.
57. Qiu YY, Zhang CB, Tu J, Zhang D. Microbubble-induced sonoporation involved in ultrasound-mediated DNA transfection in vitro at low acoustic pressures. J Biomech. 2012;45(8):1339–45.

58. Juffermans LJM, van Dijk A, Jongenelen CAM, Drukarch B, Reijerkerk A, de Vries HE, et al. Ultrasound and microbubble-induced intra- and intercellular bioeffects in primary endothelial cells. Ultrasound Med Biol. 2009;35(11):1917–27.
59. VanBavel E. Effects of shear stress on endothelial cells: possible relevance for ultrasound applications. Prog Biophys Mol Bio. 2007;93(1-3):374–83.
60. Zhong P, Cioanta I, Cocks FH, Preminger GM. Inertial cavitation and associated acoustic emission produced during electrohydraulic shock wave lithotripsy. J Acoust Soc Am. 1997;101(5):2940–50.
61. Salgaonkar VA, Datta S, Holland CK, Mast TD. Passive cavitation imaging with ultrasound arrays. J Acoust Soc Am. 2009;126(6):3071–83.
62. Farny CH, Holt RG, Roy RA. Temporal and spatial detection of Hifu-induced inertial and hot-vapor cavitation with a diagnostic ultrasound system. Ultrasound Med Biol. 2009;35(4):603–15.
63. Gyongy M, Coussios CC. Passive spatial mapping of inertial cavitation during HIFU exposure. IEEE Trans Biomed Eng. 2010;57(1):48–56.
64. Davis ME, Chen Z, Shin DM. Nanoparticle therapeutics: an emerging treatment modality for cancer. Nat Rev Drug Discov. 2008;7(9):771–82.
65. Peer D, Karp JM, Hong S, FaroKHzad OC, Margalit R, Langer R. Nanocarriers as an emerging platform for cancer therapy. Nat Nanotechnol. 2007;2(12):751–60.
66. Malam Y, Loizidou M, Seifalian AM. Liposomes and nanoparticles: nanosized vehicles for drug delivery in cancer. Trends Pharmacol Sci. 2009;30(11):592–9.
67. Allen TM, Cullis PR. Drug delivery systems: entering the mainstream. Science. 2004;303(5665):1818–22.
68. Torchilin VP. Recent advances with liposomes as pharmaceutical carriers. Nat Rev Drug Discov. 2005;4(2):145–60.
69. Andresen TL, Jensen SS, Jorgensen K. Advanced strategies in liposomal cancer therapy: problems and prospects of active and tumor specific drug release. Prog Lipid Res. 2005;44(1):68–97.
70. Yatvin MB, Weinstein JN, Dennis WH, Blumenthal R. Design of liposomes for enhanced local release of drugs by hyperthermia. Science. 1978;202(4374):1290–3.
71. Weinstein JN, Magin RL, Yatvin MB, Zaharko DS. Liposomes and local hyperthermia—selective delivery of methotrexate to heated tumors. Science. 1979;204(4389):188–91.
72. Anyarambhatla GR, Needham D. Enhancement of the phase transition permeability of DPPC liposomes by incorporation of MPPC: a new temperature-sensitive liposome for use with mild hyperthermia. J Liposome Res. 1999;9(4):491–506.
73. Needham D, Anyarambhatla G, Kong G, Dewhirst MW. A new temperature-sensitive liposome for use with mild hyperthermia: characterization and testing in a human tumor xenograft model. Cancer Res. 2000;60(5):1197–201.
74. Ponce AM, Vujaskovic Z, Yuan F, Needham D, Dewhirst MW. Hyperthermia mediated liposomal drug delivery. Int J Hyperther. 2006;22(3):205–13.
75. Needham D, Dewhirst MW. The development and testing of a new temperature-sensitive drug delivery system for the treatment of solid tumors. Adv Drug Deliver Rev. 2001;53(3):285–305.
76. Gaber MH, Hong KL, Huang SK, Papahadjopoulos D. Thermosensitive sterically stabilized liposomes—formulation and in-vitro studies on mechanism of doxorubicin release by bovine serum and human plasma. Pharm Res. 1995;12(10):1407–16.
77. Gaber MH, Wu NZ, Hong KL, Huang SK, Dewhirst MW, Papahadjopoulos D. Thermosensitive liposomes: extravasation and release of contents in tumor microvascular networks. Int J Radiat Oncol. 1996;36(5):1177–87.
78. Park SM, Kim MS, Park SJ, Park ES, Choi KS, Kim YS, et al. Novel temperature-triggered liposome with high stability: formulation, in vitro evaluation, and in vivo study combined with high-intensity focused ultrasound (HIFU). J Control Release. 2013;170(3):373–9.
79. Lindner LH, Eichhorn ME, Eibl H, Teichert N, Schmitt-Sody M, Issels RD, et al. Novel temperature-sensitive liposomes with prolonged circulation time. Clin Cancer Res. 2004;10(6):2168–78.

80. Hossann M, Wiggenhorn M, Schwerdt A, Wachholz K, Teichert N, Eibl H, et al. In vitro stability and content release properties of phosphatidylglyceroglycerol containing thermosensitive liposomes. Biochim Biophys Acta. 2007;1768(10):2491–9.
81. Mylonopoulou E, Bazan-Peregrino M, Arvanitis CD, Coussios CC. Exploitation of cavitation-enhanced heating for release of doxorubicin from thermosensitve liposomes by therapeutic ultrasound. J Acoust Soc Am. 2010;128(4):2418.
82. Peleg-Shulman T, Gibson D, Cohen R, Abra R, Barenholz Y. Characterization of sterically stabilized cisplatin liposomes by nuclear magnetic resonance. Biochim Biophys Acta. 2001;1510(1–2):278–91.
83. Torchilinl V, Papisov M. Why do polyethylene glycol-coated liposomes circulate so long?: Molecular mechanism of liposome steric protection with polyethylene glycol: Role of polymer chain flexibility. J Liposome Res. 1994;4(1):725–39.
84. Schroeder A, Sigal A, Turjeman K, Barenholz Y. Using PEGylated nano-liposomes to target tissue invaded by a foreign body. J Drug Target. 2008;16(7-8):591–5.
85. Dvorak HF, Nagy JA, Dvorak JT, Dvorak AM. Identification and characterization of the blood-vessels of solid tumors that are leaky to circulating macromolecules. Am J Pathol. 1988;133(1):95–109.
86. Schroeder A, Honen R, Turjeman K, Gabizon A, Kost J, Barenholz Y. Ultrasound triggered release of cisplatin from liposomes in murine tumors. J Control Release. 2009;137(1):63–8.
87. Al Sabbagh C, Tsapis N, Novell A, Calleja-Gonzalez P, Escoffre JM, Bouakaz A, et al. Formulation and pharmacokinetics of thermosensitive stealth(A (R)) liposomes encapsulating 5-fluorouracil. Pharm Res. 2015;32(5):1585–603.
88. Han HD, Shin BC, Choi HS. Doxorubicin-encapsulated thermosensitive liposomes modified with poly(N-isopropylacrylamide-co-acrylamide): drug release behavior and stability in the presence of serum. Eur J Pharm Biopharm. 2006;62(1):110–6.
89. Han HD, Choi MS, Hwang T, Song CK, Seong H, Kim TW, et al. Hyperthermia-induced antitumor activity of thermosensitive polymer modified temperature-sensitive liposomes. J Pharm Sci. 2006;95(9):1909–17.
90. Ta T, Convertine AJ, Reyes CR, Stayton PS, Porter TM. Thermosensitive liposomes modified with poly(N-isopropylacrylamide-co-propylacrylic acid) copolymers for triggered release of doxorubicin. Biomacromolecules. 2010;11(8):1915–20.
91. Chen KJ, Liang HF, Chen HL, Wang YC, Cheng PY, Liu HL, et al. A thermoresponsive bubble-generating liposomal system for triggering localized extracellular drug delivery. ACS Nano. 2013;7(1):438–46.
92. Chen KJ, Chaung EY, Wey SP, Lin KJ, Cheng F, Lin CC, et al. Hyperthermia-mediated local drug delivery by a bubble-generating liposomal system for tumor-specific chemotherapy. ACS Nano. 2014;8(5):5105–15.
93. Sheeran PS, Wong VP, Luois S, McFarland RJ, Ross WD, Feingold S, et al. Decafluorobutane as a phase change contrast agent for low-energy extravascular ultrasonic imaging. Ultrasound Med Biol. 2011;37(9):1518–30.
94. Sheeran PS, Luois SH, Mullin LB, Matsunaga TO, Dayton PA. Design of ultrasonically-activatable nanoparticles using low boiling point perfluorocarbons. Biomaterials. 2012; 33(11):3262–9.
95. Wang CH, Kang ST, Lee YH, Luo YL, Huang YF, Yeh CK. Aptamer-conjugated and drug-loaded acoustic droplets for ultrasound theranosis. Biomaterials. 2012;33(6):1939–47.
96. Rapoport N. Phase-shift, stimuli-responsive perfluorocarbon nanodroplets for drug delivery to cancer. Wires Nanomed Nanobi. 2012;4(5):492–510.
97. Rapoport N, Nam KH, Gupta R, Gao ZG, Mohan P, Payne A, et al. Ultrasound-mediated tumor imaging and nanotherapy using drug loaded, block copolymer stabilized perfluorocarbon nanoemulsions. J Control Release. 2011;153(1):4–15.
98. Marsh D, Seddon JM. Gel-to-inverted hexagonal (L-Beta-Hii) phase-transitions in phosphatidylethanolamines and fatty-acid phosphatidylcholine mixtures, demonstrated by P-31-Nmr spectroscopy and X-ray-diffraction. Biochim Biophys Acta. 1982;690(1):117–23.

99. Evjen TJ, Nilssen EA, Rognvaldsson S, Brandl M, Fossheim SL. Distearoylphosphatidylethanolamine-based liposomes for ultrasound-mediated drug delivery. Eur J Pharm Biopharm. 2010;75(3):327–33.
100. Graham SM, Carlisle R, Choi JJ, Stevenson M, Shah AR, Myers RS, et al. Inertial cavitation to non-invasively trigger and monitor intratumoral release of drug from intravenously delivered liposomes. J Control Release. 2014;178:101–7.
101. Graham S. Ultrasound-triggered drug release from liposomes using nanoscale cavitation nuclei. Oxford: University of Oxford; 2014.
102. Suzaki R, Takizawa T, Negishi Y, Utoguchi N, Sawamura K, Tanaka K, et al. Tumor specific ultrasound enhanced gene transfer in vivo with novel liposomal bubbles. J Control Release. 2008;125(2):137–44.
103. Suzuki R, Oda Y, Utoguchi N, Maruyama K. Progress in the development of ultrasound-mediated gene delivery systems utilizing nano- and microbubbles. J Control Release. 2011;149(1):36–41.
104. Shaw GJ, Meunier JM, Huang SL, Lindsell CJ, McPherson DD, Holland CK. Ultrasound-enhanced thrombolysis with tPA-loaded echogenic liposomes. Thromb Res. 2009;124(3): 306–10.
105. Tiukinhoy-Laing SD, Buchanan K, Parikh D, Huang SL, MacDonald RC, McPherson DD, et al. Fibrin targeting of tissue plasminogen activator-loaded echogenic liposomes. J Drug Target. 2007;15(2):109–14.
106. Tiukinhoy-Laing SD, Huang SL, Klegerman M, Holland CK, McPherson DD. Ultrasound-facilitated thrombolysis using tissue-plasminogen activator-loaded echogenic liposomes. Thromb Res. 2007;119(6):777–84.
107. Smith DAB, Vaidya SS, Kopechek JA, Huang SL, Klegerman ME, Mcpherson DD, et al. Ultrasound-triggered release of recombinant tissue-type plasminogen activator from echogenic liposomes. Ultrasound Med Biol. 2010;36(1):145–57.
108. Kopechek JA, Abruzzo TM, Wang B, Chrzanowski SM, Smith DAB, Kee PH, et al. Ultrasound-mediated release of hydrophilic and lipophilic agents from echogenic liposomes. J Ultras Med. 2008;27(11):1597–606.
109. Yin TH, Wang P, Li JG, Zheng RQ, Zheng BW, Cheng D, et al. Ultrasound-sensitive siRNA-loaded nanobubbles formed by hetero-assembly of polymeric micelles and liposomes and their therapeutic effect in gliomas. Biomaterials. 2013;34(18):4532–43.
110. Ning SC, Macleod K, Abra RM, Huang AH, Hahn GM. Hyperthermia induces doxorubicin release from long-circulating liposomes and enhances their antitumor efficacy. Int J Radiat Oncol. 1994;29(4):827–34.
111. Kinuya S, Yokoyama K, Hiramatsu T, Tega H, Tanaka K, Konishi S, et al. Combination radioimmunotherapy with local hyperthermia: increased delivery of radioimmunoconjugate by vascular effect and its retention by increased antigen expression in colon cancer xenografts. Cancer Lett. 1999;140(1–2):209–18.
112. Cope DA, Dewhirst MW, Friedman HS, Bigner DD, Zalutsky MR. Enhanced delivery of a monoclonal-antibody F(Ab')2 fragment to subcutaneous human glioma xenografts using local hyperthermia. Cancer Res. 1990;50(6):1803–9.
113. Jang SH, Wientjes MG, Lu D, Au JLS. Drug delivery and transport to solid tumors. Pharm Res. 2003;20(9):1337–50.
114. Vykhodtseva NI, Hynynen K, Damianou C. Histologic effects of high-intensity pulsed ultrasound exposure with subharmonic emission in rabbit brain in-vivo. Ultrasound Med Biol. 1995;21(7):969–79.
115. Mesiwala AH, Farrell L, Wenzel HJ, Silbergeld DL, Crum LA, Winn HR, et al. High-intensity focused ultrasound selectively disrupts the blood-brain barrier in vivo. Ultrasound Med Biol. 2002;28(3):389–400.
116. Cho CW, Liu Y, Cobb WN, Henthorn TK, Lillehei K, Christians U, et al. Ultrasound-induced mild hyperthermia as a novel approach to increase drug uptake in brain microvessel endothelial cells. Pharm Res. 2002;19(8):1123–9.

117. Chen CC, Sheeran PS, Wu SY, Olumolade OO, Dayton PA, Konofagou EE. Targeted drug delivery with focused ultrasound-induced blood-brain barrier opening using acoustically-activated nanodroplets. J Control Release. 2013;172(3):795–804.
118. Chen H, Kreider W, Brayman AA, Bailey MR, Matula TJ. Blood vessel deformations on microsecond time scales by ultrasonic cavitation. Phys Rev Lett. 2011;106(3):034301. Epub 2011/03/17.
119. Borkent BM, Gekle S, Prosperetti A, Lohse D. Nucleation threshold and deactivation mechanisms of nanoscopic cavitation nuclei. Phys Fluids. 2009;21(10).
120. Kwan JJ, Graham S, Myers R, Carlisle R, Stride E, Coussios CC. Ultrasound-induced inertial cavitation from gas-stabilizing nanoparticles. Phys Rev E Stat Nonlin Soft Matter Phys. 2015;92(2–1):023019. Epub 2015/09/19.
121. Chen Y, Yin Q, Ji XF, Zhang SJ, Chen HR, Zheng YY, et al. Manganese oxide-based multifunctionalized mesoporous silica nanoparticles for pH-responsive MRI, ultrasonography and circumvention of MDR in cancer cells. Biomaterials. 2012;33(29):7126–37.
122. Liang HD, Tang J, Halliwell M. Sonoporation, drug delivery, and gene therapy. Proc Inst Mech Eng H J Eng Med. 2010;224(2):343–61. Epub 2010/03/31.
123. Delalande A, Kotopoulis S, Postema M, Midoux P, Pichon C. Sonoporation: mechanistic insights and ongoing challenges for gene transfer. Gene. 2013;525(2):191–9. Epub 2013/04/10.
124. Nyborg WL. Ultrasonic microstreaming and related phenomena. Br J Cancer Suppl. 1982;5:156–60. Epub 1982/03/01.
125. Brujan EA, Ikeda T, Matsumoto Y. Jet formation and shock wave emission during collapse of ultrasound-induced cavitation bubbles and their role in the therapeutic applications of high-intensity focused ultrasound. Phys Med Biol. 2005;50(20):4797–809.
126. Prentice P, Cuschierp A, Dholakia K, Prausnitz M, Campbell P. Membrane disruption by optically controlled microbubble cavitation. Nat Phys. 2005;1(2):107–10.
127. Hu YX, Wan JMF, Yu ACH. Membrane perforation and recovery dynamics in microbubble-mediated sonoporation. Ultrasound Med Biol. 2013;39(12):2393–405.
128. Burgess MT, Porter TM. Acoustic cavitation-mediated delivery of small interfering ribonucleic acids with phase-shift nano-emulsions. Ultrasound Med Biol. 2015;41(8):2191–201. Epub 2015/05/17.
129. Gao D, Xu M, Cao Z, Gao J, Chen Y, Li Y, et al. Ultrasound-triggered phase-transition cationic nanodroplets for enhanced gene delivery. ACS Appl Mater Interfaces. 2015;7(24):13524–37. Epub 2015/05/29.
130. Zintchenko A, Ogris M, Wagner E. Temperature dependent gene expression induced by PNIPAM-based copolymers: potential of hyperthermia in gene transfer. Bioconjug Chem. 2006;17(3):766–72. Epub 2006/05/18.
131. Krupka TM, Solorio L, Wilson RE, Wu HP, Azar N, Exner AA. Formulation and characterization of echogenic lipid-pluronic nanobubbles. Mol Pharm. 2010;7(1):49–59.
132. Wang Y, Li X, Zhou Y, Huang PY, Xu YH. Preparation of nanobubbles for ultrasound imaging and intracellular drug delivery. Int J Pharm. 2010;384(1-2):148–53.
133. Nguyen AT, Wrenn SP. Acoustically active liposome-nanobubble complexes for enhanced ultrasonic imaging and ultrasound-triggered drug delivery. Wires Nanomed Nanobi. 2014;6(3):316–25.
134. Suzuki R, Takizawa T, Negishi Y, Hagisawa K, Tanaka K, Sawamura K, et al. Gene delivery by combination of novel liposomal bubbles with perfluoropropane and ultrasound. J Control Release. 2007;117(1):130–6.
135. Suzuki R, Namai E, Oda Y, Nishiie N, Otake S, Koshima R, et al. Cancer gene therapy by IL-12 gene delivery using liposomal bubbles and tumoral ultrasound exposure. J Control Release. 2010;142(2):245–50.
136. Negishi Y, Endo Y, Fukuyama T, Suzuki R, Takizawa T, Omata D, et al. Delivery of siRNA into the cytoplasm by liposomal bubbles and ultrasound. J Control Release. 2008;132(2):124–30.

Nanoparticles for Ultrasound-Guided Imaging of Cell Implantation

James Hartanto and Jesse V. Jokerst

1 Stem Cell Therapy

Stem cell therapy is established for many regenerative medicine applications and has shown remarkable impact in treating disease. Stem cell therapy is part of a larger component of medicine known as regenerative medicine. While the exact definition of regenerative medicine continues to evolve [1], one definition is that it creates living, functional tissues to repair or replace tissue or organ function lost due to age, disease, damage, or congenital defects [2, 3]. Cell-based therapies are particularly attractive because cells are the basic building block of tissue and can create functional tissue when added in sufficient numbers. Cell-based therapy often delivers cells directly to the site of disease, which can provide much greater efficacy than therapeutics that are delivered intravenously. Finally, stem cells can be directed into diverse lineages that are in turn tailored to specific diseases.

Cell-based therapy has been used for many applications in regenerative medicine. Examples include increasing the volume of blood being pumped by the left ventricle after myocardial infarction [4, 5], increasing the strength of muscles and the size of muscles in muscular dystrophy patients [6, 7], and decreasing joint pain in arthritis [8]. Other groups have shown that cells and tissues can be grown on support scaffolds to create entire new bulk pieces of tissue including the larynx and cornea [9]. Stem cell therapy has been described with many diverse starting materials including mesenchymal stem cells, embryonic stem cells, induced pluripotent stem cells, and adipose-derived stem cells [10]. Before, during, and after delivery of cells, researchers, physicians, and patients may have many questions about the fate of the cells including the following: (1) Are the cells alive? (2) What proteins are the

J. Hartanto • J.V. Jokerst (✉)
Department of NanoEngineering, University of California,
9500 Gilman Drive, San Diego, La Jolla, CA 92093-0448, USA
e-mail: jjokerst@ucsd.edu

J.W.M. Bulte, M.M.J. Modo (eds.), *Design and Applications of Nanoparticles in Biomedical Imaging*, DOI 10.1007/978-3-319-42169-8_14

cells expressing? (3) Where are the cells located? (4) How are the cells interacting with the surrounding tissue? (5) How has the cells' biochemical signature changed after delivery?

In vivo imaging is ideally suited to answer these questions because it can be done serially in real time, is noninvasive, and can often be carried out with high temporal and spatial resolution. The repeat nature of imaging is particularly important in stem cell therapy because the repair is a dynamic process and it is important to frequently assess the status of the therapy. Alternative techniques that do not use imaging such as biopsy are more difficult to do repeatedly over time. While no single imaging approach can yet answer all of the above questions, significant advances have been made—particularly in understanding the cell number, viability, and location.

1.1 General Approaches for Imaging in Stem Cell Therapy

There are two main approaches to stem cell imaging—direct and indirect (Fig. 1) [11]. Indirect imaging involves adding a reporter gene to the stem cell to produce a receptor, enzyme, or fluorescent/bioluminescent protein to create contrast and

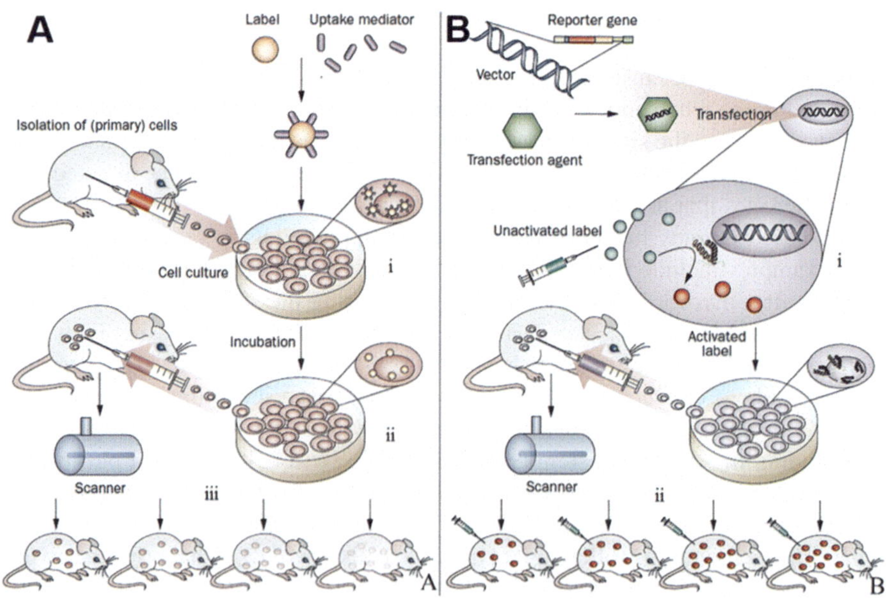

Fig. 1 Approaches to labeling stem cells. Panel **a** shows direct imaging in which an exogenous material is placed inside a cell. The cells are then purified and implanted into an animal. Over time, the amount of contrast agent per cells decreases as the cells divide. In contrast, in indirect imaging (**b**), a reporter gene is placed inside the cell. These reporter genes either produce a label or affect an injected substrate that increases contrast in the cells of interest. Because the reporter gene is an inherent part of the cells' biology, this reporter is passed to all progeny with no dilution effects. However, a substrate has to be repeatedly injected for each imaging event. Reproduced with permission from Ref. [12]

report the cells' location, number, etc. These gene products can produce signal directly such as green fluorescent protein in optical imaging or can produce a secondary reaction with an exogenous label introduced at the time of imaging. For example, the herpes simplex virus type 1 thymidine kinase (HSV1-tk) can selectively phosphorylate substrates that are tagged for imaging with positron emission tomography (PET). This phosphorylation causes selective probe accumulation only in cells that carry the reporter gene, and thus stem cells can be imaged because of the increased levels of this reporter. The luciferase/luciferin reaction in bioluminescent imaging is another example of the need to inject a substrate. The advantages of indirect imaging are that the daughter cells normally receive a copy of the reporter gene and thus the signal is "on" only when the cell is viable. The disadvantage is that this requires reprogramming of the cells' genome—something that is very difficult to clinically translate. Thus, the majority of regenerative medicine work reported in humans has used direct labels.

Cell-Based Therapy in History

Some of the first reported instances of cell-based therapy were done to counteract the aging process. The French physician Charles-Édouard Brown-Séquard was known to inject pulverized animal testicles into human subjects. Unfortunately, little long-term impact was ever noted. In 1931 Dr. Paul Niehans treated an athymic patient with cells from a bovine thyroid. Although these therapies were also failures because of immune differences between species, his work was visionary, and he described "a method of treating the whole organism on a biological basis, capable of revitalizing the human organism with trillions of cells by bringing to it those embryonic or young cells which it needs." During the 1950s and 1960s bone marrow transplantation became increasingly sophisticated and matured from grafts between identical twins to grafts between siblings as the knowledge about graft-versus-host disease increased. The first transplant between unrelated persons was in 1973, which led in time to the transplantation of entire organs. The current state of the art is *stem cell* therapy, which uses cells capable of diverse lineages to repair tissue.

In direct labeling, cells are tagged with small molecules including radioisotopes, fluorophores, and nanoparticles. These can be added during expansion in tissue culture or immediately before injection. Transfection reagents may be used to increase the efficiency of label uptake. The labels can either be on the cell surface or inside the cell. One advantage of intracellular labeling is that there is a reduced chance of the label becoming disassociated from the cell and contributing to artificially high background or erroneous signal.

Direct labeling is attractive because it is simple and straightforward to control the dose of contrast with short processing times [12]. The major limitations are dilution of the concentration of contrast with successive cell division—that is, each daughter cell only has 50 % of the amount of label as the parent cell (Fig. 1). In addition, these labels are usually "always on." They will report the presence of cells even if the cells are dead. These direct labels can also be taken up by macrophages after cells have died, which can also contribute to inaccurate cell counts.

Common examples include lipophilic fluorophores for optical imaging and (carboxy)dextran-coated super paramagnetic iron oxide (SPIO) nanoparticles such as Feridex® and Resovist® for MRI [13]. While cell loading is traditionally done ex vivo, one interesting report showed that i.v.-injected SPIO nanoparticles can accumulate in the bone marrow and label stem cells in the bone marrow in vivo through the reticuloendothelial system [14]. Fluorescent dyes have value in small animal models, but humans have too much optical scatter for direct optical labels. Radionuclides include fluorodeoxyglucose (^{18}F-FDG) in PET and ^{111}In oxine for single-photon emission computed tomography (SPECT), but one limitation of nuclear imaging methods is that the radioisotope decays over time making it difficult to perform longitudinal scans using radionuclides. More detailed descriptions of direct labels can be found elsewhere [15].

One area of imaging that has been significantly overlooked for cell tracking is ultrasound, which is somewhat surprising because ultrasound offers good temporal resolution and is widely available—features congruous with the needs of the stem cell imager. The balance of this chapter focuses on the use of ultrasound in stem cell tracking and describes the basis of contrast in ultrasound imaging as well as the types of ultrasound labels, examples of direct and indirect labeling with ultrasound, and some perspectives on future growth in the field.

2 The Rationale for Ultrasound Imaging

Ultrasound imaging offers many advantages that are useful to studying stem cell therapy. First, ultrasound is very accessible and affordable. It is by far the most common piece of imaging equipment worldwide from small rural clinics to major research university hospitals. Second, ultrasound offers spatial resolution advantages (~50 μm) that are useful to identify subtle differences in treated tissue. Third, ultrasound offers excellent temporal resolution (up to 1000 frames per second)—this is critical for instantaneous knowledge of the cell location and the cell number. Fourth, ultrasound data can be quantitative, which is critical for identifying not only the presence of the cells, but also their number. Fifth, ultrasound offers a broad portfolio of complementary imaging sequences that can be used to enhance the cell data. That is, the ultrasound can collect imaging data about the surrounding anatomy and tissue behavior that complements the functional information from the cells. This includes motion mode (M-mode) imaging that studies repetitive motion such as the heart chamber [16], Doppler imaging which monitors the direction of movement [17], and various quantification schemes that can be used to estimate organ size or cardiac behavior including the left ventricle ejection fraction [18]. However, ultrasound images can also be very difficult to interpret due to the high

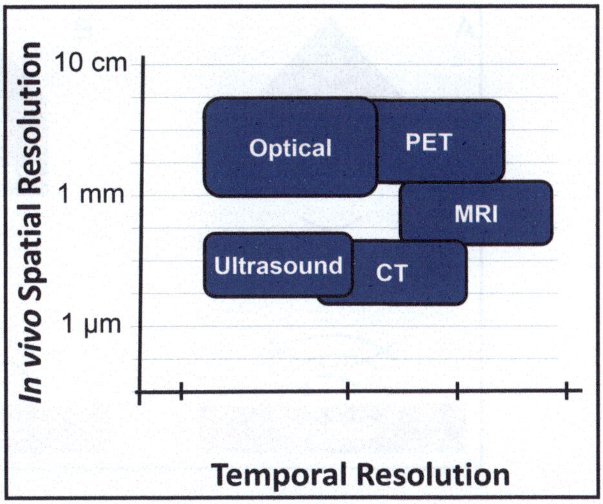

Fig. 2 Performance features of various imaging modalities. The temporal resolution (time between images) and spatial resolution (distance between points that can be distinguished) are plotted for different imaging modalities. Ultrasound offers good temporal and spatial resolution

background noise and/or complicated acoustic properties of different tissues. This is especially problematic in stem cell imaging—thus, it is critical to use either a direct or an indirect imaging technique to increase the cell-specific contrast.

The features of ultrasound and a comparison to other techniques used for stem cell tracking are shown in Fig. 2. Ultrasound is particularly powerful because of its high temporal resolution. This allows nearly instantaneous readout of the features of interest including the cell location and cell number. On the *y*-axis in Fig. 2, we plot the spatial resolution—how fine of an image can be created or the smallest distance between two objects that can be resolved. In ultrasound the spatial resolution is a function of the frequency used to create the images. At 70–100 MHz, the resolution can be as high as tens of micrometers, while at clinical frequencies (2–10 MHz), the spatial resolution is much lower—hundreds of microns to millimeters. As a trade-off, lower frequencies do offer better penetration through tissue—clinical frequencies can easily penetrate up to 25 cm into human beings depending on the tissue type, while higher preclinical frequencies (used in rodent models of human disease) are often limited to 2–3 cm of tissue. Importantly, the temporal resolution (frames per second) does not change as a function of frequency.

2.1 *Ultrasound Mechanism and Types of Ultrasound*

Ultrasound imaging in vivo is not entirely different than the approach used by bats to "see in the dark." Ultrasound imaging uses a tool called a transducer (Fig. 3a). The transducer is simply a tool that both emits and receives ultrasound pressure waves. As the emitted sound wave passes through tissue it is scattered and reflected (echoed) back to the transducer. The image is created by interpreting the backscattered sound and direction of sound as well as the speed of sound, the time of emission, the

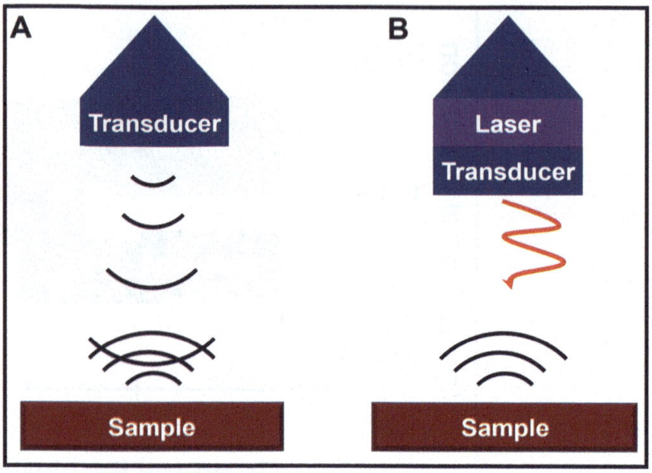

Fig. 3 Ultrasound mechanism. The mechanism of contrast in ultrasound. Traditional B-mode imaging (**a**) uses backscattered pressure waves to generate contrast. In photoacoustic imaging (**b**), an incident light is absorbed by the target tissue. Once absorbed the target heats and swells creating pressure differences that can be detected acoustically

time of arrival back at the transducer, and the angle of return. The reconstructed image thus reports the distance to the object, the size of the object, and its density (or impedance mismatch with the surrounding tissue). Most ultrasound images are two dimensions, but 3D ultrasound is possible by moving the transducer over the surface to be imaged (like a panorama shot on a mobile phone camera). Ultrasound can also be done in Doppler mode in which changes in the sound wave's pitch and phase are used to gain even more information. This is analogous to the sound differences when a siren is moving towards you and away from you. This can be used to determine blood flow or study other movement events inside the body.

In addition to Doppler ultrasound, photoacoustic ultrasound is another important subtype of ultrasound imaging. In the photoacoustic effect ultrasound waves are created due to incident light pulses on the tissue (Fig. 3b). That is, regular ultrasound is "sound in—sound out," and photoacoustic imaging is "light in—sound out." Alexander Graham Bell originally described the photoacoustic effect, but it has not been until recent years that people have used it for imaging because the transducers have become more sensitive and lasers have been developed with very short pulse lengths.

The mechanism is based on optical absorption—when the light is absorbed by the target tissue, the target heats, and swells. This thermal expansion then generates pressure waves that can be detected acoustically. The fundamental advantage of photoacoustic imaging is that it combines the high temporal and spatial resolution of ultrasound with the good contrast and spectral nature of optical imaging. Photoacoustic imaging can use both exogenous absorbers such as hemoglobin, deoxyhemoglobin, and melanin or artificial contrast agents targeted to site of interest or a cell of interest.

A third type of ultrasound imaging is contrast-enhanced ultrasound. This uses an exogenous agent such as a perfluorocarbon microbubble to artificially increase the ultrasound signal at the site of interest. Contrast-enhanced ultrasound can move beyond the anatomical images that are created with traditional imaging (bone, muscle, etc.) to imaging protein expression levels or specific stem cell types. In the following section, we discuss some of the types of ultrasound contrast agents used as well as their applications in stem cell therapy. Because of the performance features of ultrasound it is somewhat surprising that there are relatively few reports of this technology in the literature. We will present these case studies in chronological order and highlight various examples of direct and indirect imaging via ultrasound.

2.2 *Ultrasound Imaging of Cardiac Stem Cell Therapy*

Heart disease is the most common cause of death in the developed world. Stem cell therapy has shown encouraging initial results in treating heart disease [4], and yet is plagued by poor long-term efficacy because of poor cell viability after implantation [5, 19]. This is attributed to two fundamental challenges—(1) ischemia and inflammation in the treated area [19], and (2) mis-injection or implant into highly fibrotic tissue [20, 21]. Because of these barriers, one example showed that fewer than 2 % of implanted dendritic cells remain viable after 4–8 weeks [21] with cells mis-injected in 50 % of patients [22]. Ultrasound is ideally suited to solve these challenges with poor delivery. Physicians already use ultrasound to image the catheter used for delivery of cells into the cardiac tissue, but the cells have a low impedance mismatch with the surrounding tissue and thus have low contrast.

Ultrasound is particularly attractive in cardiac stem cell therapy of the heart because it can be combined with the established use of echocardiography. Echocardiography is also known as a "cardiac echo" or just an "echo," and it is used in the diagnosis and prevention of disease. Output parameters from echocardiography include the size and the shape of the ventricles and the left ventricle ejection fraction, which allows physicians to see how the heart chambers work in synchrony. This can be complemented by Doppler imaging to understand blood flow rates and directions.

In 2005, Rodriguez-Porcel and coworkers showed that ultrasound has fundamental utility in stem cell therapy [20]. Whereas prior work up to this point had performed open chest injection into the cardiac muscle, this study used ultrasound guidance in the parasternal long-axis view to deliver cells with a 28-gauge catheter. The cells were stably transfected cardiomyoblasts (plasmid–cytomegalovirus–firefly luciferase), and the target site was the anterior cardiac wall. The advantage of this approach is that the surgery is much less invasive and the surgeon can see the location of the catheter in real time. These researchers performed this work in 11 rats and then confirmed cell delivery with the bioluminescence reporter embedded in the cells. The bioluminescent signal was positively correlated with the number of cells transplanted ($R^2=0.94$, $P=0.03$).

One limitation of this work was the poor ultrasound signal from the cells. That is, the investigators could only know the location and number of the cells with downstream analysis via bioluminescence. Thus, in 2006 Bara and colleagues [23] used ultrasound not only to image the injection but also to image the cells (Fig. 4) [23]. Here, cells were labeled with a clinically approved formulation of iron oxide (CliniMACS). These materials are used for magnetic based cell separation and consisted of a 50 nm iron oxide core coated with a dextran shell that is finally annealed to a monoclonal antibody specific to the target of interest. The authors selected CD133 because it is a marker of hematopoietic stem cells. After purifying CD133+ stem cells from bone marrow aspirates, cell identify and purity were confirmed with CD34 flow cytometry.

These cells were delivered to swine models of human ischemia created via ligation/reperfusion [24]. The swine received either stem cells (5 million) or a sham injection of saline. Imaging used transesophageal echocardiography, which is a specialized type of ultrasound imaging that places a transducer in the throat of the subject to increase resolution versus ultrasound imaging via the chest wall. Figure 4 presents data including cells without any label (Fig. 4a), the ClinicMACS particles alone (Fig. 4b), as well as an animal before (Fig. 4c) and after treatment

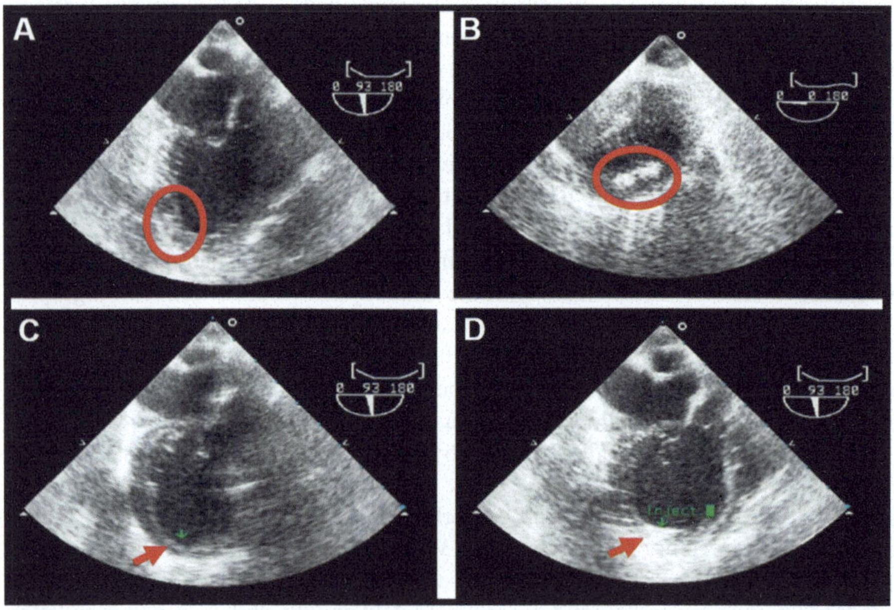

Fig. 4 Direct imaging with magnetic particles. Swine hearts are imaged at various stages of stem cell therapy. In each image, the dark oval in the center is the left ventricle. Note the clear boundary between the interior and exterior of the heart wall in **c** (between *red* and *green arrows*). Panel **a** shows injection of unlabeled cells in the area highlighted with red circle. Little signal increase is seen. When the nanoparticle contrast agent alone is injected, hyperechoic regions are seen (**b**). Panels **c** and **d** are pre- (**c**) and post- (**d**) injection images of animals treated with five million nanoparticle-labeled stem cells. Reproduced with permission from Ref. [25]

with labeled cells (Fig. 4d). The hypoechoic (bright) areas in Fig. 4b, d correspond to the increased sound backscatter due to the particles. The authors showed that the labeled cells had more signal than unlabeled cells but had challenges with long-term cell tracking. They concluded that this technique was perhaps best suited for the delivery event and that a complementary technique might be superior for long-term tracking.

In 2013, Jokerst et al. extended this to multimodal imaging with MRI and ultrasound. In contrast to the magnetic nanoparticles used by Bara [23], this group used silica nanoparticles as ultrasound contrast agents [25, 26]. These have a key advantage in that the nanoparticles can be loaded *inside* the stem cells whereas the magnetic particles were on the cell *exterior*. This is important because contrast agents on the cell exterior can easily become detached in vivo and lead to erroneous signal. The silica nanoparticles are also triple-modality agents—ultrasound signal is generated via impedance mismatch of the silica, optical imaging is enabled by an embedded fluorophore, and T1-weighted imaging is possible via chelated gadolinium (Fig. 5) [27]. Stem cells could then be labeled with these materials for both instant imaging at the time of delivery via ultrasound and long-term follow-up with MRI.

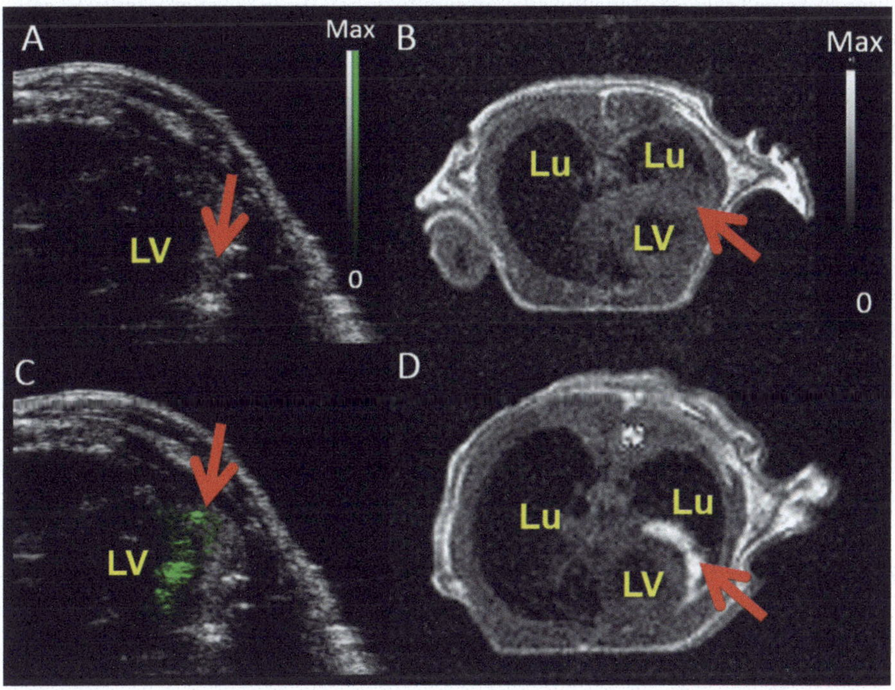

Fig. 5 Direct imaging with nanoparticles. Ultrasound (**a**) and MRI (**b**) images were collected before injection of 500,000 human mesenchymal stem cells in a rodent model. After injection, the animals were imaged again with obvious increases in signal indicating the presence of the cells on ultrasound (**c**) and MRI (**d**). Reproduced with permission from Ref. [29]

Once labeled, the nanoparticles increased the ultrasound and MRI contrast of labeled human mesenchymal stem cells 700 % and 200 %, respectively. The authors investigated the behavior of the agent on cell metabolic activity, proliferation, or pluripotency, but did not find any significant change. Electron microscopy and ultrasound imaging suggest that the mechanism of action is in vivo aggregation of the 300 nm silica nanoparticles into larger silica frameworks that amplify the ultrasound backscatter. Detection limits in cardiac tissue were 250,000 cells via MRI and 70,000 via ultrasound with cell imaging possible in animal models for 13 days after implantation (Fig. 5). This ultrasound-guided cell delivery and multimodal optical/ultrasound/MRI intracardiac cell-tracking platform could improve cell therapy in the clinic by minimizing mis-delivery or implantation into fibrotic tissue. Moreover, these materials were compatible with both preclinical and clinical ultrasound frequencies.

More recently, this same group has extended this work and developed ultrasound contrast agents that can not only produce ultrasound signal, but can also deliver pro-survival agents to increase stem cell survival and then biodegrade after the delivery and imaging tasks are complete. This was performed largely by transitioning from solid nanoparticles to mesoporous silica nanoparticles (MSNs) [28]. The MSNs still have an impedance mismatch between tissue and the silica particles, but also offer a high surface area (~1000 m^2/g) suitable for sustained release of drug [29, 30]. MSNs offer sustained release of insulin-like growth factor (IGF) in close proximity to cells at high local concentrations. They also allow imaging through multimodal approaches, which include MRI, optical, and ultrasound. Therefore, this approach allows IGF to be delivered to the cell surface membrane where the IGF receptors are located—this stimulates cell growth in the hypoxic and necrotic region of therapy.

Mesenchymal stem cells labeled with these nanoparticles had detection limits near 9000 cells with no cytotoxicity at the 250 μg/mL concentration required for labeling. Degradation studies showed that the nanoparticles clear from cells in approximately 3 weeks. The presence of IGF increased cell survival up to 40 % ($P<0.05$) versus unlabeled cells under in vitro serum-free culture conditions. The degradation time is important because MSNs have to be stable for imaging, but not so stable that they never clear from the body. One limitation of this approach is that it is not a true marker of cell viability. It continuously produces a signal whether the cell is alive or not. This problem could be counteracted by conducting further studies utilizing a reporter gene [31].

2.3 Ultrasound Tracking of Neural Progenitor Cells

Another method that can allow the utilization of ultrasound for cell tracking is by introducing microbubbles into neural progenitor cells (NPCs). Modified NPCs have previously been proposed as a way to treat genetic disorders or to target tumors, infarction, or inflammation due to their ability to disperse bioactive molecules [32–34]. In 2013, Wenjin Cui and coworkers reported that ultrasound microbubbles could be used as a means to track NPCs in vivo [35].

The authors labeled the cells using positively charged perfluorocarbon microbubbles and the negatively charged cell membrane. Transfection efficiency and cell viability were both greater than 90 %. Detection limits down to a single cell at 7 MHz were possible. The microbubble-labeled NPCs were also more resistant to ultrasound exposure showing that these were still detectable on day 7 versus the hours of in vivo stability for intravenously injected microbubbles. These results were due to the microbubbles associating themselves with the cell surfaces and being internalized. Yet the internalized microbubbles could still behave nonlinearly in ultrasound, produce harmonic signals, and be destroyed but at a higher pressure than free microbubbles. The ability to detect a single cell using microbubble-labeled cells and ultrasound is advantageous when compared to current proposed in vivo cell tracking techniques (PET and bioluminescence), which usually require thousands of cells.

One limitation of this study was that there might have been a loss of signal in vivo due to possible cell migration out of the liver. The team carried out tests to monitor the accumulation after intravenous injection in the livers of mice. Eight mice were injected with the microbubble-labeled NPCs. Four were injected with the free microbubbles. The microbubble-labeled NPCs were still visible at day 5, but the free microbubbles were gone by 8 h. The NPCs may have leaked to other areas leading to the decay of signal. This cannot be completely confirmed because the authors only studied the liver. Some microbubbles may have been released upon cell death. Nevertheless, this long lifetime is very useful for both real-time imaging of implantation and longitudinal stem cell tracking.

2.4 Indirect Imaging with Ultrasound

All of the above examples use direct imaging—that is, a nanoparticle or microbubble directly bound to the stem cell. However, indirect imaging offers many advantages including true representation of cell viability. One exciting report of an ultrasound reporter genes [36] used biogenic gas vesicles. These vesicles help bacteria maintain their proper depth in water for photosynthesis. Researchers isolated the genes responsible for these vesicles and loaded them into cells. These materials are protein-shelled compartments with typical widths of 45–250 nm and lengths of 100–600 nm. They exclude water but are gas permeable.

There are very few applications of indirect ultrasound imaging in regenerative medicine. One report in 2009 by Kuliszewski and coworkers described a method to utilize ultrasound for indirect cell imaging in vivo (Fig. 6) [37]. Here, cells were programmed to make a specific cell surface protein H-2Kk. Then microbubbles targeted to this protein were injected intravenously. Because the cell surface protein was only on the stem cells, specific signal could be obtained. The stem cells used in this application were endothelial progenitor cells (EPCs), and Kuliszewski hypothesized that the targeted microbubbles would bind to cells in vivo. The microbubbles were conjugated with a biotinylated anti-H-2Kk antibody and then injected for in vivo EPC imaging. The EPCs were implanted into rats and for ultrasound

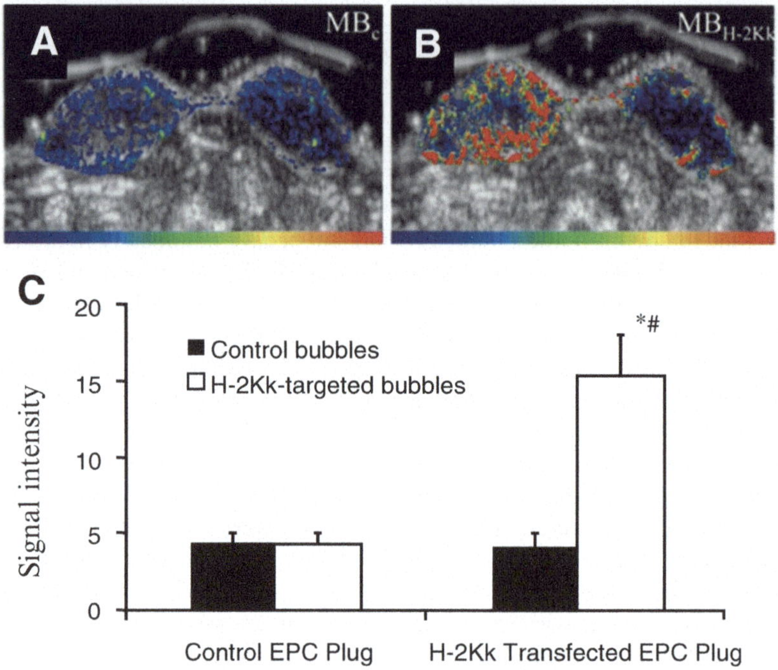

Fig. 6 Indirect imaging. (**a**) H-2Kk-transfected EPCs (*left* implant) and EPC cells (*right* implant) and treated with control, non-targeted bubbles do not show accumulation. (**b**) However, when H-2Kk-targeted bubbles are injected, the H-2kK-expressing cells specifically bind the bubbles (left implant; panel **b**). There is only background signal in the other cell types. This is further quantified in panel **c**. * and # indicate $P<0.005$. Reproduced with permission from Ref. [39]

imaging. After 1 week, imaging showed perfusion within the cell areas and a strong signal for microbubbles with H-2Kk in the H-2Kk-transfected EPC-supplemented plug (Fig. 6). Negative controls including non-targeted microbubbles and cells not expressing H-2Kk were negative. This finding shows that microbubbles targeted to an engineered cell-surface marker on EPCs can be tracked with ultrasound.

Limitations include the use of electroporation for EPC transduction. The H-2Kk gene expression was also transient. The reduction in EPC transduction was due to active cell division during culture. Another limitation is that the in vivo model of the matrigel angiogenesis may not accurately recapitulate clinical cell therapy due to inefficient cell delivery methods and a gradual decrease in the number of engrafted EPCs over time. In addition, the use of microbubbles is limited to vascular targets.

3 Photoacoustic Imaging of Stem Cell Therapy

Another useful way to track cells in vivo is to utilize photoacoustic imaging using an exogenous contrast agent. These agents produce ultrasound waves through a pressure difference caused by rapid heating from a nanosecond pulse of light on the sample. This technique complements traditional ultrasound and increases contrast. It is noninvasive and quantitative, and has quick scan times. Multiple groups have used photoacoustic imaging for cell tracking [38–40] and cancer stem cells [41, 42].

In one example, Jokerst et al. reported the use of silica-coated GNRs (SiGNRs) as a photoacoustic contrast agent and combined it with ultrasound backscatter mode imaging (Fig. 7) [38]. Stem cells were labeled with the SiGNRs, and imaged in mice, and the effect of the SiGNRS on cell viability, differentiation, and cytokine expression was measured. The silica coat played two important roles—the photoacoustic signal of the GNRs was strengthened approximately fourfold confirming Emelianov's earlier report [43]. This signal increase remained stable for at least 60 days after coating. The second important role of silica was to increase accumulation of the contrast agent in the cell more the bare gold nanorods. Transmission electron microcopy images clearly illustrate SiGNRs (Fig. 7a) inside of stem cells (Fig. 7b). Photoacoustic imaging with B-mode ultrasound imaging provides an image with clearly defined anatomic features with photoacoustic data that contain cell-specific

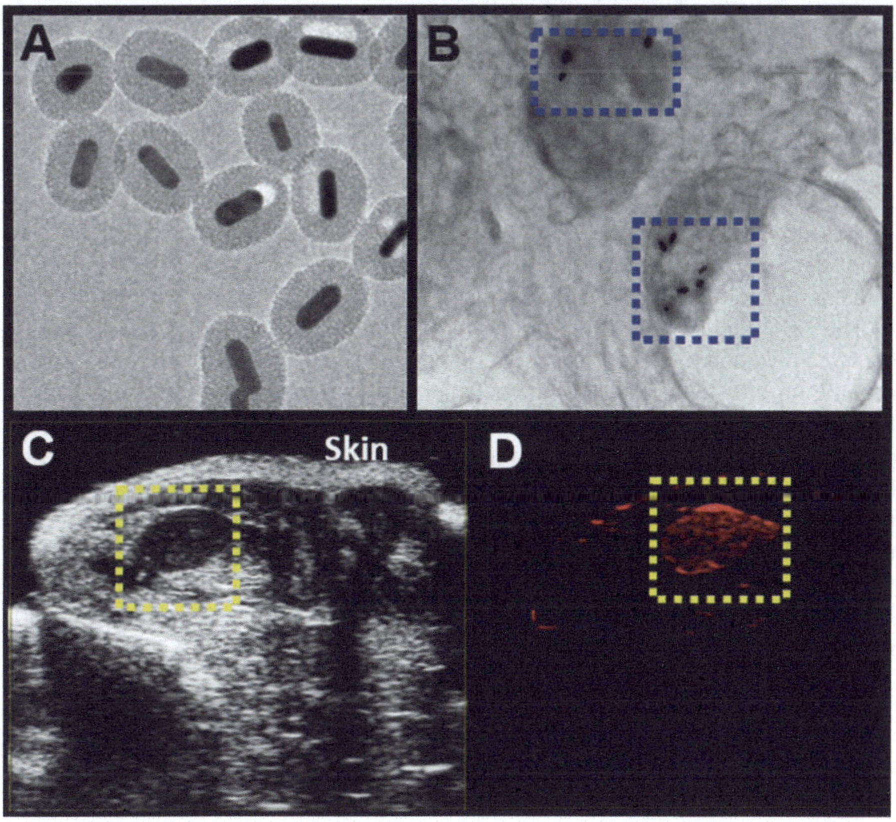

Fig. 7 Photoacoustic imaging. (**a**) Silica-coated gold nanorods are photoacoustic contrast agents that can be loaded into stem cells (**b**). (**c**) Ultrasound imaging provides details on anatomy including muscle, bone, and skin. Photoacoustic ultrasound (**d**) offers increased contrast unique to the stem cells (*dashed yellow box*). Reproduced with permission from Ref. [40]

content (Fig. 7c, d). Limitations include optical scatter due to the nature of tissue as well as inaccuracies with reconstructions. This could be solved by using a photoacoustic catheter or endoscope for deep-tissue implantation.

In another application, Nam and coworkers imaged stem cells labeled with gold nanotracers [40]. The labeled stem cells were mixed into a fibrin gel and injected into the lateral gastrocnemius muscle of a rat 5 mm beneath the skin. Ultrasound/photoacoustic images were taken and clearly show the location of the nanotracer signal of the cells indicating high sensitivity. In vivo detection limits were 4.5×10^9 nanoparticles/mL. This facilitates high-sensitivity imaging.

4 Perspectives and Conclusions

Ultrasound imaging offers many advantages to the stem cell imager. It has high temporal and spatial resolution and is also very affordable. Portable units can be purchased for below $10,000. This makes it applicable both to sophisticated research hospitals and rural or private practice clinics. As the field of regenerative medicine continues to expand, the number of people using ultrasound for cell imaging will likely increase.

We will also likely see other applications in different organ systems. Ultrasound is routinely used in many abdominal applications including imaging of the liver, spleen, and bowels. Regenerative medicine applications are likely in these areas, and ultrasound may be a very useful tool to investigators in the field. Ultrasound is also used to study joints including the knees. There are clinical trials under way using stem cells to regenerate cartilage to prevent joint pain. Ultrasound is ideally suited to understand the location and distribution of stem cells in the joints.

The community will also continue to see multimodal applications. As shown above, ultrasound and MRI are very complementary techniques in terms of tissue penetration and temporal resolution. Other groups have shown ultrasound combined with Raman imaging, which is a special optical technique with exquisite sensitivity and spectral character, but very slow temporal resolution. Thus, these two techniques can be used in tandem to minimize the relative disadvantages of each technique. Finally, researchers will continue to study ways to increase contrast in ultrasound. While photoacoustics is promising, it remains a body surface-weighted technique. Thus, magneto-acoustic or radio-frequency acoustic techniques will hopefully improve for use in humans. The interested reader is encouraged to learn more using these helpful texts [11, 44, 45].

References

1. Daar AS, Greenwood HL. A proposed definition of regenerative medicine. J Tissue Eng Regen Med. 2007;1(3):179–84.
2. Mason C, Dunnill P. A brief definition of regenerative medicine. Regen Med. 2008;3(1):1–5.

3. Lane SW, Williams DA, Watt FM. Modulating the stem cell niche for tissue regeneration. Nat Biotechnol. 2014.
4. George JC. Stem cell therapy in acute myocardial infarction: a review of clinical trials. Transl Res. 2010;155(1):10–9.
5. Bolli R, Chugh AR, D'Amario D, Loughran JH, Stoddard MF, Ikram S, et al. Cardiac stem cells in patients with ischaemic cardiomyopathy (SCIPIO): initial results of a randomised phase 1 trial. Lancet. 2011;378(9806):1847–57.
6. Gilbert PM, Havenstrite KL, Magnusson KE, Sacco A, Leonardi NA, Kraft P, et al. Substrate elasticity regulates skeletal muscle stem cell self-renewal in culture. Science. 2010;329(5995):1078–81.
7. Tedesco FS, Dellavalle A, Diaz-Manera J, Messina G, Cossu G. Repairing skeletal muscle: regenerative potential of skeletal muscle stem cells. J Clin Invest. 2010;120(1):11.
8. Wakitani S, Okabe T, Horibe S, Mitsuoka T, Saito M, Koyama T, et al. Safety of autologous bone marrow-derived mesenchymal stem cell transplantation for cartilage repair in 41 patients with 45 joints followed for up to 11 years and 5 months. J Tissue Eng Regen Med. 2011;5(2):146–50.
9. Orlando G, Baptista P, Birchall M, De Coppi P, Farney A, Guimaraes-Souza NK, et al. Regenerative medicine as applied to solid organ transplantation: current status and future challenges. Transpl Int. 2011;24(3):223–32.
10. Reya T, Morrison SJ, Clarke MF, Weissman IL. Stem cells, cancer, and cancer stem cells. Nature. 2001;414(6859):105–11.
11. Kircher MF, Gambhir SS, Grimm J. Noninvasive cell-tracking methods. Nat Rev Clin Pract. 2011;8(11):677–88.
12. Atala A, Allickson J. Translational regenerative medicine. Burlington: Elsevier Science; 2014. ScienceDirect. Restricted to UC campuses http://uclibs.org/PID/267828
13. Bulte JW, Kraitchman DL. Iron oxide MR contrast agents for molecular and cellular imaging. NMR Biomed. 2004;17(7):484–99.
14. Khurana A, Chapelin F, Beck G, Lenkov OD, Donig J, Nejadnik H, et al. Iron administration before stem cell harvest enables mr imaging tracking after transplantation. Radiology. 2013;269(1):186–97.
15. Fu Y, Kraitchman DL. Stem cell labeling for noninvasive delivery and tracking in cardiovascular regenerative therapy. Expert Rev Cardiovasc Ther. 2010;8(8):1149–60.
16. Bunce S, Moore A, Hough A. M-mode ultrasound: a reliable measure of transversus abdominis thickness? Clin Biomech. 2002;17(4):315–7.
17. Adler DD, Carson PL, Rubin JM, Quinn-Reid D. Doppler ultrasound color flow imaging in the study of breast cancer: preliminary findings. Ultrasound Med Biol. 1990;16(6):553–9.
18. Paulus WJ, Tschöpe C, Sanderson JE, Rusconi C, Flachskampf FA, Rademakers FE, et al. How to diagnose diastolic heart failure: a consensus statement on the diagnosis of heart failure with normal left ventricular ejection fraction by the Heart Failure and Echocardiography Associations of the European Society of Cardiology. Eur Heart J. 2007.
19. Nguyen PK, Lan F, Wang Y, Wu JC. Imaging: guiding the clinical translation of cardiac stem cell therapy. Circ Res. 2011;109(8):962–79.
20. Rodriguez-Porcel M, Gheysens O, Chen IY, Wu JC, Gambhir SS. Image-guided cardiac cell delivery using high-resolution small-animal ultrasound. Mol Ther. 2005;12(6):1142–7.
21. Vunjak-Novakovic G, Lui KO, Tandon N, Chien KR. Bioengineering heart muscle: a paradigm for regenerative medicine. Annu Rev Biomed Eng. 2011;13:245–67.
22. de Vries IJ, Lesterhuis WJ, Barentsz JO, Verdijk P, van Krieken JH, Boerman OC, et al. Magnetic resonance tracking of dendritic cells in melanoma patients for monitoring of cellular therapy. Nat Biotechnol. 2005;23(11):1407–13.
23. Bara C, Ghodsizad A, Niehaus M, Makoui M, Piechaczek C, Martin U, et al. In vivo echocardiographic imaging of transplanted human adult stem cells in the myocardium labeled with clinically applicable CliniMACS nanoparticles. J Am Soc Echocardiogr. 2006;19(5):563–8.
24. Kompa AR, Summers RJ. Lidocaine and surgical modification reduces mortality in a rat model of cardiac failure induced by coronary artery ligation. J Pharmacol Toxicol Methods. 2000;43(3):199–203.

25. Casciaro S, Conversano F, Ragusa A, Ada Malvindi M, Franchini R, Greco A, et al. Optimal enhancement configuration of silica nanoparticles for ultrasound imaging and automatic detection at conventional diagnostic frequencies. Invest Radiol. 45(11):715–23.
26. Martinez HP, Kono Y, Blair SL, Sandoval S, Wang-Rodriguez J, Mattrey RF, et al. Hard shell gas-filled contrast enhancement particles for colour Doppler ultrasound imaging of tumors. Med Chem Commun. 2010;1(4):266–70.
27. Jokerst JV, Khademi C, Gambhir SS. Intracellular aggregation of multimodal silica nanoparticles for ultrasound-guided stem cell implantation. Sci Transl Med. 2013;5(177):177ra35.
28. Kempen PJ, Greasley S, Parker KA, Campbell JL, Chang H-Y, Jones JR, et al. Theranostic mesoporous silica nanoparticles biodegrade after pro-survival drug delivery and ultrasound/magnetic resonance imaging of stem cells. Theranostics. 2015;5(6):631.
29. Slowing II, Trewyn BG, Giri S, Lin VSY. Mesoporous silica nanoparticles for drug delivery and biosensing applications. Adv Funct Mater. 2007;17(8):1225–36.
30. Liu J, Stace-Naughton A, Jiang X, Brinker CJ. Porous nanoparticle supported lipid bilayers (protocells) as delivery vehicles. J Am Chem Soc. 2009;131(4):1354–5.
31. Yaghoubi SS, Jensen MC, Satyamurthy N, Budhiraja S, Paik D, Czernin J, et al. Noninvasive detection of therapeutic cytolytic T cells with 18F–FHBG PET in a patient with glioma. Nat Clin Pract Oncol. 2008;6(1):53–8.
32. Brown AB, Yang W, Schmidt NO, Carroll R, Leishear KK, Rainov NG, et al. Intravascular delivery of neural stem cell lines to target intracranial and extracranial tumors of neural and non-neural origin. Hum Gene Ther. 2003;14(18):1777–85.
33. Silva GA, Czeisler C, Niece KL, Beniash E, Harrington DA, Kessler JA, et al. Selective differentiation of neural progenitor cells by high-epitope density nanofibers. Science. 2004; 303(5662):1352–5.
34. Wang L, Martin DR, Baker HJ, Zinn KR, Kappes JC, Ding H, et al. Neural progenitor cell transplantation and imaging in a large animal model. Neurosci Res. 2007;59(3):327–40.
35. Cui W, Tavri S, Benchimol MJ, Itani M, Olson ES, Zhang H, et al. Neural progenitor cells labeling with microbubble contrast agent for ultrasound imaging in vivo. Biomaterials. 2013;34(21):4926–35.
36. Shapiro MG, Goodwill PW, Neogy A, Yin M, Foster FS, Schaffer DV, et al. Biogenic gas nanostructures as ultrasonic molecular reporters. Nat Nanotechnol. 2014;9(4):311–6.
37. Kuliszewski MA, Fujii H, Liao C, Smith AH, Xie A, Lindner JR, et al. Molecular imaging of endothelial progenitor cell engraftment using contrast-enhanced ultrasound and targeted microbubbles. Cardiovasc Res. 2009;83(4):653.
38. Jokerst JV, Thangaraj M, Kempen PJ, Sinclair R, Gambhir SS. Photoacoustic imaging of mesenchymal stem cells in living mice via silica-coated gold nanorods. ACS Nano. 2012;6(7):5920–30.
39. Wang C, Ma X, Ye S, Cheng L, Yang K, Guo L, et al. Protamine functionalized single-walled carbon nanotubes for stem cell labeling and in vivo raman/magnetic resonance/photoacoustic triple-modal imaging. Adv Funct Mater. 2012;22(11):2363–75.
40. Nam SY, Ricles LM, Suggs LJ, Emelianov SY. In vivo ultrasound and photoacoustic monitoring of mesenchymal stem cells labeled with gold nanotracers. PLoS One. 2012;7(5), e37267.
41. Galanzha EI, Kim J-W, Zharov VP. Nanotechnology-based molecular photoacoustic and photothermal flow cytometry platform for in vivo detection and killing of circulating cancer stem cells. J Biophotonics. 2009;2(12):725.
42. Hu X, Wei CW, Xia J, Pelivanov I, O'Donnell M, Gao X. Trapping and photoacoustic detection of CTCs at the single cell per milliliter level with magneto-optical coupled nanoparticles. Small. 2013;9(12):2046–52.
43. Chen YS, Frey W, Kim S, Kruizinga P, Homan K, Emelianov S. Silica-coated gold nanorods as photoacoustic signal nanoamplifiers. Nano Lett. 2011;11(2):348–54.
44. Atala A, Allickson J. Translational regenerative medicine. London, UK: Academic Press; 2014.
45. Wang J, Jokerst JV. Stem cell imaging: tools to improve cell delivery and viability. Stem Cell Int. 2016. In press.

Nanoparticles for Photoacoustic Imaging of Cancer

Katheryne E. Wilson, Keerthi S. Valluru, and Jürgen K. Willmann

1 Introduction

In this chapter, we present the basics of photoacoustic imaging with an overview of the types of nanoparticles specifically used in the detection of cancer drawing from our own experience and the work of others in the field. Photoacoustic imaging is a highly complementary modality to ultrasound imaging and has potential for several clinical applications [1]. The contrast in photoacoustic imaging is based on optical absorption. Therefore, agent-free photoacoustic imaging visualizes intrinsic tissue chromophores (light absorbing molecules) such as oxygenated and deoxygenated hemoglobin, melanin, and lipid. However, for other applications, such as examining cellular expression of specific cancer markers, the use of exogenous contrast agents is required. Several types of nanoparticles (noble metal, carbon, and mixed composition) have been developed as photoacoustic contrast agents due to their strong optical absorption characteristics. Here, we review some of the notable examples for the various types of nanoparticles for photoacoustic imaging of cancer.

2 Basics of Photoacoustics

The photoacoustic (interchangeably "optoacoustic") effect was first described in 1880 by Alexander Graham Bell when he noticed that sound waves were produced from an absorbing media after being exposed to sunlight transmitted through a rapidly rotating slotted wheel [2]. Furthermore, he noticed that the emitted sound waves changed between different materials and different frequencies of light.

K.E. Wilson, Ph.D. • K.S. Valluru, M.S. • J.K. Willmann, M.D. (✉)
Department of Radiology, School of Medicine, Stanford University,
300 Pasteur Drive, Stanford, CA 94305, USA
e-mail: willmann@stanford.edu

J.W.M. Bulte, M.M.J. Modo (eds.), *Design and Applications of Nanoparticles in Biomedical Imaging*, DOI 10.1007/978-3-319-42169-8_15

These experiments provided the basis upon which modern photoacoustics are built. Formally, the photoacoustic effect is the formation of sound waves after light absorption within a media. For photoacoustic imaging, a pulsed laser is used to irradiate tissues. Upon optical absorption by either an endogenous or exogenous chromophore (a light-absorbing molecule), heat is generated and released to the immediately surrounding environment [3]. This environment undergoes localized, rapid thermal expansion. Upon contraction of the environment to the original state, a high amplitude, broadband acoustic wave is released [4–6]. These transients can be detected using standard ultrasound transducers and reconstruction techniques. As photoacoustic imaging relies on light for generating thermoelastic expansion and the ultrasound transducer for resolution, the method provides optical contrast at depths significantly deeper (up to 7 cm theoretically [5, 6]) than purely optical imaging methods. Furthermore, since optical absorption is dependent on molecular characteristics, the modality can be used as an inherently molecular methodology, which is highly complementary to the anatomical (B-mode) and functional (contrast enhanced and Doppler) imaging provided by clinical ultrasound imaging.

Photoacoustic signal is dependent on the number of photons absorbed and the ability of the media to convert that absorption to heat. Specifically, photoacoustic signal generation can be described as:

$$P = F\mu_a\gamma \tag{1}$$

Where P is the photoacoustic signal intensity, F is the fluence of laser energy in mJ/cm^2, μ_a is the optical absorption coefficient of the absorptive media in cm^{-1}, and γ is the Grüneisen parameter, which describes the heat transfer characteristics of the media. Therefore, photoacoustic signal is linearly dependent on the laser fluence irradiated onto the sample and how absorptive the sample is. Generally, two conditions must be met for the effect to occur, namely thermal and stress confinements. Thermal confinement is met when the duration of the light pulse is shorter than the thermal relaxation time of the media, meaning the pulse duration must be shorter than the time it takes for the heat generated by absorption to dissipate. Stress confinement is met when the pulse duration is shorter than the stress relaxation time. Most laser pulses less than 10 ns will fulfill these requirements for photoacoustic imaging purposes.

2.1 Spectroscopic Photoacoustic Imaging

In biological tissue, the dominant optical absorber is hemoglobin in its two oxygenation states (see Fig. 1). As photoacoustic imaging contrast is highly dependent on optical absorption, this implies photoacoustic imaging is inherently suited for imaging blood. Several applications take advantage of this property such as imaging of tumor neoangiogenesis, monitoring antiangiogenic therapy, and imaging of ischemic events, to name a few [7–11]. However, examining other endogenous or

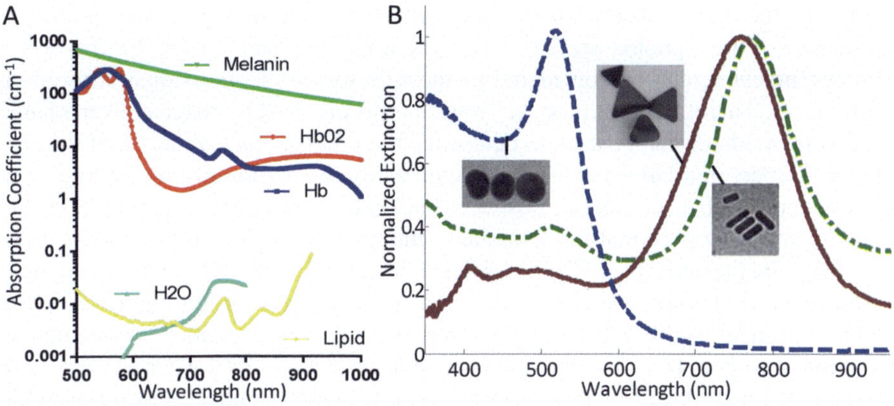

Fig. 1 Optical absorption of endogenous chromophores and exogenous nanoparticle contrast agents. (**a**) Absorption coefficients of several endogenous tissue chromophores. Melanin has the strongest optical absorption, but only occurs in the skin, and therefore hemoglobin in its two oxygenation states (HbO2—oxygenated hemoglobin, and Hb—deoxygenated hemoglobin) dominates photoacoustic images over lipid and water imaging. These spectra have local minimums in the 700–900 nm range, where (**b**) various nanoparticles have tunable peaks in their optical absorption including silver nanoplates (*brown line*) and gold nanorods (*green line*). Nanospheres (*blue line*) act as precursor nanoparticles to synthesize anisotropic shapes

exogenous absorbers beyond hemoglobin states, which is crucial to expand the cancer imaging application of photoacoustic imaging, can result in low signal to background ratios, as optical absorption of hemoglobin will dominate the photoacoustic signal production. One solution is to compare pre- and post-injection images of contrast agents to display signal only corresponding to the exogenous agents. However, using subtraction of images to remove background hemoglobin signal is prone to error due to physiological changes including tumor growth, changes in blood flow, and experimental setup, including laser fluence variations and imaging plane shifts [12]. Therefore, methods are needed to overcome these challenges and minimize background signal from hemoglobin when examining other endogenous or exogenous agents.

Spectroscopic photoacoustic imaging can be implemented to reduce the photoacoustic signal from blood and enhance the signal from the desired absorber by modeling images as linear combinations of absorbers. Optical absorption is dependent on the wavelength of light and the concentration of the photoabsorber. The absorption coefficient $\mu_a(\lambda,r)$, which is dependent on the wavelength "λ" and the spatial position "r," is defined as the product of the molar extinction coefficient ($\mathcal{E}$) and the concentration (C) of the photoabsorber as described [12, 13], providing the following Eq. (2):

$$\mu_a(\lambda,r)=C_1(r)\varepsilon_1(\lambda)+C_2(r)\varepsilon_2(\lambda)+\ldots C_n\varepsilon_n(\lambda) \tag{2}$$

Assuming the optical absorption that provides photoacoustic signal is proportional to the sum of the *n* photoabsorbers within the imaging plane, the localized absorption coefficient μ_a can be considered as the sum of the "*n*" individual absorption coefficients. Therefore, if at least "*n*" wavelengths are used to image a given plane, regression methods can be used to determine the relative concentrations of the "*n*" photoabsorbers in relative but arbitrary units within the region-of-interest based on the assumption that the photoacoustic signal (known variable as measured in the experiment) is proportional to the molar extinction coefficient "$\mathcal{E}$" (a known variable from the literature [14, 15]) times the concentration "*C*" of the individual photoabsorbers (unknown variable). The power of the regression improves with additional wavelengths and/or those that represent salient spectral characteristics of the photoabsorbers [13]. The method is most accurately applied when examining a mixture of photoabsorbers that have (1) relatively good optical absorption with hemoglobin being the standard; (2) distinctive spectral characteristics such as sharp peaks; and (3) absorption spectra that are highly distinguishable from each other.

2.2 *Label Free Spectroscopic Photoacoustic Imaging in Cancer*

Endogenous or label-free photoacoustic imaging, involves the examination of intrinsic tissue constituents such as hemoglobin, lipid, melanin, and water, which can be selectively imaged using spectroscopic photoacoustic imaging due to their strong absorption characteristics in the near infrared light region [15–19]. In cancerous tissue, angiogenesis leads to formation of new vasculature, thereby contributing to significant changes in concentration of hemoglobin within the tissue compared to healthy tissue with a subsequent increase in photoacoustic imaging signal [20, 21]. A multitude of studies taking advantage of intrinsic photoacoustic imaging contrast obtained from differences in the absorption spectra of tissue absorbers, such as oxygenated and deoxygenated hemoglobin, have been reported in the literature [22–27]. In a recent study [28], it has been demonstrated that as many as four different breast histologies, including normal ($n=82$), hyperplasia ($n=12$), ductal carcinoma in situ or DCIS ($n=96$), and invasive breast carcinoma ($n=61$), can be differentiated with reasonable accuracy based on the photoacoustic imaging contrast obtained from oxygen saturation, total hemoglobin, and lipid content in a transgenic mouse model (27 mice, $n=251$ mammary glands). With the help of a multiparametric analysis, it was shown that oxygen saturation performed best to differentiate clinically non-actionable (normal/hyperplasia) versus clinically actionable (DCIS and invasive breast carcinoma) findings. Furthermore, a reduction in lipid content was seen as the fat tissue being progressively replaced by a solid tumor during breast cancer development. However, the total hemoglobin was observed to be lower in DCIS and invasive breast carcinoma compared to normal and hyperplasic mammary glands in that study despite the general increase of vascularity in tumors. Therefore, due to complex biological processes, elucidating information concerning disease state based on those parameters only may be challenging.

Also, in instances where the blood vessels are still smaller compared to the detection sensitivity of the photoacoustic imaging system, which might often be the case in the early stages of cancer development, the intrinsic endogenous contrast may not be sufficient to detect any changes between cancer and surrounding healthy tissue. Moreover, not all cancers behave similarly, making it hard to estimate the type of change one might expect from the endogenous absorbers. For instance, will the concentration of lipid in cancer be more, or less compared to healthy tissue? Will different types of cancer show similar traits in absorption? Furthermore, the photoacoustic image contrast obtained from endogenous absorbers is not specific to the cancer or any other disease and may not be sufficient to provide a definitive diagnosis. As such, exogenous contrast agents, which can be synthesized to achieve a desired absorption, can be helpful in increasing the sensitivity and contrast-to-noise of photoacoustic imaging for cancer imaging. In addition, exogenous photoacoustic imaging agents, such as nanoparticles, that can interact with the tissue or bind to a cancer-specific biomarkers can also improve specificity, thereby allowing molecular imaging capabilities.

3 Nanoparticles for Photoacoustic Cancer Imaging

Nanoparticles are well suited for use as contrast agents, as their size is on scale with biological processes and cells, allowing visualization of the intricate processes that occur. These small sizes allow enhanced optical absorption, biodistribution, and clearance properties that can all be fine-tuned for specific applications. In the following sections, a brief introduction on general synthesis and bioconjugation methods of nanoparticles will be presented followed by key examples of use for photoacoustic imaging of cancer.

3.1 Plasmonic Noble Metal Nanoparticles

Plasmonic noble metal (e.g., gold, silver) nanoparticles constitute the most commonly researched probes for photoacoustic imaging of cancer. These particles have relatively facile and rapid (a few hours) synthesis methods with tunable size and shape parameters to allow for application-specific optimization of absorption properties that provide sharp absorption peaks within the near infrared tissue “optical window,” where endogenous tissues absorption is dominant otherwise. Furthermore, noble metal nanoparticles are easily bioconjugated to targeting moieties. However, in the case of gold nanoparticles, while they have little interaction with biological systems preventing toxicity, they are often retained in the liver and spleen. In the case of silver, which does degrade in biological environments, there is concern of potential silver toxicity. Current research is focused on overcoming these limitations because the per particle optical absorption tends to be higher in noble metal nanoparticles compared to any carbon nanotubes, polymer, or dye based nanoparticles due to the interesting phenomenon of plasmon resonance, making them promising contrast agents for photoacoustic imaging of cancer.

3.2 Plasmon Resonance

Nanoscale particles have many interesting properties that the same materials in bulk do not. One of the main reasons for this is the shift in the ratio of surface atoms to interior atoms. Surface atoms interact with the surrounding environment while interior atoms do not. This increased interaction with the environment gives rise to the superior optical absorption of noble metal nanoparticles because of surface plasmon resonance (SPR) [29–31]. Concretely, free electrons in metals are able to travel throughout the material, and in nanoscale particles, they do so free of scattering as the mean free path is greater than the size of the nanoparticle. Therefore, when an electromagnetic wave front with a wavelength larger than the diameter of the particle passes, the free electron cloud is polarized to the sides of the nanoparticle depending on the oscillating charge of the wave, resulting in coherent resonation at the same frequency. The size, shape, and surface environment of the nanoparticle all alter the SPR resonance (absorption) peaks, allowing nanoparticles to be finely tuned to specific applications in the visible and near infrared regions [29–31]. Once noble metal nanoparticles have the desired absorption characteristics, it is important to fine tune the pharmacokinetics and distribution for an optimal imaging system.

3.3 Bioconjugation

Due to their small size, unconjugated nanoparticles can be directly used as cancer contrast agents through the "enhanced permeability and retention effect" (EPR) which typically allows up to 5 % of the injected dose to accumulate nonspecifically within tumor tissues [32]. In tumors, the recruitment and growth of new blood vessels to supply nutrients occurs more quickly than typical angiogenesis [20]. Therefore, the organization and function of these rapidly grown vessels is abnormal, including poorly aligned endothelial cells, which leave large gaps of up to 800 nm in diameter between cells. Combined with ineffective lymphatic drainage, these factors make nanoparticles more likely to accumulate within tumor tissue compared to normal tissue [33, 34]. However, the efficiency of nanoparticle delivery using the EPR effect is highly variable between and within cancer types and stages, and is increased in preclinical xenograft tumor models due to a faster tumor development rate compared to native cancers [33]. Therefore, it is advantageous to conjugate molecular targets (such as antibodies, antibody fragments, peptides, aptamers, or metabolites [35]) to the surface of nanoparticles to, for example, increase receptor-mediated endocytosis into tumor cells or to allow for photoacoustic molecular imaging of a specific biomarker of the tumor vasculature. While other binding motifs (i.e., hydrophobic, electrostatic interactions, and van der Waals forces [36]) can be used to attach biomolecules to the surface of nanoparticles, covalent binding between SH, O, or N atoms through their unshared election pairs are preferred for clinical applications due to an increase in the stability of the constructs [36].

One of the typically leveraged bonds is the metal–thiol covalent bond mediated through sulfhydryl (SH) functional groups in thiols (R-SH), which is almost as strong as the gold–gold (Au–Au) bond, though the exact mechanism is still being elucidated [37, 38]. For bioconjugation purposes, the thiol interaction allows binding with cysteine amino acid residues, which have a thiol functional group allowing for direct protein conjugation. However, the number of cysteines can be limited or nonexistent in proteins, and therefore, proteins containing -NH2 groups (typically the more abundant lysine residue, occasionally arginine) could be attached by conversion to SH groups through EDC (1-ethyl-3-(3-dimethylaminopropyl) carbodiimide) or sulfo-NHS chemistry [35]. These reactions are fairly straightforward and short in duration up to a couple of hours [35]. Finally, to avoid a rapid immune response and clearance of nanoparticles triggered by nonspecific protein absorption on the surface, increased circulation time, and increased serum solubility [39], nanoparticles are typically coated with polyethylene glycol (PEG, PEGylation) using commercially available PEG-SH.

3.4 Gold Nanoparticles

Gold nanoparticles are by far the most commonly used plasmonic noble metal nanoparticle for photoacoustic imaging of cancer. Gold nanoparticles are bioinert in terms of cytotoxicity, but accumulate within the mononuclear phagocyte system [40], including the spleen, liver, and lymphatic system for up to 2–6 months (the final time point in several studies [41]); therefore, clinical use of gold nanoparticles remains elusive, as it requires more extensive biosafety and evaluation standards to elucidate whether this accumulation poses health risks, which is currently unclear. Extensive research has been dedicated to synthesizing monodisperse gold nanoparticles in a variety of sizes and shapes (spheres, rods, cubes, plates, barbells, rice, tetrapods, etc. see Fig. 2 [28, 42–44]), which allows fine-tuning of optical absorption properties for applications of photoacoustic imaging of cancer.

3.4.1 Synthesis

While "top-down" approaches (starting with bulk gold and removing material until nanoparticles remain) exist for synthesis of gold nanoparticles, "bottom-up" approaches (building nanoparticles from reduced metal ions) have become sufficiently optimized to produce relatively monodisperse particles in bulk and will be the focus here. One common method for gold nanosphere production involves reduction of gold(III) derivatives, such as $HAuCl_4$, using a weak reducing agent such as citric acid [45, 46]. The ratio of the reducing agent to the salt and reducing time determine the size of the nanoparticles, which are citrate stabilized in the solution. Particles can then be surface capped through PEGylation for purification and used for in vivo applications. Synthesis of anisotropic nanoparticles, most notably

Fig. 2 Gold nanoparticles of various size and shape with potential applications in biomedicine. Small (**a**) and large (**b**) nanospheres, (**c**) nanorods, (**d**) sharpened nanorods, (**e**) nanoshells, (**f**) nanocages/frames, (**g**) hollow nanospheres, (**h**) tetrahedra/octahedra/cubes/icosahedra, (**i**) rhombic dodecahedra, (**j**) octahedra, (**k**) concave nanocubes, (**l**) tetrahexahedra, (**m**) rhombic dodecahedra, (**n**) obtuse triangular bipyramids, (**o**) trisoctahedra, and (**p**) nanoprisms [44]—Reproduced by permission of The Royal Society of Chemistry

nanorods, begins with a similar method but using stronger reducing agents, (such as sodium borohydride) to produce small gold nanospheres (~4 nm). These small spheres act as "seeds" onto which carefully controlled epitaxial gold ion deposition is continued through the use of selective growth restriction surfactants that bind to faces of the seed with preferential crystal structures and atom spacing. In the case of gold nanorods, cetyltrimethylammonium bromide (CTAB) is used in combination with silver nitrate to block gold deposition on the {111} or {100} side faces, leaving the {110} common axis faces free for anisotropic growth {denote atomic crystal lattice structures} [47]. Similar to spheres, nanorods, and other anisotropic shapes can be stabilized through PEGylation for in vivo applications. Many other shapes (see Fig. 2) can be synthesized by varying synthesis parameters including temperature, component ratios, and surfactants. Using these synthesis methods combined with bioconjugation of targeting moieties allows fine tuning of nanoparticle composition for specific applications of photoacoustic imaging of cancer.

3.4.2 Applications

The use of nanoparticles as contrast agents for photoacoustic imaging of cancer currently remains in the preclinical research stage, though several interesting studies have been explored. For example, gold nanorods have been used to image multiple types of cancers. Jokerst et al. [48] demonstrated the use of gold nanorods to image three types of ovarian cancer tumors: 2008, HEY, and SKOV3 subcutaneously grown in mice ($n=3$). They reported a 2.3-fold photoacoustic signal increase with intratumoral injection and a 3.5-fold increase with intravenous injection of gold nanorods at excitation wavelength of 756 nm respectively. The work by Zhong et al. [49] further demonstrates the applicability of gold nanorods beyond imaging, as they reported photoacoustic therapy using gold nanorods conjugated with folic acid in mice bearing HeLa (human cervical cancer) tumors that overexpress folate receptor. Folic acid conjugation allows the nanoparticles to bind to folate receptor whose expression is minimal in healthy tissues and high in cancers such as lung, breast, kidney, brain, cervix, or ovary cancer [50, 51]. Photoacoustic therapy although similar in principle to photoacoustic imaging, destructs individual target cells with laser-induced shock waves. When a photoabsorber is irradiated with a pulsed laser, the optical energy is transformed into mechanical energy, resulting in a shock wave without the side effect of heating or cavitation [52]. In photoacoustic therapy, the generation of shock waves rely on the efficiency of photoabsorbers to convert light energy into acoustic energy, with suitable candidates reported as indocyanine green containing nanoparticles (ICG-PL-PEG), single-walled carbon nanotubes, and gold nanorods [49, 53, 54]. When compared with images of mice injected with folic acid conjugated single-walled carbon tubes, and unconjugated gold nanorods, the tumor growth rate in mice treated with folic acid-conjugated gold nanorods was observed to be significantly lower over a duration of 24 days after treatment.

Another type of gold nanoparticles, nanoshells, have been used to study colon cancers. Li et al. [55] reported the intravenous use of PEGylated gold nanoshells in mice bearing CT26.wt colon cancer tumors grown subcutaneously. When imaged for 6 h using photoacoustic microscopy, the gold nanoshells were seen to be progressively taken up by the tumor foci, while those in the vessels were cleared out, indicating an enhancement in photoacoustic signals with a contrast ratio of 6.5 between the tumor foci to the vessels. Li et al. [56] reported the use of gold nanocages for photoacoustic imaging of cerebral cortex, sentinel lymph nodes, and melanoma in small animals. When gold nanocages were injected in rats to study their brain cortex [57], 81 % enhancement in blood absorption was reported over the intrinsic contrast. In a similar study conducted in rats to image sentinel lymph nodes [58], a gradual accumulation of gold nanocages was observed leading to a gradual increase in photoacoustic signal 28 min after injection. Likewise, gold nanocages functionalized with [Nle^4, D-Phe^7]-α-melanocyte-stimulating hormone were used for active targeting and resulted in a 300 % higher photoacoustic signal enhancement because of ligand-receptor binding on melanoma cells in mice compared to PEGylated gold nanocages 6 h after injection [59]. In a recent study by Shujing et al. [60], molecular imaging and photoacoustic therapy of gastric cancer stem cells using bioconjugated gold nanostars was reported. Gold nanostars conjugated with CD44v6 monoclonal antibody were intravenously injected in orthotopic and subcutaneous mice tumor models of human gastric cancer. Their results suggest that the bioconjugated gold nanostars successfully targeted the gastric vascular system at 4 h after injection resulting in 250 % increase in photoacoustic signal compared to the signal acquired from mice injected with unconjugated PEGylated gold nanostars. Moreover, tumor growth inhibition was also reported with an extension of survivability of tumor bearing mice.

Furthermore, introduction of gold nanospheres into a biological environment leads to complex interactions, such as aggregation of nanoparticles within cells, which can be visualized due to the resulting change in optical absorption spectra. Mallidi et al. [61] reported the synthesis and application of bioconjugated gold nanospheres (20 nm dia.) that can specifically bind to the cancer biomarker endothelial growth factor receptor (EGFR) for improved differentiation of cancer from background healthy tissue. EGFR-expressing human epithelial carcinoma cells (A431, $n = 6$) and EGFR-negative human breast cancer cells (MDA-MB-435, $n = 3$) were subcutaneously inoculated in mice followed by intravenous injection of EGFR-targeted or PEGylated gold nanospheres. Results of EGFR-targeted gold nanoparticles in A431 tumors are shown in Fig. 3a, b. A wavelength shift in the

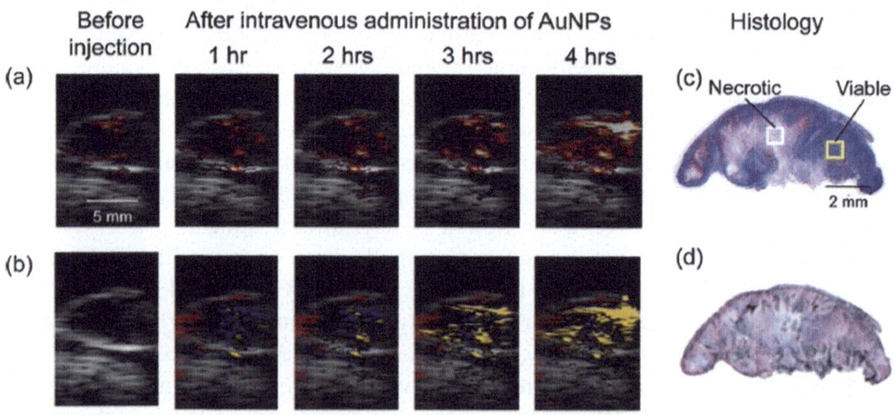

Fig. 3 Gold nanoparticles for photoacoustic imaging of cancer. (**a**) Photoacoustic and (**b**) multi-wavelength spectroscopic photoacoustic images pseudo-colored to show oxygenated- (*red*) and deoxygenated hemoglobin (*blue*), and accumulation of gold nanoparticles (*yellow*) before and 1, 2, 3, and 4 h after intravenous injection of EGFR-targeted gold nanoparticles in a subcutaneous A431 breast cancer tumor. (**c**) H&E and (**d**) silver enhancement staining of histology tumor slices showing accumulation of EGFR-targeted gold nanoparticles within the tumor. Reprinted with permission from [61].

absorption of targeted gold nanospheres in EGFR expressing tumors was observed from 520 nm (green) to 720 nm (near infrared) due to plasmon resonance coupling. Effectively, cellular internalization and aggregation of nanoparticles into lysosomal compartments brings nanoparticle, and their individual free electron clouds, into close proximity. This grouping of nanoparticles then acts as a larger nanoparticle of a different shape with one associative electron cloud that has a distinct resonance frequency and associated optical absorption, changing the overall absorption spectrum from that of individual nanoparticles. Accumulation was confirmed by silver staining (which detects the presence of noble metals by depositing ionic silver until the nanoparticle can be visualized with light microscopy) of histological sections, along with an increase in photoacoustic signal over time (4 h after injection) shown in Fig. 3c, d. In contrast, images of mice with PEGylated gold nanospheres showed no significant change in photoacoustic signals, indicating that using targeted nanoparticles increased specificity in imaging EGFR-expressing cancer cells.

Overall, gold nanoparticles have been extensively used as contrast agents for photoacoustic imaging due to their excellent optical absorption and finely tunable absorption characteristic that can be optimized for specific applications. However, the bioaccumulation of gold nanoparticles remains a limiting factor in their use clinically.

3.5 *Silver Nanoparticles*

Silver nanoparticles have been commonly used in biomedical and cosmetic purposes for their antibacterial properties [62]. Unlike gold nanoparticles, silver nanoparticles dissolve over time in biological environments making them more suitable for repeat administrations for clinical purposes compared to gold nanoparticles [62]. However, the ability of silver nanoparticles to degrade introduces new challenges for their biomedical use, including potential silver toxicity [63] through the introduction of reactive oxygen species leading to cellular oxidative stress and lipid peroxidation [64] and decreased stability which would affect their desired optical absorption characteristics.

3.5.1 Synthesis

Silver nanoparticles are synthesized very similarly to gold nanoparticles, predominantly through seed-mediated growth procedures. Reducing agents appropriate for silver particles include hydrazine, ascorbic acid, temperature increases, and light exposure. Silver seeds approximately 3 nm in diameter can be synthesized by the reduction of silver nitrate ($AgNO_3$) using sodium citrate and adding sodium borohydride ($NaBH_4$) in a dropwise fashion, followed by stabilization by bis(p-sulfonatophenyl)phenylphosphine dihydrate dipotassium (BSPP) [65]. The further growth on anisotropic shapes is accomplished similarly to that of gold nanoparticles through the addition of silver salts and reducing agents [66].

3.5.2 Applications

The use of silver nanoparticles for photoacoustic imaging of cancer has been more limited due to potential silver toxicity concerns and stability issues. However, some groups have been able to overcome these challenges by using a specific type of silver nanoparticles for imaging of pancreatic cancer xenografts. Homan et al. [66] demonstrated the synthesis of biostable sliver nanoplates (edge-length: 60.9 nm, thickness: 12.5 nm) by functionalizing them with thiol terminated methoxy-polyethylene glycol (mPEG-SH). They reported the application of PEG-passivated silver nanoplates conjugated to a monoclonal antibody to EGFR (a-EFGR) in detecting orthotopic human pancreatic cancer xenografts grown in mice. At first, cancer binding specificity of a-EFGR conjugated sliver nanoplates (SPR peak: 800 nm) was confirmed in vitro by incubating EGFR-expressing pancreatic cancer cells in three conditions: with no nanoplates, PEGylated nanoplates, and a-EGFR conjugated nanoplates, with only the last group showing any uptake. Further incubation of three different cell lines: MPanc96 (EGFR-expressing pancreatic cancer), L3.6pl (EGFR-expressing pancreatic cancer), and HPNE (non-cancerous pancreatic cells) in vitro for 24 h confirmed no cytotoxicity up to concentrations of 1 mg/mL silver nanoplates. Spectroscopic photoacoustic imaging of orthotopic pancreatic mice models up to 6 h after intravenous injection of a-EGFR conjugated nanoplates showed a heterogeneous accumulation of nanoplates in the tumor, which was also confirmed later by silver staining of histological slices, indicating that silver nanoparticles can increase the tumor contrast over surrounding tissue. Homan et al. [67] have also demonstrated an enhancement of photoacoustic imaging contrast in porcine pancreatic tissue ex vivo using another group of sliver nanoparticles called silver nanocages in a study aimed to further improve pancreatic cancer detection. The silver nanocages constitute silica nanospheres (180–520 nm) coated with a porous layer of sliver (5–60 nm) and the authors reported the possibility of using them beyond imaging for targeted drug delivery if the silica core material were to be replaced with a suitable drug-eluting material. Overall, silver nanoparticles remain a promising contrast agent for photoacoustic imaging assuming potential toxicity and stability issues can be sufficiently addressed.

3.6 Silica Coating of Noble Metal Nanoparticles

While, noble metal nanoparticles have excellent optical absorption abilities, optical absorption ultimately leads to instability of the nanoparticles after longer term imaging. Upon pulsed laser irradiation, optical absorption is converted to heat. Nanoscale particles tend to have lower melting temperatures due to the increase in ratio of surface to bulk atoms. Therefore, after prolonged imaging, anisotropically shaped noble metal nanoparticles tend to melt into more energy efficient shapes, such as spheres, altering their optical absorption peaks. Coating these particles with silica not only increases stability for prolonged or longitudinal imaging, but also provides an amplification of the photoacoustic signal due to an increase in the interfacial heat transfer between the particle and the surrounding environment. To produce silica-coated

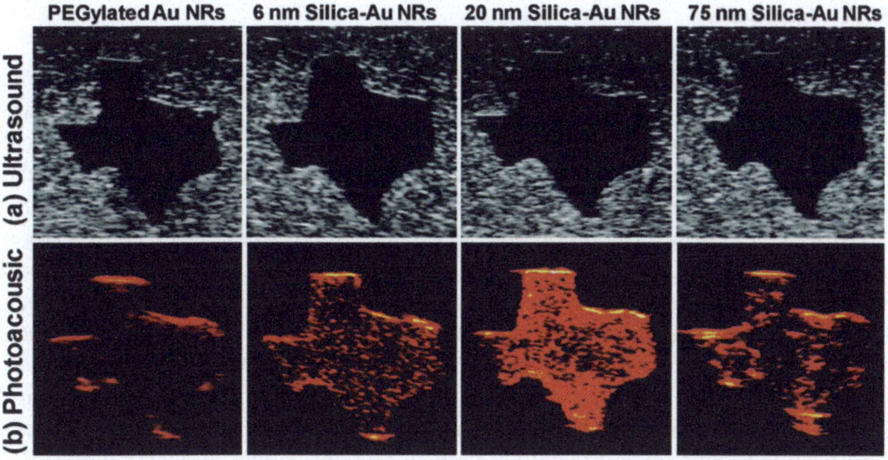

Fig. 4 (**a**) Ultrasound and (**b**) photoacoustic images (*top* to *bottom*) of inclusions containing the same number of (I) PEGylated gold nanorods and gold – silica core – shell nanorods with (II) 6 nm silica coating, (III) 20 nm silica coating, and (IV) 75 nm silica coating (left to right). Each image covers a 6 mm by 6 mm field of view. Reprinted with permission from Ref. [3]

metal nanoparticles, silica is deposited onto preprepared and PEGylated nanoparticles via the Stöber method[68]. Tetraethyl orthosilicate (TEOS) usually acts as the source of silica and the reaction is completed in an alcohol and ammonia solution [68]. The surface of silica nanoparticles can be conjugated using alkane silanes [36].

Chen et al.[3, 68] reported the development and implementation of gold nanorods coated with silica of varying thickness: 6, 20, and 75 nm for photoacoustic imaging of tissue-mimicking phantoms. When compared with signals from PEGylated gold nanorods without silica, the 20 nm silica-coated gold nanorods demonstrated the highest fold (3.8 times) increase in photoacoustic signals. The larger 75 nm silica-coated gold nanorods were reported to show a reduced photoacoustic signal (only 2.3 times higher than PEGylated) than 20 nm silica-coated nanorods, possibly due to reduced local fluence, as a consequence of increased scattering as shown in Fig. 4. Moreover, the silica-coated nanorods were reported to exhibit superior stability than PEGylated nanorods under high-fluence laser pulses. Overall, silica-coating presented a slight shift in wavelength towards the red without altering the optical absorption of gold nanoparticles increasing the usefulness of gold nanoparticles as photoacoustic imaging contrast agents through increase in stability and signal amplification.

3.7 Carbon Nanotubes

Carbon nanotubes (CNT) are graphene sheets that have been rolled into cylinders that have extremely long aspect ratios (>100) as diameters are typically a few nanometers or less and lengths up to hundreds of nanometers. CNTs have excellent stability and interesting semiconductor and optical absorption properties, and their functions in biomedical applications are currently being explored.

3.7.1 Synthesis

Carbon nanotubes represent a more challenging nanoparticle to synthesize requiring extremely high temperatures and specialized equipment. However, they are commercially available for those who would like to use them for biomedical research without large startup costs. The three main methods of carbon nanotube synthesis include arc-discharge, laser ablation, and chemical vapor deposition [68]. Arc-discharge and laser ablation are “top-down” approaches, where bulk carbon sources are vaporized at thousands of degrees Celsius into high-quality carbon nanotubes intermixed with large amounts of byproduct. The “bottom-up” approach uses small metal seeds to initiate growth through chemical vapor deposition at much lower temperatures between 500 and 1000 °C [69]. The optical absorption of CNTs is high with a very broad absorption over the visible and near infrared regions. Carbon nanotubes are hydrophobic after synthesis and, therefore, need surface modification to be compatible with biological systems. When sonicated with PEG-phospholipid compounds, the hydrophobic tails of the phospholipids interact strongly with the walls of the carbon nanotubes, leaving the PEG to interact with the aqueous solution [69, 70]. Furthermore, function linkers can be used in lace of the PEG to allow for conjugation of biologic targeting moieties. However, even after surface modification, there are concerns over cytotoxicity of the CNTs. Specifically, the concerns relate to how shapes with extremely high aspect ratios interfere with cellular endocytosis and cells physically cannot fully engulf objects of this shape [71]. However, CNTs have been used in preclinical cancer models.

3.7.2 Applications

In one such application, Zerda et al. [70] reported an eightfold increase in the photoacoustic signal in mice bearing U87MG glioblastoma tumors overexpressing $\alpha_v\beta_3$ integrin upon tail-vein injection of single-walled CNTs conjugated with cyclic RGD peptides targeting $\alpha_v\beta_3$ integrin ($n=4$), compared to photoacoustic signals from tumor bearing mice injected with non-targeted PEGylated nanotubes, as shown in Fig. 5. Three-dimensional spectroscopic photoacoustic imaging was performed before and up to four hours after injection of nanotubes followed by tumor extraction for ex vivo analysis using Raman microscope, which reported a fourfold Raman signal increase in mice injected with targeted single-walled CNTs compared to the signal from mice with non-targeted nanotubes validating the photoacoustic results. In another study, Nguyen et al. [72] demonstrated a 20-fold increase in optical absorption and a twofold increase in photoacoustic image contrast in porcine bladder tissue injected with single-walled CNTs conjugated with ICG compared to unconjugated CNTs with implications to improve bladder cancer detection. They reported the possible reasons for disproportion (10:1 ratio) between optical absorption and photoacoustic signal as nonuniform spatial distribution of ICG conjugated CNTs, the inhomogeneous absorbance of CNTs in tissue, and the precision of the mechanical movements of the PA system during signal acquisition.

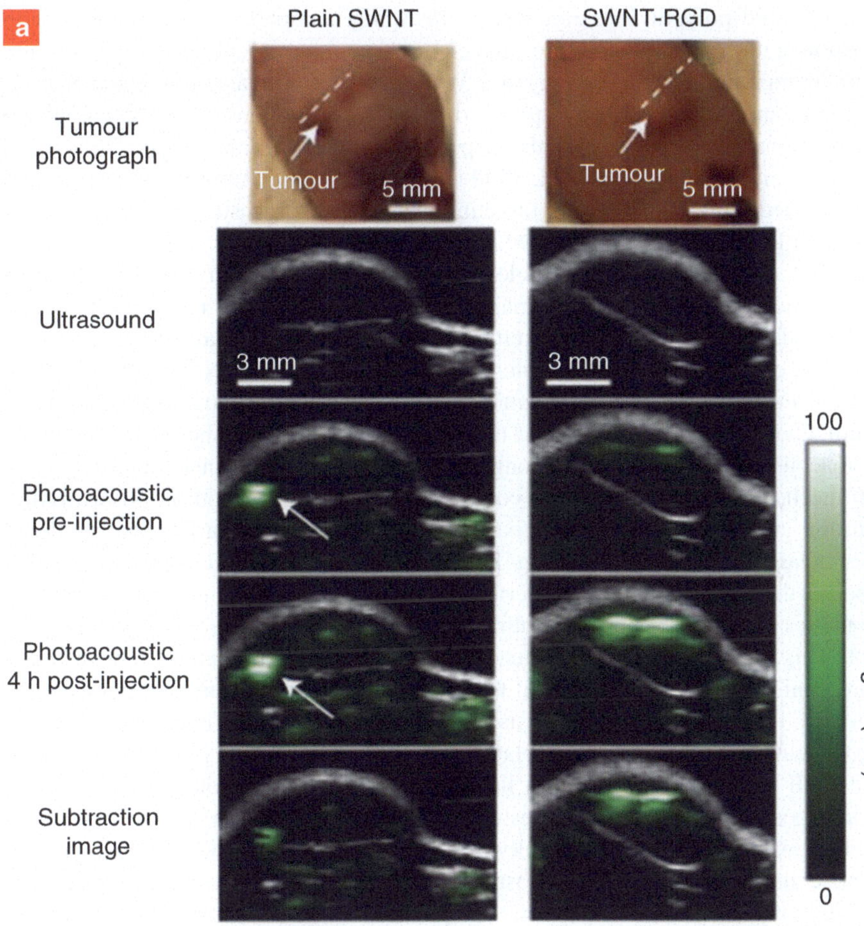

Fig. 5 Carbon nanotubes for photoacoustic imaging of cancer. Ultrasound and photoacoustic images of PEGylated and RGD peptide targeted single walled carbon nanotubes accumulated with an U87MG glioblastoma subcutaneous tumor at 4 h post intravenous injection. Targeted carbon nanotubes show increased tumor accumulation than mice injected with PEGylated carbon nanotubes (four- to sevenfold). Adapted with permission from Ref. [70]

3.8 Mixed Composition Nanoparticles

In order to achieve a complete set of desirable characteristics in nanoparticle contrast agents, including strong optical absorption, biocompatibility, appropriate size, shape, surface binding capabilities, and clearance rates, mixed composition nanoparticles have been explored. Typically, these particles comprise of a "carrier" of either polymer or biocompatible scaffold and a "payload" of optically absorbing media, either dyes or smaller nanoparticles for photoacoustic signal generation. For example, in the work by Zhong et al. [53], synthesis of indocyanine green (ICG) and phosopholipid-polyethylene glycol (ICG-PL-PEG) nanoparticles was reported with a demonstration of photoacoustic image contrast enhancement upon injection into the flank region of female Balb/c mice. With all the individual components of ICG-PL-PEG nanoparticles already approved by FDA for human use, these composite dye-based nanoparticles have a higher potential of clinical translation. The authors have later modified these ICG-PL-PEG nanoparticles by combining with folic acid and reported their use as cancer-targeting nanoprobes suggesting a reduced tumor growth after photoacoustic therapy [54].

Mixed composition nanoparticles also offer the potential for dual contrast enhancement for photoacoustic imaging and another imaging modality, typically ultrasound. Nanodroplets are an exciting field of research for dual contrast enhancement for combined ultrasound and photoacoustic imaging. Nanoscale (200 nm diameter) droplets of perfluoropentane encapsulated with a biocompatible bovine serum albumin shell in which gold nanorods (40 nm) were suspended, act as both photoacoustic contrast agents (through both thermoelastic expansion and vaporization mechanisms) and ultrasound contrast agents [73]. Upon pulsed laser irradiation, the gold nanoparticles (photoacoustic contrast agents) in the shell undergo thermoelastic expansion, emitting high frequency ultrasound transients, which vaporize the encapsulated liquid perfluoropentane into its gaseous state, converting to bubbles providing contrast for ultrasound thus providing co-registered dual contrast. This theme has been continued by replacing the gold nanorods with other, more clinically translatable optical triggers such as dyes and organic compounds. Jian et al. [74] reported the synthesis of perfluorohexane liquid nanodroplets incorporated with India ink and encapsulated with FDA approved poly-lactic-co-glycolic acid shell capable of undergoing liquid-to-gas phase transition via vaporization upon exposure to optical energy instead of ultrasound exposure. When these nanodroplets were injected intratumorally in MDA-MB-231 tumor bearing mice ($n=5$), photoacoustic signal amplitude post-injection was observed to be 20-fold higher than that of pre-injection. However, the signal enhancement reported was attributed to thermoelastic expansion alone and not phase-transition due to the laser power limitation of their photoacoustic imaging system (800 nm, 1 mJ/cm^2). To further validate the in vivo phase-transition ability of India ink nanodroplets, B-mode and contrast ultrasound imaging was performed on another group of MDA-MB-231 tumor-bearing mice ($n=5$) after exposing the tumors to a high energy laser (532 nm, 12 mJ/cm^2) for 10 s indicating a 20-fold contrast ultrasound signal enhancement post-laser exposure compared to pre-laser exposure. Hannah et al. [75] also showed similar results using ICG dye internalized to the center of the nanodroplets.

Recently, Paproski et al. [76] have synthesized a new class of nanodroplets called porphyrin nanodroplets (diameter of 185 nm) and injected them intratumorally in an HT1080 tumor in the chorioalantoic membrane of a chicken embryo. Porphyrin nanodroplets constitute liquid perfluorobutane core and a phosopholipid-porphyrin shell, which possesses strong optical absorption in the near infrared region making them suitable for photoacoustic imaging. Additionally, upon absorption of optical or ultrasound energy, porphyrin nanodroplets not only undergo phase change into a perfluorobutane gas, but also expand in size to form microbubbles that can provide ultrasound contrast.

They reported that an enhanced visualization of tumor indicating accumulation of porphyrin nanodroplets was observed with spectroscopic photoacoustic imaging throughout or at the periphery of tumor, demonstrating their usefulness in exhibiting the EPR effect. Nanodroplets have been extensively examined in proof-of-principle studies, however substantial extravascular localization in tumors has yet to be achieved leaving advancement into further preclinical animal studies a challenge.

In contrast, Huynh et al. [77] reported the reverse-conversion of porphyrin microbubbles (1–10 μm) into porphyrin nanoparticles (5–500 nm) in situ and demonstrated their applicability as photoacoustic imaging contrast agents in KB (human-epidermoid carcinoma) xenograft-bearing mice. Unlike nanodroplets that have a liquid core, porphyrin nanoparticles constitute a bacteriochlorophyll-lipid shell encapsulating perfluorocarbon gas core (824 nm peak absorbance), the same as their parent microbubbles. The nanoparticle conversion was achieved using a low-frequency, high-duty cycle ultrasound (conversion ultrasound). It was reported that conversion ultrasound ensured the delivery and retention of porphyrin nanoparticles to the tumor site bypassing the EPR, as indicated by the photoacoustic signal from the tumor, which was maintained for at least 2 h compared to the photoacoustic signal that was decreased within 30 min post-injection of porphyrin microbubbles without any conversion. Overall, mixed composition nanoparticles allow for optimization of nanoparticles for photoacoustic and other modalities, by using the optimized characteristics from each component.

4 Conclusions

Nanoparticles have been extensively used for various biomedical imaging purposes including the emerging field of photoacoustic imaging. An overview of current literature has been presented in this chapter along with background, synthesis, and applications of nanoparticles in photoacoustic imaging of cancer. Extensive explorations and reviews of nanoparticles for photoacoustic imaging in cancer can be found in the literature [78–83]. Nanoparticles that are optimized for photoacoustic imaging vary in size, shape, form, and material, but serve a common purpose of enhancing the image contrast by interacting with the tumor via enhanced permeability and retention effect either intrinsically or through surface modifications and bioconjugation with targeting moieties. Although they present excellent advantages over endogenous imaging, clinical adoption of nanoparticles has been limited due to remaining concerns over cytotoxicity and biodegradability, and the inability to accumulate sufficiently and homogenously within tumors. However, in order to leverage the molecular imaging capabilities of photoacoustic imaging and improve the diagnostic and prognostic rates of cancer detection, new classes of contrast agents that can retain their imaging benefits while minimizing the hazards, are necessary. Ongoing research explores the development and implementation of nanoparticles that constitute one or more materials, which are already FDA approved to support the eventual successful clinical translation of this promising new imaging approach.

References

1. Valluru KS, Wilson KE, Willmann JK. Photoacoustic imaging in oncology: translational preclinical and early clinical experience. Radiology. 2016;280(2):332–49.
2. Bell AG. Production of sound by radiant energy. J Franklin Inst. 1881;111(6):401–26.
3. Chen Y-S, Frey W, Kim S, Kruizinga P, Homan K, Emelianov S. Silica-coated gold nanorods as photoacoustic signal nanoamplifiers. Nano Lett. 2011;11(2):348–54.
4. Diebold GJ, Sun T, Khan MI. Photoacoustic monopole radiation in one, two, and three dimensions. Phys Rev Lett. 1991;67(24):3384–7.
5. Wang L V., Wu H-I, Masters BR. Biomedical optics, principles and imaging. J Biomed Opt. Wiley-Interscience; 2008.
6. Beard P. Biomedical photoacoustic imaging. Interface Focus. 2011;1(4):602–31.
7. Siphanto RI, Thumma KK, Kolkman RGM, Van Leeuwen TG, De Mul FFM, Van Neck JW, et al. Serial noninvasive photoacoustic imaging of neovascularization in tumor angiogenesis. Opt Express. 2005;13(1):89–95.
8. Pan D, Pramanik M, Senpan A, Allen JS, Zhang H, Wickline SA, et al. Molecular photoacoustic imaging of angiogenesis with integrin-targeted gold nanobeacons. FASEB J. 2011;25(3):875–82.
9. Flogel U, Ding Z, Hardung H, Jander S, Reichmann G, Jacoby C, et al. In vivo monitoring of inflammation after cardiac and cerebral ischemia by fluorine magnetic resonance imaging. Circulation. 2008;118(Copyright (C) 2011 U.S. National Library of Medicine.):140–8.
10. Wang B, Karpiouk A, Yeager D, Amirian J, Litovsky S, Smalling R, et al. In vivo intravascular ultrasound-guided photoacoustic imaging of lipid in plaques using an animal model of atherosclerosis. Ultrasound Med Biol. Elsevier; 2012.
11. Brannon-Peppas L, Blanchette JO. Nanoparticle and targeted systems for cancer therapy. Adv Drug Deliv Rev. 2004;56(11):1649–59.
12. Kim S, Chen Y-S, Luke GP, Emelianov SY. In vivo three-dimensional spectroscopic photoacoustic imaging for monitoring nanoparticle delivery. Biomed Opt Express. 2011;2(9):2540–50.
13. Luke GP, Nam SY, Emelianov SY. Optical wavelength selection for improved spectroscopic photoacoustic imaging. Photoacoustics. 2013;1(2):36–42.
14. Prahl S. Optical Properties Spectra. 2001.
15. Jacques SL. Optical properties of biological tissues: a review. Phys Med Biol. 2013;58(11):R37–61.
16. Guggenheim JA, Allen TJ, Plumb A, Zhang EZ, Rodriguez-Justo M, Punwani S, et al. Photoacoustic imaging of human lymph nodes with endogenous lipid and hemoglobin contrast. J Biomed Opt. 2015;20(5):050504.
17. Dogra VS, Chinni BK, Valluru KS, Joseph JV, Ghazi A, Yao JL, et al. Multispectral photoacoustic imaging of prostate cancer: preliminary ex-vivo results. J Clin Imaging Sci. 2013;3:41.
18. Ermilov SA, Khamapirad T, Conjusteau A, Leonard MH, Lacewell R, Mehta K, et al. Laser optoacoustic imaging system for detection of breast cancer. J Biomed Opt. 2009;14(2).
19. Allen TJ, Hall A, Dhillon AP, Owen JS, Beard PC. Spectroscopic photoacoustic imaging of lipid-rich plaques in the human aorta in the 740 to 1400 nm wavelength range. J Biomed Opt. 2012;17(6):612091–6120910.
20. Weis SM, Cheresh DA. Tumor angiogenesis: molecular pathways and therapeutic targets. Nat Med. 2011;17(11):1359–70.
21. Alameddine RS, Hamieh L, Shamseddine A. From sprouting angiogenesis to erythrocytes generation by cancer stem cells: evolving concepts in tumor microcirculation. Biomed Res Int. 2014;2014:1–8.
22. Wang X, Xie X, Ku G, Wang LV, Stoica G. Noninvasive imaging of hemoglobin concentration and oxygenation in the rat brain using high-resolution photoacoustic tomography. J Biomed Opt. 2006;11(2):24015.

23. Zhang HF, Maslov K, Sivaramakrishnan M, Stoica G, Wang LV. Imaging of hemoglobin oxygen saturation variations in single vessels in vivo using photoacoustic microscopy. Appl Phys Lett. 2007;90(5):5–7.
24. Chen Z, Yang S, Xing D. In vivo detection of hemoglobin oxygen saturation and carboxyhemoglobin saturation with multiwavelength photoacoustic microscopy. Opt Lett. 2012; 37(16):3414.
25. Kruger RA, Lam RB, Reinecke DR, Del Rio SP, Doyle RP. Photoacoustic angiography of the breast. Med Phys. 2010;37:6096–100.
26. Deng ZL, Yang XQ, Yu LJ, Gong H. The measurement of hemoglobin oxygen saturation using multi-wavelength photoacoustic microscopy. Saratov Fall Meet 2009 Int Sch Jr Sci Students Opt Laser Phys Biophotonics. 2010;7547.
27. Yao J, Wang L, Yang J-M, Maslov KI, Wong TTW, Li L, et al. High-speed label-free functional photoacoustic microscopy of mouse brain in action. Nat Methods. 2015;12(5):407.
28. Multiparametric spectroscopic photoacoustic imaging of breast cancer development in a transgenic mouse model [Internet]. http://www.thno.org/v04p1062.htm. Accessed 4 Sept 2014.
29. Eustis S, El-Sayed MA. Why gold nanoparticles are more precious than pretty gold: noble metal surface plasmon resonance and its enhancement of the radiative and nonradiative properties of nanocrystals of different shapes. Chem Soc Rev. 2006;35(3):209–17.
30. Jain PK, El-Sayed MA. Universal scaling of plasmon coupling in metal nanostructures: extension from particle pairs to nanoshells. Nano Lett. 2007;7(9):2854–8.
31. Aaron J, Nitin N, Travis K, Kumar S, Collier T, Park SY, et al. Plasmon resonance coupling of metal nanoparticles for molecular imaging of carcinogenesis in vivo. J Biomed Opt. 2007;12(3):34007.
32. Heneweer C, Holland JP, Divilov V, Carlin S, Lewis JS. Magnitude of enhanced permeability and retention effect in tumors with different phenotypes: 89Zr-albumin as a model system. J Nucl Med. 2011;52(4):625–33.
33. Prabhakar U, Maeda H, Jain RK, Sevick-Muraca EM, Zamboni W, Farokhzad OC, et al. Challenges and key considerations of the enhanced permeability and retention effect (EPR) for nanomedicine drug delivery in oncology. Cancer Res. 2013;73(8):2412–7.
34. Maeda H, Wu J, Sawa T, Matsumura Y, Hori K. Tumor vascular permeability and the EPR effect in macromolecular therapeutics: a review. J Control Release. 2000;65(1-2):271–84.
35. Kumar S, Aaron J, Sokolov K. Directional conjugation of antibodies to nanoparticles for synthesis of multiplexed optical contrast agents with both delivery and targeting moieties. Nat Protoc. 2008;3(2):314–20.
36. Hermanson GT. Bioconjugate techniques. Academic Press; 2013. 1200 p.
37. Sardar R, Funston AM, Mulvaney P, Murray RW. Gold nanoparticles: past, present, and future. Langmuir Am Chem Soc. 2009;25(24):13840–51.
38. Häkkinen H. The gold-sulfur interface at the nanoscale. Nat Chem. 2012;4(6):443–55.
39. Harris JM, Chess RB. Effect of pegylation on pharmaceuticals. Nat Rev Drug Discov. 2003;2(3):214–21.
40. Chow A, Brown BD, Merad M. Studying the mononuclear phagocyte system in the molecular age. Nat Rev Immunol. Nature Publishing Group, a division of Macmillan Publishers Limited. All Rights Reserved; 2011;11(11):788–98.
41. Balasubramanian SK, Jittiwat J, Manikandan J, Ong C-N, Yu LE, Ong W-Y. Biodistribution of gold nanoparticles and gene expression changes in the liver and spleen after intravenous administration in rats. Biomaterials. 2010;31(8):2034–42.
42. Sau TK, Murphy CJ. Room temperature, high-yield syntehsis of multiple shapes of gold nanoparticles in aqueous solution. J Am Chem Soc. 2004;126(28):8648–9
43. Nikoobakht B, El-Sayed MA. Preparation and growth mechanism of gold nanorods (NRs) using seed-mediated growth method. Chem Mater. 2003;15(10):1957–62.
44. Dreaden EC, Alkilany AM, Huang X, Murphy CJ, El-Sayed MA. The golden age: gold nanoparticles for biomedicine. Chem Soc Rev. 2012;41(7):2740–79.
45. Turkevich J, Stevenson PC, Hillier J. A study of the nucleation and growth processes in the synthesis of colloidal gold. Discuss Faraday Soc. 1951;11:55.

46. Daniel M-C, Astruc D. Gold nanoparticles: assembly, supramolecular chemistry, quantum-size-related properties, and applications toward biology, catalysis, and nanotechnology. Chem Rev. 2004:293–346.
47. Murphy CJ, Sau TK, Gole AM, Orendorff CJ, Gao J, Gou L, et al. Anisotropic metal nanoparticles: synthesis, assembly, and optical applications. J Phys Chem B. 2005;109(29):13857–70.
48. Jokerst JV, Cole AJ, Van de Sompel D, Gambhir SS. Gold nanorods for ovarian cancer detection with photoacoustic imaging and resection guidance via raman imaging in living mice. ACS Nano. 2012;6(11):10366–77.
49. Zhong J, Wen L, Yang S, Xiang L, Chen Q, Xing D. Imaging-guided high-efficient photoacoustic tumor therapy with targeting gold nanorods. Nanomedicine. 2015;11(6):1499–509.
50. Parker N, Turk MJ, Westrick E, Lewis JD, Low PS, Leamon CP. Folate receptor expression in carcinomas and normal tissues determined by a quantitative radioligand binding assay. Anal Biochem. 2005;338(2):284–93.
51. Zwicke GL, Mansoori GA, Jeffery CJ. Targeting of cancer nanotherapeutics. Nano Rev. 2012;1:1–11.
52. Steinhauser MO, Schmidt M. Destruction of cancer cells by laser-induced shock waves: recent developments in experimental treatments and multiscale computer simulations. Soft Matter. 2014;10(27):4778–88.
53. Zhong J, Yang S. Contrast-enhanced photoacoustic imaging using Indocyanine Green-containing nanoparticles. J Innov Opt Health Sci. 2014;7(1):1350029–1 to 1350029–7.
54. Zhong J, Yang S, Zheng X, Zhou T, Xing D. In vivo photoacoustic therapy with cancer-targeted indocyanine green-containing nanoparticles. Nanomedicine (Lond). 2013;8(6):903–19.
55. Li M-L, Wang JC, Schwartz JA, Gill-Sharp KL, Stoica G, Wang LV. In-vivo photoacoustic microscopy of nanoshell extravasation from solid tumor vasculature. J Biomed Opt. 2009;14(1):010507.
56. Li W, Brown PK, Wang LV, Xia Y. Gold nanocages as contrast agents for photoacoustic imaging. Contrast Media Mol Imaging. 2011;6(January):370–7.
57. Yang X, Skrabalak SE, Li ZY, Xia Y, Wang LV. Photoacoustic tomography of a rat cerebral cortex in vivo with Au nanocages as an optical contrast agent. Nano Lett. 2007;7(12): 3798–802.
58. Song KH, Kim C, Cobley CM, Xia Y, Wang LV. Near-infrared gold nanocages as a new class of tracers for photoacoustic sentinel lymph node mapping on a rat model. Nano Lett. 2009;9(1):183–8.
59. Kim C, Cho EC, Chen J, Song KH, Au L, Favazza C, et al. In vivo molecular photoacoustic tomography of melanomas targeted by bioconjugated gold nanocages. ACS Nano. 2010;4(8):4559–64.
60. Liang S, Li C, Zhang C, Chen Y, Xu L, Bao C, et al. CD44v6 monoclonal antibody-conjugated gold nanostars for targeted photoacoustic imaging and plasmonic photothermal therapy of gastric cancer stem-like cells. Theranostics. 2015;5(9):970–84.
61. Mallidi S, Kim S, Karpiouk A, Joshi PP, Sokolov K, Emelianov S. Visualization of molecular composition and functionality of cancer cells using nanoparticle-augmented ultrasound-guided photoacoustics. Photoacoustics. 2015;3(1):26–34.
62. Eckhardt S, Brunetto PS, Gagnon J, Priebe M, Giese B, Fromm KM. Nanobio silver: its interactions with peptides and bacteria, and its uses in medicine. Chem Rev. 2013;113(7): 4708–54.
63. Ong C, Lim JZZ, Ng C-T, Li JJ, Yung L-YL, Bay B-H. Silver nanoparticles in cancer: therapeutic efficacy and toxicity. Curr Med Chem. 2013;20(6):772–81.
64. de Lima R, Seabra AB, Durán N. Silver nanoparticles: a brief review of cytotoxicity and genotoxicity of chemically and biogenically synthesized nanoparticles. J Appl Toxicol. 2012;32(11):867–79.
65. Xue C, Metraux GS, Millstone JE, Mirkin CA. Mechanistic study of photomediated triangular silver nanoprism growth. J Am Chem Soc. 2008;130(26):8337–44.
66. Homan KA, Souza M, Truby R, Luke GP, Green C, Vreeland E, et al. Silver nanoplate contrast agents for in vivo molecular photoacoustic imaging. ACS Nano. 2012;6(1):641–50.

67. Homan K, Shah J, Gomez S, Gensler H, Karpiouk A, Brannon-Peppas L, Emelianov S. Combined ultrasound and photoacoustic imaging of pancreatic cancer using nanocage contrast agents. SPIE photons plus ultrasound: imaging and sensing. 2009. p. 71771M.
68. Chen Y-S, Frey W, Kim S, Homan K, Kruizinga P, Sokolov K, et al. Enhanced thermal stability of silica-coated? Gold nanorods for photoacoustic imaging and image-guided therapy. Opt Express. 2010;18(9):8867–78.
69. Dai H. Carbon nanotubes: synthesis, integration, and properties. Acc Chem Res. 2002;35(12): 1035–44.
70. De La Zerda A, Zavaleta C, Keren S, Vaithilingam S, Bodapati S, Liu Z, et al. Carbon nanotubes as photoacoustic molecular imaging agents in living mice. Nat Nanotechnol. 2008;3(9):557–62.
71. Smart SK, Cassady AI, Lu GQ, Martin DJ. The biocompatibility of carbon nanotubes. Carbon. 2006;44(6):1034–47.
72. Nguyen VP, Oh Y, Ha K, Oh J, Kang HW. Enhancement of high-resolution photoacoustic imaging with indocyanine green-conjugated carbon nanotubes. Jpn J Appl Phys. 2015;54(7S1):07HF04.
73. Wilson K, Homan K, Emelianov S. Biomedical photoacoustics beyond thermal expansion using triggered nanodroplet vaporization for contrast-enhanced imaging. Nat Commun. 2012;3:618.
74. Jian J, Liu C, Gong Y, Su L, Zhang B, Wang Z, et al. India ink incorporated multifunctional phase-transition nanodroplets for photoacoustic/ultrasound dual-modality imaging and photoacoustic effect based tumor therapy. Theranostics. 2014;4(10):1026–38.
75. Hannah A, Luke G, Wilson K, Homan K, Emelianov S. Indocyanine green-loaded photoacoustic nanodroplets: dual contrast nanoconstructs for enhanced photoacoustic and ultrasound imaging. ACS Nano. 2014;8(1):250–9.
76. Paproski RJ, Forbrich A, Huynh E, Chen J, Lewis JD. Porphyrin nanodroplets : sub-micrometer ultrasound and photoacoustic contrast imaging agents. Small. 2015:1–10.
77. Huynh E, Leung BYC, Helfield BL, Shakiba M, Gandier J-A, Jin CS, et al. In situ conversion of porphyrin microbubbles to nanoparticles for multimodality imaging. Nat Nanotechnol. 2015;10(4):325–32.
78. Su JL, Wang B, Wilson KE, Bayer CL, Chen Y-S, Kim S, et al. Advances in clinical and biomedical applications of photoacoustic imaging. Expert Opin Med Diagn. 2010;4(6):497–510.
79. Luke GP, Yeager D, Emelianov SY. Biomedical applications of photoacoustic imaging with exogenous contrast agents. Ann Biomed Eng. 2012;40(2):422–37.
80. Mallidi S, Larson T, Aaron J, Sokolov K, Emelianov S. Molecular specific optoacoustic imaging with plasmonic nanoparticles. Opt Express. 2007;15(11):6583–8.
81. Yang X, Stein EW, Ashkenazi S, Wang LV. Nanoparticles for photoacoustic imaging. Wiley Interdiscip Rev Nanomed Nanobiotechnol. 1(4):360–8.
82. Wilson KE, Wang TY, Willmann JK. Acoustic and photoacoustic molecular imaging of cancer. J Nucl Med. 2013;54(11):1851–4.
83. Yang X, Stein EW, Ashkenazi S, Wang LV. Nanoparticles for photoacoustic imaging. Wiley Interdiscip Rev Nanomed Nanobiotechnol. 2009;1(4):360–8.

Nanoparticles for Photoacoustic Imaging of Vasculature

Min Zhou*, Lei Li*, Junjie Yao, Richard R. Bouchard, Lihong.V. Wang, and Chun Li

1 Introduction

Photoacoustic imaging (PAI), also referred to as optoacoustic tomography, is based on the measurement of ultrasonic waves induced by biological tissues that absorb short laser pulses. In PAI, light absorption by macromolecules, nanoparticles (NPs),

*Author contributed equally with all other contributors.

M. Zhou
Department of Cancer Systems Imaging, The University of Texas M. D. Anderson Cancer Center, 1881 East Road, Houston, TX 77054, USA

Institute of Translational Medicine, Zhejiang University, Hangzhou, Zhejiang 310009, China

L. Li
Department of Electrical and Systems Engineering, Washington University in St. Louis, One Brookings Dr., St. Louis, MO 63130, USA

J. Yao
Department of Biomedical Engineering, Washington University in St. Louis, One Brookings Dr., St. Louis, MO 63130, USA

R.R. Bouchard
Department of Imaging Physics, The University of Texas M.D. Anderson Cancer Center, Houston, TX, USA

L.V. Wang (✉)
Department of Electrical and Systems Engineering, Washington University in St. Louis, One Brookings Dr., St. Louis, MO 63130, USA

Department of Biomedical Engineering, Washington University in St. Louis, One Brookings Dr., St. Louis, MO 63130, USA
e-mail: lhwang@wustl.edu

C. Li (✉)
Department of Cancer Systems Imaging, The University of Texas M. D. Anderson Cancer Center, 1881 East Road, Houston, TX 77054, USA
e-mail: cli@mdanderson.org

J.W.M. Bulte, M.M.J. Modo (eds.), *Design and Applications of Nanoparticles in Biomedical Imaging*, DOI 10.1007/978-3-319-42169-8_16

or cells in the volume of interests creates a thermally induced pressure change that is translated to ultrasonic waves, which are received by acoustic detectors to form images. The technique combines the molecular specificity of optical imaging with the depth and spatiotemporal resolution of sonography. Unlike other high-resolution optical imaging modalities, PAI relies on orders of magnitude less scattered acoustic waves, which are induced by both diffused and ballistic light, and thus can be used to image deeper biological tissues. Photoacoustic effects were demonstrated in turbid medium by Kruger in 1994 [1], in biological tissues by Oraevsky et al. in 1997 [2], and in an in vivo small animal model by Wang et al. in 2003 [3]. Since then, PAI has been widely researched and advanced toward clinical applications [4].

Because of the overwhelming light-scattering effect of biological tissues, the photoacoustic signl-to-noise ratio (SNR) decreases exponentially with depth with a decay constant that is related to effective penetration depth. Light intensity attenuation can be minimized by choosing an excitation laser wavelength within the near-infrared (NIR) region, in which biological tissues have a relatively low absorption coefficient and a low scattering coefficient. Chromophores that absorb optical energy and convert it into acoustic signal are detected by PAI. A high optical absorption in NIR region increases detection sensitivity and thus extends imaging depth. The most important chromophores in the human body are oxyhemoglobin and deoxyhemoglobin. Hemoglobin has an absorption coefficient of over 100 cm^{-1} for visible light other than red; hence, it is capable of generating strong photoacoustic signals (~10 bars). Taking advantage of this endogenous contrast agent, PAI has been successfully applied to image vascular structures and tumor angiogenesis a few millimeters under the skin [5–7]. The technique has also been used to quantify oxygen hemoglobin saturation and thus the status of hypoxia in tumors [3, 8]. Ku et al. [7] demonstrated that hemoglobin in a blood sample 5 cm deep in biological tissues could be imaged. Attempts have also been made to apply PAI to the visualization of tumor vasculature in humans [9, 10].

Imaging tissue and tumor vasculature with a high penetration depth and high resolution is difficult to achieve without the use of appropriate exogenous contrast agents. In addition, optical contrast allows spectroscopic separation of signal contributions from multiple optical absorbers, thus enabling simultaneous functional imaging (with endogenous contrast from oxyhemoglobin and deoxyhemoglobin) and molecular imaging (with exogenous contrast agents). This concept was first demonstrated by in 2008 Li et al. [8]. In this study, PAI images of a tumor-associated biomarker and tumor oxygenation status based on exogenous and endogenous contrast agents were simultaneously acquired using multiple wavelengths around 800 nm in a mouse model. In recent years, various NPs have been studied as novel contrast agents for PAI. These NPs usually possess strong and tunable optical absorption in the NIR region. In this chapter, we provide an up-to-date summary of the PAI of tissue vasculature mediated by two classes of NPs, i.e., gold nanostructures and CuS NPs. We describe the structure, preparation, and in vivo application of these nanomaterials. We also touch upon the development of multimodal imaging, which includes PAI and other imaging modalities using multifunctional NPs for tumor vasculature imaging.

2 Principle of Photoacoustic Imaging

The photoacoustic effect is the generation of acoustic waves following light absorption in a medium. Generally, a short pulse of light is absorbed by chromophores in the medium and transiently converted to heat, which causes a rapid thermoelastic expansion that generates ultrasound waves [11]. Detection of the photon-induced ultrasonic wave yields a tomographic image. In order to efficiently generate ultrasound in biological tissue, normally a short-pulsed laser (typically, several nanoseconds pulse width) is used. The amplitude of the generated ultrasound signal is proportional to the original optical energy deposition with the conversion coefficient related to the thermal and mechanical properties of the tissue. Because the absorption of any photons regardless of their scattering histories can induce PA waves, PA signals can be generated in deep tissue. Ultrasonic scattering in tissue is three orders of magnitude weaker than optical scattering in the typical ultrasonic frequency range; high spatial resolution imaging can be achieved by the detection of PA waves. Consequently, PAI can break the optical diffusion limit.

The initial pressure induced by a sufficiently short laser pulse is given by

$$p_0 = \frac{\beta T}{\kappa}, \tag{1}$$

where β is the thermal coefficient of volume expansion, T is the temperature rise, and k is the isothermal compressibility. In soft tissue, a temperature rise of 1 mK induces approximately 800 Pa pressure rise, which is above the detection sensitivity of a typical ultrasonic transducer (~77 Pa for a 50 MHz transducer) [12]. Therefore, PAI is a highly sensitive imaging modality. If all absorbed light energy is converted to heat, the initial pressure can be expressed as

$$p_0 = \Gamma \mu_a F, \tag{2}$$

where Γ is the Grueneisen parameter, μ_a is the optical absorption coefficient (cm^{-1}), and F is the optical fluence (J/cm^2). PAI converts a fractional change in optical absorption to an identical fractional change in the PA signal, tantamount to a sensitivity of 100 % in the limit of small absorption variations. In principle, any molecule has its own absorption spectrum and thus can be imaged by PAI. Thus far, PAI has spectroscopically imaged many endogenous biological molecules, such as oxyhemoglobin, deoxyhemoglobin [13–15], melanin [16, 17], water [18–20], lipids [21–23], DNA and RNA [24, 25], and cytochrome [26, 27].

At depths beyond the optical diffusion limit, PAI can provide imaging with high ultrasound resolution, which is determined by the center frequency and bandwidth of the ultrasonic detection system and the bandwidth of the detected PA signal. The higher the central frequency and the broader the bandwidth, the better the spatial resolution. Penetration depths up to 5–6 cm have been reported using a safe laser exposure in the wavelength range of 600–1500 nm [28–30]. PAI scales spatial resolution with penetration depth, as shown in Fig. 1. With a detection central frequency of 50 MHz, acoustic-resolution photoacoustic microscopy has achieved a lateral resolution of 45 μm and an imaging depth of 3 mm [31]. Reducing the detection central frequency to 5 MHz expands the imaging depth to 4 cm and relaxes the resolution to 100–560 μm, depending on the detection geometry [32–35].

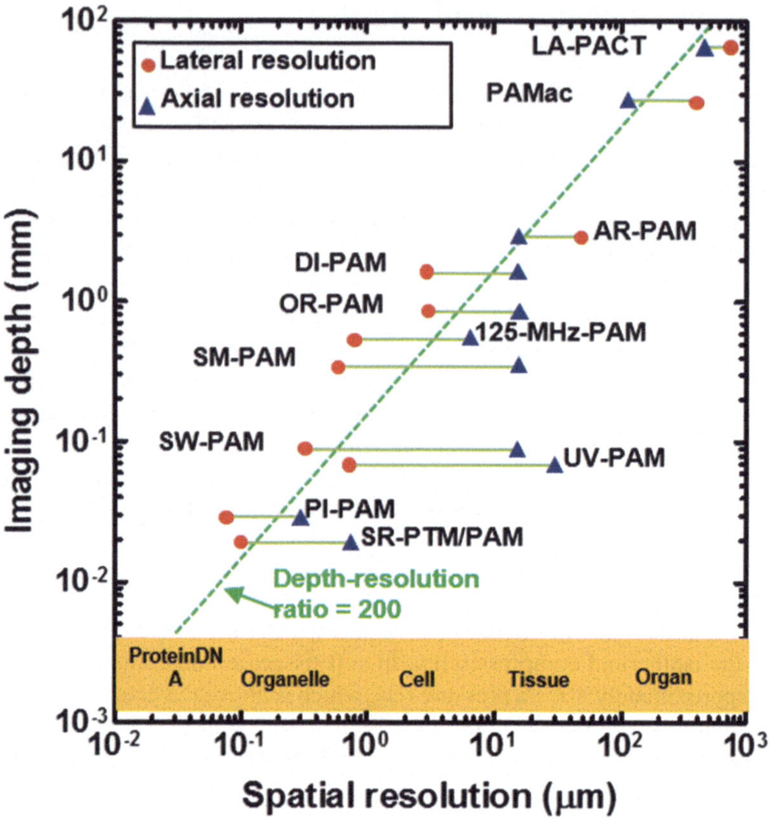

Fig. 1 Imaging depth versus spatial resolution in multi-scale PAI. The *red* circles represent lateral resolution, and the *blue* triangles denote axial resolution. LA-PACT, linear-array based PA computed tomography [80], PAMac, deep photoacoustic macroscopy [32]; AR-PAM, acoustic resolution photoacoustic microscopy (PAM) [81]; DI-PAM, double-illumination PAM [82]; OR-PAM, optical resolution PAM [14, 15, 83]; 125-MHz-PAM, PAM with a 125 MHz ultrasonic transducer [84]; SM-PAM, submicron PAM [85]; SW-PAM, subwavelength PAM [86]; UV-PAM, ultraviolet PAM [25]; PI-PAM, photo-imprint PAM [87]; SR-PTM/PAM, super-resolution photothermal/photoacoustic microscopy [88]

3 Why Nanoparticles

Without the introduction of exogenous contrast, PAI has been used in a variety of applications, including hemodynamic imaging [3, 36], blood oxygenation quantification [37], blood flow measurement [38], cancer detection [39], and tumor modeling [40, 41]. With appropriately chosen wavelengths, from the ultraviolet (UV) to visible (VIS) spectrum (250–700 nm), PAI can image cell nuclei, cytochromes, and red blood cells with high optical absorption contrast. However, UV and VIS light normally penetrates tissue only for hundreds of microns to millimeters, due to the strong optical attenuation. In the NIR wavelength region (from 700 to 1100 nm), the optical absorption of tissue is at a minimum, which is called "optical window." The optical scattering in biological tissue decreases with increasing wavelength. Therefore, NIR light suffers the least optical attenuation, which gives PAI the greatest penetration. According to Eq. (2), the low endogenous optical absorption of NIR directly leads to weak PA signals, which affects the imaging sensitivity and depth. Therefore, the employment of exogenous contrast agents can greatly improve the imaging sensitivity, contrast, and specificity. Compared with endogenous molecules, exogenous contrast agents offer several advantages. First, the optical and chemical properties of exogenous contrast agents can be specifically engineered to maximize imaging contrast and detection sensitivity, and to minimize background absorption. Second, exogenous contrast agents can be specifically conjugated with targeting agents (e.g., peptides, antibodies, aptamers) to selectively bind to cell surface receptors. Third, exogenous contrast agents, especially NPs, can be engineered with specific structures for both PAI and localized therapy. For examples, the unique optical properties of these NPs can enable controlled drug delivery by light or ultrasound to the target organs and/or to mediate heat generation for thermal therapy.

A major concern for exogenous PA contrast agents is their photostability. Organic dyes suffer from permanent oxidative photobleaching and optically induced transient changes in their absorption spectrum. They are not suitable for quantitative PA measurement or applications which require high doses of laser energy or a long exposure time. In contrast, metal NPs, such as copper and nickel chloride, silica-coated plasmonic NPs [42, 43], and absorbing pigments, such as melanin from tyrosinase-expressing cells, have much better photostability.

4 Gold Nanostructures

Gold nanostructures, especially gold nanoshells, gold nanocages, gold nanorods, and hollow gold nanospheres, exhibit a unique and tunable optical property, termed surface plasmon resonance (SPR). SPR is a collective oscillation of conduction-band electrons within the structures induced by the oscillating dipole of a resonant wavelength of light. For solid spherical particles, the resonance peaks appear at approximately 520 nm for gold, and the peak varies slightly depending on the size of the particle and the embedding medium. The SPR peaks of nanostructures can be tuned from the VIS to the NIR region (650–1100 nm) by controlling the size, shape (e.g., nanorods), and structure (e.g., hollow or core–shell structured NPs). These unique optical properties combined with its excellent biocompatibility makes gold nanostructures well suited for PAI of vasculature [44].

Gold nanoshells (AuNS) consist of a dielectric or semiconducting core (i.e., silica) surrounded by an ultrathin gold shell. These nanostructures possess photothermal properties different from solid gold nanospheres of the same size. They have been shown to be useful for a variety of potential applications in photothermal ablation therapy and molecular optical imaging. Wang et al. [45] first reported brain vasculature PAI imaging using AuNS with 100-nm silica cores, 20-nm gold shells, and with an 800-nm peak absorption. A deeply penetrating pulsed laser at 800 nm was employed to image the vasculature architecture of a rat brain. Compared to the brain PAI image based on the endogenous optical contrast, the image acquired ~20 min after three injections of the AuNS showed the brain vasculature with greater clarity. With the exogenous contrast agent, the optical absorption of the blood was increased and the contrast between the vessels and the background brain tissues was enhanced. By ~6 h after the third administration of AuNS, the optical absorption in the blood vessels decreased significantly. This was attributed to the clearance of the AuNS from the blood. The same group further reported high resolution reflection-mode (backward-mode) photoacoustic microscopy (PAM) that noninvasively imaged progressive extravasation and accumulation of AuNS within a solid tumor in vivo [46]. This study took advantage of the strong NIR absorption of AuNS that extravasated from leaky tumor vasculatures via the "enhanced permeability and retention" effect. With PAM, the three-dimensional AuNS distribution inside the tumor was visualized. Experimental results show that AuNS accumulated heterogeneously in tumors, with AuNS concentrated more in the periphery of the tumor and largely absent from the tumor core. This result is consistent with numerous other observations that drug delivery within tumor cores is ineffective because of poor blood perfusion.

Au nanocages are synthesized from Ag nanocubes by a galvanic replacement reaction [47, 48]. Similar to AuNS, the SPR absorption peak of Au nanocages can be tuned throughout the VIS and into the NIR region. Three sequential injections of Au nanocages in rats showed a gradual enhancement of the optical absorption in the cerebral cortex of brain by up to 81 % [47]. Compared with silica-cored AuNS, Au nanocages seem to have advantages for PAI due to their absorption-dominant extinction, more compact sizes (<50 nm), and larger optical absorption cross sections. Gold nanorods have also been employed as contrast enhancement agents for PAI owning to their strong absorption in the NIR range [49–55].

Hollow gold nanospheres (HAuNS) is a novel gold nanostructure consisting of only a thin Au wall with a hollow interior [56]. This type of gold nanostructures has the unique combination of being small in size (outer diameter, 40–50 nm), having spherical shape, a hollow interior, and a strong and tunable absorption band in the NIR region (Fig. 2). HAuNS are synthesized using cobalt (Co) instead of silver (Ag) NPs as sacrificial templates. The outer diameter of HAuNS is controlled by the diameter of the Co NPs, whereas the interior-cavity size is controlled by the stoichiometric ratio of $HAuCl_4$ and the reducing agents. HAuNS are coated with polyethylene glycol (PEG, MW 5000) to increase their blood circulation half-life. We evaluate the use of these PEGylated HAuNS as a contrast agent for PAI [57]. Our phantom study (Fig. 2d, e) revealed that the PAI of HAuNS was brighter than that of whole blood, aqueous solutions of $CuSO_4$ and India ink. Even though blood had optical density 2.3-fold greater than that of HAuNS in water suspension, the PA brightness of HAuNS was 2.5-fold greater than that of blood, indicating higher PA efficiency of HAuNS than that of blood.

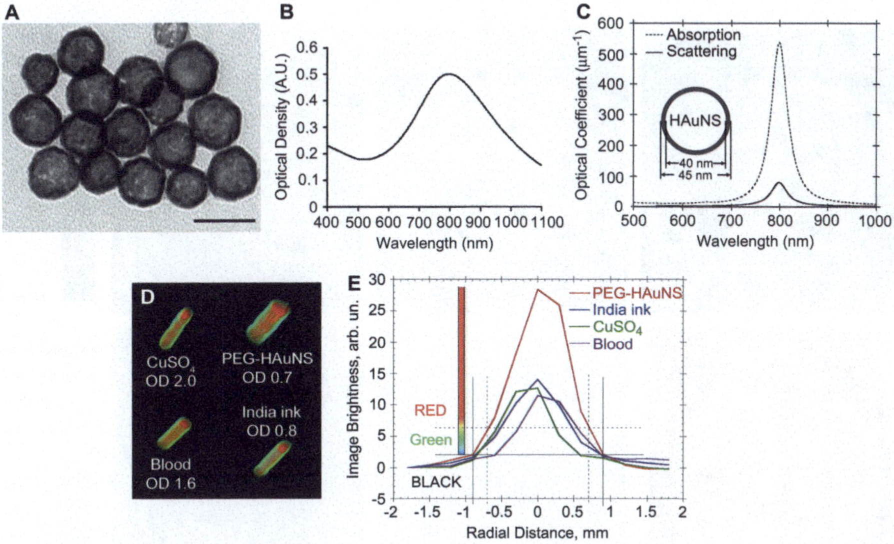

Fig. 2 (**a**) TEM image of PEG-HAuNS (Bar, 50 nm). The average outer diameter was 45 nm, and thickness of the shell was 2.5 nm. (**b**) Absorbance spectrum of PEGHAuNS in water, which peaked at 800 nm. (**c**) Theoretically calculated absorption and scattering spectra of PEG-HAuNS with water core having dielectric permittivity of 1.77. The inner core diameter was 40 nm and gold shell thickness was 2.5 nm, whose values are consistent with those shown in TEM. (**d**) Tomographic image of a tissue mimicking phantom made of poly(vinyl-chloride) plastisol with four embedded tubes filled with optically absorbing liquid media. (**e**) Quantitative analysis of the photoacoustic image brightness through the cross-section of each tube from (**d**). The color scale was designed in three steps: from *black* to *blue/green* to *red*. All tubes with different levels of brightness reflected in the object diameter. Reprinted with permission from Ref. [57]

PAI of the cerebral cortex of a mouse before contrast injection showed only large vessels (e.g., the middle cerebral artery, distributed along the mouse brain cortical surface) (Fig. 3a, arrow). This is because oxyhemoglobin and deoxyhemoglobin display weak PA signal at 800 nm. However, 5 min after the injection of PEGylated HAuNS, the PAI image revealed the brain vasculature with much greater clarity, especially for small blood vessels (Fig. 3b, c). This enhanced clarity was attributed to the strong PA signal generated with HAuNS. This is better appreciated in differential images (Fig. 3d, e) obtained by subtracting the preinjection image from the postinjection images pixel by pixel. At 2 h post-injection, the PA images remained essentially unchanged, indicating a significant amount of HAuNS circulating in the blood (Fig. 3c, e). At higher magnification, PAI revealed not only the structure of large vessels (Fig. 4a, yellow-framed picture) but also small blood vessels (Fig. 4a, green-framed pictures) 2 h after HAuNS injection. Indeed, brain blood vessels in the superficial cortex as small as ~100 μm in diameter could be clearly seen (Fig. 4a, arrows). HAuNS were confirmed to be located on the luminal

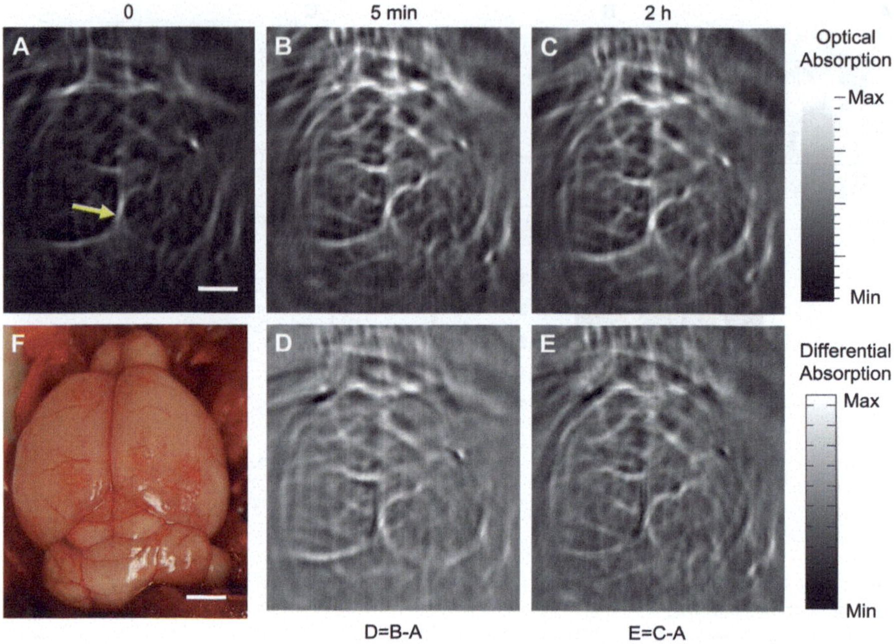

Fig. 3 Noninvasive PAI imaging of a mouse brain in vivo employing PEG-HAuNS and NIR light at a wavelength of 800 nm. Photoacoustic image acquired (**a**) before, (**b**) 5 min after, and (**c**) 2 h after the intravenous injection of PEG-HAuNS. (**d**, **e**) Differential images that were obtained by subtracting the preinjection image from the post-injection images (Image **d** = Image **b**–Image **a**; Image **e** = Image **c**–Image **a**). *Arrow*, middle cerebral artery. Bar = 2 mm. (**f**) Open-skull photograph of the mouse brain cortex obtained after the data acquisition for PAI. Bar = 2 mm. Reprinted with permission from Ref. [57]

side of the blood vessels (Fig. 4b, c). There was no particle extravasation into the brain parenchyma. This was attributed to the blood–brain barrier, which has been shown to impede the penetration of such particles. Thus, the long-circulating PEG-HAuNS enhanced the contrast between the blood vessels and the brain parenchyma [57].

Integrin αvβ3 is known to be overexpressed in the angiogenic blood vessels of solid tumors. Lu et al. [58] showed that intravenous injection of cyclic peptide c(KRGDf)-coated HAuNS targeting αvβ3 permitted PAI of orthotopically inoculated U87 glioma in nude mice (Fig. 5). Quantitative analysis confirmed that the mean PA signal ratio between the tumor and the contralateral normal brain at 24 h after c(KRGDf)-HAuNS injection was approximately twice as high as that obtained from precontrast images (without contrast agent injection). In comparison, no change in PA signal ratio between the tumor and the normal brain was observed before and after intravenous injection of the nonspecific PEG-HAuNS (Fig. 5). These data indicate that selective binding of c(KRGDf)-HAuNS to both tumor cells and tumor endothelial cells enhanced PAI of U87 glioma. c(KRGDf)-HAuNS also

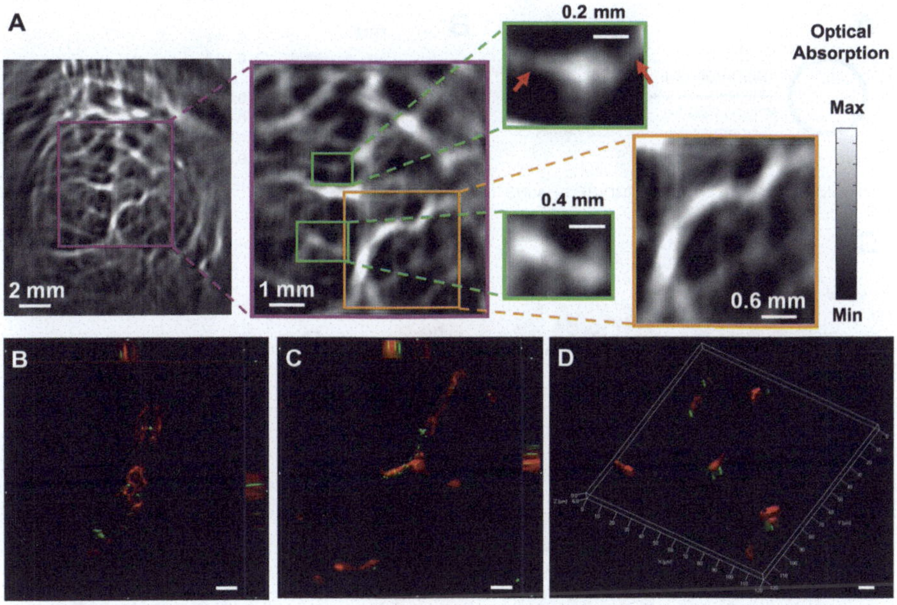

Fig. 4 (**a**) Enhanced PA signals revealed clear and detailed structure of large (*yellow*-framed picture) and small (*green*-framed picture) blood vessels in the mouse brain at higher magnification 2 h after intravenous injection of PEG-HAuNS. *Arrows* represent the small blood vessels with a diameter of about 100 μm, which can be seen in the contrast-enhanced images. (**b**–**d**) Distribution of PEG-HAuNS in brain vessels 2 h after injection (Bar = 10 μm). Brain vessels were stained with anti-CD31 antibody (*red* fluorescence), while the scattering signals of gold particles were detected under a dark field (pseudo-green). Z-stack images showed the particles located on the luminal side of brain blood vessels (**b**, **c**). In brain capillaries, three-dimensional reconstruction images show that the particles colocalized or stayed adjacent to the brain capillary endothelial cells (**d**). Reprinted with permission from Ref. [57]

mediated a selective photothermal ablation of U87 tumor, leading to a significant increase in overall survival of tumor-bearing mice compared to mice treated with laser alone or c(KRGDf)-HAuNS alone [58]. These results suggest that PAI may be used to guide photothermal ablation therapy mediated by the same nanoparticle targeted to tumor vasculature and tumor cells.

By integrating multiple functions into a NP's design, it is possible to use the same nanostructure for multimodality imaging applications. Along this line of research, Zhou et al. [59] reported dual PAI and magnetic resonance imaging (MRI) after intravenous administration of superparamagnetic iron oxide (SPIO)-containing gold nanoshells (SPIO@AuNS). Here, PAI was used to delineate tumor vasculature, while T2-weighted MRI was used to monitor therapeutic effects after photothermal therapy mediated by SPIO@AuNS. Gadolinium (Gd)-doped, gold-speckled silica NPs were synthesized as multimodal nanoparticulate contrast agents for noninvasive imaging using both PAI and T1-weighted MRI [60].

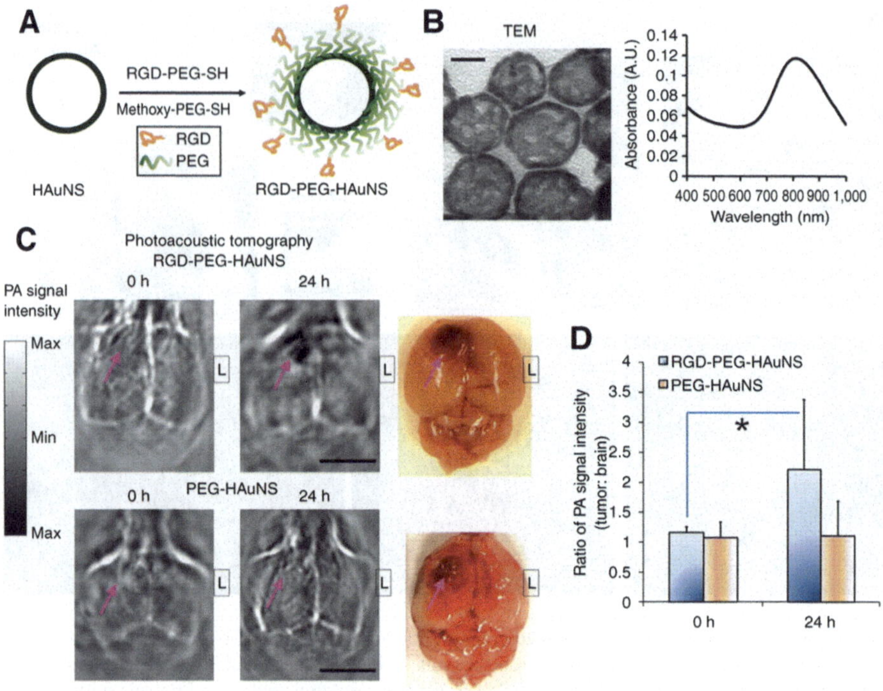

Fig. 5 (**a**) Scheme for c(KRGDf)-PEG-HAuNS bioconjugation. (**b**) c(KRGDf)-PEG-HAuNS characteristics on transmission electron microscopy (bar, 20 nm) and UV–Vis spectrum (measured in water). (**c**) PAI images of U87 human glioma in mouse brains before (0 h) and 24 hours after i.v. injection of NPs (bar, 5 mm). Photographs of corresponding mouse brains were used to confirm tumor location. *Arrows*, locations of tumors; *L* left. (**d**) PA signal intensity ratio of tumor-to-contralateral brain in mice before (0 h) and 24 h after injection of HAuNS. Data arepresented as mean=SD. c(KRGDf)-PEG-HAuNS group, $n=5$; PEG-HAuNS group, $n=4$. *, significant difference between precontrast and 24-h postcontrast groups ($p<0.05$). Reprinted with permission from Ref. [58]

5 Copper Sulfide Nanoparticles and Carbon Nanotubes

NIR light at 1064 nm lies on the second window in the NIR region that encounters less tissue attenuation by the absorption of hemoglobin and water. Among the pulsed lasers emitting NIR light, the Q-switched Nd:YAG laser, which emits laser light at 1064 nm, can provide high pulsed energy with a nanosecond pulsed widths. In fact, its second harmonic radiation at 532 nm is usually employed to pump other lasing media, such as Ti:Sapphire, to obtain tunable laser output in the NIR region. The laser energy conversion efficiency is characteristically around 10 % (~50 % of second harmonic generation and ~20 % of other lasing media). This means that with a typical laser output of ~1 J/pulse at 1064 nm from a commercial Q-switched Nd:YAG laser, only ~100 mJ/pulse maximum output can be obtained at 800 nm, which is within the first biological tissue window. With 10 times greater available laser pulse energy, a 1064-nm laser would have a much higher fluence rate than that of an 800-nm laser pulse (which is generated with a 532-nm pump and an additional lasing medium) at a depth ≤50 mm. Higher laser energy should translate to a stronger PA signal, higher SNR, and greater field-of-view. However, at 1064 nm, blood vasculature or tumor cells have little specific optical absorption that clearly distinguishes them from other normal organ structures.

Towards the goal of enhancing the sensitivity and specificity of PAI at the 1064 nm wavelength, we recently reported the use of copper sulfide NPs (CuS NPs) as a new class of contrast agent for deep-tissue PAI. CuS NPs display optical absorption tunable to 1064 nm and are much smaller (diameter <15 nm) than plasmonic Au nanostructures, which should be preferable for imaging extravascular tumor cells [61]. CuS NPs are readily synthesized in aqueous solution by reacting $CuCl_2$ and Na_2S in the presence of various stabilizers [62, 63]. Preliminary PAI studies were conducted with PEG-coated CuS NPs dispersed in 10 % polyacrylamide gel embedded in fresh chicken breast. The CuS NP targets in this phantom could be clearly visualized at ~5 cm depth from the laser irradiation surface. Figure 6 shows in vivo PAI images of a mouse brain acquired with green light at 532 nm without exogenous contrast (Fig. 6a) and with NIR light at 1064 nm using PEGylated CuS NPs as a contrast agent (Fig. 6b, c). Hemoglobin is the major chromophore in biological tissues and has strong absorption of green light at 532 nm. As has been shown before with PEG-HAuNS (Fig. 5), the superficial vascular structures of the mouse brain, such as the veins and arteries in the cerebral and temporal lobes, could be clearly visible with green light (Fig. 6a). However, green light cannot penetrate deeply because of strong tissue absorption and scattering at short wavelengths. On PAI images of the mouse brain acquired at 1064 nm, only the sagittal and transverse sinuses were visualized; blood vessels were not discernible because of the lack of contrast between blood vessels and the brain parenchyma (Fig. 6b, c). A nodule on the left cerebral cortex that was injected intracranially with 15 μL of an aqueous solution of CuS NPs (2 OD) 24 h before PAI acquisition was clearly seen, which dissolved 7 days after CuS NP injection because CuS NPs were cleared from the injection site (Fig. 6b, c). These studies confirm that CuS NPs were effective PAI contrast agent at 1064 nm and may be used for molecular PAI imaging of tumor and angiogenic blood vessels when these NPs are decorated with appropriate ligands.

The blood vessel network of a tumor regulates the supply of nutrients and oxygen to the cancer cells, affects their survival and growth, and influences the response of the tumor to therapy. Noninvasive in vivo studies of the tumor blood vasculature are therefore of interest in fundamental cancer research and the development of new drugs and other therapies. To test whether PEG-CuS NPs can be used as an exogenous contrast for PAI of tumor vasculature, we used the orthotopic 4 T1 mammary tumor model. Without contrast agent, several large vessels could be seen, probably due to light absorption by hemoglobin and water molecules at 1064 nm (Fig. 7a) [64]. However, smaller blood vessels are not discernible. At 5 min and 2 h after intravenous injection of PEG-CuS NPs (2 OD), blood vessel structures, including smaller vessels, were more clearly visualized (Fig. 7b, c), indicating increased vasculature contrast after PEG-CuS NP injection. PAI images acquired at 1064 nm

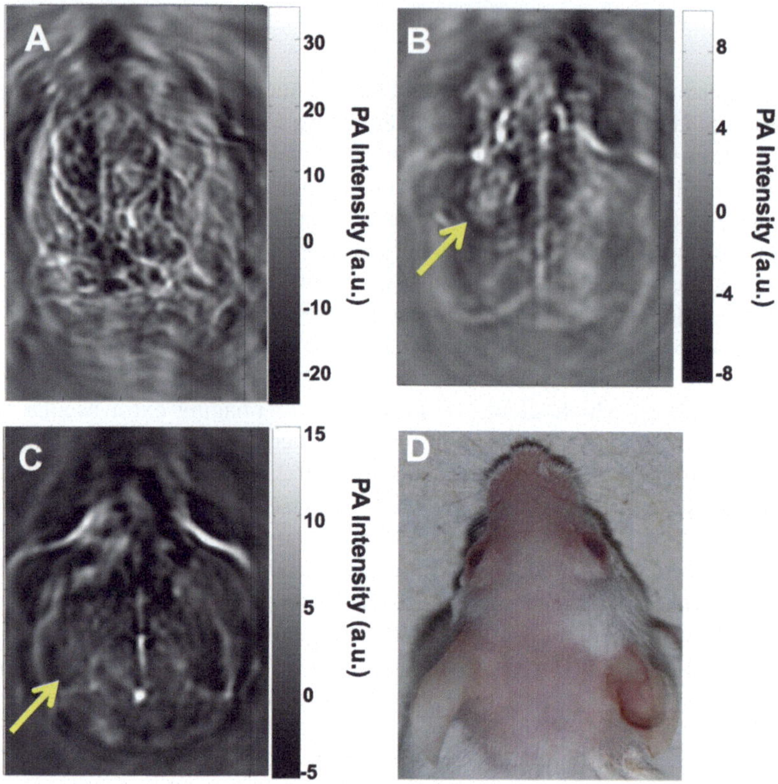

Fig. 6 Representative in vivo PAI images of a mouse brain. Images were acquired using laser light (**a**) at a wavelength of 532 nm without exogenous contrast, (**b**) at 1064 nm 24 h after intracranial injection of 15 μL of CuS NP solution, and (**c**) at 1064 nm 7 days after intracranial injection of 15 μL of CuS NP solution. (**d**) Photograph of the head of the mouse. Laser light was delivered from the top. Reprinted with permission from Ref. [61]

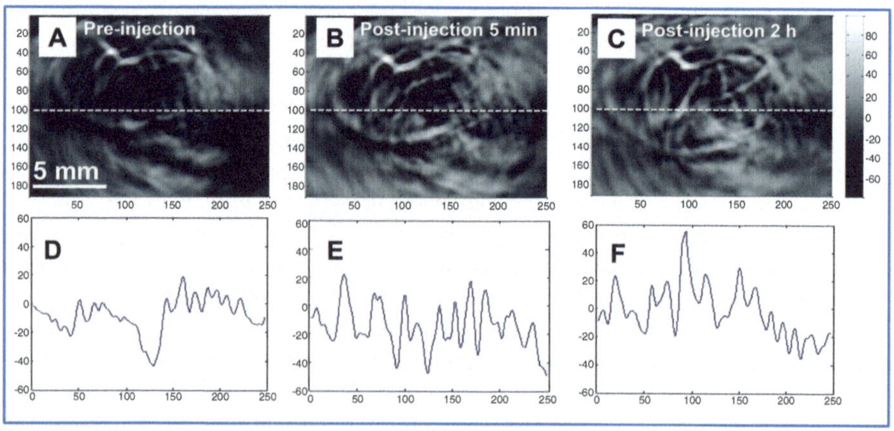

Fig. 7 Representative in vivo PAI images of a 4T1 mammary tumor grown in the mammary fatpad and corresponding PA signals ($n = 3$). Images were acquired using ns-pulsed laser light at a wavelength of 1064 nm (**a**) before CuS NP injection, (**b**) 5 min after intravenous injection of 200 μL of CuS NP solution (100 μg mL − 1, 2 OD), and (**c**) 2 h after intravenous injection of 200 μL of CuS NP solution. (**d–f**) PA signal traces correspond to the white lines in the respective PA images. Reprinted with permission from Ref. [64]

with PEG-CuS NPs suggest that this contrast agent is a promising platform for PAI of tumor blood vessels. An interesting aspect of the studies by Zhou et al. [64] was the incorporation of ^{64}Cu, a positron emitter to the same CuS NP matrix without the use of any radiometal chelator, which permitted simultaneous micro-positron emission tomography (μPET) imaging and quantification of tumor uptake of PEG-CuS NPs. Such a dual-modality imaging approach enabled by a single contrast agent is expected to provide much needed complimentary information that cannot be readily acquired with a single imaging modality.

Several other attempts have been made to increase signal intensity for PAI contrast agents. For example, conjugating small optical dye molecules to single-walled carbon nanotubes (SWNT-dye) led to sub-nanomolar sensitivity for PAI [65]. The dyes used for conjugation includes indocyanine green (ICG), methylene blue (MB), QSY_{21} (QSY), and cyanine (Cy5.5). Studies have shown that dye-containing SWNT had much higher optical absorption in the far red and NIR region than SWNT without dyes. Tumors in mice injected with RGD peptide-coated, dye-conjugated SWNT targeted integrin αvβ3 in angiogenic blood vessels, leading to a threefold increase in PA signal intensity compared to tumors in mice injected with the untargeted SWNT-dye [65]. These and other studies demonstrate that with appropriate nanoparticle design, it is possible to improve the sensitivity of PAI to picomolar concentration [51, 57, 66].

6 PA Imaging of Sentinel Lymph Nodes and Atherosclerotic Plaques

In addition to imaging vasculature and tumor blood vessels, various gold nanostructures are also studied as possible PAI contrast agents for mapping of sentinel lymph nodes (SLNs) [48, 55], ex vivo imaging of joint tissue [67], monitoring of drug release [68], and characterization of macrophages in atherosclerotic plaques [69, 70]. PAI of SLNs is worth noting in particular. The SLN, the first lymph node receiving drainage from the tumor is believed to be most likely to be positive for metastases. SLN biopsy has been used in the clinic as the preferred method for tumor staging. Before a biopsy can be taken for staging purposes, SLN must be identified. Current methods for SLN mapping typically use a blue dye (e.g., isosulfan blue or methylene blue) or radioactive colloids (e.g., technetium-99 or ^{99m}Tc). However, these approaches require either invasive techniques to visualize the blue stain or specialized facilities to deal with potentially hazardous radioactive components. Therefore, techniques such as PAI that can map SLNs without surgery or radioactivity are highly desirable. Using Au nanocages as the PAI contrast agent, Song et al. [48] have shown that SLNs could be readily imaged with PAI in the context of the surrounding vasculature and within ~3 mm below the skin surface of a rat (Fig. 8). By placing chicken breast tissue on top of the rat skin, a further study showed that the SLN could be imaged with good contrast as deep as 33 mm below the skin surface, which is significantly deeper than the ~12 mm depth of SLNs in humans.

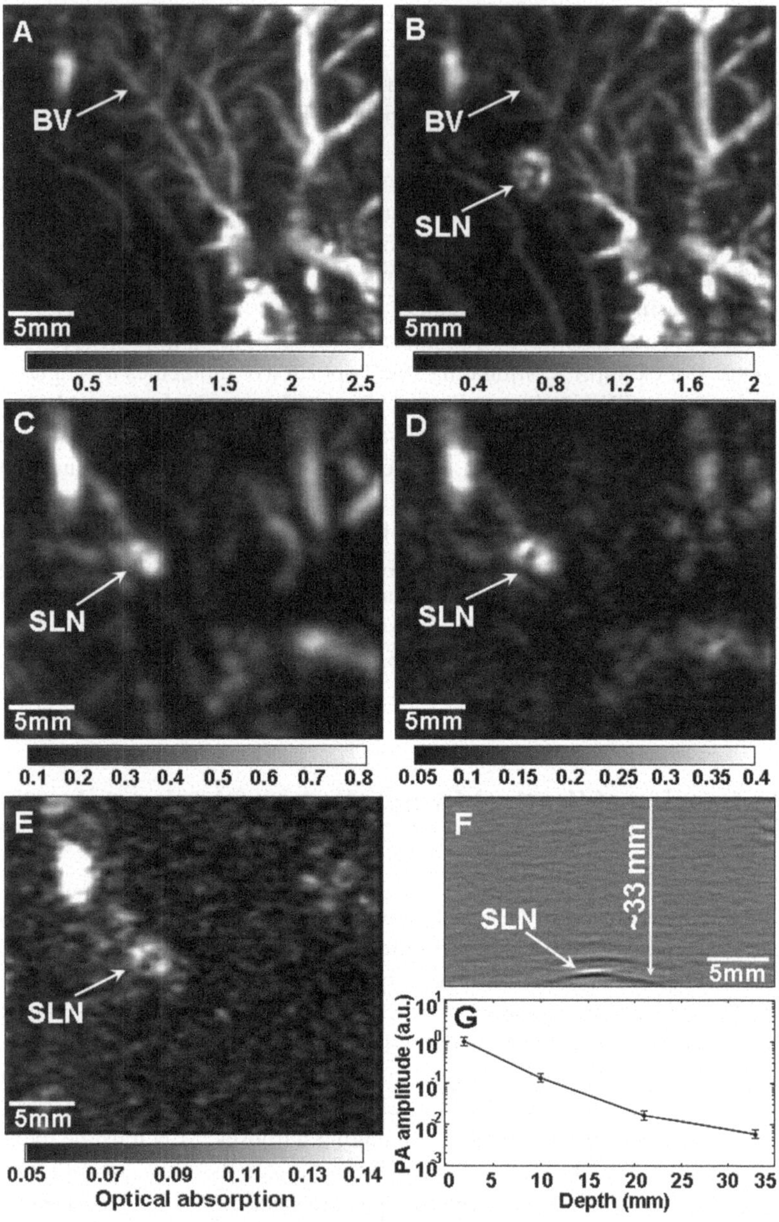

Fig. 8 Depth capability of noninvasive PA SLN mapping in another rat, demonstrating potential for clinical use. PA sagittal MAP images before (**a**) and after (**b–e**) the injection: (**a**) control image; (**b**) 28 min; (**c**) 126 min with a layer of chicken breast tissue placed on axillary region, providing a total imaging depth of 10 mm; (**d**) 165 min with the second layer of chicken breast tissue, providing a total imaging depth of 21 mm; (**e**) 226 min with the third layer of chicken breast tissue, providing a total imaging depth of 33 mm. (**f**) PA B-scan with 20 times signal average, showing the SLN located 33 mm deep. Memory of the acquisition system limited the record length in depth. (**g**) Amplitude variations of PA signals over imaging depths. Data were scaled down to 0 dB and normalized by the maximum. Error bar represents standard deviation. All images were acquired without signal averaging except the B-scan (**f**). Colorbars represent the optical absorption. *BV* blood vessel, *SLN* sentinel lymph node. Reprinted with permission from Ref. [48]

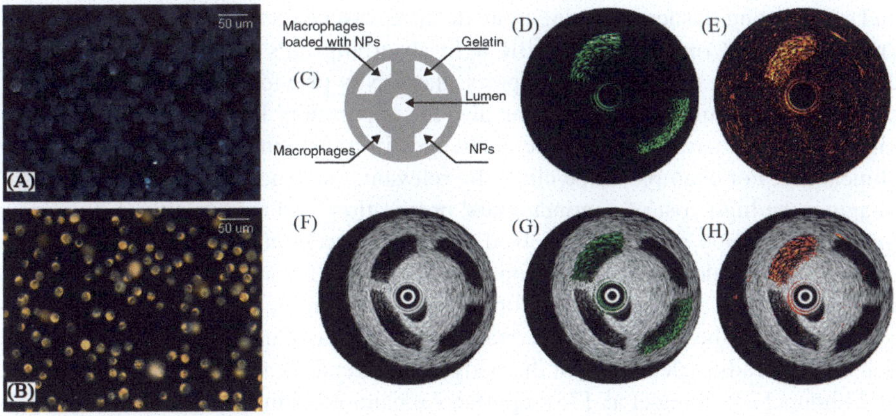

Fig. 9 Optical images of (**a**) intact murine macrophages and (**b**) murine macrophages conjugated with gold NPs. (**c**) Diagram of the tissue-mimicking phantom and (**f**) the corresponding intravascular ultrasound (IVUS) image. IVPA images of the same cross-section of the phantom taken at (**d**) 532 nm and (**e**) 680 nm. Overlaid IVUS and IVPA images of the phantom acquired at (**g**) 532 nm and (**h**) 680 nm, indicating the origin of the PA responses in the upper-left quadrants of the IVPA images. Reprinted with permission from Ref. [69]

Plasmonic gold NPs (Au-NPs) have been investigated as a contrast agent for macrophage detection in atherosclerotic plaques. The aggregation of Au-NPs internalized by macrophages causes a red shift in the particles' optical absorption spectrum. By utilizing enhanced absorption induced by plasmon resonance coupling, intravascular photoacoustic (IVPA) imaging at 680 nm can detect the signal from macrophages highlighted by aggregated Au-NPs (Fig. 9) [69, 71]. Here, the spectral difference in optical absorption between the aggregated and non-aggregated Au-NPs offers a mechanism to image macrophages. By employing multiwavelength measurements, IVPA imaging of a tissue phantom shows that aggregated NPs can be differentiated from the surrounding tissue. Thus, it is possible to identify the presence and the location of NPs associated with macrophage-rich atherosclerotic plaques.

7 Summary and Outlook

PAI imaging is a powerful and emerging noninvasive hybrid modality which provides three-dimension imaging with high spatial resolution. PAI with vasculature contrast agents can be used to perform noninvasive angiography, assess angiogenesis, quantify and measure the spacing of blood vessels, and measure blood volume and flow. PA signal amplitudes can be greatly enhanced with exogenous contrast agents, especially those leveraging NPs with strong absorption in the NIR region. These NPs usually display large absorption cross sections within the optical window of biological tissues. They have been successfully applied as intravascular contrast agents for PAI in various animal models to enhance the contrast between blood and surrounding tissues, allowing more detailed vascular structures to be imaged at greater depths. Combined with NIR light absorbing NPs decorated with ligands directed at tumor cell surface receptors, these NPs provide not only a promising platform for vasculature imaging, but also the opportunity for molecular PAI imaging. However, there are still some challenges in translating these techniques into clinical use. For example, to be clinically relevant, the penetration depth of the PAI imaging, the high cost of nanomaterials' preparation, and the toxicity of the nanomaterials currently present unresolved concerns. Recent attempts to reduce potential long-term side effects of nanoparticle-based PAI contrast agents have been made with some success by developing renally clearable NPs. For example, we reported the synthesis and characterization of ultrasmall CuS NPs (diameter <6 nm) that can be readily cleared from the body via the renal-urinary route over a period of 24 hours [72]. Song et al. [73] reported Au nanorods (dimension: $\approx$8 nm $\times$ 2 nm) embedded in biodegradable poly(lactic-co-glycolic acid) NPs (~60 nm in diameter) that could be cleared from the body over a period of 10 days. These NPs may be adapted for PAI because they all have strong absorption in the NIR region. Through synergistic combination with other imaging modalities (i.e., Raman imaging, MRI, PET), PAI may offer complementary information with greater resolution and/or detection sensitivity [74–79]. Successful translation of PAI to the clinic requires a practical PAI system that efficiently penetrates biological tissues and a contrast agent with optical absorption peaks at or near the wavelength of the laser source. It is expected that progress on both of these fronts will eventually make it possible to introduce PAI to the clinic for guiding individualized therapy for cancer patients.

Acknowledgements This work is supported in part by the John S. Dunn Foundation (CL), by the US National Institutes of Health grants DP1 EB016986 (NIH Director's Pioneer Award, LHW), R01 CA186567 (NIH Director's Transformative Research Award, LHW), and U01 NS090579 (BRAIN Initiative, LHW). L.V. Wang has a financial interest in Endra, Inc., and Microphotoacoustics, Inc., which, however, did not support this work. The other authors declare no competing financial interests.

References

1. Kruger RA. Photoacoustic ultrasound. Med Phys. 1994;21:127–31.
2. Oraevsky AA, Jacques SL, Tittel FK. Measurement of tissue optical properties by time-resolved detection of laser-induced transient stress. Appl Opt. 1997;36:402–15.
3. Wang XD, Pang YJ, Ku G, Xie XY, Stoica G, Wang LHV. Noninvasive laser-induced photoacoustic tomography for structural and functional in vivo imaging of the brain. Nat Biotechnol. 2003;21:803–6.
4. Wang LHV, Hu S. Photoacoustic tomography: in vivo imaging from organelles to organs. Science. 2012;335:1458–62.
5. Jose J, Manohar S, Kolkman RGM, Steenbergen W, van Leeuwen TG. Imaging of tumor vasculature using Twente photoacoustic systems. J Biophotonics. 2009;2:701–17.
6. Kolkman RGM, Hondebrink E, Steenbergen W, de Mul FFM. In vivo photoacoustic imaging of blood vessels using an extreme-narrow aperture sensor. IEEE J Sel Top Quant. 2003;9:343–6.

7. Ku G, Wang XD, Xie XY, Stoica G, Wang LHV. Imaging of tumor angiogenesis in rat brains in vivo by photoacoustic tomography. Appl Opt. 2005;44:770–5.
8. Li ML, Oh JT, Xie XY, Ku G, Wang W, Li C, et al. Simultaneous molecular and hypoxia imaging of brain tumors in vivo using spectroscopic photoacoustic tomography. Proc IEEE. 2008;96:481–9.
9. Ermilov SA, Khamapirad T, Conjusteau A, Leonard MH, Lacewell R, Mehta K, et al. Laser optoacoustic imaging system for detection of breast cancer. J Biomed Opt. 2009;14:024007.
10. Manohar S, Vaartjes SE, van Hespen JCG, Klaase JM, van den Engh FM, Steenbergen W, et al. Initial results of in vivo non-invasive cancer imaging in the human breast using near-infrared photoacoustics. Opt Express. 2007;15:12277–85.
11. Wang LV. Photoacoustic tomography. Scholarpedia. 2014;9:10278.
12. Yao J, Wang LV. Sensitivity of photoacoustic microscopy. Photoacoustics. 2014;2:87–101.
13. Ray A, Rajian JR, Lee YE, Wang X, Kopelman R. Lifetime-based photoacoustic oxygen sensing in vivo. J Biomed Opt. 2012;17:057004.
14. Maslov K, Zhang HF, Hu S, Wang LV. Optical-resolution photoacoustic microscopy for in vivo imaging of single capillaries. Opt Lett. 2008;33:929–31.
15. Li L, Yeh CH, Hu S, Wang LD, Soetikno BT, Chen RM, et al. Fully motorized optical-resolution photoacoustic microscopy. Opt Lett. 2014;39:2117–20.
16. Forest SE, Simon JD. Wavelength-dependent photoacoustic calorimetry study of melanin. Photochem Photobiol. 1998;68:296–8.
17. Viator JA, Komadina J, Svaasand LO, Aguilar G, Choi B, Nelson JS. A comparative study of photoacoustic and reflectance methods for determination of epidermal melanin content. J Investig Dermatol. 2004;122:1432–9.
18. Szakall M, Huszar H, Bozoki Z, Szabo G. On the pressure dependent sensitivity of a photoacoustic water vapor detector using active laser modulation control. Infrared Phys Technol. 2006;48:192–201.
19. Xu Z, Li C, Wang LV. Photoacoustic tomography of water in phantoms and tissue. J Biomed Opt. 2010;15:036019.
20. Xu Z, Zhu Q, Wang LV. In vivo photoacoustic tomography of mouse cerebral edema induced by cold injury. J Biomed Opt. 2011;16:066020.
21. Allen TJ, Hall A, Dhillon AP, Owen JS, Beard PC. Spectroscopic photoacoustic imaging of lipid-rich plaques in the human aorta in the 740 to 1400 nm wavelength range. J Biomed Opt. 2012;17:061209.
22. Wang B, Karpiouk A, Yeager D, Amirian J, Litovsky S, Smalling R, et al. Intravascular photoacoustic imaging of lipid in atherosclerotic plaques in the presence of luminal blood. Opt Lett. 2012;37:1244–6.
23. Wang P, Wang P, Wang H-W, Cheng J-X. Mapping lipid and collagen by multispectral photoacoustic imaging of chemical bond vibration. J Biomed Opt. 2012;17:096010.
24. Bugs MR, Cornelio ML. Analysis of the ethidium bromide bound to DNA by photoacoustic and FTIR spectroscopy. Photochem Photobiol. 2001;74:512–20.
25. Yao DK, Maslov K, Shung KK, Zhou QF, Wang LV. In vivo label-free photoacoustic microscopy of cell nuclei by excitation of DNA and RNA. Opt Lett. 2010;35:4139–41.
26. Di Primo C, Deprez E, Sligar SG, Hoa GHB. Origin of the photoacoustic signal in cytochrome p-450(cam): Role of the Arg186-Asp251-Lys178 bifurcated salt bridge. Biochemistry. 1997;36:112–8.
27. Zhang C, Zhang YS, Yao DK, Xia Y, Wang LV. Label-free photoacoustic microscopy of cytochromes. J Biomed Opt. 2013;18:20504.
28. Esenaliev RO, Karabutov AA, Oraevsky AA. Sensitivity of laser opto-acoustic imaging in detection of small deeply embedded tumors. IEEE J Sel Top Quantum Electron. 1999;5:981–8.
29. Ku G, Wang LV. Deeply penetrating photoacoustic tomography in biological tissues enhanced with an optical contrast agent. Opt Lett. 2005;30:507–9.
30. Kim C, Erpelding TN, Jankovic L, Pashley MD, Wang LV. Deeply penetrating in vivo photoacoustic imaging using a clinical ultrasound array system. Biomed Opt Express. 2010;1:278–84.

31. Xing W, Wang L, Maslov K, Wang LV. Integrated optical- and acoustic-resolution photoacoustic microscopy based on an optical fiber bundle. Opt Lett. 2013;38:52–4.
32. Song KH, Wang LV. Deep reflection-mode photoacoustic imaging of biological tissue. J Biomed Opt. 2007;12:060503.
33. Gamelin J, Maurudis A, Aguirre A, Huang F, Guo PY, Wang LV, et al. A real-time photoacoustic tomography system for small animals. Opt Express. 2009;17:10489–98.
34. Xia J, Chatni MR, Maslov K, Guo Z, Wang K, Anastasio M, et al. Whole-body ring-shaped confocal photoacoustic computed tomography of small animals in vivo. J Biomed Opt. 2012;17:050506.
35. Yao J, Kaberniuk AA, Li L, Shcherbakova DM, Zhang R, Wang L, et al. Multiscale photoacoustic tomography using reversibly switchable bacterial phytochrome as a near-infrared photochromic probe. Nat Meth. 2016;13:67–73.
36. Nasiriavanaki M, Xia J, Wan HL, Bauer AQ, Culver JP, Wang LV. High-resolution photoacoustic tomography of resting-state functional connectivity in the mouse brain. Proc Natl Acad Sci U S A. 2014;111:21–6.
37. Yao J, Wang L, Yang JM, Maslov KI, Wong TT, Li L, et al. High-speed label-free functional photoacoustic microscopy of mouse brain in action. Nat Methods. 2015;12:407–10.
38. Fang H, Maslov K, Wang LV. Photoacoustic doppler effect from flowing small light-absorbing particles. Phys Rev Lett. 2007;99:184501.
39. Yao J, Maslov KI, Zhang Y, Xia Y, Wang LV. Label-free oxygen-metabolic photoacoustic microscopy in vivo. J Biomed Opt. 2011;16:076003.
40. Oladipupo S, Hu S, Kovalski J, Yao J, Santeford A, Sohn RE, et al. VEGF is essential for hypoxia-inducible factor-mediated neovascularization but dispensable for endothelial sprouting. Proc Natl Acad Sci U S A. 2011;108:13264–9.
41. Oladipupo SS, Hu S, Santeford AC, Yao J, Kovalski JR, Shohet RV, et al. Conditional HIF-1 induction produces multistage neovascularization with stage-specific sensitivity to VEGFR inhibitors and myeloid cell independence. Blood. 2011;117:4142–53.
42. Agarwal A, Shao X, Rajian JR, Zhang H, Chamberland DL, Kotov NA, et al. Dual-mode imaging with radiolabeled gold nanorods. J Biomed Opt. 2011;16:051307.
43. Chen Y-S, Frey W, Kim S, Homan K, Kruizinga P, Sokolov K, et al. Enhanced thermal stability of silica-coated gold nanorods for photoacoustic imaging and image-guided therapy. Opt Express. 2010;18:8867–78.
44. You J, Zhou J, Zhou M, Liu Y, Robertson JD, Liang D, et al. Pharmacokinetics, clearance, and biosafety of polyethylene glycol-coated hollow gold nanospheres. Part Fibre Toxicol. 2014;11:26.
45. Wang YW, Xie XY, Wang XD, Ku G, Gill KL, O'Neal DP, et al. Photoacoustic tomography of a nanoshell contrast agent in the in vivo rat brain. Nano Lett. 2004;4:1689–92.
46. Li ML, Wang JC, Schwartz JA, Gill-Sharp KL, Stoica G, Wang LV. In-vivo photoacoustic microscopy of nanoshell extravasation from solid tumor vasculature. J Biomed Opt. 2009;14:010507.
47. Yang XM, Skrabalak SE, Li ZY, Xia YN, Wang LHV. Photoacoustic tomography of a rat cerebral cortex in vivo with au nanocages as an optical contrast agent. Nano Lett. 2007;7:3798–802.
48. Song KH, Kim C, Cobley CM, Xia Y, Wang LV. Near-infrared gold nanocages as a new class of tracers for photoacoustic sentinel lymph node mapping on a rat model. Nano Lett. 2009;9:183–8.
49. Agarwal A, Huang SW, O'Donnell M, Day KC, Day M, Kotov N, et al. Targeted gold nanorod contrast agent for prostate cancer detection by photoacoustic imaging. J Appl Phys. 2007;102:064701.
50. Eghtedari M, Liopo AV, Copland JA, Oraevslty AA, Motamedi M. Engineering of hetero-functional gold nanorods for the in vivo molecular targeting of breast cancer cells. Nano Lett. 2009;9:287–91.
51. Eghtedari M, Oraevsky A, Copland JA, Kotov NA, Conjusteau A, Motamedi M. High sensitivity of in vivo detection of gold nanorods using a laser optoacoustic imaging system. Nano Lett. 2007;7:1914–8.

52. Kim K, Huang SW, Ashkenazi S, O'Donnell M, Agarwal A, Kotov NA, et al. Photoacoustic imaging of early inflammatory response using gold nanorods. Appl Phys Lett. 2007;90:223901.
53. Li PC, Wang CRC, Shieh DB, Wei CW, Liao CK, Poe C, et al. In vivo photoacoustic molecular imaging with simultaneous multiple selective targeting using antibody-conjugated gold nanorods. Opt Express. 2008;16:18605–15.
54. Li PC, Wei CW, Liao CK, Chen CD, Pao KC, Wang CRC, et al. Photoacoustic imaging of multiple targets using gold nanorods. IEEE Trans Ultrason Ferroelectr Freq Control. 2007;54:1642–7.
55. Song KH, Kim C, Maslov K, Wang LV. Noninvasive in vivo spectroscopic nanorod-contrast photoacoustic mapping of sentinel lymph nodes. Eur J Radiol. 2009;70:227–31.
56. Melancon MP, Zhou M, Li C. Cancer theranostics with near-infrared light-activatable multimodal nanoparticles. Acc Chem Res. 2011;44:947–56.
57. Lu W, Huang Q, Geng KB, Wen XX, Zhou M, Guzatov D, et al. Photoacoustic imaging of living mouse brain vasculature using hollow gold nanospheres. Biomaterials. 2010;31:2617–26.
58. Lu W, Melancon MP, Xiong C, Huang Q, Elliott A, Song S, et al. Effects of photoacoustic imaging and photothermal ablation therapy mediated by targeted hollow gold nanospheres in an orthotopic mouse xenograft model of glioma. Cancer Res. 2011;71:6116–21.
59. Zhou M, Singhana B, Liu Y, Huang Q, Mitcham T, Wallace MJ, et al. Photoacoustic- and magnetic resonance-guided photothermal therapy and tumor vasculature visualization using theranostic magnetic gold nanoshells. J Biomed Nanotech. 2015;11:1442–50.
60. Sharma P, Brown SC, Bengtsson N, Zhang QZ, Walter GA, Grobmyer SR, et al. Gold-speckled multimodal nanoparticles for noninvasive bioimaging. Chem Mater. 2008;20:6087–94.
61. Ku G, Zhou M, Song SL, Huang Q, Hazle J, Li C. Copper sulfide nanoparticles as a new class of photoacoustic contrast agent for deep tissue imaging at 1064 nm. ACS Nano. 2012;6:7489–96.
62. Zhou M, Song S, Zhao J, Tian M, Li C. Theranostic CuS nanoparticles targeting folate receptors for PET image-guided photothermal therapy. J Mater Chem B. 2015;3:8939–48.
63. Zhou M, Zhang R, Huang M, Lu W, Song S, Melancon MP, et al. A chelator-free multifunctional [^{64}Cu]CuS nanoparticle platform for simultaneous micro-PET/CT imaging and photothermal ablation therapy. J Am Chem Soc. 2010;132:15351–8.
64. Zhou M, Ku G, Pageon L, Li C. Theranostic probe for simultaneous in vivo photoacoustic imaging and confined photothermolysis by pulsed laser at 1064 nm in 4T1 breast cancer model. Nanoscale. 2014;6:15228–35.
65. de la Zerda A, Bodapati S, Teed R, May SY, Tabakman SM, Liu Z, et al. Family of enhanced photoacoustic imaging agents for high-sensitivity and multiplexing studies in living mice. ACS Nano. 2012;6:4694–701.
66. Bouchard L-S, Anwar MS, Liu GL, Hann B, Xie ZH, Gray JW, et al. Picomolar sensitivity MRI and photoacoustic imaging of cobalt nanoparticles. Proc Natl Acad Sci U S A. 2009;106:4085–9
67. Chamberland DL, Agarwal A, Kotov N, Fowlkes JB, Carson PL, Wang X. Photoacoustic tomography of joints aided by an Etanercept-conjugated gold nanoparticle contrast agent—an ex vivo preliminary rat study. Nanotechnology. 2008;19:095101.
68. Lee HJ, Liu Y, Zhao J, Zhou M, Bouchard RR, Mitcham T, et al. In vitro and in vivo mapping of drug release after laser ablation thermal therapy with doxorubicin-loaded hollow gold nanoshells using fluorescence and photoacoustic imaging. J Controlled Release. 2013;172:152–8.
69. Wang B, Yantsen E, Larson T, Karpiouk AB, Sethuraman S, Su JL, et al. Plasmonic intravascular photoacoustic imaging for detection of macrophages in atherosclerotic plaques. Nano Lett. 2009;9:2212–7.
70. Rouleau L, Berti R, Ng VW, Matteau-Pelletier C, Lam T, Saboural P, et al. VCAM-1-targeting gold nanoshell probe for photoacoustic imaging of atherosclerotic plaque in mice. Contrast Media Mol Imaging. 2013;8:27–39.
71. Wang B, Joshi P, Sapozhnikova V, Amirian J, Litovsky SH, Smalling R, et al. Intravascular photoacoustic imaging of macrophages using molecularly targeted gold nanoparticles. Proc SPIE 7564. 2010;75640A.

72. Zhou M, Li J, Liang S, Sood AK, Liang D, Li C. CuS nanodots with ultrahigh efficient renal clearance for positron emission tomography imaging and image-guided photothermal therapy. ACS Nano. 2015;9(7):7085–96.
73. Song J, Yang X, Jacobson O, Huang P, Sun X, Lin L, et al. Ultrasmall gold nanorod vesicles with enhanced tumor accumulation and fast excretion from the body for cancer therapy. Adv Mater. 2015;27:4910–7.
74. Yang K, Hu L, Ma X, Ye S, Cheng L, Shi X, et al. Multimodal imaging guided photothermal therapy using functionalized graphene nanosheets anchored with magnetic nanoparticles. Adv Mater. 2012;24:1868–72.
75. Qin H, Zhou T, Yang S, Chen Q, Xing D. Gadolinium(III)-gold nanorods for MRI and photoacoustic imaging dual-modality detection of macrophages in atherosclerotic inflammation. Nanomedicine. 2013;8:1611–24.
76. Lovell JF, Jin CS, Huynh E, Jin H, Kim C, Rubinstein JL, et al. Porphysome nanovesicles generated by porphyrin bilayers for use as multimodal biophotonic contrast agents. Nat Mater. 2011;10:324–32.
77. Kim J-W, Galanzha EI, Shashkov EV, Moon H-M, Zharov VP. Golden carbon nanotubes as multimodal photoacoustic and photothermal high-contrast molecular agents. Nat Nanotechnol. 2009;4:688–94.
78. Zhang YM, Jeon M, Rich LJ, Hong H, Geng JM, Zhang Y, et al. Non-invasive multimodal functional imaging of the intestine with frozen micellar naphthalocyanines. Nat Nanotechnol. 2014;9:631–8.
79. Kircher MF, de la Zerda A, Jokerst JV, Zavaleta CL, Kempen PJ, Mittra E, et al. A brain tumor molecular imaging strategy using a new triple-modality MRI-photoacoustic-Raman nanoparticle. Nat Med. 2012;18:829–34.
80. Guo L, Lei L, Liren Z, Jun X, Wang LV. Multiview Hilbert transformation for full-view photoacoustic computed tomography using a linear array. J Biomed Opt. 2015;20:066010.
81. Zhang HF, Maslov K, Stoica G, Wang LHV. Functional photoacoustic microscopy for high-resolution and noninvasive in vivo imaging. Nat Biotechnol. 2006;24:848–51.
82. Yao J, Maslov KI, Puckett ER, Rowland KJ, Warner BW, Wang LV. Double-illumination photoacoustic microscopy. Opt Lett. 2012;37:659–61.
83. Zhu L, Li L, Gao L, Wang LV. Multiview optical resolution photoacoustic microscopy. Optica. 2014;1:217–22.
84. Zhang C, Maslov K, Yao J, Wang LV. In vivo photoacoustic microscopy with 7.6-μm axial resolution using a commercial 125-MHz ultrasonic transducer. J Biomed Opt. 2012;17:116016.
85. Zhang C, Maslov K, Hu S, Chen R, Zhou Q, Shung KK, et al. Reflection-mode submicron-resolution in vivo photoacoustic microscopy. J Biomed Opt. 2012;17:0205011.
86. Zhang C, Maslov K, Wang LV. Subwavelength-resolution label-free photoacoustic microscopy of optical absorption in vivo. Opt Lett. 2010;35:3195–7.
87. Yao J, Wang L, Li C, Zhang C, Wang LV. Photoimprint photoacoustic microscopy for three-dimensional label-free subdiffraction imaging. Phys Rev Lett. 2014;112:014302.
88. Nedosekin DA, Galanzha EI, Dervishi E, Biris AS, Zharov VP. Super-resolution nonlinear photothermal microscopy. Small. 2014;10:135–42.

Nanoparticles for Cardiovascular Imaging with CT

Peter Chhour, Rabe'e Cheheltani, Pratap C. Naha, Harold I. Litt, Victor A. Ferrari, and David P. Cormode

1 Introduction

Cardiovascular diseases (CVD) are the leading cause of death, accounting for nearly 1 in 3 deaths in the developed world today. In the USA, approximately 35.1 % of adults have at least one cardiovascular disease, and this number is expected to rise to 43.9 % by 2030 [1]. Cardiovascular disease is a broad definition that covers several pathologies including stroke, high blood pressure, heart failure, and coronary artery disease (CAD). Nearly half of the total deaths from CVD arise from CAD which includes angina pectoris, atherosclerosis of the coronary arteries, myocardial infarction, and ischemic heart disease [2]. The majority of imaging studies presented here focuses on these coronary artery disease-related pathologies.

P. Chhour
Department of Radiology, University of Pennsylvania, Philadelphia, PA, USA

Department of Bioengineering, University of Pennsylvania, Philadelphia, PA, USA

R. Cheheltani • P.C. Naha
Department of Radiology, University of Pennsylvania, Philadelphia, PA, USA

H.I. Litt • V.A. Ferrari
Department of Radiology, University of Pennsylvania, Philadelphia, PA, USA

Division of Cardiovascular Medicine, University of Pennsylvania,
3400 Spruce Street, 1 Silverstein, Philadelphia, PA 19104, USA

D.P. Cormode (✉)
Department of Radiology, University of Pennsylvania, Philadelphia, PA, USA

Department of Bioengineering, University of Pennsylvania, Philadelphia, PA, USA

Division of Cardiovascular Medicine, University of Pennsylvania,
3400 Spruce Street, 1 Silverstein, Philadelphia, PA 19104, USA
e-mail: david.cormode@uphs.upenn.edu

J.W.M. Bulte, M.M.J. Modo (eds.), *Design and Applications of Nanoparticles in Biomedical Imaging*, DOI 10.1007/978-3-319-42169-8_17

In the clinic, cardiovascular diseases are typically diagnosed and monitored using an X-ray based imaging technique known as computed tomography. Imaging using this technique is done together with injection of iodinated contrast agents, which make the blood visible, thus delineating the heart and coronary arteries. These iodinated agents have a number of drawbacks such as swift clearance, lack of specificity, and causing adverse effects in some patients [3]. These drawbacks and the advent of the use of nanotechnology in medicine (nanomedicine) have spurred the development of a raft of new nanoparticle based contrast agents for CT [4]. As we explain below, these agents offer advantages such as long blood circulation times, high payloads, multifunctionality, site-specific imaging, and potentially improved biocompatibility.

In this chapter, we first provide background on the characteristics of coronary artery disease. We next describe the principles of computed tomography and the contrast agents currently used for this technique in the clinic. We then discuss novel nanoparticle contrast agents for CT. Agents based on a range of elements are covered (iodine, gold, bromine, bismuth, tantalum, and others). We describe agents designed to provide vascular contrast and agents that are targeted to specific features of interest within blood vessels. Examples featuring novel CT technologies are included, such as spectral CT. We explain certain studies in detail via the use of figures.

2 Coronary Artery Disease

Coronary artery disease is the progression of atherosclerosis in the coronaries such that it leads to clinical manifestations, i.e., ischemic heart disease and myocardial infarction. Atherosclerosis is prevalent in areas of high oscillatory flow and low endothelial shear stress, indicating that initiation of atherosclerosis may be due to localized mechanical stresses [5, 6]. These local flow conditions allow for the accumulation of circulating low density lipoproteins (LDL) into the intima. Oxidation of these LDL particles leads to the activation of the endothelium, increasing the expression of surface adhesion molecules [7, 8]. Increased expression of vascular cell adhesion molecule (VCAM) and intercellular cell adhesion molecules (ICAM) promote the adherence of circulating leukocytes [9–11]. Subsequently, a chemokine gradient drives the migration of these adhered leukocytes into intima. Inflammatory monocytes play a key role in the progression of atherosclerosis and have been found to be recruited into the intima through chemokine gradients for MCP-1 (CCR2 receptor), fractalkine (CX3CR1 receptor), and CCL5 (CCR5 receptor) [12–14]. In the intima, monocytes differentiate into phenotypical macrophages and begin engulfing oxidized LDL particles. These lipid particles accumulate in the macrophages, eventually becoming foam cells, which are thus named due to their appearance under microscopy [15]. Foam cells promote the destabilization of the surrounding extracellular matrix (ECM) through the release of proteolytic enzymes, including matrix metalloproteinases (MMP) and cathepsins [16]. The chronic buildup of foam cells and other cells, such as altered smooth muscle cells migrating from the media, increases the chances of plaque rupture [17]. The progression of atherosclerosis from initiation through to plaque rupture offers a number of targets for imaging that may provide improved diagnoses or increase our knowledge of the disease. Including imaging modalities such as magnetic resonance imaging (MRI), fluorescence and positron emission tomography, there have been reports of imaging oxidized LDL for endothelial dysfunction [18], cell adhesion molecules for endothelial activation [19], monocytes/macrophages for inflammation [20], proteases for ECM disruption [21], integrins for angiogenesis [22], and fibrin for thrombosis [23]. Calcification of atherosclerotic plaque provides a useful clinical target for CT imaging. Coronary artery calcium scoring measures the area of individual calcified plaques and multiples it by a factor based on the highest attenuation value found in the area. Coronary artery calcium score is strongly associated with future clinical events but the presence of non-calcified and partially calcified plaques diminishes its predictive value [24, 25].

The buildup of plaques may eventually lead to rupture, which releases the contents of the plaque into the circulation. The contents of the plaque such as necrotic cells and cell debris induce the formation of a thrombus, which can occlude the artery [26]. Such an occlusion occurring in the coronary artery can result in an acute myocardial infarction. The myocardium downstream of the occlusion may become ischemic, which can lead to necrosis. If the occlusion is resolved, complications involved in the wound healing process of the injured myocardium present additional targets for imaging [27, 28]. For example, several groups have published studies focused on imaging myocyte necrosis and apoptosis, which occurs under ischemic conditions [29, 30]. Monocytes are recruited to the infarct site as part of the wound healing process, but the benefit of their presence in the injured myocardium is debated [31]. Therefore, there is interest in monitoring the recruitment of monocytes into the infarct site to elucidate their role [32, 33]. Similarly to the progression of coronary artery disease, proteases [34], extracellular matrix [35], and angiogenesis [36] can be informative imaging targets after a myocardium infarct.

Imaging plays an important role in the diagnosis and study of cardiovascular disease and each noninvasive imaging modality has its own advantages and weaknesses. Plaques in the coronary arteries are relatively small, requiring high spatial resolution for imaging. Additionally, cardiac and respiratory motions present a challenge for image acquisition. Magnetic resonance imaging provides excellent soft tissue contrast, but suffers from long image acquisition times, preventing imaging of the coronary arteries from being performed in a single breath hold to limit respiratory motion. Moreover, MRI of the coronary arteries lacks information on plaque calcification and cannot quantify the extent of the disease [37]. MRI is not routinely used for coronary artery imaging in clinical practice. Positron emission tomography can provide functional cardiac information with the use of radioisotope contrast agents [38]. Radiolabeled contrast agents typically have short-half-lives, which increases the complexity of their distribution from production centers. The use of PET for coronary artery disease continues to expand but still remains extremely limited clinically [39]. 18F-fluorodeoxyglucose (18F-FDG) has been used as a macrophage marker and subsequent marker of inflammation in atherosclerosis [40].

However, the low spatial resolution of PET has caused concern about reliability when quantifying 18F-FDG uptake in the coronary plaque [41]. 18F-sodium fluoride has also been explored as a PET marker for arterial calcification [42]. For both contrast agents, low spatial resolution, motion, and off-target uptake typically limit PET studies to evaluation of plaque only in larger arteries such as the carotids or thoracic aorta. As we explain below, the modality of choice for clinical imaging of coronary artery disease is computed tomography.

3 Computed Tomography

Computed tomography has unique advantages for visualizing the coronary arteries that make it an excellent modality for this application. CT relies on X-rays that are produced by accelerating electrons from a cathode towards an anode, which is typically made of tungsten. The impacting electrons, depending on their energy, produce X-rays in the form of bremsstrahlung and characteristic radiation. In bremsstrahlung radiation, the incoming electron path is diverted by interaction with the nuclei, converting lost kinetic energy into an X-ray. In the case of characteristic radiation, the incoming electron ejects an inner-orbital electron of the atom. An electron from an outer orbital fills the vacancy while releasing a characteristic X-ray. For imaging applications, X-rays from the generating source interact with the object of interest in two primary mechanisms, Compton scattering and the photoelectric effect. For Compton scattering, the incoming X-ray transfers a portion of its energy to an outer shell electron which causes the X-ray to scatter in a different direction. For CT energy levels, the primary form of interaction between X-rays and the subject is the photoelectric effect. The incoming X-ray ejects an inner shell electron of the atom, typically of the k or l shell. An outer shell electron fills the inner-shell vacancy and produces a characteristic X-ray. The photoelectric effect occurs when the energy of the incoming X-ray is higher than the binding energy of the inner-shell electron, also known as the k-edge or l-edge energy. The photoelectric effect produces a significant jump in mass attenuation, improving contrast for imaging. For CT contrast agents, imaging with peak voltages higher than the k-edge energy of the X-ray attenuating agent improves contrast enhancement between the agent and surrounding tissue. X-ray attenuation is quantified by the Hounsfield scale which defines Hounsfield units $(\mathrm{HU}) = 1000 \times (\mu - \mu_{\mathrm{water}})/(\mu_{\mathrm{water}} - \mu_{\mathrm{air}})$, where μ is the linear attenuation coefficient of the corresponding material. The Hounsfield scale standardizes the attenuation of a material in reference to the attenuation of water (HU = 0) and air (HU = −1000).

CT is the current clinical gold standard for noninvasive detection of coronary stenosis. The coronary arteries in humans are approximately 4–5 mm in diameter proximally, tapering rapidly to less than 1 mm, mandating high spatial resolution for imaging. A variety of adjustable factors including slice thickness, reconstruction parameters, and field of view can affect spatial resolution in CT; most modern scanners are capable of producing resolution of 0.5 mm or less in each dimension [43]. Moreover, cardiac and respiratory motion can be overcome by the fast temporal resolution of CT and a volume coverage allowing imaging of the entire heart to be acquired in a single breath hold. Increases in the number of detectors and improvement in gantry rotation speed in CT scanners has considerably improved the temporal resolution over the past two decades. Modern clinical 64-detector CT systems are capable of sub-200 ms resolution [44], while the newest CT scanners are capable of sub-100 ms temporal resolution [45]. Synchronization of the acquisition to the electrocardiogram (ECG) signal can be used to reduce artifacts from heart motion [46]. ECG synchronization can be performed using two methods, prospective triggering and retrospective gating. Prospective triggering acquires images at a predefined portion of the R-R interval, while retrospective gating acquires images throughout the cardiac cycle, allowing for image reconstruction at any phase of the heartbeat, but at the cost of a higher radiation dose.

CT can quickly image the coronary arteries and exclude coronary artery disease from patients presenting with chest pain [47–49]. CT allows for post-processing of cross-sectional scans to create improved visual information, such as that seen in the 3D renderings of coronary arteries shown in Fig. 1 [50]. The applications of cardiac CT for plaque characterization include calcium scoring, determination of noncalcified plaque volume, and extent of stenosis [51]. These parameters can be assessed with CT angiography (CTA), a noninvasive CT scan using contrast agents to visualize arterial flow. The plaque characteristics that can be derived from CTA can be used to identify patients at high risk of a significant cardiac event. High risk plaques features that can be visualized with CTA include stenosis ≥ 50 %, positive remodeling, low HU, napkin ring sign, and spotty calcium [52]. A novel application of cardiac CT angiography is noninvasive determination of the hemodynamic significance of an observed stenosis. Fractional flow reserve (FFR) describes the pressure difference across an arterial stenosis as compared to the pressure in the absence of a stenosis

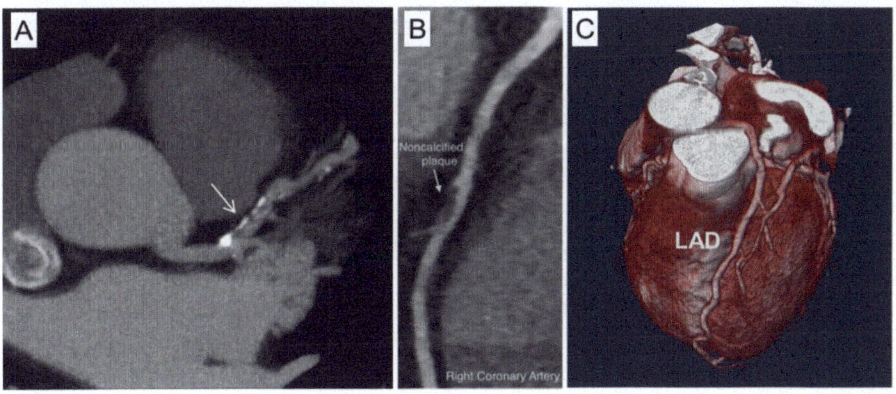

Fig. 1 Clinical CT scans and 3D reconstructions of human hearts displaying high spatial resolution and anatomical features of cardiovascular disease. (**a**) Cross-sectional CT image of the heart. *White arrow* indicates regions of calcified plaques in the coronary artery. (**b**) CT scan showing region of non-calcified plaques in the right coronary artery. (**c**) 3D reconstruction of the heart from CT images, with the left arterial descending (LAD) coronary artery in proximal view

and can be measured directly with catheter angiography [53]. However, advancements in computer modeling has allowed calculations of fluid dynamics from CTA scans for accurate determination of coronary flow and pressure without the need for an invasive procedure. CTA-derived computational models of FFR have been shown to have higher accuracy of determining the impact of stenoses than CTA based measurements of stenosis dimensions alone [54]. The myocardium can also be visualized with high resolution allowing for detection of abnormal morphology in the chambers of the heart and valves [55]. Moreover, the degree of injured myocardium after an infarct can be determined due to the changes in the thickness and/or scar formation of the myocardium [56].

4 Contrast Agents for CT

CT contrast agents play a key role in cardiovascular CT scans; the use of contrast agents in CT can improve the delineation between different soft tissues and is critical for highlighting blood vessels. The primary mechanism for X-ray attenuation in CT is the photoelectric effect. At CT energy levels (25–140 keV), incoming X-rays have enough energy to interact with and eject inter-shell electrons (k or l shell) of the atom. Ejection of a k-shell electron produces a significant increase in X-ray attenuation and allows for increased separation between tissue and contrast media. For this reason, CT contrast agents are typically high atomic number elements such as iodine, barium, gold, bismuth, and lanthanides that have l and/or k edges that are at much higher energies than tissue and are closer to diagnostic X-ray energies. However, CT suffers from low sensitivity to contrast agents, typically requiring millimolar local concentrations to produce contrast that can be distinguished from the background [57].

Barium sulfate is an agent used for imaging the gastrointestinal tract. The agent is typically swallowed and allows visualization of the digestive system [58]. While barium sulfate is limited to imaging of the GI tract, iodine based agents are widely used intravenously for a variety of applications. Iodinated contrast agents (ICA) can be used to visualize the vasculature as well as the organs such as the brain, heart, liver, and kidneys [59–61]. Clinical ICA are small molecules that are based on a tri-iodinated benzene ring. These types of structures reduce osmolality and toxicity compared with injections of iodide ion based agents. These iodinated rings are typically functionalized with amides and alcohol groups to provide water solubility, and can be monomers or dimers. These agents can be further categorized into ionic and non-ionic formulations which yield different properties in terms of toxicity and clearance from the blood stream. Regardless of the molecular structure of these agents, clinically approved ICA suffer from rapid renal clearance from the circulation after intravenous injection, on the order of minutes [62]. Given this rapid clearance, large amounts of contrast agent are often used for prolonged or multiple CT scans, with doses of 150 ml being used routinely [63]. Injecting large volumes of ICA increases the chance of severe renal impairment, known as contrast induced nephropathy, in specific patient populations [64]. Contrast-induced nephropathy is more likely in patients with preexisting renal dysfunction [65], who are now 26 % of the US population over 65 [66]. This proportion is expected to increase in the coming years, since renal dysfunction is common in diabetics and the prevalence of diabetes is predicted to rapidly rise [67, 68]. Iodinated contrast agents also can commonly cause adverse reactions in patients ranging from mild symptoms including nausea and vomiting to severe reactions such as anaphylaxis [69].

These issues with agents have created interest in developing new contrast agents for CT. Nanotechnology has been the primary source of new contrast agents for CT, since nanoparticle platforms allow for increased payload delivery, longer circulation times, targeted agents, and straightforward integration of multiple properties (e.g., drugs or additional contrast media) [70–74]. New agents are also being sought that have less adverse effects than current clinical agents. Researchers are able to control the size, shape, and surface properties of nanoparticles, which can drastically alter the biological interactions of these agents. For instance, while small molecules are rapidly cleared through renal filtration, a nanoparticle can be precisely synthesized to a size >10 nm, reducing renal clearance and therefore increasing circulation time. Additionally, control of the surface properties of the particle can allow for increased circulation time [75–78], or specific targeting to tissue [79–81]. With control over these properties, nanotechnology offers an effective solution to designing contrast agents with desirable properties for specific applications.

5 Nanoparticles for CT Imaging of Cardiovascular Disease

5.1 Iodinated Contrast Agents

As mentioned, iodinated small molecule contrast agents are the most widely used class of agent in CT scanning. While different formulations (ionic vs non-ionic, dimer vs monomer) have slightly differing biocompatibility properties, they continue to suffer from rapid clearance and risk of adverse reactions [69]. Researchers have turned to nanotechnology to improve the properties of these iodinated agents. Incorporation of iodinated agents into nanoparticles such as nano-emulsions, micelles, liposomes, dendrimers have resulted in improvements in circulation time and biocompatibility. The structures of some of the nanoparticle agents used as CT contrast agents are displayed in Fig. 2. Long blood circulation time of contrast agents may help to improve the assessment of cardiovascular health. A long circulating agent also can decrease the total dose necessary for an imaging procedure, potentially reducing the burden of the agent on the patient. The use of a blood pool contrast agent for CT may provide information about coronary artery disease and progression, allowing visualization of the extent of stenosis in an artery and well as calcified and non-calcified plaques [25]. Use of a contrast agent also allows for assessment of cardiac functions such as ejection fraction and stroke volume [43]. In addition, contrast agents are needed to highlight blood vessels during stent placement procedures.

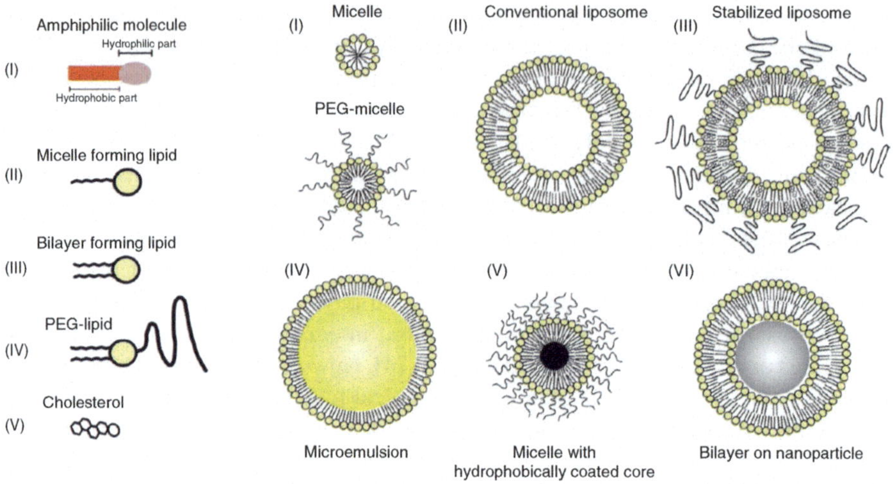

Fig. 2 Schematic representation of nanoparticle platforms. PEG functionalized nanoparticles are less likely to be taken up by cells of the mononuclear phagocytic system, leading to longer circulation half-lives. Reproduced with permissions from Ref. [82]

To incorporate iodinated agents into nanoparticles, nano-emulsions have been formed with a hydrophobic core of iodinated oil. Nano-emulsions can be formed via simple synthetic methods, which produce stable iodinated particles and have been an attractive choice for iodinated contrast agents. An emulsion is formed with a suitable amphiphilic surfactant encapsulating an iodinated oil core. An iodinated oil called Lipiodol (also known as Ethiodol) is FDA approved and consists of iodine combined with ethyl esters of poppy seed oil [83]. Because Lipiodol is hydrophobic, it has limited clinical uses (it is used as an agent for lymphangiography and trans-arterial chemoembolization) [84, 85]. To expand the applications of Lipiodol as a blood pool agent, Pluronic F127 and a polyethylene glycol (PEG) diamine were used to encapsulate Lipiodol into a nano-emulsion [86]. These particles, 150 nm in diameter, were shown to be thermodynamically stable and produce vascular CT contrast in the ventricles and major arteries for up to 4 h.

Nano-emulsions have also been reported where a mixture of phospholipids and cholesterol was used to surround the hydrophobic core, which consisted of iodinated triglycerides. A nano-emulsion formed with endogenous oils and lipids was demonstrated as an effective contrast agent for hepatic imaging and is available as a commercial product known as Fenestra LC® [87, 88]. The iodinated triglycerides making up the core of this nano-emulsion are rapidly metabolized by the liver [89]. To provide longer circulation as a blood pool contrast agent, the agent has been reformulated with PEG included at the surface of the particle to reduce uptake by hepatocytes, which is known as Fenestra VC® [90, 91]. As a blood pool agent, Fenestra VC® has been used to study cardiac function in mice, quantifying ejection fraction, stroke volume, and cardiac output [92]. In this study, Fenestra VC® provided vascular contrast for 3 h with a nearly 500 HU value difference between the blood and myocardium. Moreover, with the use of Fenestra VC®, Detombe et al. studied cardiac function and ventricular remodeling in a mouse model of myocardial infarction. The group was able to assess systolic and diastolic volumes and ejection fraction noninvasively over the course of 4 weeks after myocardial infarction [93]. Both studies utilized cardiac gating with micro-CT to quantify cardiac function accurately.

While Fenestra formulations have been used successfully for cardiac applications, relatively poor inclusion of iodine by weight means that very large volume administrations of the agent are required (up to 40 % of total blood volume). To overcome this issue, researchers have used denser iodinated oils to improve iodine content in the formulations. Using, a dense tri-iodinated oil, de Vries et al. formed nano-emulsions using the amphiphilic polymer, poly(butadiene)-b-poly(ethylene oxide) (PBD-PEO) [94]. The agent was able to provide strong vascular contrast in the chambers of the heart of up to 220 HU directly after administration. Attenuation of the liver and spleen remained at high levels over the three hour observation period suggesting the mononuclear phagocytic system as the main clearance pathway for the agent. Novel approaches to increasing the density of iodinated oils continue to be explored [95, 96]. For instance, Attia et al. reported agents that contain up to 65 % wt. of iodine via use of a hexa-iodinated monoglyceride and a PEGylated nonionic surfactant [91]. This formulation showed clear delineation of the major arteries and heart chambers post-injection with a long circulation half-life of 6.1 h.

Micelles are similar to nano-emulsions, in that they are self-assemblies of amphiphilic molecules where the hydrophilic headgroups point outwards and the hydrophobic tails point inwards (Fig. 2i). The main difference between emulsions and micelles is that there is not an oil core for micelles. Therefore, as CT contrast agents, iodinated molecules are typically linked to the hydrophobic tails of the amphiphiles, concentrating the iodinated moieties into the center of the micelle. This allows for encapsulation of high iodinated payloads [97]. There have been several reports of iodinated micelles used as blood pool contrast agents [98]. Torchilin et al. reported the synthesis of micelles using iodine containing poly-L-lysine (MPEG-iodolysine). The micelles were approximately 80 nm in diameter with 33.8 % iodine content. The agent was injected intravenously and allowed for CT imaging of the aorta, heart, liver, and spleen for up to 3 h in rats [99].

While nano-emulsions and micelles can carry hydrophobic contrast media, liposomes are versatile platforms that can carry both aqueous and hydrophobic payloads. Liposomes are synthesized through the use of amphiphilic lipids or polymers that self-assemble into bilayers. This architecture allows for encapsulation of hydrophilic agents in the core as well as hydrophobic agents in the bilayer. One of the first liposomal iodinated contrast agents was synthesized using L-α-phosphatidylcholine and was loaded with a number of different aqueous iodinated agents [100]. The agent was injected intravenously as a blood pool contrast agent. Uptake of the liposome was seen in the liver and spleen in a dose dependent manner with maximum attenuation of 240 HU seen at 1 h. Because these early liposomes provided strong contrast of the liver and spleen, it was primarily used to visualize tumors in these organs [101, 102].

With improvements to liposome design, most notably the modification of amphiphilic lipids/polymers with PEG, the diagnostic applications of liposomal contrast agents expanded [103]. The addition of PEG to liposomal formulations allows for significantly increased blood circulation times improving their utility as a blood pool agents. One of the earliest reports of PEGylated liposomes, PEG-phosphatidylethanolamine (PEG-PE) liposomes, found that the addition of PEG increased the blood time circulation to up to 5 h as compared to 30 min for non-PEGylated formulations. Iodinated PEGylated liposomes have been successful in providing strong vascular contrast of up to 900 HU in the aorta, as reported in a study by Mukundan et al. [104–106].

With the use of liposomes, a significant issue is the leakage of internal payloads through the lipid bilayer. Hydrophilic payloads have been found to diffuse through bilayers due to chemical gradients. To overcome this issue, some groups have added cholesterol to liposome formulations, which reduces the permeability of the liposome and thereby decreases payload leakage [107, 108]. Additionally, direct iodination of the lipids used to form the liposome can be used to prevent internal leakage, since covalently binding iodine onto lipid constrains the contrast generating material in the bilayer [109].

Moreover by incorporating iodinated molecules into the lipid bilayer, the aqueous core can carry a separate payload including drugs for theranostics applications, another imaging agent for multimodal imaging, or more of the same agent for increased payload delivery. A study by Kweon et al. reported liposomes that were simultaneously loaded with iodinated contrast into both the core and lipid bilayer (Fig. 3a) [110]. For lipid preparation, 1,2-dimyristoyl-sn-glycero-3-phosphocholine

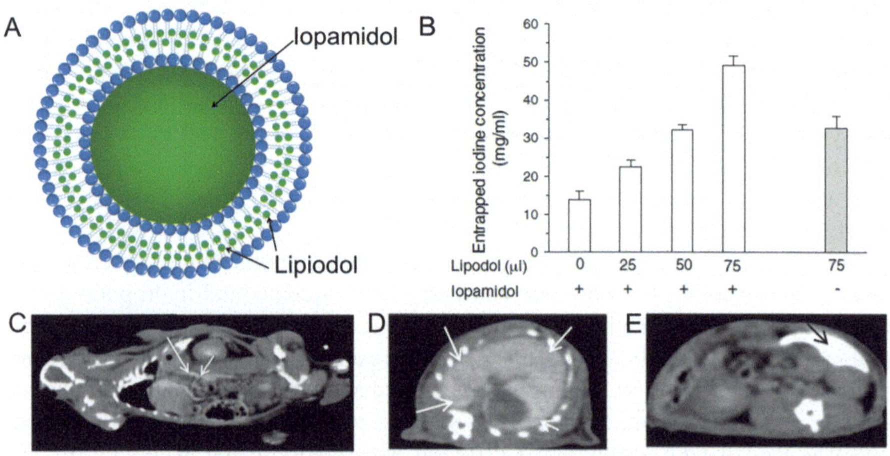

Fig. 3 (**a**) Schematic depiction of a liposomal formulation where Lipiodol is incorporated into the lipid bilayer and iopamidol is loaded into the core. (**b**) Iodine content of liposomes loaded with increasing concentrations of Lipiodol in the bilayer and with and without iopamidol in the core. The highest iodine loading was seen with inclusion of both Lipiodol and iopamidol as compared to Lipiodol alone. (**c**) Enhancement of aorta post-injection of Lipiodol and iopamidol containing liposomes indicated by white arrows. (**d, e**) Accumulation of iodinated liposomes in the (**d**) liver and (**e**) spleen. Reproduced with permission from Ref. [110]

(DMPC), cholesterol, and Lipiodol (iodinated oil) were mixed together before freeze-drying. As mentioned above, cholesterol increases the rigidity of the lipid membrane decreasing leakiness of the liposomes. These lipids prepared together were used to encapsulate iopamidol (a water soluble iodinated agent) into the core of the liposomes. This dual loading procedure produced particles about 280 nm in diameter, and a solution whose iodine concentration was 49.2 mg/ml (Fig. 3b). Inclusion of both Lipiodol and iopamidol was found to improve iodine content in the nanoparticle solution over either payload alone. Intravenous injection of the liposome produced strong contrast in the aorta after injection and was observed for up to 2.5 h (Fig. 3c). Because of the lack of PEG, high uptake was seen in the liver and spleen, resulting in attenuation values of up to 684 HU for the spleen (Fig. 3d, e). This study demonstrates the versatility of liposomes for contrast agent delivery.

Dendrimers are another platform that has been used to develop contrast agents for computed tomography. Dendrimers are polymeric molecules formed through sequential branched growth steps termed "generations." Each generation exponentially increases the number of branches, isolating the core from the external environment. Additionally, the last generation of the branched sequence can be functionalized to confer specific properties to the dendrimer. The branched structure provides excellent molecular stability and can provide many sites for functionalization at the ends of the branches [111]. Dendrimers synthesized using poly(amido amine) (PAMAM) allow for functionalization through amine groups at the available termini of the molecule. By conjugating triiodobenzenepropanoic acid (DMAA-IPA) onto a PAMAM dendrimer, Yordanov et al. were able to incorporate up to 33 % iodine content by weight [112]. Large dendrimers are often quickly cleared through the mononuclear phagocytic system; however, Fu et al. reported dendrimer-based CT contrast agents with improved circulation times. These dendrimers were formed via the use of a large PEG core (6000–12,000 MW) and poly-L-lysine for branch generation. The amine termini of these dendrimers were functionalized with triiodophthalamide molecules to provide CT contrast [113].

5.2 *Targeted Iodinated Agents*

Nanotechnology offers several platforms upon which to develop contrast agents for CT as outlined above. Blood pool agents are useful to help visualize several pathologies in coronary artery disease and myocardial infarction, however specific targeting of contrast agents can allow single mechanisms or processes in cardiovascular disease to be probed. The progression of coronary artery disease occurs in many stages, providing numerous targets to study. Targeting can be accomplished by attaching ligands such as antibodies or peptides to the surface of the particle. Moreover, the use of native biological material such as lipoproteins (e.g., LDL, HDL) can provide targeting for cardiovascular disease.

Pan et al. utilized poly(styrene-b-acrylic acid) to produce nano-emulsions with high payloads of Lipiodol (up to 37 % wt of iodine) under 100 nm in diameter [114]. Furthermore, the authors conjugated an anti-fibrin monoclonal antibody onto the particle using a modified avidin–biotin linker. Fibrin is generated during the coagulation cascade and is a key element in thrombus formation. It is a very high density target and therefore suitable for molecular imaging via CT (high density targets are needed due to the low sensitivity of CT). A fibrin targeted agent may reveal information on plaque rupture and subsequent thrombus formation. Using a fibrin clot phantom, the agent found a CNR value of 98 ± 12 compared to the surrounding water. From rat experiments, the blood half-life of the agent was found to be 56 min, although in vivo thrombus targeting experiments were not reported.

Detecting the macrophage content within a plaque is hypothesized to be a marker of plaque stability, since increased macrophage population has been found to be characteristic of vulnerable plaques [115]. Macrophages are another high density target, due to their high populations in plaques and ability to take up very large amounts of agent. Several types of iodinated contrast agents have been developed to noninvasively image macrophages in plaques. Hyafil et al. utilized a formulation for macrophage imaging termed N1177, which consists of ethyl-3,5-bis(acetylamino)-2,4,6-triiodobenzoate contained in an amphiphilic tri-block copolymer [116]. The emulsion was prepared through milling and further stabilized with PEG. The particles had an average diameter of 259 nm. The authors hypothesized that N1177 is taken up by macrophages in the circulation and subsequently infiltrate the plaque allowing for detection. Two hours after intravenous injection using the balloon-injury rabbit model of atherosclerosis, specific enhancement of the arterial plaque could be identified in CT with a maximum enhancement of 14 HU. Histology confirmed the localization of N1177 particles in macrophages within plaques. The same group published a follow-up study focusing on N1177 where they quantified inflammation through macrophage targeting [117].

The use of N1177 as a specific macrophage marker was used to investigate atherosclerotic plaque rupture in a study by Van Herck et al. [118]. The authors hypothesized that after plaque rupture, increased influx of macrophages to the rupture site would allow for detection of the rupture. The authors initially investigated cytotoxicity of macrophages using N1177 in vitro which they found no effect on viability or secretion of cytokines by J774 macrophages (Fig. 4a, b). In this case, the atherosclerosis model used was rabbits kept on hypercholesterolemic diets long term (12–15 months) to establish plaques. As a control, an I.V. injection of N1177 showed slight enhancement of the aortic lumen but returned to pre-scan levels within 2 h, demonstrating that N1177 did not significantly enhance non-ruptured plaques. This result was contrary to the findings of Hyafil et al. [116], but the models of atherosclerosis used in the two studies differed considerably. On the other hand, when the authors induced plaque rupture mechanically and injected N1177, the attenuation of the ruptured plaque was increased significantly (34 HU, $P < 0.001$) as compared to pre-ruptured plaque (Fig. 4c). Histology on the plaques showed increased staining of RAM-11 (a macrophage marker) in ruptured plaques. Additionally, N1177 was found in macrophage rich areas of the ruptured plaques but not in non-ruptured plaques using time-of-flight-static secondary ion mass spectrometry. These results demonstrate that N1177 uptake in macrophages could be used to rapidly detect and assess plaque rupture. Other groups have also focused on macrophage specific targeting. For example, Ding et al. developed iodinated oil

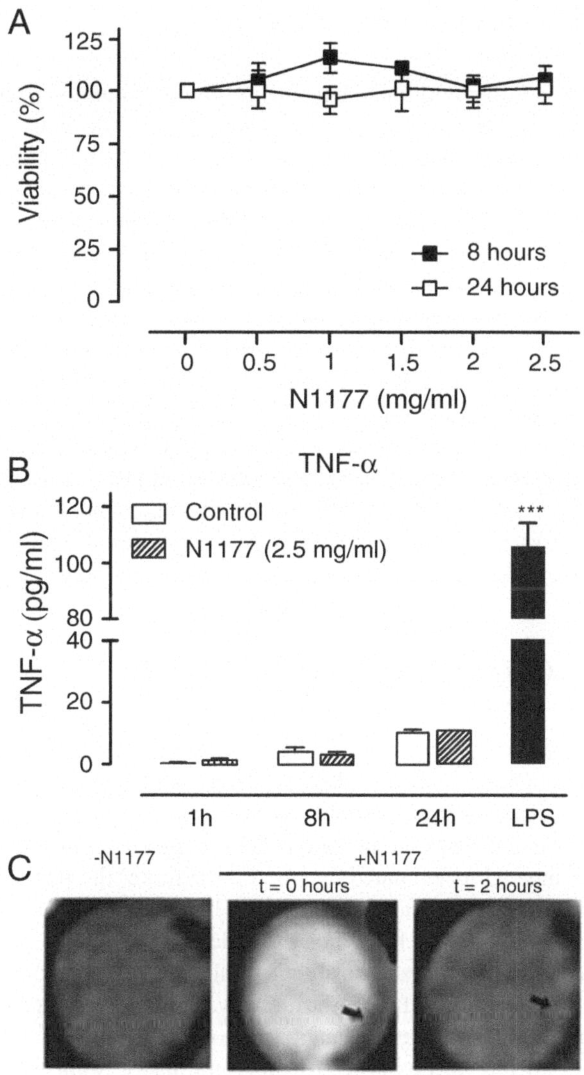

Fig. 4 (**a**) Viability of J774 macrophages after treatment with N177. (**b**) TNF-α release after treatment with N1177. (**c**) CT images of rabbit aortas after mechanical plaque rupture without N1177, immediately after injection of N1177 and 2 h post-injection. Significant increases in attenuation were seen after rupture of plaque. Reproduced with permission from Ref. [118]

nano-emulsions containing quantum dots for dual modal imaging of plaque macrophages with CT and fluorescence [119].

Low-density lipoprotein is an attractive platform for targeted imaging due to its evasion of the immune system, biocompatibility, biodegradability, defined size and innate targeting. LDL infiltration in leaky endothelium is a key stage in early atherosclerosis development. A number of studies have attempted to use LDL as nanoparticle platform to carry imaging agents for detection of plaque for multimodalities [120–122]. For iodinated contrast agents, a study by Hill et al. demonstrated the feasibility of using LDL containing iodinated triglycerides to target liver cells (HepG2) that overexpress the LDL-receptor [123]. This in vitro study resulted in visible enhancement of the cells, but would require further testing in vivo.

5.3 Bromine Based Agents

In addition to iodinated agents, bromide containing agents have been explored as CT contrast agents for several decades [124]. The most widely used formulations are based on perfluorooctylbromide (PFOB), which is a brominated fluorocarbon. While the CT contrast generated by bromine is relatively poor, necessitating very high doses, the excellent biocompatibility and inert nature of PFOB motivated developing CT contrast agents on this compound. These fluorocarbons are emulsified with phospholipids to produce stable nanoparticles that are capable of attenuating X-rays. Early studies demonstrated PFOB as an effective CT contrast agent for enhancement of the liver, spleen and vasculature [125, 126]. Other studies have looked at the potential of PFOB emulsions as a multimodal imaging contrast agent for ultrasound or MRI in addition to CT [127, 128]. A study by Li et al. included superparamagnetic iron nanoparticles (Fe_3O_4) in a PFOB emulsification to produce tri-modal contrast agent for CT, MRI, and ultrasound [129]. The magnetic-PFOB nanoparticles were approximately 200 nm in diameter and interestingly demonstrated higher echogenicity than PFOB particles alone. Additionally, contrast enhancement of the liver was seen after intravenous injection for both CT and MR imaging.

5.4 Inorganic Blood Pool Agents

While iodinated small molecules are the contrast agents used for X-ray computed tomography clinically, there has been significant preclinical interest in the development of inorganic nanoparticles as imaging agents. Inorganic nanoparticles can provide improved X-ray attenuation at CT energy ranges due to higher k-edge energy levels as compared to iodinated agents. Furthermore, the payloads of inorganic nanoparticles can be much higher than for iodine based agents, since the density of inorganic materials is often very high. For example, Brown et al. reported a solid bismuth core CT contrast agent (core density 9.8 g/ml) whose 76 nm cores carried a payload of 6 million Bi atoms each [130]. Additionally, the use of inorganic nanocrystals may alleviate biocompatibility complications associated with iodinated agents as outlined above.

Of the various types of inorganic nanocrystals that have been proposed as CT contrast agents, gold nanoparticles (AuNP) have been the focus of most work. Gold nanoparticle syntheses allow for control of size, shape, and functionality of the particles. This precise control allows for specific tailoring of these particles for a given application. AuNP can be included in various nanotechnology platforms, i.e., micelles, liposomes, dendrimers, and as ligand stabilized crystalline nanoparticles [131, 132]. Gold nanoparticles are also well known for their biocompatible properties and have been used in a variety of biomedical applications such as drug delivery, imaging, biosensors, and photothermal ablation [133–137].

As a CT contrast agent, the physical parameters of the particle as well as the chemical properties can drastically influence the particle's biological properties in terms of clearance and biocompatibility. One of the first AuNP developed as an X-ray contrast agent was a 1.9 nm spherical formulation, and was shown to provide strong enhancement of the major vessels. However, due to the small size of the agent, the particles were rapidly cleared through the renal system as evidenced by high attenuation in the bladder in images acquired 10 min after injection [138]. Increasing the size and/or modifying the surface functionality of AuNP has been shown to improve circulation time, as found in studies where gold nanoparticles whose core size was around 10 nm. After coating with PEG, the nanoparticles had an overall size of 38 nm. These nanoparticles were not renally cleared and provided vascular contrast over a period of 12–24 h [139].

A study by Kim et al. examined the effectiveness of AuNP as blood pool contrast agents after surface modification with PEG [140]. AuNP 30 nm in diameter were synthesized through the reduction of gold chloride by sodium citrate using a modified Turkevich method [141, 142]. Ligand exchange was performed with 5000 methoxy-PEG-thiol for stabilization and to provide a long circulation half-life (Fig. 5a). Cytotoxicity of the AuNP-PEG particles were evaluated in HepG2 (epithelial liver cell line). The particles were well tolerated up to 1000 μg/ml of Au (Fig. 5b). After intravenous injection of AuNP-PEG, the agent yielded strong contrast in the heart and major arterial vessels (>100 HU). Figure 5c shows a 3D reconstruction of

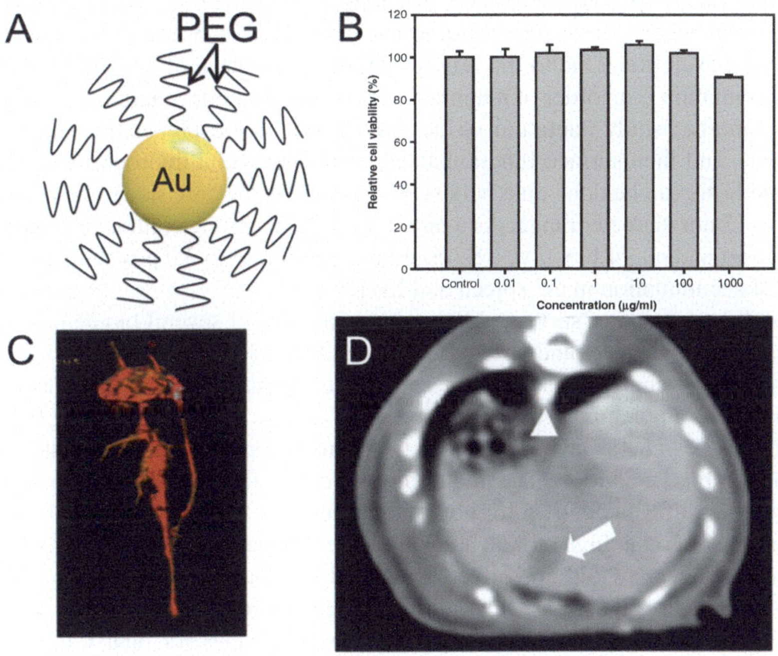

Fig. 5 (**a**) Schematic of PEG coated gold nanoparticles for blood pool imaging. (**b**) Cytotoxicity results for HepG2 cells treated with increasing concentrations of PEG coated AuNP. (**c**) 3D reconstruction of the heart and major vessels after injection of AuNP-PEG from CT images. (**d**) Cross section scans of rat hepatoma model, 12 h post injection of AuNP-PEG. Attenuation of aorta indicated by *arrow head* and hepatoma by *white arrow*. Reproduced with permission from Ref. [140]

the heart and vasculature, derived from images acquired at 10 min post injection. A roughly 100 HU increase in attenuation in the heart and blood vessels was sustained for 4 h post-injection, attesting to their long-circulation half-life provided by the PEG coating. The attenuation of the liver and spleen was found to increase over the 24 h period, indicating the major clearance pathways for the PEG coated AuNP. Due to the uptake of the particles by Kupffer cells and hepatocytes, the authors demonstrated visualization of a hepatoma in vivo in a rat model. Figure 5d shows cross section scans 12 h post injection, with strong attenuation in the aorta (white arrow head) and delineation of the hepatoma (white arrow). Other studies have focused on different AuNP structures (e.g., nanorods [80, 143]), coatings (such as gum-arabic [144, 145]) or incorporated gold into other platforms such as dendrimers [146–149].

While AuNP have desirable properties for X-ray imaging, the cost of gold may be an issue for widespread adoption and scale up. Other high atomic number elements have been explored as CT contrast agents. For example, studies have shown bismuth nanoparticles as a possible CT blood pool contrast agent [150]. Naha et al. reported the development of dextran coated bismuth-iron nanoparticles for in vivo blood contrast imaging with both CT and MRI [151].

Tantalum is another element that has been recently explored as a candidate for CT contrast agents due to the low toxicity of its oxide and strong X-ray absorption [152, 153]. A group from General Electric have published on several tantalum oxide formulations where the overall nanoparticle size is below 5 nm, thus allowing for swift renal excretion. Due to the relative natural abundance of tantalum compared to gold, tantalum is an order of magnitude cheaper, which may facilitate widespread use. In another study, tantalum oxide nanocrystals were prepared using reverse emulsions and then surface functionalized with PEG and a fluorophore tag [154]. This synthesis can be done on a bulk scale and allows the core size to be controlled in the 6–15 nm diameter range. As a blood pool agent, strong contrast enhancement of the heart and major blood vessels could be seen for up to 3 h post-injection, with eventual accumulation in the spleen and liver.

Gadolinium chelates serve as MRI contrast agents for several biomedical applications including cancer and vascular imaging [155, 156]. Despite the toxicity concerns over non-chelated gadolinium and tissue retention, studies have evaluated gadolinium as a potential CT contrast agent [157, 158]. Due to its unique properties for MRI, researchers have developed gadolinium nanoparticles for dual modal imaging with CT and MRI. The addition of gold to gadolinium chelates has shown to provide significant contrast for both modalities [73, 159, 160].

A number of other high atomic number elements have shown promise as effective CT contrast agents. Gold and silver core–shell nanoparticles have been shown to produce stronger CT contrast than iodinated agents [161, 162]. A report by Chou et al. demonstrated the use of iron/platinum nanoparticles for dual CT and MRI molecular imaging after intravenous injection [163]. Several studies have reported the use of tungsten nanoparticles as CT contrast agents that provide better enhancement compared to traditional iodinated agents [164–166]. Moreover, ytterbium based nanoparticles have received attenuation as potential CT agents due to favorable X-ray absorption spectra and relative natural abundance compared to other potential elements mentioned [167, 168]. Ytterbium based nanoparticles also present multimodal capabilities through upconversion fluorescence imaging. Upconverting fluorescent probes allow for excitation at NIR-wavelengths for deep tissue penetration with reduced background tissue absorption. Liu et al. synthesized ytterbium based particles coated in PEG for use as a CT contrast agent [169]. Xing et al. synthesized tri-modal nanoparticles for upconversion fluorescence, MR and CT imaging [170]. The group synthesized gadolinium-based upconverting nanoparticles (doped with yttrium, ytterbium, erbium, and thulium) that were also surface coated with gold nanoparticles. In vivo imaging of these trimodal particles was demonstrated with subcutaneous injection of the particles at tumor bearing locations in mice.

5.5 *Targeted Inorganic CT Contrast Agents*

The synthesis of inorganic nanoparticles presents an opportunity for relatively straightforward functionalization of the particle surface for targeting capabilities. Common methods for targeting involve conjugation of peptides complimentary to a specific target [171, 172]. The use of antibodies conjugated to inorganic nanoparticles to create targeted CT contrast agents has been demonstrated in several studies [173, 174]. While several methods have been developed for targeted tumor imaging with CT, targeted CT cardiovascular imaging with inorganic nanoparticles is less well explored. Chhour et al. demonstrated the use AuNP to specifically label monocytes and track their recruitment to developing atherosclerotic plaques in mice with CT [175]. The group found significantly increased attenuation in the plaques 5 days after injection of the labeled cells. Due to CT's low sensitivity to contrast agents, a large amount of contrast generating material needs to be delivered to a site for imaging, as compared to MRI or PET.

The advent of multicolor or spectral CT has allowed different elements to be distinguished in CT images due to the characteristic X-ray attenuation profiles of the elements [176]. Spectral CT utilizes novel, photon-counting detectors to discern the energies of incident X-rays. By separating the transmitted X-rays into multiple energy bins, X-ray attenuation in the field of view can be correlated to specific elements (since the boundaries of the bins are usually set to the k-edges of the elements expected to be found in the subject). Discerning multiple material types in CT scans presents the opportunity for better delineation between contrast agent and high attenuating tissue. For instance, early investigations have involved discriminating barium/calcifications and ytterbium/bone using spectral CT [177, 178]. A study by Cormode et al. examined the composition of atherosclerotic plaques using spectral CT [179]. The purpose of the study was to examine the use of spectral CT to characterize macrophage burden, calcification and artery stenosis in a single CT scan. Gold nanoparticles incorporated into a coating similar to high density lipoprotein (Au-HDL), 7.2 nm in overall diameter, were used to target macrophages in atherosclerotic plaques (Fig. 6a) [180]. As an initial demonstration of spectral CT, a phantom consisting of different concentrations of Au-HDL, iodine, and calcium phosphate were scanned (Fig. 6b). By using spectral CT with energy bins tuned to the k-edges of the elements involved, gold, iodine, and $Ca_3(PO_4)_2$, concentration gradients could

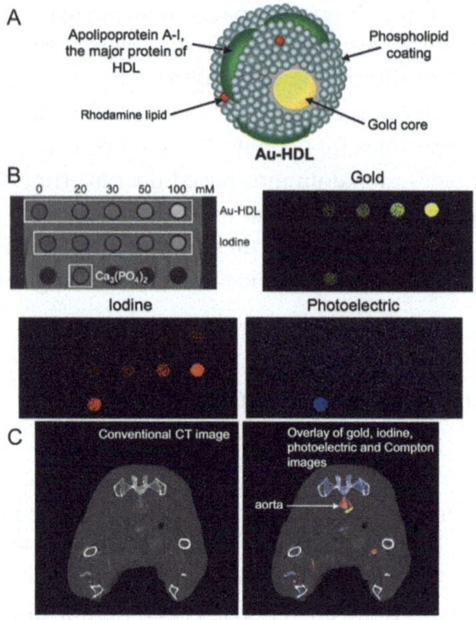

Fig. 6 (**a**) Schematic depiction of the structure of Au-HDL. (**b**) Spectral CT phantom imaging of concentration gradients of Au-HDL and an iodinated contrast agent. $Ca_3(PO_4)_2$ was located in the third row of the phantom. Via materials basis decomposition of the spectral imaging data, attenuation from Au-HDL, iodine, and $Ca_3(PO_4)_2$ was distinguishable. (**c**) Injection of Au-HDL and 24 h later injection of a vascular phase iodinated agent. Conventional CT image (*left*) compared to an overlay of gold, iodine, photoelectric, and Compton images from a spectral CT scan (*right*). Reproduced with permission from Ref. [179]

be distinguished. The photoelectric effect was used to visualize $Ca_3(PO_4)_2$. For in vivo studies, Au-HDL was intravenously injected to allow macrophage targeting and after 24 h an additional vascular phase iodinated contrast agent was injected. A conventional CT scan image is seen in Fig. 6c. Comparatively, the scan with multicolor CT revealed the aorta, which had been strongly enhanced by the iodinated agent (red color), was distinguishable from Au-HDL localized in a nearby plaque (gold color) (Fig. 6c). Electron microscopy and confocal microscopy confirmed that the Au-HDL agent was targeted to macrophages within the plaque. Pan et al. studied the use a bismuth based nanoparticle conjugated to a fibrin specific antibody, which the authors termed NanoK [181]. The NanoK particles were found to have a hydrodynamic diameter between 150 and 230 nm. Anti-fibrin NanoK were in situ incubated with balloon induced thrombuses using temporary snares for 30 min. Arterial circulation was reestablished for 30 min to wash unbound particles and then the animal was euthanized. Using spectral CT, the NanoK particles could be distinguished from calcified regions at the area of induced clot injury.

6 Conclusions

Current clinical contrast agents for X-ray computed tomography provide important diagnostic information for evaluation of cardiovascular disease but suffer from a number of drawbacks. New contrast agents are being designed utilizing nanotechnology, offering solutions to improving blood circulation time, biocompatibility, and allowing for specific targeting. Several strategies have been developed to synthesize iodinated contrast agents for blood pool imaging including nano-emulsions, liposomes, micelles, and dendrimers. Similarly, a variety of inorganic elements have been synthesized for use as CT contrast agents. Gold nanoparticles have been most frequently reported as inorganic nanoparticles for CT contrast agents, however other elements including bismuth, tantalum, and ytterbium continue to be explored. Each technology has its own advantages and weaknesses in terms of particle stability, clearance time, and X-ray attenuation properties that make each platform distinct. With a variety of platforms and formulations, nanoparticle technology will continue to improve the diagnostic capabilities of cardiac CT. Nanoparticle based agents seem to have clear advantages for blood pool imaging and we expect that the next decade will see efforts to translate nanoparticle blood pool agents to the clinic. The field of targeted CT agents for cardiovascular agents is relatively nascent, but we expect that considerable progress will be made in this area, both in terms of expanding the number of targets that can be imaged and with regards to improvement in nanoparticle technology. Advances in CT imaging technology promise to aid in these efforts. The advent of spectral (multicolor) CT allows detection of contrast media without pre-injection scans and allows identification of multiple materials in a single scan. Developments in computing and algorithms such as iterative reconstruction will increase the sensitivity of CT via suppression of image noise. Doubtless, unforeseen innovations and new technologies will arise. Overall, we expect to see the field of nanoparticle contrast agents for cardiovascular CT imaging to continue to flourish over the next decade and beyond.

Acknowledgments This work was supported by R01 HL131557 (D.P.C.), R00 EB012165 (D.P.C.), W. W. Smith Charitable Trust (D.P.C.), and the T32 HL007954 (P.C.). We also thank the University of Pennsylvania for startup funding.

References

1. Mozaffarian D, Benjamin EJ, Go AS, Arnett DK, Blaha MJ, Cushman M, et al. Heart disease and stroke statistics- 2015 update: a report from the American heart association. Circulation. 2015;131(4):e29–322.
2. Go AS, Mozaffarian D, Roger VL, Benjamin EJ, Berry JD, Blaha MJ, et al. Heart disease and stroke statistics—2014 update: a report from the American heart association. Circulation. 2014;129(3):e28–292.
3. Lumbroso P, Dick CE. X-ray attenuation properties of radiographic contrast media. Med Phys. 1987;14(5):752–8.

4. Lee N, Choi SH, Hyeon T. Nano-sized CT contrast agents. Adv Mater. 2013;25(19):2641–60.
5. Chatzizisis YS, Coskun AU, Jonas M, Edelman ER, Feldman CL, Stone PH. Role of endothelial shear stress in the natural history of coronary atherosclerosis and vascular remodeling: molecular, cellular, and vascular behavior. J Am Coll Cardiol. 2007;49(25):2379–93.
6. Stone PH, Coskun AU, Yeghiazarians Y, Kinlay S, Popma JJ, Kuntz RE, et al. Prediction of sites of coronary atherosclerosis progression: in vivo profiling of endothelial shear stress, lumen, and outer vessel wall characteristics to predict vascular behavior. Curr Opin Cardiol. 2003;18(6):458–70.
7. Skalen K, Gustafsson M, Rydberg EK, Hulten LM, Wiklund O, Innerarity TL, et al. Subendothelial retention of atherogenic lipoproteins in early atherosclerosis. Nature. 2002;417(6890):750–4.
8. Srinivasan SR, Vijayagopal P, Dalferes Jr ER, Abbate B, Radhakrishnamurthy B, Berenson GS. Low density lipoprotein retention by aortic tissue. Contribution of extracellular matrix. Atherosclerosis. 1986;62(3):201–8.
9. Cybulsky MI, Iiyama K, Li H, Zhu S, Chen M, Iiyama M, et al. A major role for vcam-1, but not icam-1, in early atherosclerosis. J Clin Invest. 2001;107(10):1255–62.
10. Huo Y, Hafezi-Moghadam A, Ley K. Role of vascular cell adhesion molecule-1 and fibronectin connecting segment-1 in monocyte rolling and adhesion on early atherosclerotic lesions. Circ Res. 2000;87(2):153–9.
11. Ramos CL, Huo Y, Jung U, Ghosh S, Manka DR, Sarembock IJ, et al. Direct demonstration of p-selectin- and vcam-1-dependent mononuclear cell rolling in early atherosclerotic lesions of apolipoprotein e-deficient mice. Circ Res. 1999;84(11):1237–44.
12. Boisvert WA, Rose DM, Johnson KA, Fuentes ME, Lira SA, Curtiss LK, et al. Up-regulated expression of the cxcr2 ligand kc/gro-alpha in atherosclerotic lesions plays a central role in macrophage accumulation and lesion progression. Am J Pathol. 2006;168(4):1385–95.
13. Boring L, Gosling J, Cleary M, Charo IF. Decreased lesion formation in ccr2-/- mice reveals a role for chemokines in the initiation of atherosclerosis. Nature. 1998;394(6696):894–7.
14. Combadiere C, Potteaux S, Rodero M, Simon T, Pezard A, Esposito B, et al. Combined inhibition of ccl2, cx3cr1, and ccr5 abrogates ly6c(hi) and ly6c(lo) monocytosis and almost abolishes atherosclerosis in hypercholesterolemic mice. Circulation. 2008;117(13):1649–57.
15. Shashkin P, Dragulev B, Ley K. Macrophage differentiation to foam cells. Curr Pharm Des. 2005;11(23):3061–72.
16. Libby P. Inflammation in atherosclerosis. Nature. 2002;420(6917):868–74.
17. Moreno PR, Falk E, Palacios IF, Newell JB, Fuster V, Fallon JT. Macrophage infiltration in acute coronary syndromes. Implications for plaque rupture. Circulation. 1994;90(2):775–8.
18. Briley-Saebo KC, Shaw PX, Mulder WJ, Choi SH, Vucic E, Aguinaldo JG, et al. Targeted molecular probes for imaging atherosclerotic lesions with magnetic resonance using antibodies that recognize oxidation-specific epitopes. Circulation. 2008;117(25):3206–15.
19. Nahrendorf M, Jaffer FA, Kelly KA, Sosnovik DE, Aikawa E, Libby P, et al. Noninvasive vascular cell adhesion molecule-1 imaging identifies inflammatory activation of cells in atherosclerosis. Circulation. 2006;114(14):1504–11.
20. Nahrendorf M, Zhang H, Hembrador S, Panizzi P, Sosnovik DE, Aikawa E, et al. Nanoparticle PET-CT imaging of macrophages in inflammatory atherosclerosis. Circulation. 2008;117(3):379–87.
21. Nahrendorf M, Waterman P, Thurber G, Groves K, Rajopadhye M, Panizzi P, et al. Hybrid in vivo FMT-CT imaging of protease activity in atherosclerosis with customized nanosensors. Arterioscler Thromb Vasc Biol. 2009;29(10):1444–51.
22. Winter PM, Neubauer AM, Caruthers SD, Harris TD, Robertson JD, Williams TA, et al. Endothelial alpha(v)beta3 integrin-targeted fumagillin nanoparticles inhibit angiogenesis in atherosclerosis. Arterioscler Thromb Vasc Biol. 2006;26(9):2103–9.
23. Johansson LO, Bjornerud A, Ahlstrom HK, Ladd DL, Fujii DK. A targeted contrast agent for magnetic resonance imaging of thrombus: implications of spatial resolution. J Magn Reson Imaging. 2001;13(4):615–8.

24. Tota-Maharaj R, Al-Mallah MH, Nasir K, Qureshi WT, Blumenthal RS, Blaha MJ. Improving the relationship between coronary artery calcium score and coronary plaque burden: addition of regional measures of coronary artery calcium distribution. Atherosclerosis. 2015;238(1):126–31.
25. Hou ZH, Lu B, Gao Y, Jiang SL, Wang Y, Li W, et al. Prognostic value of coronary CT angiography and calcium score for major adverse cardiac events in outpatients. JACC Cardiovasc Imaging. 2012;5(10):990–9.
26. Galis ZS, Sukhova GK, Lark MW, Libby P. Increased expression of matrix metalloproteinases and matrix degrading activity in vulnerable regions of human atherosclerotic plaques. J Clin Invest. 1994;94(6):2493–503.
27. Pfeffer MA, Braunwald E. Ventricular remodeling after myocardial infarction. Experimental observations and clinical implications. Circulation. 1990;81(4):1161–72.
28. Sutton MG, Sharpe N. Left ventricular remodeling after myocardial infarction: pathophysiology and therapy. Circulation. 2000;101(25):2981–8.
29. Sosnovik DE, Nahrendorf M, Panizzi P, Matsui T, Aikawa E, Dai G, et al. Molecular mri detects low levels of cardiomyocyte apoptosis in a transgenic model of chronic heart failure. Circ Cardiovasc Imaging. 2009;2(6):468–75.
30. Hofstra L, Liem IH, Dumont EA, Boersma HH, van Heerde WL, Doevendans PA, et al. Visualisation of cell death in vivo in patients with acute myocardial infarction. Lancet. 2000;356(9225):209–12.
31. Nahrendorf M, Pittet MJ, Swirski FK. Monocytes: protagonists of infarct inflammation and repair after myocardial infarction. Circulation. 2010;121(22):2437–45.
32. Flogel U, Ding Z, Hardung H, Jander S, Reichmann G, Jacoby C, et al. In vivo monitoring of inflammation after cardiac and cerebral ischemia by fluorine magnetic resonance imaging. Circulation. 2008;118(2):140–8.
33. Tsujioka H, Imanishi T, Ikejima H, Kuroi A, Takarada S, Tanimoto T, et al. Impact of heterogeneity of human peripheral blood monocyte subsets on myocardial salvage in patients with primary acute myocardial infarction. J Am Coll Cardiol. 2009;54(2):130–8.
34. Panizzi P, Swirski FK, Figueiredo JL, Waterman P, Sosnovik DE, Aikawa E, et al. Impaired infarct healing in atherosclerotic mice with ly-6c(hi) monocytosis. J Am Coll Cardiol. 2010;55(15):1629–38.
35. Helm PA, Caravan P, French BA, Jacques V, Shen L, Xu Y, et al. Postinfarction myocardial scarring in mice: molecular MR imaging with use of a collagen-targeting contrast agent. Radiology. 2008;247(3):788–96.
36. van den Borne SW, Isobe S, Verjans JW, Petrov A, Lovhaug D, Li P, et al. Molecular imaging of interstitial alterations in remodeling myocardium after myocardial infarction. J Am Coll Cardiol. 2008;52(24):2017–28.
37. Yonezawa M, Nagata M, Kitagawa K, Kato S, Yoon Y, Nakajima H, et al. Quantitative analysis of 1.5-t whole-heart coronary MR angiograms obtained with 32-channel cardiac coils: a comparison with conventional quantitative coronary angiography. Radiology. 2014;271(2):356–64.
38. Health QO. Positron emission tomography for the assessment of myocardial viability: an evidence-based analysis. Ont Health Technol Assess Ser. 2010;10(16):1–80.
39. Sarikaya I. Cardiac applications of PET. Nucl Med Commun. 2015;36(10):971–85.
40. Mehta NN, Torigian DA, Gelfand JM, Saboury B, Alavi A. Quantification of atherosclerotic plaque activity and vascular inflammation using [18-f] fluorodeoxyglucose positron emission tomography/computed tomography (FDG-PET/CT). J Vis Exp. 2012;63:e3777.
41. Huet P, Burg S, Le Guludec D, Hyafil F, Buvat I. Variability and uncertainty of 18F-FDG PET imaging protocols for assessing inflammation in atherosclerosis: suggestions for improvement. J Nucl Med. 2015;56(4):552–9.
42. Derlin T, Richter U, Bannas P, Begemann P, Buchert R, Mester J, et al. Feasibility of 18F-sodium fluoride PET/CT for imaging of atherosclerotic plaque. J Nucl Med. 2010;51(6):862–5.
43. Bardo DM, Brown P. Cardiac multidetector computed tomography: basic physics of image acquisition and clinical applications. Curr Cardiol Rev. 2008;4(3):231–43.

44. Otero HJ, Steigner ML, Rybicki FJ. The "post-64" era of coronary CT angiography: understanding new technology from physical principles. Radiol Clin North Am. 2009;47(1): 79–90.
45. Flohr TG, McCollough CH, Bruder H, Petersilka M, Gruber K, Suss C, et al. First performance evaluation of a dual-source CT (DSCT) system. Eur Radiol. 2006;16(2):256–68.
46. Dewey M, Teige F, Laule M, Hamm B. Influence of heart rate on diagnostic accuracy and image quality of 16-slice CT coronary angiography: comparison of multisegment and halfscan reconstruction approaches. Eur Radiol. 2007;17(11):2829–37.
47. Hoffmann U, Pena AJ, Cury RC, Abbara S, Ferencik M, Moselewski F, et al. Cardiac CT in emergency department patients with acute chest pain. Radiographics. 2006;26(4):963–78. discussion 79-80.
48. Lee NJ, Litt H. Cardiac CT angiography for evaluation of acute chest pain. Int J Cardiovasc Imaging. 2015.
49. Goldberg A, Litt HI. Evaluation of the patient with acute chest pain. Radiol Clin North Am. 2010;48(4):745–55.
50. Achenbach S, Daniel WG. Cardiac imaging in the patient with chest pain: coronary CT angiography. Heart. 2010;96(15):1241–6.
51. Nasis A, Mottram PM, Cameron JD, Seneviratne SK. Current and evolving clinical applications of multidetector cardiac CT in assessment of structural heart disease. Radiology. 2013;267(1):11–25.
52. Puchner SB, Liu T, Mayrhofer T, Truong QA, Lee H, Fleg JL, et al. High-risk plaque detected on coronary CT angiography predicts acute coronary syndromes independent of significant stenosis in acute chest pain: results from the Romicat-II trial. J Am Coll Cardiol. 2014;64(7):684–92.
53. Pijls NH, De Bruyne B, Peels K, Van Der Voort PH, Bonnier HJ, Bartunek JKJJ, et al. Measurement of fractional flow reserve to assess the functional severity of coronary-artery stenoses. N Engl J Med. 1996;334(26):1703–8.
54. Taylor CA, Fonte TA, Min JK. Computational fluid dynamics applied to cardiac computed tomography for noninvasive quantification of fractional flow reserve scientific basis. J Am Coll Cardiol. 2013;61(22):2233–41.
55. Bittencourt MS, Achenbach S, Marwan M, Seltmann M, Muschiol G, Ropers D, et al. Left ventricular thrombus attenuation characterization in cardiac computed tomography angiography. J Cardiovasc Comput Tomogr. 2012;6(2):121–6.
56. Grude M, Juergens KU, Wichter T, Paul M, Fallenberg EM, Muller JG, et al. Evaluation of global left ventricular myocardial function with electrocardiogram-gated multidetector computed tomography: comparison with magnetic resonance imaging. Invest Radiol. 2003;38(10):653–61.
57. Galper MW, Saung MT, Fuster V, Roessl E, Thran A, Proksa R, et al. Effect of computed tomography scanning parameters on gold nanoparticle and iodine contrast. Invest Radiol. 2012;47(8):475–81.
58. Oliva MR, Erturk SM, Ichikawa T, Rocha T, Ros PR, Silverman SG, et al. Gastrointestinal tract wall visualization and distention during abdominal and pelvic multidetector CT with a neutral barium sulphate suspension: comparison with positive barium sulphate suspension and with water. JBR-BTR. 2012;95(4):237–42.
59. Bae KT. Intravenous contrast medium administration and scan timing at CT: considerations and approaches. Radiology. 2010;256(1):32–61.
60. Weininger M, Barraza JM, Kemper CA, Kalafut JF, Costello P, Schoepf UJ. Cardiothoracic CT angiography: current contrast medium delivery strategies. Am J Roentgenol. 2011; 196(3):W260–72.
61. Fleischmann D, Kamaya A. Optimal vascular and parenchymal contrast enhancement: the current state of the art. Radiol Clin North Am. 2009;47(1):13–26.
62. Bourin M, Jolliet P, Ballereau F. An overview of the clinical pharmacokinetics of X-ray contrast media. Clin Pharmacokinet. 1997;32(3):180–93.

63. Watchorn J, Miles R, Moore N. The role of CT angiography in military trauma. Clin Radiol. 2013;68(1):39–46.
64. Goldenberg I, Matetzky S. Nephropathy induced by contrast media: pathogenesis, risk factors and preventive strategies. CMAJ. 2005;172(11):1461–71.
65. Thomsen HS, Morcos SK. Contrast media and the kidney: European society of urogenital radiology (ESUR) guidelines. Br J Radiol. 2003;76(908):513–8.
66. Jun YW, Choi JS, Cheon J. Shape control of semiconductor and metal oxide nanocrystals through nonhydrolytic colloidal routes. Angew Chem Int Ed Engl. 2006;45(21):3414–39.
67. Boyle JP, Thompson TJ, Gregg EW, Barker LE, Williamson DF. Projection of the year 2050 burden of diabetes in the us adult population: dynamic modeling of incidence, mortality, and prediabetes prevalence. Popul Health Metr. 2010;8:29.
68. Ritz E, Orth SR. Nephropathy in patients with type 2 diabetes mellitus. N Engl J Med. 1999;341(15):1127–33.
69. Pasternak JJ, Williamson EE. Clinical pharmacology, uses, and adverse reactions of iodinated contrast agents: a primer for the non-radiologist. Mayo Clin Proc. 2012;87(4):390–402.
70. Cormode DP, Naha PC, Fayad ZA. Nanoparticle contrast agents for computed tomography: a focus on micelles. Contrast Media Mol Imaging. 2014;9(1):37–52.
71. Chhour P, Gallo N, Cheheltani R, Williams D, Al-Zaki A, Paik T, et al. Nanodisco balls: control over surface versus core loading of diagnostically active nanocrystals into polymer nanoparticles. ACS Nano. 2014;8(9):9143–53.
72. Jarzyna PA, Gianella A, Skajaa T, Knudsen G, Deddens LH, Cormode DP, et al. Multifunctional imaging nanoprobes. Wires Nanomed Nanobi. 2010;2(2):138–50.
73. van Schooneveld MM, Cormode DP, Koole R, van Wijngaarden JT, Calcagno C, Skajaa T, et al. A fluorescent, paramagnetic and pegylated gold/silica nanoparticle for MRI, CT and fluorescence imaging. Contrast Media Mol Imaging. 2010;5(4):231–6.
74. Mieszawska AJ, Kim Y, Gianella A, van Rooy I, Priem B, Labarre MP, et al. Synthesis of polymer-lipid nanoparticles for image-guided delivery of dual modality therapy. Bioconjug Chem. 2013;24(9):1429–34.
75. Satomi T, Nagasaki Y, Kobayashi H, Tateishi T, Kataoka K, Otsuka H. Physicochemical characterization of densely packed poly(ethylene glycol) layer for minimizing nonspecific protein adsorption. J Nanosci Nanotechnol. 2007;7(7):2394–9.
76. Moghimi SM, Hunter AC, Andresen TL. Factors controlling nanoparticle pharmacokinetics: an integrated analysis and perspective. Annu Rev Pharmacol Toxicol. 2012;52:481–503.
77. Almeida JP, Chen AL, Foster A, Drezek R. In vivo biodistribution of nanoparticles. Nanomedicine (Lond). 2011;6(5):815–35.
78. Alexis F, Pridgen E, Molnar LK, Farokhzad OC. Factors affecting the clearance and biodistribution of polymeric nanoparticles. Mol Pharm. 2008;5(4):505–15.
79. Laakkonen P, Porkka K, Hoffman JA, Ruoslahti E. A tumor-homing peptide with a targeting specificity related to lymphatic vessels. Nat Med. 2002;8(7):751–5.
80. Popovtzer R, Agrawal A, Kotov NA, Popovtzer A, Balter J, Carey TE, et al. Targeted gold nanoparticles enable molecular CT imaging of cancer. Nano Lett. 2008;8(12):4593–6.
81. Liu J, Wong HL, Moselhy J, Bowen B, Wu XY, Johnston MR. Targeting colloidal particulates to thoracic lymph nodes. Lung Cancer. 2006;51(3):377–86.
82. Mulder WJ, Strijkers GJ, van Tilborg GA, Griffioen AW, Nicolay K. Lipid-based nanoparticles for contrast-enhanced MRI and molecular imaging. NMR Biomed. 2006;19(1):142–64.
83. Bhattacharya S, Novell JR, Winslet MC, Hobbs KE. Iodized oil in the treatment of hepatocellular carcinoma. Br J Surg. 1994;81(11):1563–71.
84. Lim JS, Choi J, Song J, Chung YE, Lim SJ, Lee SK, et al. Nanoscale iodized oil emulsion: a useful tracer for pretreatment sentinel node detection using CT lymphography in a normal canine gastric model. Surg Endosc. 2012;26(8):2267–74.
85. Ahrar K, Gupta S. Hepatic artery embolization for hepatocellular carcinoma: technique, patient selection, and outcomes. Surg Oncol Clin N Am. 2003;12(1):105–26.

86. Kong WH, Lee WJ, Cui ZY, Bae KH, Park TG, Kim JH, et al. Nanoparticulate carrier containing water-insoluble iodinated oil as a multifunctional contrast agent for computed tomography imaging. Biomaterials. 2007;28(36):5555–61.
87. Weichert JP, Longino MA, Bakan DA, Spigarelli MG, Chou TS, Schwendner SW, et al. Polyiodinated triglyceride analogs as potential computed tomography imaging agents for the liver. J Med Chem. 1995;38(4):636–46.
88. Henning T, Weber AW, Bauer JS, Meier R, Carlsen JM, Sutton EJ, et al. Imaging characteristics of DHOG, a hepatobiliary contrast agent for preclinical microCT in mice. Acad Radiol. 2008;15(3):342–9.
89. Weber SM, Peterson KA, Durkee B, Qi C, Longino M, Warner T, et al. Imaging of murine liver tumor using microCT with a hepatocyte-selective contrast agent: accuracy is dependent on adequate contrast enhancement. J Surg Res. 2004;119(1):41–5.
90. Willekens I, Lahoutte T, Buls N, Vanhove C, Deklerck R, Bossuyt A, et al. Time-course of contrast enhancement in spleen and liver with Exia 160, Fenestra LC, and VC. Mol Imaging Biol. 2009;11(2):128–35.
91. Attia MF, Anton N, Chiper M, Akasov R, Anton H, Messaddeq N, et al. Biodistribution of X-ray iodinated contrast agent in nano-emulsions is controlled by the chemical nature of the oily core. ACS Nano. 2014;8(10):10537–50.
92. Badea CT, Fubara B, Hedlund LW, Johnson GA. 4-d micro-CT of the mouse heart. Mol Imaging. 2005;4(2):110–6.
93. Detombe SA, Ford NL, Xiang F, Lu X, Feng Q, Drangova M. Longitudinal follow-up of cardiac structure and functional changes in an infarct mouse model using retrospectively gated micro-computed tomography. Invest Radiol. 2008;43(7):520–9.
94. de Vries A, Custers E, Lub J, van den Bosch S, Nicolay K, Grull H. Block-copolymer-stabilized iodinated emulsions for use as CT contrast agents. Biomaterials. 2010;31(25):6537–44.
95. Hallouard F, Briancon S, Anton N, Li X, Vandamme T, Fessi H. Iodinated nano-emulsions as contrast agents for preclinical X-ray imaging: impact of the free surfactants on the pharmacokinetics. Eur J Pharm Biopharm. 2013;83(1):54–62.
96. Li X, Anton N, Zuber G, Zhao M, Messaddeq N, Hallouard F, et al. Iodinated alpha-tocopherol nano-emulsions as non-toxic contrast agents for preclinical X-ray imaging. Biomaterials. 2013;34(2):481–91.
97. Trubetskoy VS. Polymeric micelles as carriers of diagnostic agents. Adv Drug Deliv Rev. 1999;37(1-3):81–8.
98. Trubetskoy VS, Gazelle GS, Wolf GL, Torchilin VP. Block-copolymer of polyethylene glycol and polylysine as a carrier of organic iodine: design of long-circulating particulate contrast medium for X-ray computed tomography. J Drug Target. 1997;4(6):381–8.
99. Torchilin VP, Frank-Kamenetsky MD, Wolf GL. Ct visualization of blood pool in rats by using long-circulating, iodine-containing micelles. Acad Radiol. 1999;6(1):61–5.
100. Ryan PJ, Davis MA, DeGaeta LR, Woda B, Melchior DL. Liposomes loaded with contrast material for image enhancement in computed tomography. Work in progress. Radiology. 1984;152(3):759–62.
101. Havron A, Seltzer SE, Davis MA, Shulkin P. Radiopaque liposomes: a promising new contrast material for computed tomography of the spleen. Radiology. 1981;140(2):507–11.
102. Seltzer SE, Davis MA, Adams DF, Shulkin PM, Landis WJ, Havron A. Liposomes carrying diatrizoate. Characterization of biophysical properties and imaging applications. Invest Radiol. 1984;19(2):142–51.
103. Klibanov AL, Maruyama K, Torchilin VP, Huang L. Amphipathic polyethyleneglycols effectively prolong the circulation time of liposomes. FEBS Lett. 1990;268(1):235–7.
104. Kao CY, Hoffman EA, Beck KC, Bellamkonda RV, Annapragada AV. Long-residence-time nano-scale liposomal iohexol for X-ray-based blood pool imaging. Acad Radiol. 2003;10(5):475–83.
105. Pannu HK, Thompson RE, Phelps J, Magee CA, Fishman EK. Optimal contrast agents for vascular imaging on computed tomography: iodixanol versus iohexol. Acad Radiol. 2005;12(5):576–84.

106. Mukundan Jr S, Ghaghada KB, Badea CT, Kao CY, Hedlund LW, Provenzale JM, et al. A liposomal nanoscale contrast agent for preclinical CT in mice. Am J Roentgenol. 2006;186(2):300–7.
107. Carruthers A, Melchior DL. Studies of the relationship between bilayer water permeability and bilayer physical state. Biochemistry. 1983;22(25):5797–807.
108. Seltzer SE, Blau M, Herman LW, Hooshmand RL, Herman LA, Adams DF, et al. Contrast material-carrying liposomes: biodistribution, clearance, and imaging characteristics. Radiology. 1995;194(3):775–81.
109. Elrod DB, Partha R, Danila D, Casscells SW, Conyers JL. An iodinated liposomal computed tomographic contrast agent prepared from a diiodophosphatidylcholine lipid. Nanomedicine. 2009;5(1):42–5.
110. Kweon S, Lee HJ, Hyung WJ, Suh J, Lim JS, Lim SJ. Liposomes coloaded with iopamidol/lipiodol as a res-targeted contrast agent for computed tomography imaging. Pharm Res. 2010;27(7):1408–15.
111. Simon GH, Fu Y, Berejnoi K, Fournier LS, Lucidi V, Yeh B, et al. Initial computed tomography imaging experience using a new macromolecular iodinated contrast medium in experimental breast cancer. Invest Radiol. 2005;40(9):614–20.
112. Yordanov AT, Lodder AL, Woller EK, Cloninger MJ, Patronas N, Milenic D, et al. Novel iodinated dendritic nanoparticles for computed tomography (ct) imaging. Nano Lett. 2002;2(6):595–9.
113. Fu Y, Nitecki DE, Maltby D, Simon GH, Berejnoi K, Raatschen HJ, et al. Dendritic iodinated contrast agents with peg-cores for CT imaging: synthesis and preliminary characterization. Bioconjug Chem. 2006;17(4):1043–56.
114. Pan D, Williams TA, Senpan A, Allen JS, Scott MJ, Gaffney PJ, et al. Detecting vascular biosignatures with a colloidal, radio-opaque polymeric nanoparticle. J Am Chem Soc. 2009;131(42):15522–7.
115. Robbins CS, Hilgendorf I, Weber GF, Theurl I, Iwamoto Y, Figueiredo JL, et al. Local proliferation dominates lesional macrophage accumulation in atherosclerosis. Nat Med. 2013;19(9):1166–72.
116. Hyafil F, Cornily JC, Feig JE, Gordon R, Vucic E, Amirbekian V, et al. Noninvasive detection of macrophages using a nanoparticulate contrast agent for computed tomography. Nat Med. 2007;13(5):636–41.
117. Hyafil F, Cornily JC, Rudd JH, Machac J, Feldman LJ, Fayad ZA. Quantification of inflammation within rabbit atherosclerotic plaques using the macrophage-specific CT contrast agent N1177: a comparison with 18F-FDG PET/CT and histology. J Nucl Med. 2009;50(6):959–65.
118. Van Herck JL, De Meyer GR, Martinet W, Salgado RA, Shivalkar B, De Mondt R, et al. Multi-slice computed tomography with n1177 identifies ruptured atherosclerotic plaques in rabbits. Basic Res Cardiol. 2010;105(1):51–9.
119. Ding J, Wang Y, Ma M, Zhang Y, Lu S, Jiang Y, et al. Ct/fluorescence dual-modal nanoemulsion platform for investigating atherosclerotic plaques. Biomaterials. 2013;34(1):209–16.
120. Glickson JD, Lund-Katz S, Zhou R, Choi H, Chen IW, Li H, et al. Lipoprotein nanoplatform for targeted delivery of diagnostic and therapeutic agents. Mol Imaging. 2008;7(2):101–10.
121. Song L, Li H, Sunar U, Chen J, Corbin I, Yodh AG, et al. Naphthalocyanine-reconstituted ldl nanoparticles for in vivo cancer imaging and treatment. Int J Nanomedicine. 2007;2(4):767–74.
122. Allijn IE, Leong W, Tang J, Gianella A, Mieszawska AJ, Fay F, et al. Gold nanocrystal labeling allows low-density lipoprotein imaging from the subcellular to macroscopic level. ACS Nano. 2013;7(11):9761–70.
123. Hill ML, Corbin IR, Levitin RB, Cao W, Mainprize JG, Yaffe MJ, et al. In vitro assessment of poly-iodinated triglyceride reconstituted low-density lipoprotein: initial steps toward CT molecular imaging. Acad Radiol. 2010;17(11):1359–65.
124. Caride VJ, Sostman HD, Twickler J, Zacharis H, Orphanoudakis SC, Jaffe CC. Brominated radiopaque liposomes: contrast agent for computed tomography of liver and spleen: a preliminary report. Invest Radiol. 1982;17(4):381–5.

125. Mattrey RF, Long DM, Peck WW, Slutsky RA, Higgins CB. Perfluoroctylbromide as a blood pool contrast agent for liver, spleen, and vascular imaging in computed tomography. J Comput Assist Tomogr. 1984;8(4):739–44.
126. Bruneton JN, Falewee MN, Francois E, Cambon P, Philip C, Riess JG, et al. Liver, spleen, and vessels: preliminary clinical results of CT with perfluorooctylbromide. Radiology. 1989;170(1 Pt 1):179–83.
127. Behan M, O'Connell D, Mattrey RF, Carney DN. Perfluorooctylbromide as a contrast agent for CT and sonography: preliminary clinical results. Am J Roentgenol. 1993;160(2): 399–405.
128. Mattrey RF. Perfluorooctylbromide: a new contrast agent for CT, sonography, and MR imaging. Am J Roentgenol. 1989;152(2):247–52.
129. Li A, Zheng Y, Yu J, Wang Z, Yang Y, Wu W, et al. Superparamagnetic perfluorooctylbromide nanoparticles as a multimodal contrast agent for us, MR, and CT imaging. Acta Radiol. 2013;54(3):278–83.
130. Brown AL, Naha PC, Benavides-Montes V, Litt HI, Goforth AM, Cormode DP. Synthesis, X-ray opacity, and biological compatibility of ultra-high payload elemental bismuth nanoparticle X-ray contrast agents. Chem Mater. 2014;26(7):2266–74.
131. Mieszawska AJ, Mulder WJ, Fayad ZA, Cormode DP. Multifunctional gold nanoparticles for diagnosis and therapy of disease. Mol Pharm. 2013;10(3):831–47.
132. Thakor AS, Jokerst J, Zavaleta C, Massoud TF, Gambhir SS. Gold nanoparticles: a revival in precious metal administration to patients. Nano Lett. 2011;11(10):4029–36.
133. Wang B, Yantsen E, Larson T, Karpiouk AB, Sethuraman S, Su JL, et al. Plasmonic intravascular photoacoustic imaging for detection of macrophages in atherosclerotic plaques. Nano Lett. 2009;9(6):2212–7.
134. von Maltzahn G, Park JH, Agrawal A, Bandaru NK, Das SK, Sailor MJ, et al. Computationally guided photothermal tumor therapy using long-circulating gold nanorod antennas. Cancer Res. 2009;69(9):3892–900.
135. Qian X, Peng XH, Ansari DO, Yin-Goen Q, Chen GZ, Shin DM, et al. In vivo tumor targeting and spectroscopic detection with surface-enhanced raman nanoparticle tags. Nat Biotechnol. 2008;26(1):83–90.
136. Lee SE, Sasaki DY, Park Y, Xu R, Brennan JS, Bissell MJ, et al. Photonic gene circuits by optically addressable sirna-au nanoantennas. ACS Nano. 2012;6(9):7770–80.
137. Naha PC, Chhour P, Cormode DP. Systematic in vitro toxicological screening of gold nanoparticles designed for nanomedicine applications. Toxicol In Vitro. 2015;29(7): 1445–53.
138. Hainfeld JF, Slatkin DN, Focella TM, Smilowitz HM. Gold nanoparticles: a new X-ray contrast agent. Br J Radiol. 2006;79(939):248–53.
139. Cai QY, Kim SH, Choi KS, Kim SY, Byun SJ, Kim KW, et al. Colloidal gold nanoparticles as a blood-pool contrast agent for X-ray computed tomography in mice. Invest Radiol. 2007;42(12):797–806.
140. Kim D, Park S, Lee JH, Jeong YY, Jon S. Antibiofouling polymer-coated gold nanoparticles as a contrast agent for in vivo X-ray computed tomography imaging. J Am Chem Soc. 2007;129(24):7661–5.
141. Jana NR, Gearheart L, Murphy CJ. Evidence for seed-mediated nucleation in the chemical reduction of gold salts to gold nanoparticles. Chem Mater. 2001;13(7):2313–22.
142. Turkevich J, Stevenson P, Hillier J. A study of the nucleation and growth processes in the synthesis of colloidal gold. Discuss Faraday Soc. 1951;11:55–75.
143. Huang P, Bao L, Zhang C, Lin J, Luo T, Yang D, et al. Folic acid-conjugated silica-modified gold nanorods for X-ray/CT imaging-guided dual-mode radiation and photo-thermal therapy. Biomaterials. 2011;32(36):9796–809.
144. Kattumuri V, Katti K, Bhaskaran S, Boote EJ, Casteel SW, Fent GM, et al. Gum arabic as a phytochemical construct for the stabilization of gold nanoparticles: in vivo pharmacokinetics and X-ray-contrast-imaging studies. Small. 2007;3(2):333–41.

145. Chanda N, Upendran A, Boote EJ, Zambre A, Axiak S, Selting K, et al. Gold nanoparticle based X-ray contrast agent for tumor imaging in mice and dog: a potential nano-platform for computer tomography theranostics. J Biomed Nanotechnol. 2014;10(3):383–92.
146. Peng C, Li K, Cao X, Xiao T, Hou W, Zheng L, et al. Facile formation of dendrimer-stabilized gold nanoparticles modified with diatrizoic acid for enhanced computed tomography imaging applications. Nanoscale. 2012;4(21):6768–78.
147. Guo R, Wang H, Peng C, Shen MW, Pan MJ, Cao XY, et al. X-ray attenuation property of dendrimer-entrapped gold nanoparticles. J Phys Chem C. 2010;114(1):50–6.
148. Liu H, Xu Y, Wen S, Chen Q, Zheng L, Shen M, et al. Targeted tumor computed tomography imaging using low-generation dendrimer-stabilized gold nanoparticles. Chemistry (Easton). 2013;19(20):6409–16.
149. Ye K, Qin J, Peng Z, Yang X, Huang L, Yuan F, et al. Polyethylene glycol-modified dendrimer-entrapped gold nanoparticles enhance CT imaging of blood pool in atherosclerotic mice. Nanoscale Res Lett. 2014;9(1):529.
150. Rabin O, Manuel Perez J, Grimm J, Wojtkiewicz G, Weissleder R. An X-ray computed tomography imaging agent based on long-circulating bismuth sulphide nanoparticles. Nat Mater. 2006;5(2):118–22.
151. Naha PC, Zaki AA, Hecht E, Chorny M, Chhour P, Blankemeyer E, et al. Dextran coated bismuth-iron oxide nanohybrid contrast agents for computed tomography and magnetic resonance imaging. J Mater Chem B Mater Biol Med. 2014;2(46):8239–48.
152. Bonitatibus Jr PJ, Torres AS, Goddard GD, FitzGerald PF, Kulkarni AM. Synthesis, characterization, and computed tomography imaging of a tantalum oxide nanoparticle imaging agent. Chem Commun (Camb). 2010;46(47):8956–8.
153. Bonitatibus Jr PJ, Torres AS, Kandapallil B, Lee BD, Goddard GD, Colborn RE, et al. Preclinical assessment of a Zwitterionic tantalum oxide nanoparticle X-ray contrast agent. ACS Nano. 2012;6(8):6650–8.
154. Oh MH, Lee N, Kim H, Park SP, Piao Y, Lee J, et al. Large-scale synthesis of bioinert tantalum oxide nanoparticles for X-ray computed tomography imaging and bimodal image-guided sentinel lymph node mapping. J Am Chem Soc. 2011;133(14):5508–15.
155. Zhou Z, Lu ZR. Gadolinium-based contrast agents for magnetic resonance cancer imaging. Wiley Interdiscip Rev Nanomed Nanobiotechnol. 2013;5(1):1–18.
156. Mohs AM, Lu ZR. Gadolinium(iii)-based blood-pool contrast agents for magnetic resonance imaging: status and clinical potential. Expert Opin Drug Deliv. 2007;4(2):149–64.
157. Karagiannis P, Niumsawatt V, Rozen WM. An alternative contrast medium for computed tomographic angiography: gadolinium. Plast Reconstr Surg. 2014;133(6):900e–1e.
158. Gierada DS, Bae KT. Gadolinium as a CT contrast agent: assessment in a porcine model. Radiology. 1999;210(3):829–34.
159. Zeng C, Shi X, Wu B, Zhang D, Zhang W. Colloids containing gadolinium-capped gold nanoparticles as high relaxivity dual modality contrast agents for CT and MRI. Colloids Surf B. 2014;123:130–5.
160. Alric C, Taleb J, Le Duc G, Mandon C, Billotey C, Le Meur-Herland A, et al. Gadolinium chelate coated gold nanoparticles as contrast agents for both X-ray computed tomography and magnetic resonance imaging. J Am Chem Soc. 2008;130(18):5908–15.
161. Huo D, He J, Li H, Yu H, Shi T, Feng Y, et al. Fabrication of au@ag core-shell NPs as enhanced CT contrast agents with broad antibacterial properties. Colloids Surf B. 2014;117:29–35.
162. Huo D, Ding J, Cui YX, Xia LY, Li H, He J, et al. X-ray CT and pneumonia inhibition properties of gold-silver nanoparticles for targeting MRSA induced pneumonia. Biomaterials. 2014;35(25):7032–41.
163. Chou SW, Shau YH, Wu PC, Yang YS, Shieh DB, Chen CC. In vitro and in vivo studies of fept nanoparticles for dual modal CT/MRI molecular imaging. J Am Chem Soc. 2010;132(38):13270–8.
164. Zhou Z, Kong B, Yu C, Shi X, Wang M, Liu W, et al. Tungsten oxide nanorods: an efficient nanoplatform for tumor CT imaging and photothermal therapy. Sci Rep. 2014;4:3653.

165. Jakhmola A, Anton N, Anton H, Messaddeq N, Hallouard F, Klymchenko A, et al. Poly-epsilon-caprolactone tungsten oxide nanoparticles as a contrast agent for X-ray computed tomography. Biomaterials. 2014;35(9):2981–6.
166. Mongan J, Rathnayake S, Fu Y, Wang R, Jones EF, Gao DW, et al. In vivo differentiation of complementary contrast media at dual-energy CT. Radiology. 2012;265(1):267–72.
167. Xing H, Bu W, Ren Q, Zheng X, Li M, Zhang S, et al. A NaYbF4: Tm3+ nanoprobe for CT and NIR-to-NIR fluorescent bimodal imaging. Biomaterials. 2012;33(21):5384–93.
168. Liu Y, Liu J, Ai K, Yuan Q, Lu L. Recent advances in ytterbium-based contrast agents for in vivo X-ray computed tomography imaging: promises and prospects. Contrast Media Mol Imaging. 2014;9(1):26–36.
169. Liu Y, Ai K, Liu J, Yuan Q, He Y, Lu L. A high-performance ytterbium-based nanoparticulate contrast agent for in vivo X-ray computed tomography imaging. Angew Chem Int Ed Engl. 2012;51(6):1437–42.
170. Xing H, Bu W, Zhang S, Zheng X, Li M, Chen F, et al. Multifunctional nanoprobes for upconversion fluorescence, MR and CT trimodal imaging. Biomaterials. 2012;33(4):1079–89.
171. Chanda N, Kattumuri V, Shukla R, Zambre A, Katti K, Upendran A, et al. Bombesin functionalized gold nanoparticles show in vitro and in vivo cancer receptor specificity. Proc Natl Acad Sci U S A. 2010;107(19):8760–5.
172. Wang H, Zheng L, Peng C, Shen M, Shi X, Zhang G. Folic acid-modified dendrimer-entrapped gold nanoparticles as nanoprobes for targeted CT imaging of human lung adencarcinoma. Biomaterials. 2013;34(2):470–80.
173. Eck W, Nicholson AI, Zentgraf H, Semmler W, Bartling S. Anti-cd4-targeted gold nanoparticles induce specific contrast enhancement of peripheral lymph nodes in X-ray computed tomography of live mice. Nano Lett. 2010;10(7):2318–22.
174. Hainfeld JF, O'Connor MJ, Dilmanian FA, Slatkin DN, Adams DJ, Smilowitz HM. Micro-CT enables microlocalisation and quantification of her2-targeted gold nanoparticles within tumour regions. Br J Radiol. 2011;84(1002):526–33.
175. Chhour P, Naha PC, O'Neill SM, Litt HI, Reilly MP, Ferrari VA, et al. Labeling monocytes with gold nanoparticles to track their recruitment in atherosclerosis with computed tomography. Biomaterials. 2016;87:93–103.
176. Danad I, Fayad ZA, Willemink MJ, Min JK. New applications of cardiac computed tomography: dual-energy, spectral, and molecular CT imaging. JACC Cardiovasc Imaging. 2015;8(6):710–23.
177. Pan D, Schirra CO, Senpan A, Schmieder AH, Stacy AJ, Roessl E, et al. An early investigation of ytterbium nanocolloids for selective and quantitative "multicolor" spectral CT imaging. ACS Nano. 2012;6(4):3364–70.
178. Feuerlein S, Roessl E, Proksa R, Martens G, Klass O, Jeltsch M, et al. Multienergy photon-counting k-edge imaging: potential for improved luminal depiction in vascular imaging. Radiology. 2008;249(3):1010–6.
179. Cormode DP, Roessl E, Thran A, Skajaa T, Gordon RE, Schlomka JP, et al. Atherosclerotic plaque composition: analysis with multicolor CT and targeted gold nanoparticles. Radiology. 2010;256(3):774–82.
180. Cormode DP, Skajaa T, van Schooneveld MM, Koole R, Jarzyna P, Lobatto ME, et al. Nanocrystal core high-density lipoproteins: a multimodality contrast agent platform. Nano Lett. 2008;8(11):3715–23.
181. Pan D, Roessl E, Schlomka JP, Caruthers SD, Senpan A, Scott MJ, et al. Computed tomography in color: Nanok-enhanced spectral CT molecular imaging. Angew Chem Int Ed Engl. 2010;49(50):9635–9.

Molecular Imaging with Spectral CT Nanoprobes

Dipanjan Pan, Anne H. Schmieder, Angana SenPan, Xiaoxia Yang, Samuel A. Wickline, Ewald Roessl, Roland Proksa, Carsten O. Schirra, and Gregory M. Lanza

1 Spectral Computed Tomography Imaging Technology

Traditional computerized tomography (CT) exploits the attenuation of photons traversing through an imaged object and impinging on a sensitive detector (or film) to create images. CT systems typically employ scintillation-based detector arrays that produce light when individual photons are absorbed. On a pixel-by-pixel basis, the integrated intensity of signal detected by the photodiode corresponds to the overall loss of signal intensity due to the intervening object represented. Despite the broad energy distribution of photons emitted in an X-ray beam and the known energy dependence of photon attenuation, traditional integrating detector measurements eliminate all inherent spectral information that might be used to further characterize the imaged object.

The two predominant physical causes of X-ray attenuation effects are the photoelectric process and Compton scattering [1]. From these factors, which can be differentiated with two measurements at distinct photon energies, the concept of "dual-energy" CT evolved and is now implemented in clinical practice. Dual-energy CT systems can be realized in two different ways, source-based and detection-based spectral separation. Source-based dual-energy CT is achieved utilizing conventional detector principles under X-ray exposure of different tube voltages.

D. Pan
Department of Bioengineering, University of Illinois at Urbana, Urbana, IL, USA

A.H. Schmieder • A. SenPan • X. Yang • G.M. Lanza, M.D., Ph.D. (✉)
Division of Cardiology, Department of Medicine, Washington University School of Medicine, CORTEX Building, Suite 101, 4320 Forest Park Avenue, Saint Louis, MO 63108, USA
e-mail: greg.lanza@mac.com

S.A. Wickline • E. Roessl • R. Proksa
Philips Research Hamburg, Hamburg, Germany

C.O. Schirra
Philips Research Brazil, São Paulo, Brazil

J.W.M. Bulte, M.M.J. Modo (eds.), *Design and Applications of Nanoparticles in Biomedical Imaging*, DOI 10.1007/978-3-319-42169-8_18

The kVp-switching method acquires X-ray projections at alternating tube voltages in order to obtain spectrally resolved measurements [2–5]. Another clinically available solution is dual-source CT which mounts two "source-detector" pairs on a single gantry with each pair operating at a different tube voltage. Detection-based spectral CT uses a single X-ray tube and instead a stack of detector-layers of which each layer is sensitive to different photon-energies. This "dual-layer" approach allows one to acquire both spectral measurements on the same projection path simultaneously in a single rotation of the gantry [6–12].

Advancements of the dual-layer CT energy-integrated detector technology were further extended into the development of photon-counting detectors, leading to the next generation of CT instrumentation often referred to as spectral or multicolored CT [13–17]. Photon-counting detectors produce a digital measure of photon energy through pulse height analysis that eliminates electronic noise. Photon-counting detectors simultaneously discriminate the energy of counted photons by assigning measured pulses into energy bins based on preselected thresholds. In principle, a multiplicity of spectral determinations characterizing multiple tissues and contrast agents within each voxel from the same X-ray beam can be made.

Electrons for all elements of the periodic table are distributed into energy shells, with the K-shell electrons closest to the nucleus possessing the lowest energy. When photons are iso-energetic with a K-shell electron, the photon can be fully attenuated, deploying its energy to free an electron on the K-shell. This ionization causes a discontinuity in the photoelectric attenuation cross-section, and is characteristic for each individual element. Elements between iodine and bismuth on the periodic chart possess K-shell electron energies within typical photon energy bandwidth of clinical CT, and each could serve a K-edge contrast agent. In particular, gold, gadolinium, ytterbium, tantalum, or bismuth are among the metals with K-edges well inside the X-ray beam energy bandwidth and promise the highest signal-to-noise levels [18, 19]. Iodine, which is a stalwart X-ray contrast element for traditional CT, is on the lower edge of the bandwidth. While useful for small animal preclinical spectral CT imaging, the low K-edge energy signal detected will be significantly compromised by photon starvation and scattering occurring as the X-ray beam passes through patient torsos. Some new CT contrast agents are being developed with iodine alone or as part of a hybrid multimodality approach, however the present chapter is focused on K-edge agents incorporating nontraditional radiopaque elements.

2 Spectral CT Contrast Agents: Challenges

Spectral CT contrast agents must be designed with metals with suitable K-shell energy electrons in order to absorb photons emitted in the X-ray beam. However, another significant challenge in the development of CT and spectral CT probes is the requisite high concentrations of metal that must be present within a voxel for detectability. In comparison with nuclear imaging, ultrasound, and even MRI, CT contrast imaging is very insensitive. High dosages of heavy metals are required to elicit effective CT contrast inherently bringing patient long-term safety into consideration.

This is particularly relevant since CT contrast probes will likely be used in many patients to rule-out disease and in others to serial assess and guide clinical management. Proving long-term human safety following exposure to heavy metals will be challenging unless the metal component of CT contrast agents are essentially bioeliminated over a reasonable time frame. A "what goes in must come out" philosophy regarding heavy metals should be embraced in the development of these pharmaceuticals.

Because the human renal clearance threshold for nanoparticles or their breakdown products is around 6–8 nm, the use of poorly degradable solid particles in excess of this size could ultimately result in prolonged or even indefinite retention of such contrast agents internally [20]. While some metal particles, such as iron oxides used for MRI imaging, are known to be metabolized slowly with the iron reentering the iron metabolic pool, other solid oxide or sulfide crystalline particles are unlikely to be metabolized and bioeliminated adequately. Even gold, which was used as a colloidal suspension for various medical applications since ancient times and in the more recent past for rheumatoid arthritis, may not be as inert as anticipated [21]. Large and or frequent dosages of gold particles (>8 nm) may clear the body very slowly or not at all. Proving that this or similar situations with other elements is safe to regulatory agencies will be a translational challenge that the field must recognize.

Rodent models are often used to study the pharmacology and safety of drugs of all types, including nanomedicines. In man, rabbits, and larger mammals, solid nanoparticles 8–10 nm or greater are above the renal threshold for glomerular clearance and must be metabolized into smaller constituent parts for elimination through the bile, urine, or respiration. However, the clearance mechanism of nanoparticles in rodents differs markedly from man and larger mammals. Particles of 250 nm or greater can pass directly into the biliary system of rats and mice within a few minutes of intravenous injection. Bulte et al. illustrated this phenomenon by acquiring an MR cholangiogram using proton and fluorine MR imaging of paramagnetic perfluorocarbon nanoparticles in rats [22] (Fig. 1). This MRI demonstration corroborated experimental

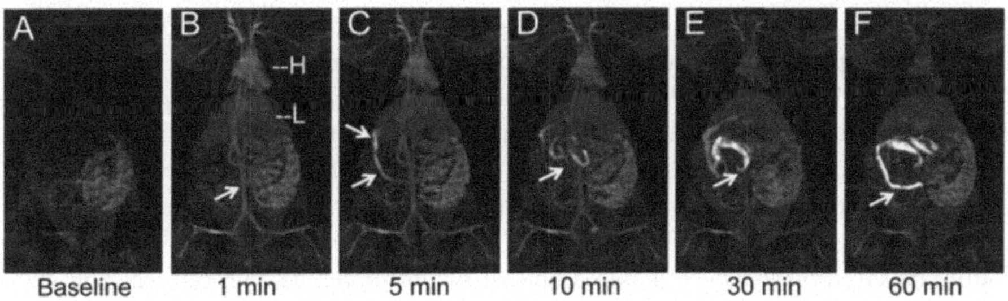

Fig. 1 3D time-of-flight (TOF) MR cholangiogram in a rat following intravenous injection of gadolinium-functionalized perfluorooctylbromide nanoparticles (NP). (**a**) Baseline image showing no evidence of vasculature or common bile duct. (**b**) 1 min post-injection, the blood pool as seen in the heart, aorta (*arrow*), and peripheral vasculature have strong T1-weighted positive contrast from the NP that are still constrained within the vasculature. *H* heart, *L* liver. (**c**) Within 5 min, the NP are rapidly excreted through the common bile duct (*arrows*), reflecting the rapid shunt of contrast from the liver into the small intestine. (**d**). After 10 min the small intestine contains most of the contrast, that is passed on to the large intestines at 30 min (**e**) and 60 min (**f**). Reproduced with permission from Ref. [22]

investigations of particle elimination dating to the 1950s and possibly decades earlier. In 1958, Hampton showed that intravenous particles in rodents distributed into Kupffer cells, hepatic cells, and bile within the liver [23]. This report cited and confirmed earlier reports of Müllendorf in 1916 and Weatherford in 1956 (*cited in* [23]) using colloidal dyes and India ink, respectively. Hampton demonstrated that particles phagocytosed by Kupffer cells were retained indefinitely; whereas, particles endocytosed by hepatocytes were transported rapidly into the biliary system, as imaged by Bulte. Interestingly, Hampton performed retrograde biliary infusions of Thorotrast™ colloidal particles and used transmission electron microscopy (TEM) to reveal that those particles transited in reverse through hepatocytes and into the space of Disse. While some of the Thoratrast™ became sequestered in Kupffer cells, much of it passed into the circulatory system. Similar evidence of nanoparticle excretion into rodent bile and feces include silica particles from 50 to 200 nm [24] citrate-coated silver particles (~8 nm) [25], and iron oxide core high-density lipoproteins (~10 nm) [26]. In larger non-rodent mammals, Juhlin [27] reported a few years later that fluorescent spherical hydrophilic particles of methyl methacrylate injected intravenously in rabbits did not transit effectively into the bile. Particles in the 20–110 nm produced very minimal biliary concentrations and those greater than 60 nm were not excreted into the biliary tree. While rodent models offer a wealth of research opportunity, species difference in biliary nanoparticle excretion must be considered to avoid incorrect conclusions regarding nanoparticle pharmacokinetics, pharmacodynamics, metabolism, bioelimination, and ultimately safety.

3 Spectral CT Contrast Agents

Many CT contrast agents have been created using metals with K-edge values within the X-ray energy bandwidth and the vast majority have crystalline cores that exceed 10 nm. While ultimately these types of agents may prove safe and effective, the scope of the present review generally considered small crystalline nanoparticles (<10 nm) and large degradable nanoparticles encapsulating small molecule organometallic complexes or small solid nanoparticles (<10 nm).

From a spectral CT perspective, two independent blood pool contrast agents imaged simultaneously may have important medical utility; however, the use of a single blood pool agent will likely have limited differential benefit over current iodine based approaches. For new contrast technologies to gain clinical and economic traction, they must address important unmet functional or biochemical (i.e., molecular) imaging needs. In some situations, this may be achieved through in situ labeling of cells, e.g., macrophages in inflammatory imaging, and in other situations by ligand-directed targeting. Regardless, the development of a molecular imaging agent should begin with a clear understanding of the unmet need, which then defines parameters for pharmaceutical design.

4 Bismuth

4.1 Organobismuth Nanocolloid (NanoK)

Emergency departments in the USA assess more than 8 million patients annually with complaints of chest pain or shortness of breath [28]. While a presumptive diagnosis of life-threatening acute coronary syndrome (ACS) may be apparent from the patient's initial presentation, the vast majorities of cases are equivocal and require the potential of cardiac involvement to be excluded. Consequently, individuals are retained for close observation and testing. For patient without cardiovascular symptoms, there is inconvenience, family stress, and high healthcare cost. Patients admitted for possible ACS require continuous cardiac telemetry, serial ECGs, and repeated cardiac troponin assays over 12–24 h. If cardiac infarction and unstable angina are excluded, then noninvasive cardiac stress testing is typically performed.

Coronary CT angiography has become increasingly refined and offers the potential to recognize patients with low likelihood of disease but the negative and positive predictive values do not meet the standard set by cardiac catheterization [29]. Moreover, coronary calcium, due to its attenuating and blooming artifacts, further complicates the clarity of the angiogram in the region of potentially critical disease [30]. Spectral CT contrast imaging offers an opportunity to target and detect intraluminal thrombus associated with acutely ruptured plaque while discriminating between attenuation artifacts of calcium deposits, which have low (K-edge energies (4.4 keV) (Fig. 2).

An initial approach to this medical imaging problem was reported by Pan et al. [31] using nanocolloids comprised of high concentrations of an organobismuth compound, bismuth neodecanoate, commixed in sorbitan sesquioleate. The phospholipid-encapsulated "soft" nanocolloid was 20% w/v aqueous suspension with hydrodynamic diameters between 180 and 250 nm, a negative electrophoretic potential ranging from −20 to −27 mV, and 1.06 g/ml. The particle core was 12–14 wt% bismuth. The biocompatible outer lipid membrane can be functionalized for ligand-direct (peptide and antibody) homing to microthrombus (Fig. 3).

In vitro studies with the bismuth nanocolloid compared cross-sectional CT (conventional) and spectral CT images of phantoms containing serially diluted bismuth particles and a reference calcium acetate suspension in water. This illustrated the potential of K-edge imaging to segment calcium from bismuth nanocolloid despite the X-ray attenuation similarity of the materials at 60 keV, (Ca: 1060HU, Bi: 1050HU). As expected the attenuation of the bismuth formulation varied linearly with the element concentration ($R^2=0.999$).

Additionally, fibrin-targeted bismuth nanocolloid revealed excellent delineation and signal enhancement on spectral CT images of fibrin clot phantoms embedded with calcium. The control clot treated with targeted nonmetallic nanoparticles had negligible contrast, exhibiting only the highly attenuating calcium. The specific targeting of rhodamine-labeled bismuth particles to fibrin on human carotid endarterectomy specimens was corroborated microscopically with immunohistochemistry

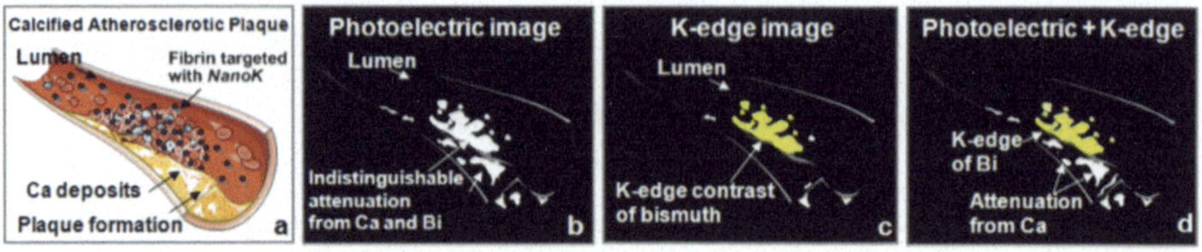

Fig. 2 Example of spectral CT using a coronary vascular example to illustrate the discrimination of calcium and spectral CT contrast agents; (**a**) coronary artery with a partial occluding thrombus emerging from rupture of the unstable intimal cap. (**b**) fibrin targeted with bismuth nanocolloid (BiNC) with in classic CT (photoelectric) image and attenuation due to calcium are seen (**c**) spectral CT (K-edge image) discriminates the fibrin-targeted BiNC from the calcium deposits, (**d**) integration of the classic CT X-ray image with the spectral CT molecular imaging result acquired simultaneously, reveals the signal from the fibrin-bound bismuth resolved from the atherosclerotic calcium deposits. Reproduced with permission from Ref. [31]

[31] and exemplified how the K-edge agent targeting was constrained to intravascular rather than intramural fibrin deposits. Dynamic imaging of NanoK with a clinical multidetector CT revealed that blood pool background was no longer detectable after 15–30 min, which would permit optimal molecular imaging of intraluminal thrombus after 1–2 h. The antibody-targeted fibrin bismuth agent concentrated in situ over the acute thrombus in 30 min, remained bound to the clot during 90 min continuous blood circulation, and was detected with a first generation spectral CT scanner as a partial-occlusive thrombus within a 1.41 mm diameter artery, equivalent to a small coronary artery in humans. The bismuth-enhanced clot was clearly differentiated within the vessel and distinct from adjacent attenuation effects of the femur. Superimposition of the simultaneously acquired K-edge image with the traditional CT image spatially oriented and localized the intravascular lesion relative to the rabbit's skeletal anatomy and demonstrated the complementarity of spectral CT molecular imaging and traditional CT [31] (Fig. 4).

Importantly, whole-body bioelimination of bismuth following intravenous injection was studied in adult male BALB/c mice in a 2-week pilot study. Nearly all of the metal was cleared from the mice within 14 days, and most was bioeliminated by the first week. The residual biodistribution on day 14 into primary particle clearance organs (i.e., liver, spleen, kidney) was assessed using inductively coupled plasma optical emission spectrometry in a second cohort of mice ($n=3$), which revealed less than 10 ppb (i.e., the lower detection limit). A third cohort of mice assessed the in vivo impact of bismuth nanocolloid and saline on liver (ALT, AST, albumin) and renal (BUN, CR, Na, K, Cl) function was assessed on days 1, 7, and 14 post-injection. All tests in both groups remained within accepted normal limits for each parameter and no statistical differences between groups were noted at any time point. The overarching conclusions from this early work were: (1) that the nanocolloid offered stable, detectable bismuth contrast for imaging, (2) that the organobismuth complexes could be bioeliminated, and (3) that the nanosystem acute toxicology was favorable [31].

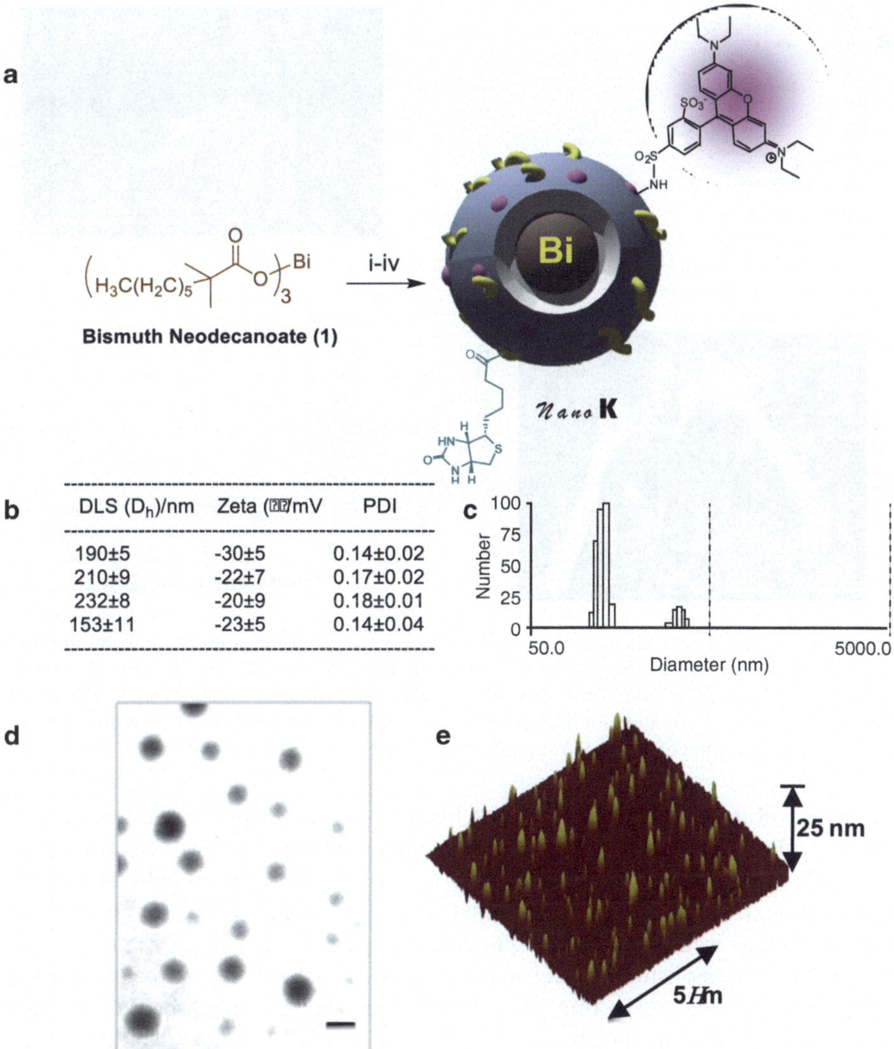

DLS (D_h)/nm	Zeta ([illegible]/mV	PDI
190±5	-30±5	0.14±0.02
210±9	-22±7	0.17±0.02
232±8	-20±9	0.18±0.01
153±11	-23±5	0.14±0.04

Fig. 3 Synthesis and physicochemical characterization of NanoK: (**a**) Schematic describing the preparation of bismuth-enriched K-edge nanocolloid (NanoK (Bi)): (**b**) characterization table for three replicates of NanoK; (**c**) hydrodynamic particle size distribution from dynamic light scattering (DLS); (**d**) anhydrous state TEM images (staining: uranyl acetate; scale bar: 100 nm; (**e**) Atomic force microscopy image (deposited on glass substrate). Reproduced with permission from Ref. [31]

4.2 *Bismuth Sulfate: Alginate Microcapsules*

The concept and benefit of small molecule bismuth contrast entrapped within larger particles or microcapsules was instituted by Barnett et al. [32]. These investigators addressed the challenge of pancreatic islet cells transplantation by entrapping the

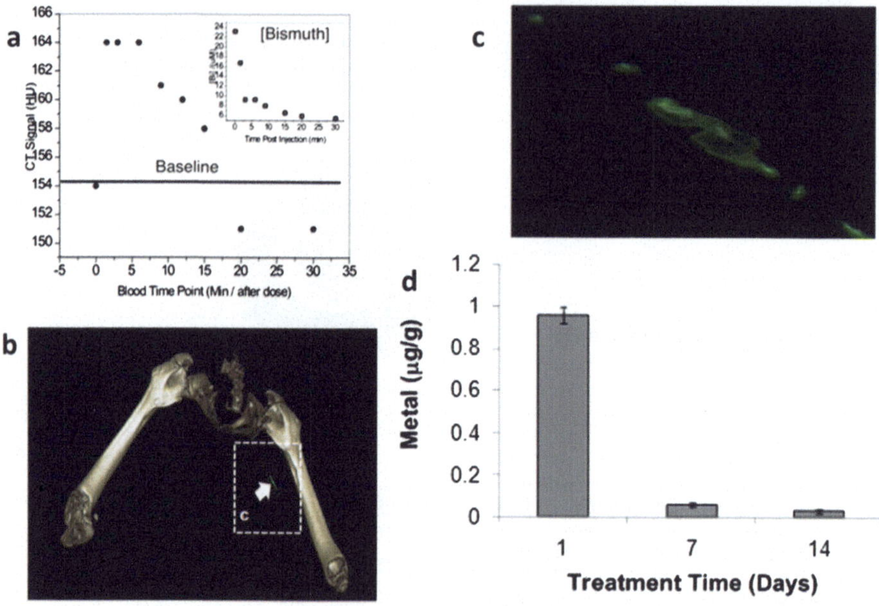

Fig. 4 (**a**) CT blood pool signal in rabbits following IV injection of NanoK. Inset shows the concentration of bismuth (ICP) in blood versus time post injection. Note that the background signal is at baseline in less than 30 min; (**b**, **c**) targeting in situ clot (thrombus) in rabbits (*arrow* indicates thrombus); (Scale: maximum clot diameter = 1.41 mm; minimum diameter = 1.25 mm). (**d**) Two weeks clearance profile of bismuth from mice. Reproduced with permission from Ref. [31]

cells and within a permeability selective bismuth-alginate capsule. The semipermeable capsule accommodated the influx of glucose and the responsive efflux of insulin, while excluding the penetration of humoral and cell-mediated immune factors. In this context, the bismuth-doped alginate offers radiopaque contrast to both guide implantation as well as support longitudinal monitoring capsule persistence. Upon the eventual metabolism of the capsule, the alginate and bismuth are metabolized or eliminated via urinary excretion. Traditional CT imaging provided very good spatial localization of the implants, however, spectral CT could provide quantitative estimates of bismuth content. Estimates of bismuth content can be converted into more precise assessment of microcapsule numbers and at least qualitative projections of islet numbers the adequacy of their insulin response to glucose challenge. Functional longitudinal monitoring afforded by quantitative spectral CT would clearly facilitate improved clinical management of these pancreatic surrogates.

4.3 *Bismuth Sulfide Nanodots*

Large-scale synthesis of bismuth sulfide (Bi_2S_3) nanodots was reported by Lu [33]. In contradistinction to the report from the Weissleder lab [34] using large crystalline bismuth particles, these nanodots had uniform particle sizes approximately 2–3 nm, well below the renal clearance threshold. Although current production methods at that time for hydrophobic Bi_2S_3 nanoparticles lacked effective surface modification methods, Ai et al. coated the nanodots with either poly (vinylpyrrolidone) (PVP) for stability and biocompatibility. PVP-bismuth nanodots were compared in vivo to Iobitridol™, an X-ray contrast molecule with 45.6 wt% iodine. The intravenously administered nanodots circulated for 1 h and accumulated rapidly in the liver and spleen. Little bismuth was observed clearing via the kidneys into the bladder. Since these tiny particles were quickly sequestered into the mononuclear phagocyte system (MPS), one might infer that they aggregated in circulation. Subsequent histology showed no gross changes to clearance organ microanatomy, which is consistent with the expected low toxicity of nonionized bismuth. The Iobitridol™ control produced a very brief blood pool contrast then cleared through the kidney into the urine within 3 min. While prolonged blood pool contrast can be beneficial, particularly for techiques like MRI, a typical clinical CT imaging sequence is concluded in seconds, diminishing this advantage.

4.4 Dendrimeric Bismuth Sulfide

An alternative approach to coated minute bismuth nanoparticles was offered by the Shi laboratory, which reported the development of dendrimer-bismuth sulfide [35]. The agent was produced by reacting Bi^{3+} with generation 4 poly(amidoamine) dendrimers (G4.NGlyOH) followed by exposure to hydrogen sulfide to generate dendrimer coated-Bi_2S_3 nanoparticles, sized 5.2–5.7 nm. As expected, the X-ray contrast of dendrimer-bismuth sulfide was greater than that of iodine on an equimolar metal basis. The complex had minimal cellular cytotoxicity and good hemocompatibility. When injected subcutaneously in a rabbit thigh, the localized complex was readily appreciated with CT. However, when administered systemically in mice, the enhancement seen within the pulmonary vein and aorta was only slightly greater than an equimolar dosage of Omnipaque™, a commercially available iodine contrast agent. No pharmacokinetic, biodistribution, and bioelimination data of the dendrimer-bismuth sulfide were provided.

5 Gold

Gold nanoparticles and nanoshells have been extensively studied as optical, photoacoustic, and more recently X-ray contrast agents. Indeed the use of gold for optical contrast, particularly photoacoustic contrast agents, precedes later reports of its use for CT contrast nanoparticles, although it is long known to be X-ray attenuating. Many of the gold-based agents used for photoacoustic imaging are relatively large and exceed the renal threshold for clearance, but some are comprised of small gold nanoparticles alone or encased in larger nanoparticles.

5.1 Gold Nanobeacons

One important indication for nanotechnology has been the need for image-guided sentinel lymph node biopsy (SLNB) in patients with breast cancer and melanoma to stage metastases and avoid more invasive biopsy. Invasive SLNB can lead to seroma formation, lymphedema, sensory nerve injury, and limitation in the range of motion. Alternative noninvasive methods to identify sentinel lymph nodes (SLNs) in conjunction with either minimally invasive percutaneous fine-needle biopsy or molecular techniques could offer new prospects for noninvasive axillary staging of breast cancer. Pan et al. demonstrated the use of gold nanobeacons with varying gold nanoparticle content in the context of photoacoustic imaging; these gold nanobeacons were demonstrated to be highly effective for the same application with spectral CT [36–38]. The initial gold nanobeacons were designed as the encapsulation of octanethiol coated metallic gold nanoparticles (2–4 nm) suspended in vegetable and encapsulated with phospholipid with a nominal hydrodynamic diameter of 154 ± 10 nm. The polydispersity and zeta potential were measured to be 0.08 ± 0.03 and −47 ± 7 mV, respectively. Gold content, as determined by inductively coupled emission mass spectrometry, was 1080 μg/g of the 20 % colloid suspension (of GNB_{160}). For SLN mapping, GNB_{160} was compared with a polymer-encapsulated gold nanobeacons (GNB_{290}) developed through the self-assembly of amphiphilic di-block copolymer in aqueous media to entrap high payloads of gold. The same octanethiol coated AuNPs (2 w/v%) were suspended in polysorbate and microfluidized with a PS-*b*-PAA dispersion to obtain the GNB_{290} particles (particle size: 289 ± 24 nm, polydispersity index: 0.15 ± 0.04, gold: 134 μg/g or ~71,493 gold atoms per nanobeacon. While the original GNB was effective for SLN imaging, GNB_{290} was not due to its weight, inhibiting uptake and migration through the lymphatics following intradermal injection into the forepaw. Consequently, a smaller lipid-encapsulated (~90 nm) gold nanobeacon (GNB_{90}) was produced by suspending octanethiol-functionalized, coated AuNPs (2 w/v% of inner matrix) in polysorbate then microfluidizing the mixture with phospholipid-based surfactant. The GNB_{90} particle size was 92 ± 12 nm with a polydispersity of 0.35 ± 0.05. The gold content was only 1.56 μg/g or approximately 9 gold metal atoms per GNB_{90}. This particle and smaller payload rapidly transited from the forepaw intradermal injection site through the lymphatics into the SLN within a few minutes. The extent of lymph node uptake of the GNB_{90} particle far exceeded the original signal appreciated with GNB_{160} nanocolloid. For SLN imaging, it was evident that smaller particles with less gold/particle yielded more PA signal.

SLN mapping using GNB_{90} was next evaluated with spectral CT using an analogous experiment. As shown, the magnitude of the spectral CT image was dramatic, reflecting the high mass of gold accumulated rapidly into the node. The "soft" nature of the colloid, the incorporation of very small gold particles, and the very limited gold dose required for SLN imaging all contributed to what could be a translatable safe and effective formulation for this application [39] (Fig. 5).

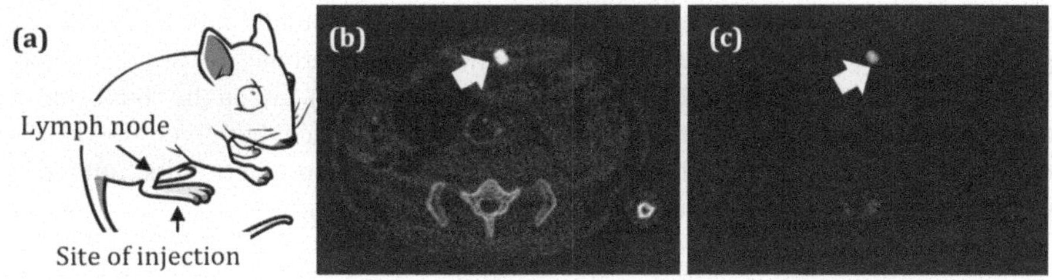

Fig. 5 In vivo noninvasive spectral CT imaging of SLN. (**a**) A cartoon illustrating the site of injection and the area of interest. 150 ml of nanobeacons were injected intradermally. (**b**) regional SLN were clearly contrasted on conventional CT. (**c**) K-edge contrast of accumulated gold nanoparticles in the SLN was selectively imaged with spectral CT. Reproduced with permission from Ref. [39]

5.2 High-Density Lipoprotein Gold Nanoparticles

The quest for distinguishing vulnerable versus stable atherosclerotic plaque prior to rupture remains a daunting challenge and unmet need in cardiovascular medicine. Since the early work of Benson [40] and Constantinides [41], the acute formation of thrombus following atherosclerotic plaque rupture has been well recognized as the etiology of unstable angina, myocardial infarction, transient ischemic attacks and stroke [42, 43]. Sensitive detection and differentiation of vulnerable versus stable atherosclerotic plaques in vessels with mild severity stenoses remains limited. As the PROSPECT trial showed, the current best invasive imaging technologies, i.e., angiograms and intravascular ultrasound, predict future regions of plaque rupture poorly, even in high-risk patients presenting with myocardial infarction [44]. While still inadequately understood, the risk of plaque rupture is primarily related to the composition of plaque, and considerable noninvasive imaging effort has been directed to characterizing the atherosclerotic inflammatory component.

In this regard, Cormode et al. developed a gold enriched HDL nanoparticle mimetic as a relatively specific approach to labeling and identifying macrophage-rich plaques in the aortas of apoprotein E knock-out mice ($APO^{e-/e-}$) using micro-CT [45]. As previously discussed plaque is often laden with calcium deposits, which like gold has marked X-ray attenuation that can be differentiated with spectral CT. Moreover, spectral CT was employed to differentiate the uptake of the gold HDL particles in $APO^{e-/e-}$ mice from the iodine contrast enhancement of the aortic lumen [46, 47]. Although these first-generation photon-counting scanners had very slow image acquisition speeds unable to support live animal studies, this seminal report offered the first in vivo illustration of multi-spectral imaging potential in a pathological animal model. Moreover, in contradistinction to earlier versions of this gold-HDL approach, the nanosystem used in this study incorporated gold particles of 3 nm with the overall diameter of the HDL particle around 7 nm, basically one gold particle per HDL particle. The dose of gold used was around 500 mg/kg, somewhat higher than a typical iodine contrast load (370 mg/kg). Given the small size of the gold particles, they have the potential for renal clearance, but given the dosage of particles required for in situ macrophage labeling, the long-term bioelimination of the metal will need to be better characterized.

5.3 Pegylated Dendrimer Entrapped Gold Nanoparticles

Peng and Shi [48] recently reported on the development of a polyethylene glycol (PEG)ylated dendrimer containing gold nanoparticles as a vehicle to achieve prolonged blood pool circulation. Indeed, prolonged blood pool contrast, particularly if it is vascular constrained has merit, but one must also be cognizant that the speed of photon-counting is already approaching clinical CT acquisition rates and current rapid imaging of approved iodinated agents with CT presents a high barrier to overcome. Peng and Shi synthesized and characterized amine-terminated poly(amidoamine) dendrimers (generation 5) that were partially modified by PEG monomethylether and gold salt. The dendrimers were 2–4 nm, had low acute toxicity up to 100 μM, and offered a circulatory half-life between 31 and 42 h depending on the gold density. Subsequently, Shi et al. reported the tripeptide arginylglycylaspartic acid (RGD) functionalization of this dendrimeric-gold construct and demonstrated its binding and potential X-ray benefits in vitro [49]. Clearly, this recent work is still evolving and seeks to take advantage of the metal concentration benefits accrued by targeting to increase imaging detectability of occult pathologies.

5.4 Gold-Loaded Polymeric Micelles

The Tsourkas lab [50] followed an approach similar to that described by Pan et al. in development of the polymer-encapsulated gold nanobeacons (GNB_{290}) [37]. These investigators took advantage of the higher X-ray attenuation properties of gold versus iodine and constructed PEG-b-poly(e-caprolactone) particles between 25 nm and 150 nm that entrapped 1.9 nm gold particles to augment CT-guided radiation therapy and radiosensitization. Radiosensitization of gold stems primarily from photoelectrons generated in the kilovolt energy range and theoretically from Auger electrons within the megavolt realm [51]. Radiation therapy in the presence of gold particles increases the extent of DNA fragmentation and treatment efficacy [52, 53]. This therapeutic response was corroborated with a 75 nm version of this nanotechnology.

Similar to the gold dendrimers described above, these polymeric gold particles had extended circulatory half-lives with enhanced blood pool contrast extending for 24 h. The particles were passively entrapped within the tumor periphery, as has been described before for most nanoparticles. In combination with radiation therapy median survival of tumor-bearing mice was increased by 1.7-fold. Issues of acute toxicity and clinical pathology changes in mice receiving the nanoparticles were unremarkable. Issue of inflammation and biocompatibility associated with the "suture-like" polymer, the bioelimination of gold, and other long-term safety concerns will need to be further elucidated. Once again, this technology points to contrast designs that avoid large solid particle bioelimination issues by encapsulating numerous small particles with potentially filterable into the urine.

6 Tantalum Oxide

Other metals within the X-ray bandwidth include tantalum, which is chemically inert and offers good radiopacity. Although the metal oxide is water insoluble, Bonitatibus et al. [54] developed a coated water-soluble version 6 nm particle size. While the desirable X-ray properties of tantalum have been known, these investigators provide the first work to permit the nanocrystal to be suitable for intravenous injection at high concentration without aggregation. The tantalum oxide core was coated with (2-diethylphosphato-ethyl)triethoxysilane to form a homogeneous product concept stable at least for 6 months. Tantalum had greater X-ray attenuation versus iodine across the X-ray spectrum, particularly for the higher peak kilovoltages (kVp) in which realm the augmenting K-edge benefit that iodine enjoys with typical CT X-ray beams is missing. Using dynamic contrast imaging over 30 s, the higher and more persistent attenuation achieved with the tantalum oxide was appreciated versus iodine. Although the difference in contrast between the metals at baseline was modest, the rapid loss of iodinated contrast from the blood pool over 30 s led to progressively less iodine contrast relative to the more slowly clearing tantalum particles. This proof of concept study demonstrated yet another potential translatable approach to CT and K-edge imaging. Metal bioelimination, safety, and optimized efficacy for a specific unmet clinical are pending.

Along a similar line of investigation, the Hyeon and Choi laboratories [55] reported a silane modified PEG-coated tantalum oxide nanoparticle that was further functionalized with a fluorescent dye and applied to SLN mapping and blood pool imaging. The low end of the nanocrystal particle size distribution by TEM was 6 nm while the other particles had greater size reaching 15 nm with hydrodynamic particle size averaging just under 20 nm. For blood pool imaging, high dose (840 mg/kg) of the tantalum produced strong blood pool contrast with MPS clearance into the liver and spleen. For SLN mapping, the enhanced CT contrast obtained was corroborated with optical imaging for the bimodal agent. No acute safety problems were reported, but the issues of metal bioelimination, safety, and specific clinical application remain to be elucidated, particularly for the larger particle population.

7 Ytterbium

Ytterbium (Yb) based nanomaterials are particularly interesting from a K-edge imaging perspective. Yb has a K-edge at 61 keV, which is in the middle of the higher energy X-ray spectrum used clinically. At this spectral location the flux of photons impinging on the detector at somewhat higher and lower energies affords better resolution of the K-edge discontinuity than that observed on either extreme end of the bandwidth. Ytterbium is a lanthanide metal that is highly abundant at much lower cost than other relevant metals, particularly gold. However, as a lanthanide similar to gadolinium, unsuspected safety issues, including the potential for nephrogenic systemic fibrosis, will need to be carefully assessed.

The Lu laboratory [56] reported a Yb-based nanoparticulate that had prolonged circulating time, low acute toxicity, and in addition to CT contrast, offered optical up-conversion luminescence (near infra-red (NIR) to visible or NIR to NIR). To extend its multimodal functionality, they included gadolinium for magnetic resonance imaging (MRI) as well as to contribute to the overall the X-ray attenuation. The initial basic particle, $NaYbF4:Er$, stabilized with oleic acid was noted to have considerable size and shape variation. Particle size and shape control was conveniently resolved through Gd-doping at 20+ mol%. Nevertheless, particle sizes ranged from 20 to 50 nm as determined by TEM. The particles provided marked CT contrast and were cleared into the liver and spleen. For MRI, the r1 relaxivity attributable to the gadolinium was 0.41 $mM^{-1}\ s^{-1}$, approximately a $\log_{10}$ order below a typical MR blood pool contrast agent, e.g., Gd-DTPA, being ~4.5 $mM^{-1}\ s^{-1}$.

Pursuing a "soft" nanoparticle approach, Pan et al. [57] developed an ytterbium nanocolloid wherein a trivalent ytterbium complex suspended in polysorbate was encapsulated within a phospholipid membrane through microfluidization. The overall particle size was 240 nm with low polydispersity (0.2), a slightly negative zeta potential (−12 mV) and was designed for vascular constrained application, such as detection of ruptured plaque. The 20 % (v/v) nanocolloid suspension (~10^{12} particles/ml) contained 0.41 mg/ml of metallic Yb, which equated to approximately 1200 K Yb atoms/nanoparticle and was stable at 4 °C for months. Spectral CT imaging of the Yb^{3+} nanocolloid in the blood pool was clearly shown in the heart, perhaps the first K-edge image of Yb^{3+} produced (Fig. 6). However, largish nanoparticles provide considerably less blood pool contrast than small molecules or tiny particle on an equimolar basis, which is an advantage for vascular-targeted molecular imaging. The particles followed a typical biexponential pharmacokinetic profile with a beta-elimination $t_{1/2}$ of 5 h. Biodistribution of the nanocolloid was primarily in the liver and spleen acutely and whole body bioelimination was progressive over 1-week, with about 90 % elimination of the injected dose at 7 days. Similar to the bismuth neodecanoate nanocolloids, the Yb-based formulation could be further developed for detection of microthrombus in ruptured plaques or other vascular molecular imaging applications. However, in contradistinction to bismuth, which has a high K-edge (89.5 keV), the Yb^{3+}-based nanocolloid K-edge benefits in terms of contrast sensitivity and radiation dose from being more central within the photon energy spectrum of the X-ray beam.

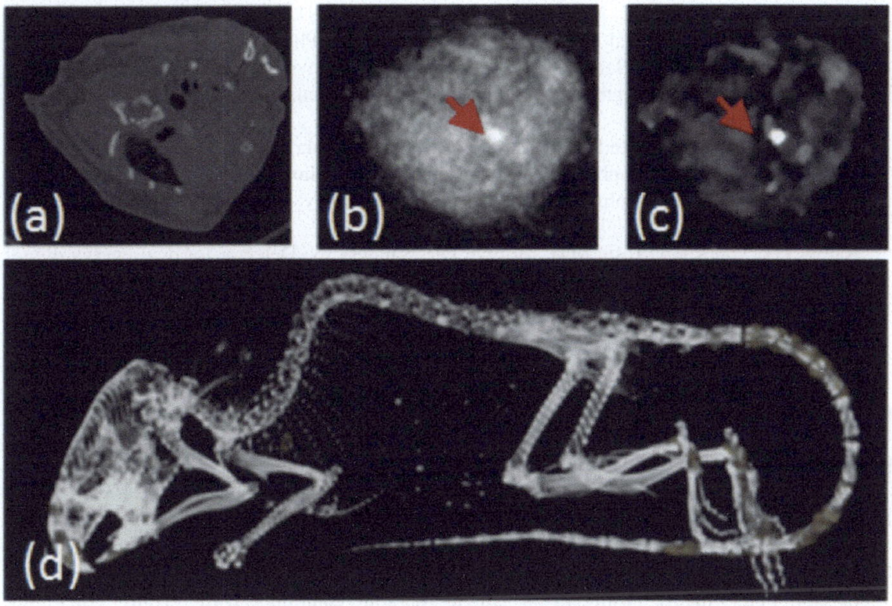

Fig. 6 Blood pool imaging in mouse after bolus injection of non-targeted Yb (2.46 mg/kg as a 20 vol% nanocolloid). The upper row shows a representative cross section through the heart. While change in contrast is only subtle in the conventional CT image (**a**), K-edge imaging using spectral CT and iterative reconstruction yields selective Yb signal at high signal to noise ratio (**b**, **c**, *red arrow*). The conventional CT and Yb images were used to compose a gradient rendered 3D image of the mouse (**d**) Reproduced with permission from Ref. [57]

8 Conclusions

Beyond iodinated systems, the recent increase in the development of CT contrast agents has coincided with the development of spectral CT. While these agents could all be envisioned for use with current clinical and dual energy CT, the incremental improvement over approved iodinated contrast agents, particularly for blood pool imaging, may be an economic barrier to translation. However, recent advancements in photon-counting detectors now place spectral CT clearly on the path to the clinic, offering higher resolution with lower radiation exposure as basic advantages. Spectral CT molecular imaging could be the "icing on the cake" that drives expansion of this capability by addressing intractable medical imaging issues unresolved today, such as the issue of coronary ruptured plaque in patients with possible ACS.

References

1. Alvarez RE, Macovski A. Energy-selective reconstructions in X-ray computerized tomography. Phys Med Biol. 1976;21(5):733–44.
2. Kelcz F, Joseph PM, Hilal SK. Noise considerations in dual energy CT scanning. Med Phys. 1979;6(5):418–25.

3. Lehmann LA, Alvarez RE, Macovski A, Brody WR, Pelc NJ, Riederer SJ, et al. Generalized image combinations in dual KVP digital radiography. Med Phys. 1981;8(5):659–67.
4. Brody WR, Cassel DM, Sommer FG, Lehmann LA, Macovski A, Alvarez RE, et al. Dual-energy projection radiography: initial clinical experience. AJR Am J Roentgenol. 1981;137(2):201–5.
5. Asaga T, Chiyasu S, Mastuda S, Mastuura H, Kato H, Ishida M, et al. Breast imaging: dual-energy projection radiography with digital radiography. Radiology. 1987;164(3):869–70.
6. Brooks RA, Di Chiro G. Split-detector computed tomography: a preliminary report. Radiology. 1978;126(1):255–7.
7. Barnes GT, Sones RA, Tesic MM, Morgan DR, Sanders JN. Detector for dual-energy digital radiography. Radiology. 1985;156(2):537–40.
8. Sones RA, Barnes GT. Noise correlations in images acquired simultaneously with a dual-energy sandwich detector. Med Phys. 1989;16(6):858–61.
9. Gauntt DM, Barnes GT. X-ray tube potential, filtration, and detector considerations in dual-energy chest radiography. Med Phys. 1994;21(2):203–18.
10. Alvarez RE. Active energy selective image detector for dual-energy computed radiography. Med Phys. 1996;23(10):1739–48.
11. Stevens GM, Pelc NJ. Depth-segmented detector for X-ray absorptiometry. Med Phys. 2000;27(5):1174–84.
12. Alvarez RE, Seibert JA, Thompson SK. Comparison of dual energy detector system performance. Med Phys. 2004;31(3):556–65.
13. Campbell M, Heijne EHM, Meddeler G, Pernigotti E, Snoeys W. A readout chip for a 64×64 pixel matrix with 15-bit single photon counting. IEEE Trans Nucl Sci. 1998;45(3):751–3.
14. Fischer P, Helmich A, Lindner M, Wermes N, Blanquart L. A photon counting pixel chip with energy windowing. IEEE Trans Nucl Sci. 2000;47(3):881–4.
15. Llopart X, Campbell M, Dinapoli R, Segundo DS, Pemigotti E. Medipix2: a 64-k pixel readout chip with 55 mu m square elements working in single photon counting mode. IEEE Trans Nucl Sci. 2002;49(5):2279–83.
16. Bronnimann C, Baur R, Eikenberry EF, Fischer P, Florin S, Horisberger R, et al. A pixel detector for the protein crystallography beamline at the SLS. Nucl Instrum Meth A. 2002;477(1–3): 531–5.
17. Locker M, Fischer P, Krimmel S, Kruger H, Lindner M, Nakazawa K, et al. Single photon counting X-ray imaging with Si and CdTe single chip pixel detectors and multichip pixel modules. IEEE Trans Nucl Sci. 2004;51(4):1717–23.
18. Roessl E, Brendel B, Engel KJ, Schlomka JP, Thran A, Proksa R. Sensitivity of photon-counting based K-edge imaging in X-ray computed tomography. IEEE Trans Med Imaging. 2011;30(9):1678–90.
19. Schirra CO, Brendel B, Anastasio MA, Roessl E. Spectral CT: a technology primer for contrast agent development. Contrast Media Mol Imaging. 2014;9(1):62–70.
20. Soo Choi H, Liu W, Misra P, Tanaka E, Zimmer JP, Itty Ipe B, et al. Renal clearance of quantum dots. Nat Biotech. 2007;25(10):1165–70.
21. Khlebtsov N, Dykman L. Biodistribution and toxicity of engineered gold nanoparticles: a review of in vitro and in vivo studies. Chem Soc Rev. 2011;40(3):1647–71.
22. Bulte JWM, Schmieder AH, Keupp J, Caruthers SD, Wickline SA, Lanza GM. MR cholangiography demonstrates unsuspected rapid biliary clearance of nanoparticles in rodents: Implications for clinical translation. Nanomedicine. 2014;10(7):1385–8.
23. Hampton JC. An electron microscope study of the hepatic uptake and excretion of submicroscopic particles injected into the blood stream and into the bile duct. Acta Anat. 1958;32:262–91.
24. Cho EC, Kim C, Zhou F, Cobley CM, Song KH, Chen J, et al. Measuring the optical absorption cross sections of Au-Ag nanocages and Au nanorods by photoacoustic imaging. J Phys Chem C Nanomater Interfaces. 2009;113(21):9023–8.
25. Park K, Park EJ, Chun IK, Choi K, Lee SH, Yoon J, et al. Bioavailability and toxicokinetics of citrate-coated silver nanoparticles in rats. Arch Pharm Res. 2011;34(1):153–8.

26. Skajaa T, Cormode DP, Jarzyna PA, Delshad A, Blachford C, Barazza A, et al. The biological properties of iron oxide core high-density lipoprotein in experimental atherosclerosis. Biomaterials. 2011;32(1):206–13.
27. Juhlin L. Excretion of intravenously injected solid particles in bile. Acta Physiol Scand. 1960;49:224–30.
28. Amsterdam EA, Kirk JD, Bluemke DA, Diercks D, Farkouh ME, Garvey JL, et al. Testing of low-risk patients presenting to the emergency department with chest pain. A scientific statement from the American Heart Association. Circulation. 2010;122:756–76.
29. Miller JM, Rochitte CE, Dewey M, Arbab-Zadeh A, Niinuma H, Gottlieb I, et al. Diagnostic performance of coronary angiography by 64-row CT. N Engl J Med. 2008;359(22):2324–36.
30. Arbab-Zadeh A, Miller JM, Rochitte CE, Dewey M, Niinuma H, Gottlieb I, et al. Diagnostic accuracy of computed tomography coronary angiography according to pre-test probability of coronary artery disease and severity of coronary arterial calcification: the CORE-64 (Coronary Artery Evaluation Using 64-Row Multidetector Computed Tomography Angiography) International Multicenter Study. J Am Coll Cardiol. 2012;59(4):379–87.
31. Pan D, Roessl E, Schlomka JP, Caruthers SD, Senpan A, Scott MJ, et al. Computed tomography in color: NanoK-enhanced spectral CT molecular imaging. Angew Chem Int Ed Engl. 2010;49(50):9635–9.
32. Barnett BP, Kraitchman DL, Lauzon C, Magee CA, Walczak P, Gilson WD, et al. Radiopaque alginate microcapsules for X-ray visualization and immunoprotection of cellular therapeutics. Mol Pharm. 2006;3(5):531–8.
33. Ai K, Liu Y, Liu J, Yuan Q, He Y, Lu L. Large-scale synthesis of Bi(2)S(3) nanodots as a contrast agent for in vivo X-ray computed tomography imaging. Adv Mater. 2011;23(42):4886–91.
34. Rabin O, Manuel Perez J, Grimm J, Wojtkiewicz G, Weissleder R. An X-ray computed tomography imaging agent based on long-circulating bismuth sulphide nanoparticles. Nat Mater. 2006;5(2):118–22.
35. Fang Y, Peng C, Guo R, Zheng L, Qin J, Zhou B, et al. Dendrimer-stabilized bismuth sulfide nanoparticles: synthesis, characterization, and potential computed tomography imaging applications. Analyst. 2013;138(11):3172–80.
36. Pan D, Pramanik M, Senpan A, Yang X, Song KH, Scott MJ, et al. Molecular photoacoustic tomography with colloidal nanobeacons. Angew Chem Int Ed Engl. 2009;48(23):4170–3.
37. Pan D, Pramanik M, Senpan A, Ghosh S, Wickline SA, Wang LV, et al. Near infrared photoacoustic detection of sentinel lymph nodes with gold nanobeacons. Biomaterials. 2010;31(14):4088–93.
38. Pan D, Pramanik M, Senpan A, Allen JS, Zhang H, Wickline SA, et al. Molecular photoacoustic imaging of angiogenesis with integrin-targeted gold nanobeacons. FASEB J. 2011;25(3):875–82.
39. Schirra CO, Senpan A, Roessl E, Thran A, Stacy AJ, Wu L, et al. Second generation gold nanobeacons for robust K-edge imaging with multi-energy CT. J Mater Chem. 2012;22(43):23071–7.
40. Benson R. Present status of coronary artery disease. Arch Pathol Lab Med. 1926;2:876–916.
41. Constantinides P. Plaque fissuring in human coronary thrombosis. J Atheroscler Res. 1966;6:1–17.
42. Brown BG, Gallery CA, Badger RS, Kennedy JW, Mathey D, Bolson EL, et al. Incomplete lysis of thrombus in the moderate underlying atherosclerotic lesion during intracoronary infusion of streptokinase for acute myocardial infarction: quantitative angiographic observations. Circulation. 1986;73(4):653–61.
43. Fuster V, Badimon L, Cohen M, Ambrose JA, Badimon JJ, Chesebro J. Insight into the pathogenesis of acute ischemic syndromes. Circulation. 1988;77:1213–20.
44. Sanidas EA, Mintz GS, Maehara A, Cristea E, Wennerblom B, Iniguez A, et al. Adverse cardiovascular events arising from atherosclerotic lesions with and without angiographic disease progression. JACC Cardiovasc Imaging. 2012;5(3 Suppl):S95–105.
45. Cormode DP, Skajaa T, van Schooneveld MM, Koole R, Jarzyna P, Lobatto ME, et al. Nanocrystal core high-density lipoproteins: a multimodality contrast agent platform. Nano Lett. 2008;8(11):3715–23.

46. Cormode DP, Roessl E, Thran A, Skajaa T, Gordon RE, Schlomka J-P, et al. Atherosclerotic plaque composition: analysis with multicolor CT and targeted gold nanoparticles. Radiology. 2010;256:774–82.
47. Bulte JW. Science to practice: can CT be performed for multicolor molecular imaging? Radiology. 2010;256(3):675–6.
48. Peng C, Zheng L, Chen Q, Shen M, Guo R, Wang H, et al. PEGylated dendrimer-entrapped gold nanoparticles for in vivo blood pool and tumor imaging by computed tomography. Biomaterials. 2012;33(4):1107–19.
49. Zhu J, Fu F, Xiong Z, Shen M, Shi X. Dendrimer-entrapped gold nanoparticles modified with RGD peptide and alpha-tocopheryl succinate enable targeted theranostics of cancer cells. Colloids Surf B: Biointerfaces. 2015;133:36–42.
50. Al Zaki A, Joh D, Cheng Z, De Barros AL, Kao G, Dorsey J, et al. Gold-loaded polymeric micelles for computed tomography-guided radiation therapy treatment and radiosensitization. ACS Nano. 2014;8(1):104–12.
51. Douglass M, Bezak E, Penfold S. Monte Carlo investigation of the increased radiation deposition due to gold nanoparticles using kilovoltage and megavoltage photons in a 3D randomized cell model. Med Phys. 2013;40(7):071710.
52. Hainfeld JF, Dilmanian FA, Zhong Z, Slatkin DN, Kalef-Ezra JA, Smilowitz HM. Gold nanoparticles enhance the radiation therapy of a murine squamous cell carcinoma. Phys Med Biol. 2010;55(11):3045–59.
53. Chithrani DB, Jelveh S, Jalali F, van Prooijen M, Allen C, Bristow RG, et al. Gold nanoparticles as radiation sensitizers in cancer therapy. Radiat Res. 2010;173(6):719–28.
54. Bonitatibus Jr PJ, Torres AS, Goddard GD, FitzGerald PF, Kulkarni AM. Synthesis, characterization, and computed tomography imaging of a tantalum oxide nanoparticle imaging agent. Chem Commun (Camb). 2010;46(47):8956–8.
55. Oh MH, Lee N, Kim H, Park SP, Piao Y, Lee J, et al. Large-scale synthesis of bioinert tantalum oxide nanoparticles for X-ray computed tomography imaging and bimodal image-guided sentinel lymph node mapping. J Am Chem Soc. 2011;133(14):5508–15.
56. Liu Y, Ai K, Liu J, Yuan Q, He Y, Lu L. A high-performance ytterbium-based nanoparticulate contrast agent for in vivo X-ray computed tomography imaging. Angew Chem Int Ed Engl. 2012;51(6):1437–42.
57. Pan D, Schirra CO, Senpan A, Schmieder AH, Stacy AJ, Roessl E, et al. An early investigation of ytterbium nanocolloids for selective and quantitative "multicolor" spectral CT imaging. ACS Nano. 2012;6(4):3364–70.

Theranostic Gold Nanoparticles for CT Imaging

Tamar Dreifuss, Eran Barnoy, Menachem Motiei, and Rachela Popovtzer

1 Introduction

One of the major aims in nanomedicine is the ability to perform multiple functions using the same nanovehicle, that is, the ability to achieve both therapeutic and diagnostic imaging capabilities using only a single nanoparticle. The unique physical and optical properties of gold nanoparticles (GNPs), along with the well-known biosafety of gold [1], make GNPs ideal candidates for various biomedical applications, including imaging, therapy, and diagnostic systems. So far, GNPs have been utilized for various therapy applications, such as drug delivery, phototherapy, and radiotherapy, and as a contrast-enhancing agent for computed tomography (CT) and other imaging modalities [2–11]. The wide applications of GNPs and their potential for clinical implementation, as well as the high flexibility in terms of size and shape and the ability to attach multiple types of ligands to their surfaces, have led to varying schemes for developing multifunctional GNPs, with multiple capabilities within a single platform [2, 12, 13]. Compared with other methods, treatment plans involving the use of multifunctional nanoparticles hold the promise of more accurately targeted treatment, with a higher likelihood of a successful outcome. It is this multipronged approach on which many studies have been anchored, since it addresses many issues usually associated with the most aggressive aspects of disease, including multidrug resistance, radioresistance, and recurrence of tumors. Therefore, GNPs have high potential to become the next generation theranostic agents for cancerous diseases. A schematic diagram of potential multifunctional gold nanoparticles for theranostic applications is presented in Fig. 1. This chapter discusses the main approaches for the use of GNPs for both therapy and imaging applications, focusing on multifunctional gold nanoparticles for theranostic applications involving CT imaging.

T. Dreifuss • E. Barnoy • M. Motiei • R. Popovtzer (✉)
Faculty of Engineering and the Institutes of Nanotechnology and Advanced Materials, Bar-Ilan University, Ramat-Gan, Israel
e-mail: rachela.popovtzer@biu.ac.il

J.W.M. Bulte, M.M.J. Modo (eds.), *Design and Applications of Nanoparticles in Biomedical Imaging*, DOI 10.1007/978-3-319-42169-8_19

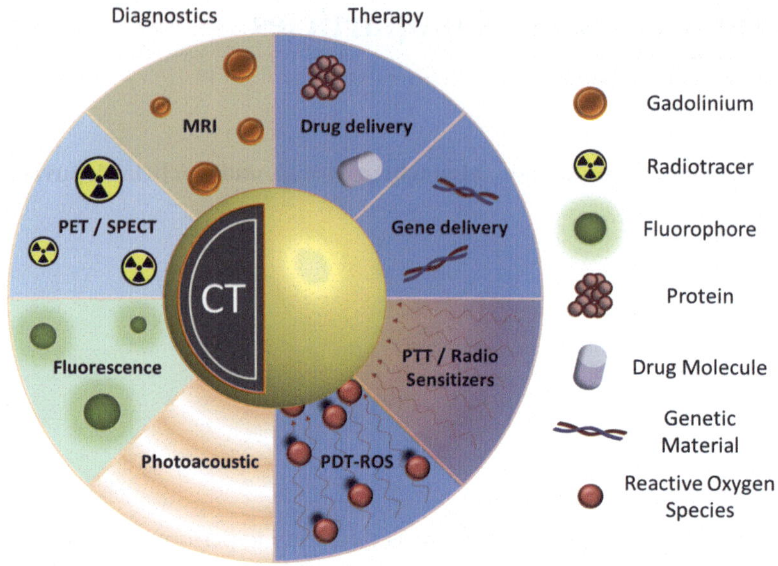

Fig. 1 Schematic diagram of multifunctional gold nanoparticles (GNPs). Multifunctional GNPs can be prepared by either combining metals with different functionalities or combining functional small-molecules through different surface engineering strategies

2 Gold Nanoparticles as CT Contrast Agents

2.1 Physical Properties

X-ray based CT is among the most convenient imaging/diagnostic tools in hospitals today in terms of availability, efficiency, and cost. CT provides superior visualization of bone structures due to the inherent contrast between electron-dense bones and the more permeable surrounding soft tissues. CT, however, is limited in distinguishing between different soft tissues that have similar densities [14]. CT contrast agents were introduced in order to improve vascular contrast and to enable better delineation of soft tissue structures with similar or identical contrast properties.

Different tissues provide different degrees of X-ray attenuation, according to Eq. (1):

$$I = I_0 e^{-\ x} \tag{1}$$

where I_0 is the incident X-ray intensity, I is the transmitted X-ray intensity, x is the thickness of the absorber medium, and μ is the mass attenuation coefficient. The most dominant factor impacting the mass attenuation coefficient is the photoelectric effect, which is proportional to the third power of the atomic number of the material (Z^3). Therefore, in order to provide good contrast in CT images, the key factor in the selection of CT contrast agents is high atomic number materials.

Present CT contrast agents are predominantly based on iodine containing molecules, which are effective in absorbing X-rays ($Z=53$). However, the atomic number of gold ($Z=79$) is much higher than that of iodine, and therefore gold can induce stronger X-ray attenuation [15]. The ability of gold to induce high CT contrast was first demonstrated, inadvertently, by Wilhelm Roentgen, in the first medical imaging (Fig. 2). In addition, while iodine containing molecules allow only very short imaging times due to rapid clearance by the kidneys, and are nonspecifically targeted because they cannot be conjugated to most biological components or cancer markers, GNPs can be designed so as to overcome biological barriers and to remain confined to the intravascular space for prolonged times [16–18]. In addition, their strong binding affinity toward thiol, disulfide, and amine groups, enables binding with various targeting agents. Therefore, GNPs are ideal candidates for CT contrast agents. Another advantage of GNPs is their ability to be fabricated in a variety of shapes and sizes, using simple wet-laboratory techniques. Note that for CT imaging the total amount of gold per unit volume (voxel) is the only important parameter regardless of the shape of the particles.

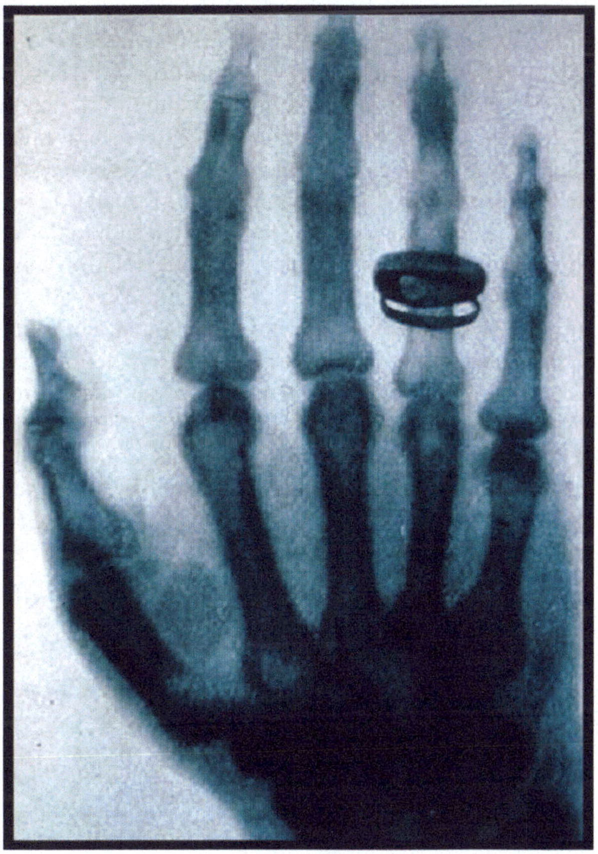

Fig. 2 First ever medical X-ray image (1895) taken by Roentgen. "Hand with Ring" print of Wilhelm Roentgen's first "medical" X-ray, taken on 22 December 1895. It dramatically showed the bones of his wife's fingers; however the real size of her finger's soft tissue could be garnered from the clearly visible gold ring on her finger. Radiology Centennial, Inc. copyrighted in 1993

The high-Z nanoparticle contrast agents could also address the important issue of relatively high radiation exposure of CT. The new generation CT contrast agents that are based on high atomic number materials, such as gold and bismuth, have a great potential not only because of their ability to produce contrast higher than conventional iodine-based contrast agents, but even more importantly, because of the possibility to lower the overall radiation exposure to patients.

Low-to-middle X-ray photon energy (25–120 keV) is used for diagnostic radiology, producing significant contrast between bone and other tissues, resulting in high quality CT images. However, since most of this energy is being absorbed, it exposes the patient to a high dose of radiation. Since the higher energy photons in the energy spectrum produced by the X-ray tubes have a much lower interaction cross section for soft tissue than the nanoparticles, it is possible that filtering lower energies of the X-ray spectrum will lead to lower absorbed radiation dose to the patient. The uptake pattern of these particles can be visualized as distinct contrast relative to their soft tissue background [14]. Therefore, high-Z nanoparticles as contrast agents may permit CT imaging at lower patient doses, with better sensitivity and good specificity.

Spectral CT (popularly referred to as multicolor CT) is a new promising technology that enhances the traditional CT images by using simultaneously different energy settings, which enable the differentiation of materials, based on their energy-related attenuation profiles. The concept of spectral CT exploits the K-edge discontinuity in the photo-electric component of X-ray absorption [19]. Gold has K-edge values within the X-ray energy bandwidth of CT, and thus is an excellent candidate as contrast agents for spectral CT [20]. Several studies have demonstrated that spectral CT can accurately distinguish GNP-based contrast agents from an iodinated contrast agent, tissue, or calcium-rich matter [21–23].

Another important parameter, and indeed a major challenge in the implementation of high-Z contrast agents—lies in the relatively low sensitivity of the CT to the contrast agents' concentration. For example, for MRI contrast agents, proton relaxation must be altered, requiring agents that can perturb the local magnetic field around the proton. The perturbing field in MRI of a super-paramagnetic particle is effective at up to 50-times its diameter, and therefore influences water protons in several cell layers around its location [24]. This phenomenon results in sensitivity of the MRI to micromolar contrast agents' concentration [25]. By contrast, CT contrast agents lack such an amplification ability. CT is sensitive to millimolar contrast agent concentrations; therefore, in order to induce sufficient contrast in the desired organ, a much larger amount of high-Z molecules is needed, since the CT contrast is linearly proportional to the total amount of the high-Z molecules in a voxel.

2.2 GNPs as Blood Pool, Passive and Active Contrast Agents

As blood pool contrast agents, GNPs extended the blood circulation time from several minutes (with the clinically used iodine compounds) to hours or even days [26–29] and show stronger X-ray attenuation than the currently used iodine-based compounds (under the same clinically relevant conditions) [26, 30]. By taking advantage of the progressive permeation through transendothelial pores on tumor blood vessels (the EPR effect), GNPs of a certain size range can passively accumulate on tumors. This nonselective "passively targeting" approach was demonstrated in several studies which showed significant CT contrast enhancement caused by accumulation of GNPs within the tumor and in areas surrounding it [9, 31].

Conjugation of GNPs to a variety of ligands, including antibodies, peptides, aptamers, or small molecules that possess high affinity toward unique molecular signatures found on diseased cells, produces active targeting agents, which can selectively accumulate on specific cells or tissues. Molecularly targeted nanoparticles reach tumor tissues through the EPR effect (as in passive targeting). However, the active targeting has additive values; the nanoparticles home selectively into specific tumors and remain at the tumor site for extended durations, thereby increasing the local accumulation of the nanoparticles at sites of interest. Studies have shown that molecularly targeted GNPs can enhance the visibility of small tumors in CT, and that active tumor targeting is more efficient than passive targeting [9, 32–34]. This specific interaction between antigen and antibody or receptor and its ligand was shown to be an effective strategy to improve the amount and residence time of contrast agents in tumors, as well as to provide specific molecular knowledge regarding the findings. This new approach of molecularly targeted CT contrast agents has changed the concept of CT from diagnosis based on anatomical structures to diagnosis according to molecular markers.

3 GNP as Multimodal Imaging Contrast Agents

The idea of multimodality imaging, with two or more imaging modalities, has brought new perceptions to the fields of clinical and preclinical imaging. The integration between different individual imaging techniques allows each to compensate for the limitations of the other, thus to achieve synergism and improve the quality and accuracy of the diagnostic tool or research platform. GNPs can be utilized as multimodal imaging contrast agents, for combining CT with additional imaging modalities, as described below.

3.1 CT and MRI

For the development of nanoparticle-based contrast agents for dual mode CT/MR imaging, both CT and MR imaging elements should be incorporated. As the most commonly used MRI contrast agents are based on chelates of gadolinium, gadolinium (Gd)-chelated Au nanoparticles have been designed for CT/MR dual mode imaging in several studies [35–37]. For example, Wen et al. [35] have reported the utilization of gadolinium-loaded dendrimer-entrapped gold nanoparticles (GdeAu DENPs) for dual mode CT/MR imaging of the heart, liver, kidney, and bladder of rat and mouse. They showed that the GdeAu DENPs have enhanced X-ray attenuation due to the entrapped AuNPs, as well as the complexed Gd(III) ions, which enable efficient dual mode CT/MR imaging. They further showed that the GdeAu DENPs are non-cytotoxic at an Au concentration up to 50 mM, and can be cleared from major organs within 24 h. In another study [36], folic acid-modified multifunctional dendrimer-entrapped gold nanoparticles loaded with gadolinium have been used for targeted CT/MR imaging of cancer in vitro and in vivo via FA receptor-mediated active targeting pathway.

Triple-modal imaging with a single nanostructure has also gained attention in recent years. Hu et al. [37] have developed hybrid gold–gadolinium nanoclusters (NCs), which are suitable for triple-modal NIRF/CT/MRI imaging. The hybrid NCs have been effectively accumulated in tumor tissues and quickly cleared by renal excretion.

3.2 CT and Nuclear Imaging

The incorporation of CT with nuclear imaging (PET and SPECT) enables both functional and anatomical information in a single setting. GNPs can be radiolabeled with different radioactive tracers in order to supplement CT with nuclear imaging. Indeed, coupling of gold to a radiotracer was investigated in a number of studies. A recent study [38] reported the integration of ^{64}Cu into the structures of gold nanoparticles. In this study, PET/CT image of mouse breast cancer model showed high passive tumor targeting and contrast ratios after intravenous injection of those ^{64}CuAuNPs. In another study [39], gold nanocages were prepared and radiolabeled with ^{64}Cu, for evaluating the pharmacokinetics and in vivo cancer targeting capability in a murine EMT-6 breast cancer model. Furthermore, Su et al. [40] have represented theranostic iodine-125-labeled cRGD-gold nanoparticles, which act as tumor-targeted radiosensitizer and SPECT/CT imaging agent. The introduction of ^{125}I on the GNP serves as both a therapeutic factor and a radiotracer for in vivo nuclear imaging. It is important to note that in these aforementioned studies, the use of gold was not aimed to enhance CT intensity, but to earn stability and high loading capabilities of gold nanoparticles, as well as their pharmacokinetics and biodistribution properties.

3.3 CT and Fluorescent Imaging

While optical imaging methods suffer in spatial resolution and offer only a limited penetration depth, they are very sensitive, fast, and readily distinguish between several sources with great soft tissue contrast on cellular and even molecular scales [41–43]. Near-infrared radiation (NIR) allows for relatively deep light penetration into biological tissues with minimal autofluorescence, providing higher signal-to-noise ratios than visible light [41, 43, 44]. A property often used in CT-fluorescence dual-modal GNP constructs is that GNPs possess very efficient surface-energy transfer properties that can significantly affect the fluorescence of molecules near their surface, both with quenching and with fluorescence enhancement [42, 43, 45–57]. Combining the strong spatial resolution and unlimited penetration depth of CT with the sensitivity available from fluorescence is a promising and widely explored dual-modal imaging technique [42, 43].

Zhang et al. [43] packed gold nanoparticles with a fluorescent dye into polymer micelles. The dye used possessed the unusual property of aggregation-induced emission, so that the aggregated dye was able to produce a signal that overcame the quenching effects of the GNPs. The result was a construct with a strong fluorescent signal, as well as being an efficient CT contrast agent in tumor sites. Sun et al. [58] modified GNPs with glycol chitosan polymers and an NIR fluorescent dye adsorbed using a peptide sequence cleavable by a matrix metalloproteinase (MMP). Prior to cleavage, the GNPs quenched approximately 95 % of the fluorescence compared to the fluorophore alone. Fluorescence was regained proportionally to the concentration of environmental MMP. GNPs allowed easy imaging of the tumor in CT, while the NIR fluorescence signaled strongly starting an hour post injection. Luo et al. [44] attached the NIR dye ICG to gold nanorods coated with mesoporous silica. The layer of silica protected the ICG from GNP quenching. By injecting the constructs into mouse tumors, in addition to the efficient CT images, they witnessed NIR fluorescence up to 12 h post injection. Song et al. [41] coated gold nanospheres with mesoporous silica, and inserted the dye IR-783 into the silica pores. The silica shell was prepared to prevent quenching. In combination, these constructs proved to be strong CT contrast agents with significant fluorescent abilities lasting several hours.

3.4 CT and Photoacoustic Imaging

In photoacoustic (PA) imaging, short laser pulses are delivered into the subject, resulting in local heating which leads to thermoelastic expansion and thus wideband ultrasonic emission. GNPs strongly absorb light and thus are potential contrast agents for PA. The strategy of using GNPs as dual mode CT/PA contrast enhancers has been investigated by Chanda et al. [59]. They used cinnamon-coated gold nanoparticles (Cin-AuNPs) and showed that these particles underwent significant accumulation in breast cancer and prostate cancer cells, facilitating detectable PA signals. In addition, quantitative analysis of CT values in a phantom model revealed that the Cin-AuNPs have reasonable attenuation efficiency. Another recent study, published by Jing et al. [60], has utilized Prussian blue-coated gold nanoparticles (Au@PB NPs) for simultaneous PA/CT bimodal imaging and photothermal ablation of cancer, applying the theranostic approach. The Au@PB NPs were found to be an excellent photoabsorbing agent for both PA imaging and photothermal therapy due to a high photostability and high molar extinction coefficient in NIR region. Moreover, a remarkable enhancement of CT values in tumor sites was observed due to the Au@PB NPs.

4 Therapeutic Gold Nanoparticles

Therapeutic nanoparticles can overcome various limitations of conventional drug delivery systems, such as nonspecific targeting, lack of water solubility, poor oral bioavailability, and low therapeutic indices [16, 61]. GNPs are highly suitable for drug delivery systems, due to their strong binding affinity toward thiol, disulfide, and amine groups, which enables binding with various targeting agents and therapeutic moieties. In addition, by taking advantage of the inherent unique characters of gold itself, GNPs can be exploited to photothermal therapy (PTT) and radiotherapy in which the GNP itself has the therapeutic value, as will be discussed below.

4.1 Drug Delivery

Gold is consistently used in various forms for biological applications owing to ease of fabrication and surface modification, as well as gold's inherent biological compatibility [2, 3, 62, 63]. Using simple wet-laboratory techniques, GNPs have been fabricated in a variety of shapes and sizes, and used as the core or the shell for polymer–metal [62, 63] and metal–metal [64, 65] hybrid nanoparticles. Therapeutics, such as proteins, peptides, oligonucleotides, or small drug molecules, can be attached to the gold nanoparticle surface directly, via amine or thiol groups [63], or indirectly, using a molecule such as bovine serum albumin [66] or the penta-peptide with the amino acid sequence CALNN [67].

GNPs have been widely investigated as drug delivery vehicles for cancer treatment, while implementing both passive and active targeting approaches. For example, a drug delivery system was developed [68] by tethering doxorubicin onto the surface of GNPs with a poly(ethylene glycol) spacer via an acid-labile linkage. The doxorubicin-tethered GNPs (DOX-Hyd@AuNPs) achieved enhanced drug accumulation and retention in multidrug resistant MCF-7/ADR cancer cells when it was compared with free doxorubicin. It released doxorubicin in response to the pH of acidic organelles following endocytosis, and significantly enhanced the cytotoxicity of doxorubicin. Therefore, the DOX-Hyd@AuNPs considerably inhibited the growth of multidrug-resistant MCF-7/ADR cancer cells. In another study [69], the antitumor efficacy of GNPs conjugated non-covalently with three conventional chemotherapy drugs (doxorubicin, cisplatin, and capecitabine) was evaluated in vitro. The cellular proliferation rates in the presence of the anticancer drugs delivered by the GNPs were found to be significantly lower than those of cells exposed to free drugs. In addition, these GNPs were also efficient when tested with chemotherapy resistant cells.

To improve the therapeutic outcomes, the active tumor targeting approach is implemented by coupling targeting agents to drug-loaded GNPs. Prabaharan et al. [70] have synthesized folate (FA)-conjugated amphiphilic GNPs with a poly(L-aspartate-doxorubicin)-b-poly(ethylene glycol) monolayer (Au-P(LA-DOX)-b-PEG-OH/FA) that interacts with the FA receptor. They showed that the presence of FA enables active tumor targeting abilities, as the cellular uptake of the

Au-P(LA-DOX)-b-PEG-OH/FA micelles into cells overexpressing the FA receptor was higher than that of micelles without folate. In addition, Conjugation of DOX onto the NPs and the pH-sensitive properties of its linkage to the NP, greatly reduce the chance of premature drug release during circulation in the bloodstream, offering a potential DOX carrier. In another study [71], Paciotti et al. developed a colloidal GNP that targets the delivery of tumor necrosis factor (TNF) to solid tumors. TNF is an antitumor cytokine that is prone to systemic toxicity when administrated as a free drug. While conjugated to the GNP (which was coated also with PEG-THIOL), it was less toxic and more effective in reducing tumor burden, since maximal antitumor responses in tumor bearing mice were achieved at lower doses of drug. In this study, the TNF served both as antitumor and targeting ligand.

4.2 Gene Delivery

As DNA is not efficiently translocated through the cell membrane [72], designing effective synthetic vectors for transporting nucleic acids, including plasmid DNA, small interfering RNA, or antisense oligonucleotides, is also a challenge which can be addressed using gold nanoparticles [73–76]. Ghosh et al. [77] have demonstrated that coating GNPs with lysine-based headgroups produces effective transfection vectors, without any observed cytotoxicity. They further showed that DNA delivery efficiency strongly depends on the structure of headgroups and their concomitant ability to condense DNA. Importantly, these amino acid-based nanoparticles were responsive to intracellular glutathione levels, during in vitro transfection, providing insight into their mode of activity, as well as a tool for controlled release and concomitant expression of DNA. Guo et al. [78] reported charge-reversal functional GNP as carrier for gene delivery. In this study, Lamin A/C, an important nuclear envelope protein, was effectively silenced by lamin A/C-siRNA delivered by the charge-reversal functional GNPs, whose knockdown efficiency was better than that of commercial Lipofectamine 2000. The charge reversion under an acidic environment facilitated the escape of GNP/nucleic acid complexes from endosome/lysosome and the release of functional nucleic acids into the cytoplasm. Another example of using GNPs for gene silencing was recently presented by Conde et al. [79], who synthesized a library of multifunctional GNPs, aimed to deliver siRNA and specifically silence the c-myc protooncogene. The efficiency of the GNPs was evaluated using a hierarchical approach including three biological systems of increasing complexity (cultured cells, invertebrate, and mammal). After fine-tuning of multiple structural and functional moieties, the most active formulation was selected, consisting of siRNA, which is either covalently or ionically linked to the nanoparticles. Another interesting study [80] has reported proof of concept for the selective release of two different DNA oligonucleotides from two different gold nanorods. DNA was loaded onto the nanorods via thiol conjugation, and selective release was specifically induced by melting of gold nanorods via laser irradiation at the nanorods' longitudinal surface plasmon resonance peaks. The released oligonucleotides remained functional.

4.3 Photothermal Therapy

In the case of photothermal therapy (PTT), GNPs are advantageous because of their ability to strongly absorb light (orders of magnitude higher than organic dye molecules) [3], and then convert the absorbed light into heat via a nonradiative processes. This process occurs on a time scale on the order of picoseconds, leading to intense localized heating and irreparable damage to the cell [81, 82]. In many cases gold by itself has served as the photosensitizer. On the other hand, GNPs enable delivery of exogeneous photosensitizers to cells in high enough concentrations [83]. Moreover, the ability to easily add targeting moieties to the GNP surface enables preferential uptake into the malignant cells, for selective heating of the targeted cells, while leaving the neighboring cells intact.

Consequently, GNPs have been most frequently exploited for PTT, in comparison to options involving other metal, dye–polymer or carbon-based nanoparticles. The exceptionally high extinction coefficient of gold (1×10^{19} M^{-1} cm^{-1} for 20 nm gold nanoparticles) [84] is orders of magnitude higher than that of strongly absorbing organic dyes (e.g., Coomassie blue, 4.3×10^4 M^{-1} cm^{-1}) [85], which makes gold an ideal PTT mediator. Also, the localized surface plasmon resonance of GNPs is highly dependent on the morphology of the nanoparticles and can be easily tuned during the fabrication process [84]. Typically, the absorption of GNPs is tuned to be in the range between 600 and 1000 nm, the so-called therapeutic window in which the interaction of light with biological tissues is low, that is, attenuation and scattering effects are at a minimum.

Nanoparticle-mediated PTT is especially well suited for treating cancer owing to the increased heat sensitivity associated with cancer cells, as well as a tumor's inability to efficiently dissipate heat as a result of its poor vasculature networks. To date, various cancer cell lines, including breast [5], epithelial [86, 87] and colon [12] cancers, have been successfully treated by GNP-mediated PTT, both in vitro and in vivo. Recently, a study that builds on work by Qian et al. [88] has successfully achieved controlled surface engineering of gold nanoparticles, using multiple ligands [PEG 5K, (RGD)4 and NLS (nuclear localization signal) peptides] for efficient delivery into cell nuclei [89]. The nanoparticles were then utilized in PTT studies. One motivation for this study [89] was to investigate the likely advantages of directly heating the nucleus of cancer cells, advantages stemming from the fact that the nucleus has a smaller target volume, a lower heat capacity and an increased likelihood of causing irreparable damage to the cell's DNA. A stark reduction in the survival curves of the cells that had GNPs located predominantly in the nucleus, as opposed to cells that had nanoparticles located predominantly in the cytosol, was observed for cells that were treated under the same conditions, that is, the same GNP concentration and treatment time. Another recent study involving both in vitro and in vivo PTT studies was reported by Yuan et al. [90]. They detailed the fabrication of surfactant-free gold nanostars and their use for both in vitro and in vivo studies of nanoparticle tracking, via plasmon enhanced two-photon photoluminescence, and for PTT of breast cancer cells. Yuan and his colleagues found that 3 min of treatment yielded successful cell kill, and 5 min of treatment gave the best results.

Control measurements involving cells that were exposed to the laser, but had not been incubated with nanostars, showed that these cells were still viable after 5 min of illumination. For the in vivo studies, PEGylated nanostars were injected into CD-1 nude mice that were previously implanted with small dorsal window chambers. After 10 min of PTT, visual confirmation of photo-ablation was indicated by the release of the gold nanoparticles out of the ruptured vessels; scarring occurred 1 week later. One major research area aimed at improving GNP-mediated PTT is the utilization of gold clusters or aggregates [91–94], as the corresponding maximum absorption is red-shifted toward the near-IR range of the "therapeutic window," resulting in less concern regarding unwanted interaction with the surrounding healthy tissues. It should be noted that, despite the multitude of promising outcomes, the variability among the results has continued to encourage further studies aimed at elucidating the mechanisms of heat generation and thermolysis (e.g., shockwaves, micro- and nano-bubbles, and nanoparticle destruction [93, 95, 96]).

4.4 Radiation Therapy

The use of targeted radiosensitizers in cancer treatment is intended to increase tumor sensitivity to radiation, while sparing normal healthy tissue the effects of x-radiation, thereby increasing the therapeutic window. Lately, several studies have suggested that the administration of GNPs into tumors may improve radiosensitivity. GNPs are an ideal candidate to be used as radiosensitizing agents, since high atomic number (Z) materials increase radiation sensitivity owing to their greater absorption of photons and release of secondary energy in the form of photoelectrons, auger electrons, and X-rays into surrounding tissue [30, 97]. The close proximity of GNP to nuclear DNA increases the probability of creating DNA strand breaks, the primary mechanism of radiation-induced cytotoxicity [98, 99]. When GNPs specifically target cancer cells (e.g., via antibody–antigen interaction), large amounts of gold atoms accumulate in the tumor, leading to an enhancement of the radiation effect on the tumor. Chattopadhyay et al. [100] tested the effect of molecularly targeted GNPs on tumor radiosensitization both in vitro and in vivo. In vitro, they combined X-ray exposure with targeted AuNPs (Au-T) and non-targeted AuNPs (Au-P), and found that DNA damage was increased by 1.7- and 3.3-fold in comparison to X-radiation alone, for Au-T and Au-P, respectively. In vivo, the combination of Au-T and X-radiation resulted in regression of human breast cancer tumors by 46 % as compared to treatment with X-radiation. No significant normal tissue toxicity was observed. The mechanism underlying cell death in GNP-induced radiation sensitivity is unclear and several theories have been proposed. The main suggested hypothesis includes an increase in apoptosis, increased generation of intracellular reactive oxygen species or direct DNA damage causing DNA double strand break [98]. Roa et al. [101] found that GNP may sensitize radiotherapy by regulating the cell cycle. They showed that it induced the acceleration of the G0/G1 phase and cell accumulation in the G2/M phase, which were accompanied by down-regulation of p53 and cyclin expression and upregulation of cyclin B1 and cyclin E.

4.5 Photodynamic Therapy

Photodynamic therapy (PDT) in cancer treatment involves the uptake of a photosensitizer by cancer tissue followed by photoirradiation [102]. Photoirradiation in the appropriate wavelength excites the photosensitizer to produce reactive oxygen species (ROS), which causes damage to tumor cells. GNPs hold the promise to be highly efficient as photosensitizer carriers, and their use for PDT applications was investigated in several studies. Wieder et al. [103] presented a comparison between silica and gold nanoparticle systems. They argued that photosensitizer-coated GNP is preferable over an encapsulated photosensitizer, in that the generated singlet oxygen would not need to diffuse out of the porous particle structure to elicit cell kill. Hone et al. [104] reported the preparation of GNPs coated with a phthalocyanine (Pc) photosensitizer and their use as delivery vehicle for photodynamic therapy. The nanoparticles, consisted of three components (photosensitizer/gold/phase transfer reagent), were shown to generate singlet oxygen with enhanced quantum yields as compared to the free Pc. In addition, the association of the transfer reagent promoted the solubility of the surface-bound hydrophobic sensitizer in polar solvents, facilitating their systemic injection. In another study, Cheng et al. [105] have developed PEGylated gold nanoparticle conjugates with a reversible PDT drug adsorption, which act as a water-soluble and biocompatible "cage" that allows the delivery of the hydrophobic drug to its site of PDT action. This delivery mode greatly improved the transport of the PDT drug to the tumor relative to conventional drug administration.

5 The Theranostic Approach: Combined Imaging and Therapy

Co-delivery of the above therapeutic and imaging functions by specifically tailored theranostic gold nanosystems can considerably improve medical treatment, particularly in oncology. Diagnosis, treatment, and monitoring response to treatment can be reached with a single approach [5], thus enhancing personalized medicine. We present here prominent recent publications that combine different therapeutic strategies with CT imaging. The data is summarized in Table 1.

5.1 Combining Drug Delivery with CT Imaging

Chemotherapy success requires delivery of drugs to the tumor cells at concentrations above a therapeutic threshold. This is affected not only by the accumulation level in the tumor but also by drug distribution in the tumor stroma, which can often vary among different tumor regions, disease stages, and patients [106]. This issue can be addressed by image-guided drug delivery, which enables monitoring drug distribution in the tumor, as well as in the whole body.

Table 1 Prominent recent studies that combine different therapeutic and diagnostic strategies with CT imaging

Theranostic strategy	Nanoparticle type	Therapeutic agent	Targeting agent	Reference
Drug delivery and CT	Aptamer-conjugated GNP	Dox	PSMA aptamer	[107]
Drug delivery and CT	Dendrimer-entrapped GNP	α-TOS	Folic acid	[108]
Drug delivery and CT	Dendrimer-entrapped GNP	α-TOS	RGD peptide	[113]
PTT and CT	PEG-protected GNR	–	–	[5]
PTT and CT	GNR	–	–	[114]
PTT and CT	Polymer grafted GNP	–	–	[115]
PTT and bimodal SPECT/CT	125I-labeled GNR	–	Folate	[116]
PTT and bimodal PA/CT	PLA microcapsules containing GNPs	–	–	[117]
PTT and bimodal PA/CT	Au@PB core–shell nanoparticle	Prussian blue	–	[60]
PTT and multimodal SERS/CT/TPL	Gold nanostar	–	–	[118]
PTT and multimodal MR/CT/thermal imaging	Fe3O4@Au core–shell nanostar	–	Hyaluronic acid	[119]
RT and CT	PEG-modified GNP	–	–	[120]
RT and CT	Nanogel encapsulating PEGylated GNPs	–	–	[121]
RT and CT	Gold-loaded polymeric micelles	–	–	[122]
RT and bimodal SPECT/CT	125I-labeled GNP	125I	RGD	[40]
RT and bimodal SPECT/CT	^{131}I-labeled GNP	–	C225	[123]
Bimodal RT/PTT and CT	Silica-modified GNR	–	Folic acid	[124]
Multimodal drug delivery/PTT/RT and CT	HGNP	Dox	–	[125]

For this purpose, several studies were conducted in which drugs were loaded onto or into gold nanoparticles for visualization by CT imaging. Kim et al. [107] reported a multifunctional GNP for targeted molecular CT imaging and therapy of prostate cancer. For targeting, the GNP surface was functionalized with a prostate-specific membrane antigen (PSMA) RNA aptamer that binds to PSMA. In vitro experiments showed that the PSMA aptamer-conjugated GNP induced more than fourfold greater CT intensity for a targeted LNCaP cell than that of a nontargeted PC3 cell. Moreover, after loading the targeted GNP with doxorubicin (approximately 615 Dox molecules per GNP), drug release experiments showed that approximately 35 % of Dox was released within 1 h, and the particle potency against targeted LNCaP cells was significantly higher than against nontargeted PC3 cells. Another multifunctional theranostic GNP for targeted CT imaging and drug delivery was developed by Zhu et al. [108]. They reported dendrimer-entrapped GNPs (Au DENPs) covalently linked with α-tocopheryl succinate (α-TOS), which can induce apoptosis of various cancer cells, inhibit the cell cycle, and disrupt the necessary autocrine signaling pathways of tumor growth [109, 110], while not affecting the proliferation of most normal cells [111, 112]. The practical use of free α-TOS is limited due to poor water solubility and bioavailability. In this study, the GNP platform has been found to improve these limitations, without compromising α-TOS therapeutic activity (there was an even higher therapeutic efficacy than that of free α-TOS). Au DENPs were also modified with folic acid (FA) for targeting to cancer cells overexpressing FA receptors (FAR). In vivo experiments with a xenografted tumor model demonstrated that CT images and corresponding CT values of the tumor site did not show an enhancement at 2 h post-injection of both targeted and nontargeted Au DENPs. However, at 24 h post injection, significant CT enhancement in the tumor was observed for mice injected with targeted Au DENPs, compared to that injected with non-targeted Au DENPs (Fig. 3a). In the therapeutic aspect, the tumor growth rate of mice injected with targeted Au DENPs was much slower than that of mice treated with saline, free α-TOS, and non-targeted Au DENPs (Fig. 3b). Following this work, Zhu et al. [113] reported multifunctional Au DENPs, also modified with α-TOS, but with another targeting ligand: arginine–glycine–aspartic acid (RGD) peptide. These multifunctional Au DENPs were able to target cancer cells overexpressing $\alpha_v\beta_3$ integrin and specifically inhibit the growth of the cancer cells. Likewise, via the specific RGD-mediated targeting, $\alpha_v\beta_3$ integrin-overexpressing cancer cells were specifically detected by CT imaging.

5.2 *Combining Photothermal Therapy with CT Imaging*

The potential of simultaneous utilization of GNPs for CT imaging along with the emerging and highly promising treatment, photothermal therapy, has been widely studied in recent years. Maltzahn et al. [5] developed PEG–protected gold nanorods (PEG-NRs) for both photothermal heat generation and tumor CT imaging. After intratumoral administration, Maltzahn et al. fused PEG-NR biodistribution data derived via noninvasive X-ray computed tomography, with four-dimensional computational

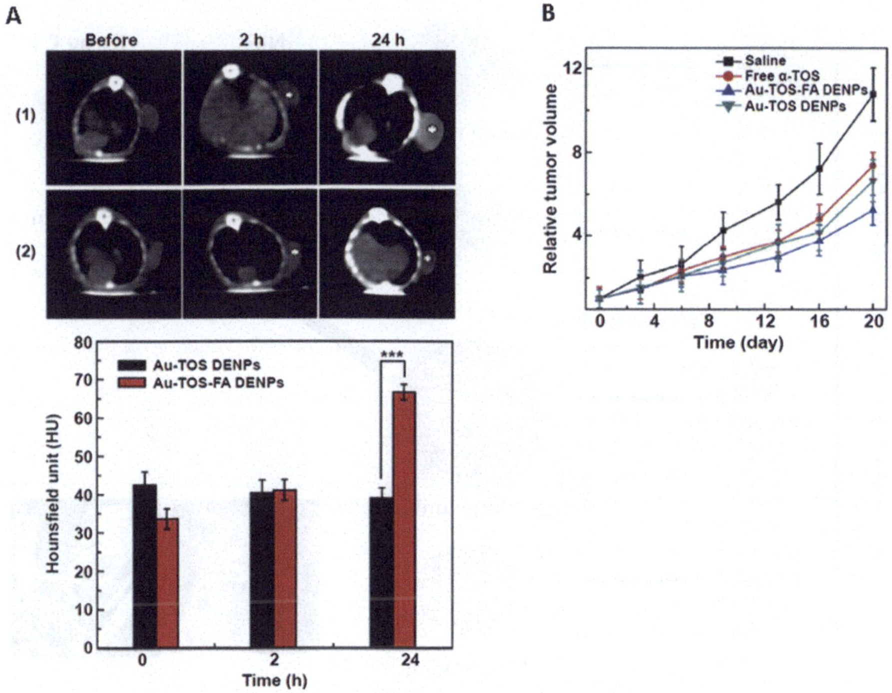

Fig. 3 (**a**) Representative transverse CT images (*top*) and CT values (*bottom*) of U87MG tumor xenografts in nude mice before and after intravenous injection of Au-TOS-FA (1) and Au-TOS (2) DENPs for 2 and 24 h, respectively. The white star indicates the location of the tumor. (**b**) The growth of U87MG xenografted tumors after various treatments. Reprinted with permission from Ref. [108]

heat transport modeling to predict photothermal heating during irradiation. The PEG-NRs exhibited ~2-fold higher X-ray absorption than a clinical iodine contrast agent, and enabled destruction of all irradiated human xenograft tumors in mice. Recently, Qin et al. [114] developed gold nanorods (GNRs) as a theranostic platform for CT imaging and PTT of inflammatory macrophages. In this study, in vitro experiments showed that the GNRs exhibited a significant cell-killing efficacy of macrophages, even at relatively low concentrations of GNRs and low NIR powers. In addition, in vivo therapeutic experiments demonstrated that the GNRs could image and use PTT of inflammatory macrophages in the femoral artery restenosis of an apolipoprotein E knockout (Apo E) mouse model. In another study, Deng et al. [115] presented a gold assembly structure as a potential photothermal therapeutic and CT contrast agent. Two different amphiphilic polymers were anchored onto the GNP, yielding a smart hybrid building block for self-assembly. By varying the ratio of these mixed polymer brushes, a series of amphiphilic nanocrystals was obtained, while an increase in the hydrophobic/hydrophilic ratio resulted in different assembly structures, from a normal monolayer vesicle to large compound micelle (LCM). Due to the strictly packed structure of LCM, the LSPR peaks of GNPs could be tuned to NIR region, for enhanced PTT. In vitro and In vivo studies showed efficient treatment of optimum gold assemblies as a PTT and CT contrast agent (Fig. 4).

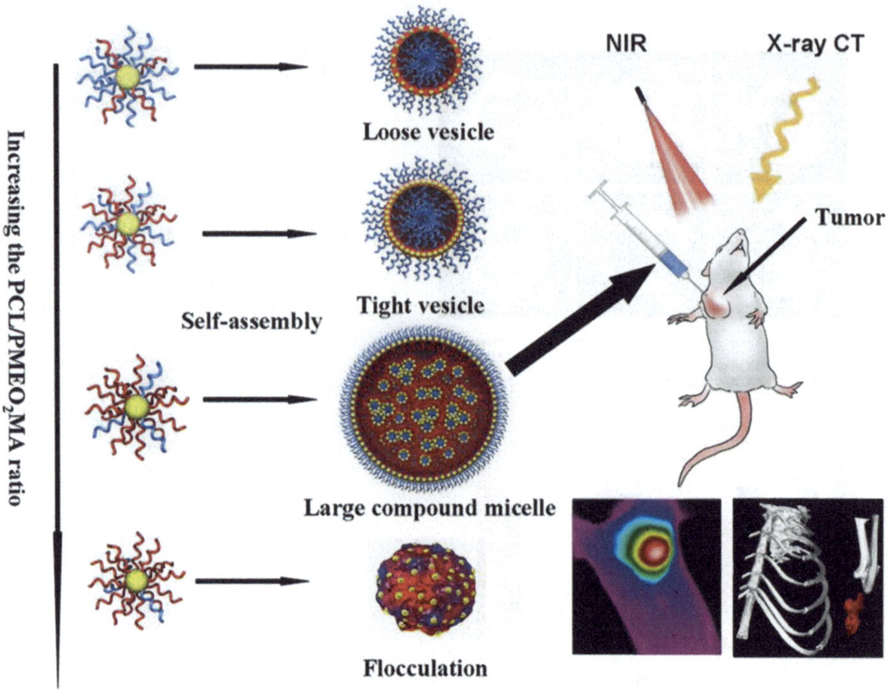

Fig. 4 Schematic representation of assemblies composed by polymer grafted GNPs and their potential application in photothermal therapy and CT imaging of cancer. Reprinted with permission from Ref. [115]

Multifunctional GNPs can serve also for a combination of PTT and multimodal imaging for theranostic applications. For example, incorporation of a radioactive tracer on GNP for dual mode SPECT/CT imaging and PTT was presented by Jang et al. [116]. Likewise, simultaneous photoacoustic/CT bimodal imaging and PTT of cancer using GNPs was presented recently in formulations of graphene oxide modified PLA microcapsules containing gold nanoparticles [117], and core–shell nanoparticles of Au@Prussian blue (Au@PB) [60]. Additional recent studies have demonstrated gold nanostars (GNS) as theranostic probes for PTT and multimodal imaging of cancer [118, 119].

5.3 *Combining Radiation Therapy with CT Imaging*

Image-guided RT (IGRT) relies on the ability to localize tumors in order to optimize the therapeutic outcomes. GNPs may represent an efficient theranostic adjuvant for CT-guided RT treatment, due to their radiosensitive properties and the ability to enhance CT contrast. Accordingly, they are currently being studied in this theranostic role, and have thus far shown great potential in clinical applications. Joh et al. [120]

have investigated the therapeutic and diagnostic value of PEG-modified GNPs (P-GNPs) combined with RT for sarcoma. Their experimental techniques included intravenous administration of the P-GNP, CT imaging of mice with engrafted sarcoma tumors for RT treatment planning, and precision-targeted RT delivered using a Small Animal Radiation Research Platform (SARRP). The results demonstrated that P-GNP in conjunction with RT enhanced CT imaging, caused increased RT-induced DNA damage, and led to significantly reduced clonogenic survival of tumor cells, compared to RT alone, while minimizing radiation dose to normal tissues. Jølck et al. [121] have developed a nanogel based on colloidal gold in a sucrose acetate isobutyrate gelating matrix, for IGRT. They investigated the induced CT contrast and the stability of the nanogel over a period of 12 weeks. It was demonstrated that once injected, the nanogel remained at the injection site and provided excellent CT-contrast as visualized by micro-CT imaging. In addition, in vitro and in vivo experiments showed that the nanogel was stable over time both in terms of geometry, volume, and GNP content. Minor degradation was observed after 12 weeks, suggesting an acceptable degradation profile for image guidance in patients undergoing fractioned RT, as the treatment is typically conducted within a 6 weeks time frame. Another nanoplatform that could simultaneously generate a CT image contrast and serve as a radiosensitizer was developed by Zaki et al. [122]. They prepared gold-loaded polymeric micelles (GPMs), which provided blood pool contrast for up to 24 h following intravenous injection, and improved the delineation of tumor margins via CT. In combination of GPMs with radiation, tumor-bearing mice exhibited a 1.7-fold improvement in the median survival time compared with mice receiving radiation alone (Fig. 5).

Multifunctional theranostic GNPs for combination of RT with dual-mode SPECT/CT were also presented by Su et al. [40], who used targeted ^{125}I-labeled cRGD-GNPs, and by Kao et al. [123], who used ^{131}I-labeled EGFR-targeted GNPs.

5.4 Combining Multimodal Therapy with CT Imaging

Today, treating patients with a broad-spectrum approach is preferable among most clinicians. Multimodal therapy (MMT) combines different therapeutic strategies to improve treatment outcomes. GNPs have been investigated in several studies as both CT contrast agents and MMT enhancers. Huang et al. [124] have designed a multifunctional nanoprobe of folic acid-conjugated silica-modified GNR with the aim of investigating the feasibility of their use as X-ray/CT imaging-guided targeting dual-mode RT and PTT. The prepared nanoprobes showed highly selective targeting, enhanced RT and PTT effects on MGC803 gastric cancer cells, and also exhibited strong X-ray attenuation for in vivo X-ray and CT imaging. In another study, Park et al. [125] have reported multifunctional hollow GNPs (HGNP) designed for triple combination therapy and CT imaging. The shell thickness of HGNPs can tune the surface plasmon resonance to the near infrared light, resulting in a photothermal ablation of tumors, whereas the hollow cavity within a HGNP is able to accommodate a high payload of chemotherapeutic agents. This study utilized HGNPs for CT imaging and combined drug delivery (of doxorubicin), thermal

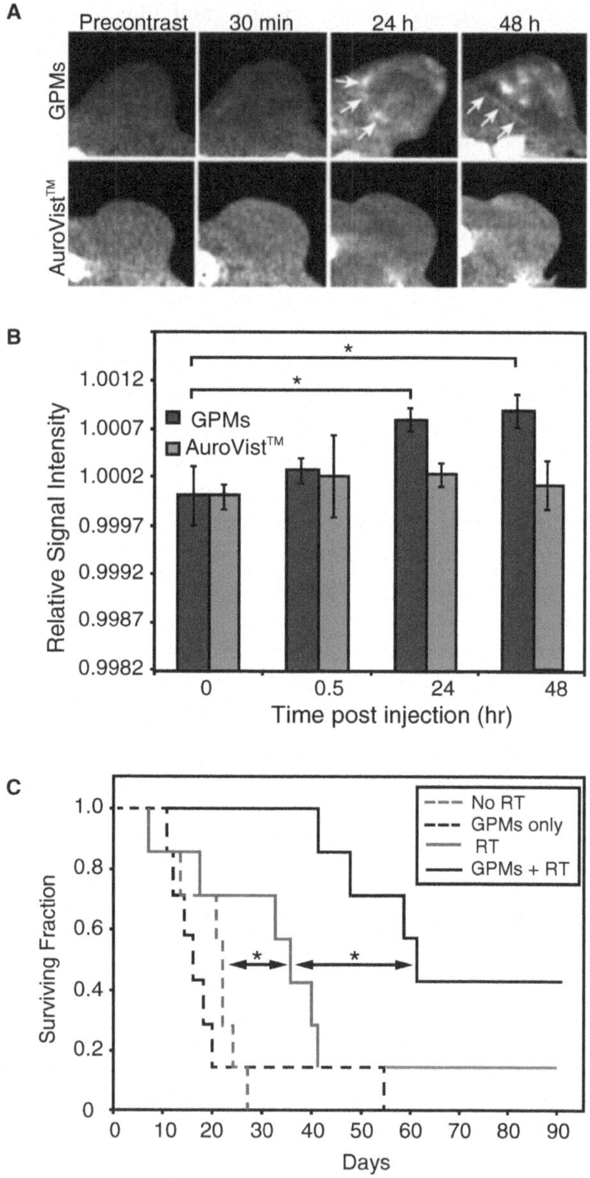

Fig. 5 (**a**) Representative CT images in the axial plane prior to injection (precontrast) and 30 min, 24 h, and 48 h postinjection of GPMs or AuroVist (individual 1.9 nm GNPs). Tumor boundaries are indicated by *white arrows*. (**b**) Quantitative analysis of CT images. (**c**) Kaplan-Meier survival analysis for tumor-bearing mice receiving no treatment (*dotted gray line*), GPMs only (*dotted black line*), irradiation only (*solid gray line*), or irradiation 24 h after retro-orbital injection of GPMs (*solid black line*). Reprinted with permission from Ref. [122]

therapy, and radiotherapy. Synthesized Dox-HGNPs exhibited better response to radiation for CT imaging and cancer therapy. It was shown also that Dox-HGNPs were capable of triggering Dox release by NIR laser, converting light to heat for photothermal ablation and enhancing radiosensitization of the tumor. In vivo experiments showed that the triple treatment effects of Dox, heat, and radiation dramatically enhanced tumor growth delay by a factor of 4.3.

6 Conclusions and Future Perspectives

The aforementioned studies, both for imaging and therapy, demonstrate that GNPs are highly favorable candidates for various therapeutic and diagnostic implementations, each one separately, or together in a multimodal approach. Furthermore, the unique physicochemical properties of GNPs, along with their biocompatibility and low toxicity, have led to excellent performance in the few clinical trials involving GNPs [126]. In the theranostic field, which is based on the "all-in-one" approach, the broad possible range of structures, hybridizations, coatings, and capacities marks GNPs as the next generation of theranostic agents. The concept of theranostics with multifunctional GNPs has a great potential to enhance the medical area towards personalized medicine. However, a broad and thorough evaluation of the long-term toxicity of GNPs is still required in order to translate them to the clinic.

References

1. Connor EE, Mwamuka J, Gole A, Murphy CJ, Wyatt MD. Gold nanoparticles are taken up by human cells but do not cause acute cytotoxicity. Small. 2005;1:325–7. doi:10.1002/smll.200400093.
2. Boisselier E, Astruc D. Gold nanoparticles in nanomedicine: preparations, imaging, diagnostics, therapies and toxicity. Chem Soc Rev. 2009;38:1759. doi:10.1039/b806051g.
3. Jain S, Hirst DG, O'Sullivan JM. Gold nanoparticles as novel agents for cancer therapy. Br J Radiol. 2012;85:101–13. doi:10.1259/bjr/59448833.
4. Sean Norman R, Stone JW, Gole A, Murphy CJ, Sabo-Attwood TL. Targeted photothermal lysis of the pathogenic bacteria, pseudomonas aeruginosa, with gold nanorods. Nano Lett. 2008;8:302–6. doi:10.1021/nl0727056.
5. Von Maltzahn G, Park J-H, Agrawal A, Bandaru NK, Das SK, Sailor MJ, et al. Computationally guided photothermal tumor therapy using long-circulating gold nanorod antennas. Cancer Res. 2009;69:3892–900. doi:10.1158/0008-5472.CAN-08-4242.
6. Minati L, Antonini V, Torrengo S, Serra MD, Boustta M, Leclercq X, et al. Sustained in vitro release and cell uptake of doxorubicin adsorbed onto gold nanoparticles and covered by a polyelectrolyte complex layer. Int J Pharm. 2012;438:45–52. doi:10.1016/j.ijpharm.2012.08.057.
7. Rand D, Ortiz V, Liu Y, Derdak Z, Wands JR, Tatíček M, et al. Nanomaterials for X-ray imaging: gold nanoparticle enhancement of X-ray scatter imaging of hepatocellular carcinoma. Nano Lett. 2011;11:2678–83. doi:10.1021/nl200858y.
8. Popovtzer R, Agrawal A, Kotov NA, Popovtzer A, Balter J, Carey TE, et al. Targeted gold nanoparticles enable molecular CT imaging of cancer. Nano Lett. 2008;8:4593–6. doi:10.1021/nl8029114.
9. Popovtzer R. Targeted gold nanoparticles enable molecular CT imaging of cancer: an in vivo study. Int J Nanomed. 2011;6:2859. doi:10.2147/IJN.S25446.
10. Jakobsohn K, Motiei M, Sinvani M, Popovtzer R. Towards real-time detection of tumor margins using photothermal imaging of immune-targeted gold nanoparticles. Int J Nanomed. 2012;7:4707–13. doi:10.2147/IJN.S34157.
11. Shilo M, Reuveni T, Motiei M, Popovtzer R. Nanoparticles as computed tomography contrast agents: current status and future perspectives. Nanomedicine. 2012;7:257–69. doi:10.2217/nnm.11.190.
12. Huang H-C, Barua S, Sharma G, Dey SK, Rege K. Inorganic nanoparticles for cancer imaging and therapy. J Control Release. 2011;155:344–57. doi:10.1016/j.jconrel.2011.06.004.

13. Liu D, Wang Z, Jiang X. Gold nanoparticles for the colorimetric and fluorescent detection of ions and small organic molecules. Nanoscale. 2011;3:1421–33. doi:10.1039/c0nr00887g.
14. Yu S, Watson AD, Bismuth B. Metal-based X-ray contrast media. 1999.
15. Xu C, Tung GA, Sun S. Size and concentration effect of gold nanoparticles on X-ray attenuation as measured on computed tomography. Chem Mater. 2008;20:4167–9. doi:10.1021/cm8008418.
16. Petros RA, DeSimone JM. Strategies in the design of nanoparticles for therapeutic applications. Nat Rev Drug Discov. 2010;9:615–27. doi:10.1038/nrd2591.
17. Hallouard F, Anton N, Choquet P, Constantinesco A, Vandamme T. Iodinated blood pool contrast media for preclinical X-ray imaging applications—a review. Biomaterials. 2010;31:6249–68. doi:10.1016/j.biomaterials.2010.04.066.
18. Allkemper T, Bremer C, Matuszewski L, Ebert W, Reimer P. Contrast-enhanced blood-pool MR angiography with optimized iron oxides: effect of size and dose on vascular contrast enhancement in rabbits. Radiology. 2002;223:432–8. doi:10.1148/radiol.2232010241.
19. Schirra CO, Brendel B, Anastasio MA, Roessl E. Spectral CT: a technology primer for contrast agent development. Contrast Media Mol Imaging 9:62–70. doi: 10.1002/cmmi.1573.
20. Pan D, Roessl E, Schlomka J-P, Caruthers SD, Senpan A, Scott MJ, et al. Computed tomography in color: NanoK-enhanced spectral CT molecular imaging. Angew Chem Int Ed. 2010;49:9635–9. doi:10.1002/anie.201005657.
21. Cormode DP, Roessl E, Thran A, Skajaa T, Gordon RE, Schlomka J-P, et al. Atherosclerotic plaque composition: analysis with multicolor CT and targeted gold nanoparticles. Radiology. 2010;256:774–82. doi:10.1148/radiol.10092473.
22. Schirra CO, Senpan A, Roessl E, Thran A, Stacy AJ, Wu L, et al. Second generation gold nanobeacons for robust K-edge imaging with multi-energy CT. J Mater Chem. 2012;22:23071–7. doi:10.1039/C2JM35334B.
23. Roessl E, Cormode D, Brendel B, Jürgen Engel K, Martens G, Thran A, et al. Preclinical spectral computed tomography of gold nano-particles. Nucl Instruments Methods Phys Res Sect A Accel Spectrometers Detect Assoc Equip. 2011;648:S259–64. doi:10.1016/j.nima.2010.11.072.
24. Wolf GL. Magnetic resonance imaging and the future of cardiac imaging. Am J Cardiol. 1989;64:E60–3. doi:10.1016/0002-9149(89)90736-4.
25. Skotland T, Iversen T-G, Sandvig K. New metal-based nanoparticles for intravenous use: requirements for clinical success with focus on medical imaging. Nanomed Nanotechnol Biol Med. 2010;6:730–7. doi:10.1016/j.nano.2010.05.002.
26. Kojima C, Umeda Y, Ogawa M, Harada A, Magata Y, Kono K. X-ray computed tomography contrast agents prepared by seeded growth of gold nanoparticles in PEGylated dendrimer. Nanotechnology. 2010;21:245104. doi:10.1088/0957-4484/21/24/245104.
27. Kim D, Park S, Lee JH, Jeong YY, Jon S. Antibiofouling polymer-coated gold nanoparticles as a contrast agent for in vivo X-ray computed tomography imaging. J Am Chem Soc. 2007;129:7661–5. doi:10.1021/ja071471p.
28. Cai Q-Y, Kim SH, Choi KS, Kim SY, Byun SJ, Kim KW, et al. Colloidal gold nanoparticles as a blood-pool contrast agent for X-ray computed tomography in mice. Invest Radiol. 2007;42:797–806. doi:10.1097/RLI.0b013e31811ecdcd.
29. Peng C, Zheng L, Chen Q, Shen M, Guo R, Wang H, et al. PEGylated dendrimer-entrapped gold nanoparticles for in vivo blood pool and tumor imaging by computed tomography. Biomaterials. 2012;33:1107–19. doi:10.1016/j.biomaterials.2011.10.052.
30. Hainfeld JF, Slatkin DN, Focella TM, Smilowitz HM. Gold nanoparticles: a new X-ray contrast agent. Br J Radiol. 2006;79:248–53. doi:10.1259/bjr/13169882.
31. Ghaghada KB, Badea CT, Karumbaiah L, Fettig N, Bellamkonda RV, Johnson GA, et al. Evaluation of tumor microenvironment in an animal model using a nanoparticle contrast agent in computed tomography imaging. Acad Radiol. 2011;18:20–30. doi:10.1016/j.acra.2010.09.003.
32. Hainfeld JF, O'Connor MJ, Dilmanian FA, Slatkin DN, Adams DJ, Smilowitz HM. Micro-CT enables microlocalisation and quantification of Her2-targeted gold nanoparticles within tumour regions. Br J Radiol. 2014.

33. Chanda N, Kattumuri V, Shukla R, Zambre A, Katti K, Upendran A, et al. Bombesin functionalized gold nanoparticles show in vitro and in vivo cancer receptor specificity. Proc Natl Acad Sci U S A. 2010;107:8760–5. doi:10.1073/pnas.1002143107.
34. Wang H, Zheng L, Peng C, Shen M, Shi X, Zhang G. Folic acid-modified dendrimer-entrapped gold nanoparticles as nanoprobes for targeted CT imaging of human lung adencarcinoma. Biomaterials. 2013;34:470–80. doi:10.1016/j.biomaterials.2012.09.054.
35. Wen S, Li K, Cai H, Chen Q, Shen M, Huang Y, et al. Multifunctional dendrimer-entrapped gold nanoparticles for dual mode CT/MR imaging applications. Biomaterials. 2013;34:1570–80. doi:10.1016/j.biomaterials.2012.11.010.
36. Chen Q, Li K, Wen S, Liu H, Peng C, Cai H, et al. Targeted CT/MR dual mode imaging of tumors using multifunctional dendrimer-entrapped gold nanoparticles. Biomaterials. 2013;34:5200–9. doi:10.1016/j.biomaterials.2013.03.009.
37. Hu D-H, Sheng Z-H, Zhang P-F, Yang D-Z, Liu S-H, Gong P, et al. Hybrid gold–gadolinium nanoclusters for tumor-targeted NIRF/CT/MRI triple-modal imaging in vivo. Nanoscale. 2013;5:1624. doi:10.1039/c2nr33543c.
38. Zhao Y, Sultan D, Detering L, Cho S, Sun G, Pierce R, et al. Copper-64-alloyed gold nanoparticles for cancer imaging: improved radiolabel stability and diagnostic accuracy. Angew Chem Int Ed. 2014;53:156–9. doi:10.1002/anie.201308494.
39. Wang Y, Liu Y, Luehmann H, Xia X, Brown P, Jarreau C, et al. Evaluating the pharmacokinetics and in vivo cancer targeting capability of au nanocages by positron emission tomography imaging. ACS Nano. 2012;6:5880–8. doi:10.1021/nn300464r.
40. Su N, Dang Y, Liang G, Liu G. Iodine-125-labeled cRGD-gold nanoparticles as tumor-targeted radiosensitizer and imaging agent. Nanoscale Res Lett. 2015;10:160. doi:10.1186/s11671-015-0864-9.
41. Song J-T, Yang X-Q, Zhang X-S, Yan D-M, Wang Z-Y, Zhao Y-D. Facile synthesis of gold nanospheres modified by positively charged mesoporous silica, loaded with near-infrared fluorescent dye, for in vivo X-ray computed tomography and fluorescence dual mode imaging. ACS Appl Mater Interfaces. 2015;7:17287–97. doi:10.1021/acsami.5b04359.
42. Sun I-C, Eun D-K, Koo H, Ko C-Y, Kim H-S, Yi DK, et al. Tumor-targeting gold particles for dual computed tomography/optical cancer imaging. Angew Chem. 2011;123:9520–3. doi:10.1002/ange.201102892.
43. Zhang J, Li C, Zhang X, Huo S, Jin S, An F-F, et al. In vivo tumor-targeted dual-modal fluorescence/CT imaging using a nanoprobe co-loaded with an aggregation-induced emission dye and gold nanoparticles. Biomaterials. 2015;42:103–11. doi:10.1016/j.biomaterials.2014.11.053.
44. Luo T, Huang P, Gao G, Shen G, Fu S, Cui D, et al. Mesoporous silica-coated gold nanorods with embedded indocyanine green for dual mode X-ray CT and NIR fluorescence imaging. Opt Express. 2011;19:17030. doi:10.1364/OE.19.017030.
45. Lee D-E, Koo H, Sun I-C, Ryu JH, Kim K, Kwon IC. Multifunctional nanoparticles for multimodal imaging and theragnosis. Chem Soc Rev. 2012;41:2656–72. doi:10.1039/c2cs15261d.
46. Lee S, Cha E-J, Park K, Lee S-Y, Hong J-K, Sun I-C, et al. A near-infrared-fluorescence-quenched gold-nanoparticle imaging probe for in vivo drug screening and protease activity determination. Angew Chemie. 2008;120:2846–9. doi:10.1002/ange.200705240.
47. Razgulin A, Ma N, Rao J. Strategies for in vivo imaging of enzyme activity: an overview and recent advances. Chem Soc Rev. 2011;40:4186–216. doi:10.1039/c1cs15035a.
48. Cheng W, Chen Y, Yan F, Ding L, Ding S, Ju H, et al. Ultrasensitive scanometric strategy for detection of matrix metalloproteinases using a histidine tagged peptide-Au nanoparticle probe. Chem Commun (Camb). 2011;47:2877–9. doi:10.1039/c0cc04441e.
49. Park SY, Lee SM, Kim GB, Kim Y-P. Gold nanoparticle-based fluorescence quenching via metal coordination for assaying protease activity. Gold Bull. 2012;45:213–9.
50. Tira DS, Focsan M, Ulinici S, Maniu D, Astilean S. Rhodamine B-coated gold nanoparticles as effective "turn-on" fluorescent sensors for detection of Zinc II ions in water. Spectrosc Lett. 2013;47:153–9. doi:10.1080/00387010.2013.782557.
51. Hutter E, Maysinger D. Gold-nanoparticle-based biosensors for detection of enzyme activity. Trends Pharmacol Sci. 2013;34:497–507. doi:10.1016/j.tips.2013.07.002.

52. Mohamed MB, Volkov V, Link S, El-sayed MA. The "lightning" gold nanorods: fluorescence enhancement of over a million compared to the gold metal. Chem Phys Lett. 2000;317:517–23.
53. Geddes C, Lakowicz J. Metal-enhanced fluorescence. J Fluoresc. 2002;12:121–9.
54. Ming T, Zhao L, Yang Z, Chen H, Sun L, Wang J, et al. Strong polarization dependence of plasmon-enhanced fluorescence on single gold nanorods. Nano Lett. 2009;9:3896–903. doi:10.1021/nl902095q.
55. Abadeer N, Brennan M, Wilson W, Murphy C. Distance and plasmon wavelength dependent fluorescence of molecules bound to silica-coated gold nanorods. ACS Nano. 2014;8:8392–406.
56. Kang K, Wang J. Conditionally activating optical contrast agent with enhanced sensitivity via gold nanoparticle plasmon energy transfer: feasibility study. J Nanobiotechnology. 2014;12:56. doi:10.1186/s12951-014-0056-2.
57. Kang KA, Wang J. Smart dual-mode fluorescent gold nanoparticle agents. Wiley Interdiscip Rev Nanomed Nanobiotechnol. 2014;6:398–409. doi:10.1002/wnan.1267.
58. Sun I-C, Eun D-K, Koo H, Ko C-Y, Kim H-S, Yi DK, et al. Tumor-targeting gold particles for dual computed tomography/optical cancer imaging. Angew Chem Int Ed Engl. 2011;50:9348–51. doi:10.1002/anie.201102892.
59. Chanda N, Shukla R, Zambre A, Mekapothula S, Kulkarni RR, Katti K, et al. An effective strategy for the synthesis of biocompatible gold nanoparticles using cinnamon phytochemicals for phantom CT imaging and photoacoustic detection of cancerous cells. Pharm Res. 2011;28:279–91. doi:10.1007/s11095-010-0276-6.
60. Jing L, Liang X, Deng Z, Feng S, Li X, Huang M, et al. Prussian blue coated gold nanoparticles for simultaneous photoacoustic/CT bimodal imaging and photothermal ablation of cancer. Biomaterials. 2014;35:5814–21. doi:10.1016/j.biomaterials.2014.04.005.
61. Cho K, Wang X, Nie S, Chen ZG, Shin DM. Therapeutic nanoparticles for drug delivery in cancer. Clin Cancer Res. 2008;14:1310–6. doi:10.1158/1078-0432.CCR-07-1441.
62. Beija M, Li Y, Duong HT, Laurent S, Elst LV, Muller RN. Polymer–gold nanohybrids with potential use in bimodal MRI/CT: enhancing the relaxometric properties of Gd(iii) complexes. J Mater Chem. 2012;22:21382. doi:10.1039/c2jm34999j.
63. Xie M, Ding L, You Z, Gao D, Yang G, Han H. Robust hybrid nanostructures comprising gold and thiol-functionalized polymer nanoparticles: facile preparation, diverse morphologies and unique properties. J Mater Chem. 2012;22:14108. doi:10.1039/c2jm31228j.
64. Han S-Y, Guo Q-H, Xu M-M, Yuan Y-X, Shen L-M, Yao J-L, et al. Tunable fabrication on iron oxide/Au/Ag nanostructures for surface enhanced Raman spectroscopy and magnetic enrichment. J Colloid Interface Sci. 2012;378:51–7. doi:10.1016/j.jcis.2012.04.047.
65. Guo X, Zhang Q, Sun Y, Zhao Q, Yang J. Lateral etching of core–shell Au@Metal nanorods to metal-tipped Au nanorods with improved catalytic activity. ACS Nano. 2012;6:1165–75. doi:10.1021/nn203793k.
66. Tkachenko AG, Xie H, Liu Y, Coleman D, Ryan J, Glomm WR, et al. Cellular trajectories of peptide-modified gold particle complexes: comparison of nuclear localization signals and peptide transduction domains. Bioconjug Chem. 2004;15:482–90. doi:10.1021/bc034189q.
67. Nativo P, Prior IA, Brust M. Uptake and intracellular fate of surface-modified gold nanoparticles. ACS Nano. 2008;2:1639–44. doi:10.1021/nn800330a.
68. Wang F, Wang YC, Dou S, Xiong MH, Sun TM, Wang J. Doxorubicin-tethered responsive gold nanoparticles facilitate intracellular drug delivery for overcoming multidrug resistance in cancer cells. ACS Nano. 2011;5:3679–92. doi:10.1021/nn200007z.
69. Tomuleasa C, Soritau O, Orza A, Dudea M, Petrushev B, Mosteanu O, et al. Gold nanoparticles conjugated with cisplatin/doxorubicin/capecitabine lower the chemoresistance of hepatocellular carcinoma-derived cancer cells. J Gastrointest Liver Dis. 2012;21:188–96.
70. Prabaharan M, Grailer JJ, Pilla S, Steeber DA, Gong S. Gold nanoparticles with a monolayer of doxorubicin-conjugated amphiphilic block copolymer for tumor-targeted drug delivery. Biomaterials. 2009;30:6065–75. doi:10.1016/j.biomaterials.2009.07.048.
71. Paciotti GF, Myer L, Weinreich D, Goia D, Pavel N, McLaughlin RE, et al. Colloidal gold: a novel nanoparticle vector for tumor directed drug delivery. Drug Deliv. 2004;11:169–83. doi:10.1080/10717540490433895.

72. Sokolova V, Epple M. Inorganic nanoparticles as carriers of nucleic acids into cells. Angew Chemie Int Ed. 2008;47:1382–95. doi:10.1002/anie.200703039.
73. Han G, Ghosh P, De M, Rotello VM. Drug and gene delivery using gold nanoparticles. NanoBiotechnology. 2007;3:40–5. doi:10.1007/s12030-007-0005-3.
74. Chen C-C, Lin Y-P, Wang C-W, Tzeng H-C, Wu C-H, Chen Y-C, et al. DNA-gold nanorod conjugates for remote control of localized gene expression by near infrared irradiation. J Am Chem Soc. 2006;128:3709–15. doi:10.1021/ja0570180.
75. Kim J-H, Yeom J-H, Ko J-J, Han MS, Lee K, Na S-Y, et al. Effective delivery of anti-miRNA DNA oligonucleotides by functionalized gold nanoparticles. J Biotechnol. 2011;155:287–92. doi:10.1016/j.jbiotec.2011.07.014.
76. Shan Y, Luo T, Peng C, Sheng R, Cao A, Cao X, et al. Gene delivery using dendrimer-entrapped gold nanoparticles as nonviral vectors. Biomaterials. 2012;33:3025–35. doi:10.1016/j.biomaterials.2011.12.045.
77. Ghosh PS, Kim C-K, Han G, Forbes NS, Rotello VM. Efficient gene delivery vectors by tuning the surface charge density of amino acid-functionalized gold nanoparticles. ACS Nano. 2008;2:2213–8. doi:10.1021/nn800507t.
78. Guo S, Huang Y, Jiang Q, Sun Y, Deng L, Liang Z, et al. Enhanced gene delivery and siRNA silencing by gold nanoparticles coated with charge-reversal polyelectrolyte. ACS Nano. 2010;4:5505–11. doi:10.1021/nn101638u.
79. Conde J, Ambrosone A, Sanz V. Design of multifunctional gold nanoparticles for in vitro and in vivo gene silencing. ACS Nano. 2012;6(9):8316–24.
80. Wijaya A, Schaffer SB, Pallares IG, Hamad-Schifferli K. Selective release of multiple DNA oligonucleotides from gold nanorods. ACS Nano. 2009;3:80–6. doi:10.1021/nn800702n.
81. Link S, El-Sayed MA. Spectral properties and relaxation dynamics of surface plasmon electronic oscillations in gold and silver nanodots and nanorods. J Phys Chem B. 1999;103:8410–26. doi:10.1021/jp9917648.
82. El-Sayed MA. Some interesting properties of metals confined in time and nanometer space of different shapes. Acc Chem Res. 2001;34:257–64. doi:10.1021/ar960016n.
83. Curry T, Epstein T, Smith R, Kopelman R. Photothermal therapy of cancer cells mediated by blue hydrogel nanoparticles. Nanomedicine. 2013;8:1577–86. doi:10.2217/nnm.12.190.
84. Link S, El-Sayed MA. Shape and size dependence of radiative, non-radiative and photothermal properties of gold nanocrystals. Int Rev Phys Chem. 2010.
85. Chial HJ, Thompson HB, Splittgerber AG. A spectral study of the charge forms of Coomassie blue G. Anal Biochem. 1993;209:258–66. doi:10.1006/abio.1993.1117.
86. Choi J, Yang J, Bang D, Park J, Suh J-S, Huh Y-M, et al. Targetable gold nanorods for epithelial cancer therapy guided by near-IR absorption imaging. Small. 2012;8:746–53. doi:10.1002/smll.201101789.
87. El-Sayed IH, Huang X, El-Sayed MA. Selective laser photo-thermal therapy of epithelial carcinoma using anti-EGFR antibody conjugated gold nanoparticles. Cancer Lett. 2006;239:129–35. doi:10.1016/j.canlet.2005.07.035.
88. Qian W, Murakami M, Ichikawa Y, Che Y. Highly efficient and controllable PEGylation of gold nanoparticles prepared by femtosecond laser ablation in water. J Phys Chem C. 2011;115:23293–8. doi:10.1021/jp2079567.
89. Curry TY. Nanoparticles for biomedical applications: photothermal therapy and nuclear delivery. 2013.
90. Yuan H, Khoury CG, Wilson CM, Grant GA, Bennett AJ, Vo-Dinh T. In vivo particle tracking and photothermal ablation using plasmon-resonant gold nanostars. Nanomed Nanotechnol Biol Med. 2012;8:1355–63. doi:10.1016/j.nano.2012.02.005.
91. Zharov VP, Galitovskaya EN, Johnson C, Kelly T. Synergistic enhancement of selective nanophotothermolysis with gold nanoclusters: potential for cancer therapy. Lasers Surg Med. 2005;37:219–26. doi:10.1002/lsm.20223.
92. Nam J, Won N, Jin H, Chung H, Kim S. pH-Induced aggregation of gold nanoparticles for photothermal cancer therapy. J Am Chem Soc. 2009;131:13639–45. doi:10.1021/ja902062j.

93. Hühn D, Govorov A, Gil PR, Parak WJ. Photostimulated Au nanoheaters in polymer and biological media: characterization of mechanical destruction and boiling. Adv Funct Mater. 2012;22:294–303. doi:10.1002/adfm.201101134.
94. Murthy AK, Stover RJ, Borwankar AU, Nie GD, Gourisankar S, Truskett TM, et al. Equilibrium gold nanoclusters quenched with biodegradable polymers. ACS Nano. 2013;7:239–51. doi:10.1021/nn303937k.
95. Letfullin RR, Joenathan C, George TF, Zharov VP. Laser-induced explosion of gold nanoparticles: potential role for nanophotothermolysis of cancer. Nanomedicine. 2006;1:473–80. doi:10.2217/17435889.1.4.473.
96. Coronado EA, Encina ER, Stefani FD. Optical properties of metallic nanoparticles: manipulating light, heat and forces at the nanoscale. Nanoscale. 2011;3:4042–59. doi:10.1039/c1nr10788g.
97. Hainfeld JF, Slatkin DN, Smilowitz HM. The use of gold nanoparticles to enhance radiotherapy in mice. Phys Med Biol. 2004;49:N309–15. doi:10.1088/0031-9155/49/18/N03.
98. De Jong WH, Hagens WI, Krystek P, Burger MC, Sips AJ, Geertsma RE. Particle size-dependent organ distribution of gold nanoparticles after intravenous administration. Biomaterials. 2008;29:1912–9. doi:10.1016/j.biomaterials.2007.12.037.
99. Hainfeld JF, Dilmanian FA, Slatkin DN, Smilowitz HM. Radiotherapy enhancement with gold nanoparticles. J Pharm Pharmacol. 2008;60:977–85. doi:10.1211/jpp.60.8.0005.
100. Chattopadhyay N, Cai Z, Kwon YL, Lechtman E, Pignol JP, Reilly RM. Molecularly targeted gold nanoparticles enhance the radiation response of breast cancer cells and tumor xenografts to X-radiation. Breast Cancer Res Treat. 2013;137:81–91. doi:10.1007/s10549-012-2338-4.
101. Roa W, Zhang X, Guo L, Shaw A, Hu X, Xiong Y, et al. Gold nanoparticle sensitize radiotherapy of prostate cancer cells by regulation of the cell cycle. Nanotechnology. 2009;20:375101. doi:10.1088/0957-4484/20/37/375101.
102. Bechet D, Couleaud P, Frochot C, Viriot M-L, Guillemin F, Barberi-Heyob M. Nanoparticles as vehicles for delivery of photodynamic therapy agents. Trends Biotechnol. 2008;26:612–21. doi:10.1016/j.tibtech.2008.07.007.
103. Wieder ME, Hone DC, Cook MJ, Handsley MM, Gavrilovic J, Russell DA. Intracellular photodynamic therapy with photosensitizer-nanoparticle conjugates: cancer therapy using a "Trojan horse". Photochem Photobiol Sci. 2006;5:727–34. doi:10.1039/b602830f.
104. Hone D, Walker P, Evans-Gowing R. Generation of cytotoxic singlet oxygen via phthalocyanine-stabilized gold nanoparticles: a potential delivery vehicle for photodynamic therapy. Langmuir. 2002:2985–7. doi: 10.1021/la0256230.
105. Cheng Y, Samia AC, Meyers JD, Panagopoulos I, Fei B, Burda C. Highly efficient drug delivery with gold nanoparticle vectors for in vivo photodynamic therapy of cancer. J Am Chem Soc. 2008;130:10643–7. doi:10.1021/ja801631c.
106. Bao G, Mitragotri S, Tong S. Multifunctional nanoparticles for drug delivery and molecular imaging. Annu Rev Biomed Eng. 2013;15:253–82. doi:10.1146/annurev-bioeng-071812-152409.
107. Kim D, Jeong YY, Jon S. A drug-loaded aptamer-gold nanoparticle bioconjugate for combined CT imaging and therapy of prostate cancer. ACS Nano. 2010;4:3689–96. doi:10.1021/nn901877h.
108. Zhu J, Zheng L, Wen S, Tang Y, Shen M, Zhang G, et al. Targeted cancer theranostics using alpha-tocopheryl succinate-conjugated multifunctional dendrimer-entrapped gold nanoparticles. Biomaterials. 2014;35:7635–46. doi:10.1016/j.biomaterials.2014.05.046.
109. Won Y-W, Yoon S-M, Lim KS, Kim Y-H. Self-assembled nanoparticles with dual effects of passive tumor targeting and cancer-selective anticancer effects. Adv Funct Mater. 2012;22:1199–208. doi:10.1002/adfm.201101979.
110. Won Y-W, Yoon S-M, Sonn CH, Lee K-M, Kim Y-H. Nano self-assembly of recombinant human gelatin conjugated with α-tocopheryl succinate for Hsp90 inhibitor, 17-AAG, delivery. ACS Nano. 2011;5:3839–48. doi:10.1021/nn200173u.
111. Gogvadze V, Norberg E, Orrenius S, Zhivotovsky B. Involvement of Ca2+ and ROS in α-tocopheryl succinate-induced mitochondrial permeabilization. Int J Cancer. 2010;127:1823–32. doi:10.1002/ijc.25204.

112. Prasad KN, Kumar B, Yan X-D, Hanson AJ, Cole WC. α-tocopheryl succinate, the most effective form of vitamin E for adjuvant cancer treatment: a review. J Am Coll Nutr. 2003;22:108–17. doi:10.1080/07315724.2003.10719283.
113. Zhu J, Fu F, Xiong Z, Shen M, Shi X. Dendrimer-entrapped gold nanoparticles modified with RGD peptide and alpha-tocopheryl succinate enable targeted theranostics of cancer cells. Colloids Surf B Biointerfaces. 2015;133:36–42. doi:10.1016/j.colsurfb.2015.05.040.
114. Qin J, Peng Z, Li B, Ye K, Zhang Y, Yuan F, et al. Gold nanorods as a theranostic platform for in vitro and in vivo imaging and photothermal therapy of inflammatory macrophages. Nanoscale. 2015;7:13991–4001. doi:10.1039/C5NR02521D.
115. Deng H, Zhong Y, Du M, Liu Q, Fan Z, Dai F, et al. Theranostic self-assembly structure of gold nanoparticles for NIR photothermal therapy and X-Ray computed tomography imaging. Theranostics. 2014;4:904–18. doi:10.7150/thno.9448.
116. Jang B, Park S, Kang SH, Kim JK, Kim S-K, Kim I-H, et al. Gold nanorods for target selective SPECT/CT imaging and photothermal therapy in vivo. Quant Imaging Med Surg. 2012;2:1–11. doi:10.3978/.issn.2223-4292.2012.01.03.
117. Jin Y, Wang J, Ke H, Wang S, Dai Z. Graphene oxide modified PLA microcapsules containing gold nanoparticles for ultrasonic/CT bimodal imaging guided photothermal tumor therapy. Biomaterials. 2013;34:4794–802. doi:10.1016/j.biomaterials.2013.03.027.
118. Liu Y, Ashton JR, Moding EJ, Yuan H, Register JK, Fales AM, et al. A plasmonic gold nanostar theranostic probe for in vivo tumor imaging and photothermal therapy. Theranostics. 2015;5:946–60. doi:10.7150/thno.11974.
119. Li J, Hu Y, Yang J, Wei P, Sun W, Shen M, et al. Hyaluronic acid-modified Fe3O4@Au core/shell nanostars for multimodal imaging and photothermal therapy of tumors. Biomaterials. 2015;38:10–21. doi:10.1016/j.biomaterials.2014.10.065.
120. Joh DY, Kao GD, Murty S, Stangl M, Sun L, Al Zaki A, et al. Theranostic gold nanoparticles modified for durable systemic circulation effectively and safely enhance the radiation therapy of human sarcoma cells and tumors. Transl Oncol. 2013;6:722–31. doi:10.1593/tlo.13433.
121. Jølck RI, Binderup T, Hansen AE, Scherman JB, Munch AF, Rosenschold P, Kjaer A, et al. Injectable colloidal gold in a sucrose acetate isobutyrate gelating matrix with potential use in radiation therapy. Adv Healthc Mater. 2014;3:1680–7. doi:10.1002/adhm.201300668.
122. Al Zaki A, Joh D, Cheng Z, De Barros ALB, Kao G, Dorsey J, et al. Gold-loaded polymeric micelles for computed tomography-guided radiation therapy treatment and radiosensitization. ACS Nano. 2014;8:104–12. doi:10.1021/nn405701q.
123. Kao HW, Lin YY, Chen CC, Chi KH, Tien DC, Hsia CC, et al. Evaluation of EGFR-targeted radioimmuno-gold-nanoparticles as a theranostic agent in a tumor animal model. Bioorg Med Chem Lett. 2013;23:3180–5. doi:10.1016/j.bmcl.2013.04.002.
124. Huang P, Bao L, Zhang C, Lin J, Luo T, Yang D, et al. Folic acid-conjugated silica-modified gold nanorods for X-ray/CT imaging-guided dual-mode radiation and photo-thermal therapy. Biomaterials. 2011;32:9796–809. doi:10.1016/j.biomaterials.2011.08.086.
125. Park J, Park J, Ju EJ, Park SS, Choi J, Lee JH, et al. Multifunctional hollow gold nanoparticles designed for triple combination therapy and CT imaging. J Control Release. 2015;207:77–85. doi:10.1016/j.jconrel.2015.04.007.
126. Pillai G. Nanomedicines for cancer therapy: an update of FDA approved and those under various stages of development. SOJ Pharm Pharm Sci. 2014;1:13.

Bismuth-Based Nanoparticles for CT Imaging

Songping D. Huang

1 Introduction

The search for suitable materials that can enhance X-ray image contrast began almost immediately after Röntgen discovered X-rays and demonstrated their use in medical imaging in 1895. After noting that sheets of different metals such as copper, silver, lead, gold, and platinum show varying degrees of transparency, Röntgen recognized that the X-ray transparency of a substance primarily depends on its density [1]. Theoretical and experimental studies show that when an X-ray beam traverses a medium, the beam intensity will be reduced due to both absorption and deflection of photons by the medium. The degree of X-ray attenuation obeys the following equation [2, 3]:

$$I = I_0 e^{-\mu x} \tag{1}$$

where I is the transmitted beam intensity, I_0 is the incident beam intensity, and x is the thickness of the medium. The mass attenuation coefficient, μ, expressed in cm^2/g, is best described by the following formula:

$$\mu = \frac{\rho Z^4}{AE^3} \tag{2}$$

where ρ is the density, Z is the atomic number, A is the atomic mass, and E is the X-ray energy. Therefore, X-ray attenuation is high with low-energy X-rays and with materials of high atomic number Z. All known X-ray contrast agents contain a heavy element that effectively contributes almost all the X-ray attenuation produced by such materials [4]. Based on this reason, many compounds with metallic elements of high atomic numbers were examined for potential applications in X-ray

S.D. Huang (✉)
Department of Chemistry and Biochemistry, Kent State University, Kent, OH 44240, USA
e-mail: shuang1@kent.edu

J.W.M. Bulte, M.M.J. Modo (eds.), *Design and Applications of Nanoparticles in Biomedical Imaging*, DOI 10.1007/978-3-319-42169-8_20

imaging. Consequently, the development of metal-based X-ray imaging agents predates the advent of the iodinated organic counterparts [5–7]. For instance, a colloidal silver suspension was used in urinary tract radiography as early as 1905. Two other preparations containing AgI or Ag_2O were later used as contrast agents in X-ray retrograde pyelography. A bismuth subnitrate emulsion was used as the contrast agent to obtain the very first contrast-enhanced coronary angiogram from a human cadaver. As early as 1897, the first radiograph of a bismuth-enhanced segment of the gastrointestinal (GI) tract was obtained using a bismuth subnitrate suspension. A year later bismuth-containing X-ray imaging agents were developed for X-ray examinations of the entire GI tract. However, after the aqueous suspensions of $BaSO_4$ was introduced in 1910 as a contrast agent for GI imaging, the use of bismuth compounds for this purpose failed to continue due to the higher costs of such compounds at that time and potential toxicity of bismuth at high dose levels. Interestingly, $BaSO_4$ has remained in clinical use for over 100 years with little or no change in the formulation. A variety of compounds or complexes containing other heavy metals such as cesium, tin, zirconium, tantalum, tungsten, and rare earth metals were also investigated for potential X-ray imaging applications [5–7]. On the other hand, the use of thorium oxide (ThO_2) as an X-ray contrast agent, first began about some 80 years ago [8], but would later result in tragic consequences [9, 10]. Thorium (Z=90) is the second heaviest naturally occurring element after uranium (Z=92). A suspension of ThO_2 showed superb image quality with virtually no acute toxicity or any immediate side effects. This formulation was introduced into clinical use as an X-ray imaging agent (Thorotrast®) in 1928 and quickly gained widespread applications for imaging cerebral arteries, liver, spleen, lymph nodes, and other organs. However, thorium has a naturally occurring radioactive isotope that is an *alpha*-emitter with an extremely long radioactive half-life (4.08 MeV, $t_{1/2} = 1.41 \times 10^{10}$ year). This problem, compounded with the unusually long biological half-life of the ThO_2 colloidal formulations ($t_{1/2}$=ca.22 years), would later cause liver cancer and leukemia in millions of patients who were injected with Thorotrast® in Europe, North America, and Japan [9].

Modern intravenous X-ray imaging agents, first appeared in 1954, are all based on the 1,3,5-triiodobenzene structural platform [11–13]. Because of the superior water solubility, versatile synthetic chemistry and high tolerance by the human body, such iodinated organic compounds have dominated the clinical application of X-ray imaging agents with ionic or covalent derivatives of tri-iodinated benzene compounds. The iodinated organic compounds are characterized by a nonspecific extracellular distribution and a short imaging time of approximately 1.5 h due to their rapid clearance by the kidney [13]. The rapid pharmacokinetics of iodinated organic CT imaging agents has thus limited their microvascular and targeting performance [11, 12]. From the viewpoint of contrast enhancement, use of iodine as an X-ray imaging agent is not optimal because its X-ray attenuation effect is rather low compared with the typical metal-based counterparts [2, 3]. The current use of iodinated X-ray imaging agents is mainly based on their superior safety and cost, rather than on their X-ray attenuation effects [13, 14]. In this regard, bismuth-based materials represent a unique opportunity for the development of a new generation of X-ray imaging agents.

2 Physical, Chemical, and Biological Properties of Bismuth

Bismuth (Z=83) is a soft gray metal that occurs in nature commonly as bismite (Bi_2O_3), bismutite ($(BiO)_2CO_3$), or bismuthinite (Bi_2S_3). The metallic form of this element is usually obtained as a byproduct from copper, lead, tin, molybdenum and tungsten smelting. Bismuth is a relatively rare element with the abundance in the Earth's crust estimated to be about 0.2 ppm, approximately about twice as abundant as gold. When freshly produced, it is a brittle metal with a silvery white color, but often exhibits a pink tint after exposure to air due to surface oxidation. Among metals, bismuth is the most diamagnetic element, has the lowest thermal conductivity, and exhibits the highest Hall effect. Because of its wide temperature range between the melting point (T_{melt}=271 °C) and boiling point (T_{boil}=1560 °C), and low neutron absorption cross section, liquid bismuth is used as a coolant in nuclear reactors. The density of bismuth, d=9.78 g/cm^3 is ~86 % of the density for lead, d=11.32 g/cm^3, making bismuth useful as a nontoxic replacement for lead in ceramic glazes, fishing sinkers, food-processing equipment, free-machining brasses for plumbing applications, lubricating greases, and ammunition for waterfowl hunting [15]. There is only one naturally occurring isotope of bismuth, $^{209}_{83}Bi$ that has long been considered as the heaviest stable isotope. In 2003, however, $^{209}_{83}Bi$ was found to decay via *alpha*-emission with a half-life to be 1.9×10^{19} years, more than a billion times longer than the current estimated age of the universe [16]. For all practical purposes, bismuth is still treated as nonradioactive.

Bismuth has two oxidation states, Bi(III) and Bi(V). The latter is a powerful oxidant in aqueous solution with a redox potential of $E°$=2.03 V for Bi(V)/Bi(III). Under the physiological conditions, Bi(III) is the most common and the only stable form of the element. The inorganic Bi(III) salts of sulfate, perchlorate, nitrate, as well as halides, except fluoride, all appear to be soluble in water, but hydrolyze to form insoluble hydroxide or oxyhydroxide products in neutral solution [15]. In acidic solution, such hydrolytic reactions are complex, and products formed are sensitive to the pH value, as well as the presence of different counter ions. These products are commonly referred to as "basic," "oxy-", or "sub-" salts with the formation of the bismuthyl ion (BiO^+) or various polynuclear clusters of low aqueous solubility [17]. As a heavy metal, the toxicity of bismuth is much lower than that of its heavy-metal neighbors, such as mercury, thallium, lead, and polonium [18]. Because of its low solubility in the blood, bismuth is readily cleared from urine, and exhibits no long-term carcinogenic, mutagenic or teratogenic effects. Its biological half-life for whole-body retention is about 5 days, although it may remain in the kidney for years in patients treated with bismuth compounds [19].

The use of bismuth compounds in medicine can be traced back to the Middle Ages, although the first full account of such use did not appear until 1786 for the treatment of dyspepsia [19]. A decade later, bismuth compounds were used as antisyphilitic drugs. In the late nineteenth century, a variety of bismuth compounds mixed with iodoform were promoted as a surgical wound dressing owing to its antimicrobial and antibacterial properties. The use of various bismuth compounds, such as subnitrate, subgallate, subcitrate, subcarbonate and subsalicylate to treat syphilis, hypertension, infections, skin conditions and gastrointestinal disorders lasted until 1950s after penicillin and several other antibiotics had become widely available [18]. Currently, the only continued medicinal application of bismuth is for the treatment of diarrhea and dyspepsia using bismuth subsalicylate (Pepto-Bismol®; the Procter & Gamble Company, Cincinnati, OH, USA), or bismuth subcitrate (De-Nol®; Gist Brocades, Delft, The Netherlands) [18, 19]. On the other hand, bismuth oxychloride (BiOCl) found in nature as the mineral bismoclite has been used as a pigment in cosmetics since the ancient Egyptian Times. Owing to its layered structure, BiOCl refracts light chromatically, resulting in an iridescent appearance resembling nacre of pearl [20].

3 Development of Bismuth-Based Nanoparticles for CT Imaging

Bismuth compounds have high X-ray attenuation efficiency, and thus can provide sufficient contrast enhancement at low dosage levels when they are employed as CT imaging agents. Furthermore, the biocompatible and nontoxic nature of bismuth represents an extremely attractive feature for many molecular compounds of bismuth to function as effective CT imaging agents. However, the very first requirement of an effective CT imaging agent is the high solubility of the compound, as a high concentration is necessary for an intravenous administration to produce sufficient contrast in CT imaging due to the intrinsic low contrast sensitivity of the X-ray CT modality. Additionally, the typical ionic, covalent or coordination complexes of bismuth may bind to proteins or other biomolecules or undergo transmetallation reactions with endogenous ions, such as Ca^{2+}, Mg^{2+}, Mn^{2+}, and Zn^{2+}, thus altering their pharmacological characteristics when administrated intravenously.

Use of bismuth-based nanoparticles, rather than bismuth-containing ionic or covalent compounds, to develop CT imaging agents have potential to circumvent the high-osmolality or poor-solubility limitations of its ionic or covalent compounds. It should be noted that thus far the small-molecule platform has not yet afforded clinically useful bismuth-containing CT contrast agents. On the other hand, there has been an important paradigm shift to use various nanoparticles for the development of new-generation imaging agents in different imaging modalities including X-ray CT imaging [21, 22]. Additionally, particulate-imaging agents can have a greater blood circulation half-life than the molecular counterparts, thus increasing the time window for imaging [23]. Furthermore, surface-modifications via ionic or covalent attachment of a targeting agent on nanoparticles provide possibilities to develop organ-specific imaging agents [24, 25].

3.1 *Polymer-Coated Bi_2S_3 Nanoparticles for CT Imaging*

Weissleder and coworkers were the first to explore the application of nanoparticles for CT imaging in small animals [26]. In 2006, they reported on the development of a polymer-coated Bi_2S_3 nanoparticle preparation as an injectable CT contrast agent.

They developed a two-step synthetic method to prepare homogeneous preparations more suitable for parenteral use than those previously reported. Nanocrystals of bismuth sulfide were first grown by precipitation from the reaction between bismuth citrate and sodium sulfide in the presence of a thiol surfactant. The nanocrystals were then coated with polyvinylpyrrolidone (PVP) to afford PVP-coated Bi_2S_3 nanoparticle (BPNPs).

The TEM images revealed that such BPNPs have a quasi-rectangular platelet shape with the length ranging from 10 nm to about 50 nm per side. The thickness of the crystals is 3–4 nm. The mean hydrodynamic diameter of BPNPs was measured to be 30 ± 10 nm by laser-light scattering (Fig. 1).

The CT opacity measurements showed that that a BPNP preparation of 0.55 M in bismuth concentration is equivalent in X-ray absorption to a 2.36 M iodine solution. After intravenous administration of 250 µl BPNPs (~57 µmol Bi3+) into Balb/c

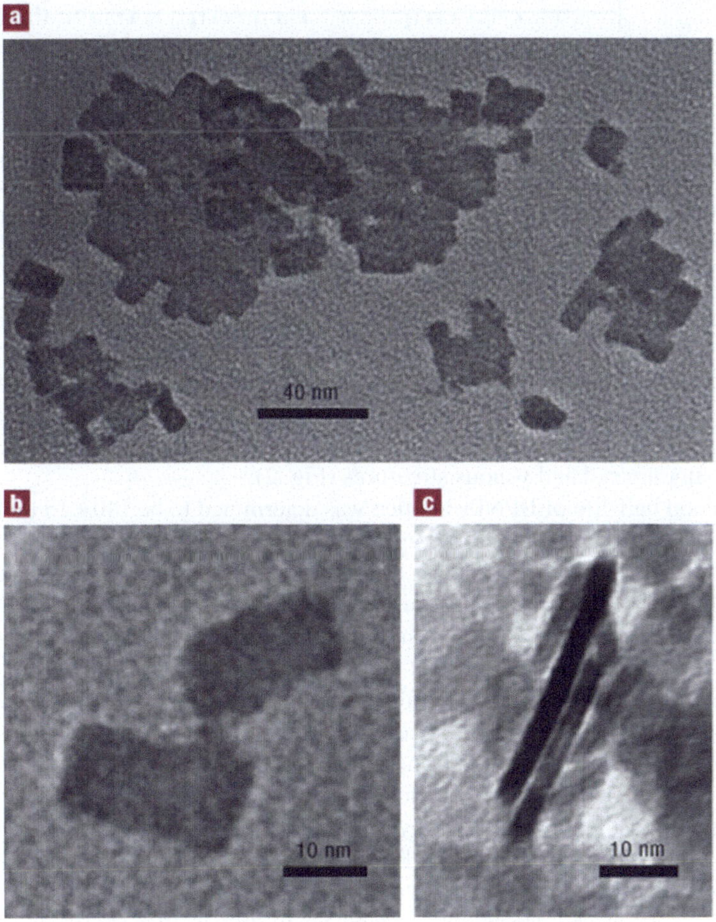

Fig. 1 BPNP characterization. TEM characterization of the Bi_2S_3 nanoparticles

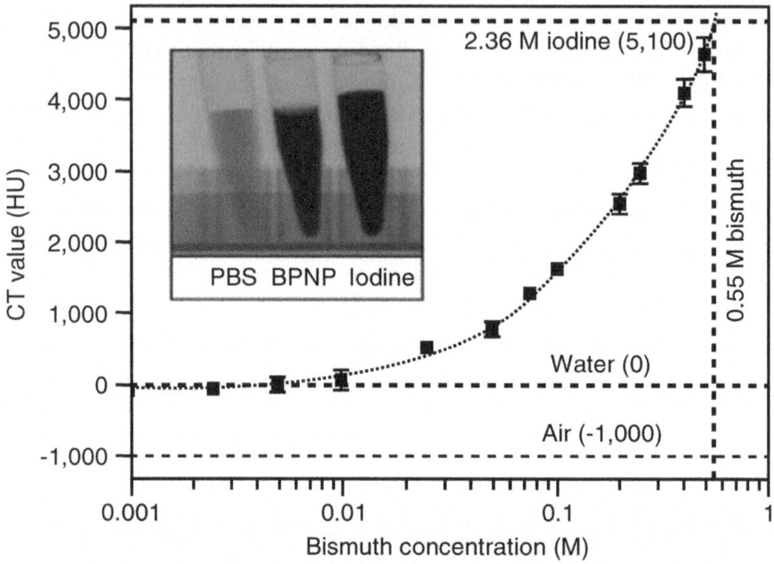

Fig. 2 X-ray absorption of BPNPs. Experimental CT opacity of bismuth solutions as a function of concentration. The horizontal dashed lines indicate the opacities of air, water and 2.36 M (300 mg × ml^{-1}) iodine contrast agent, for comparison. Inset: X-ray fluoroscopy of tubes containing PBS, 0.5 M Bi_2S_3 nanoparticle suspension (BPNPs) and 2.36 M iodine contrast agent (Iopromide). The error bars represent the standard deviation in the CT value of a set of 30 × 30 × 1 (900) voxels on a planar square aligned with the center of the sample receptacle

mice, serial CT imaging was performed to determine vascular enhancement and half-life. It was found that the blood signal increases from −27 ± 77 HU to 530 ± 150 HU at its peak. The enhancement provides clear delineation of the cardiac ventricles and all major arterial and venous structures (Fig. 2).

The blood half-life of BPNPs in mice was determined to be 140 ± 15 min, which is significantly longer than for commercial iodinated preparations (<10 min). Furthermore, their extended imaging studies at 12–24 h after intravenous administration showed that the BPNPs were distributed to organs containing phagocytic cells (liver, spleen, lymph nodes). They found that liver signal intensity increased from −22 ± 77 HU to 740 ± 210 HU at 24 h (Fig. 3d), which probably reflects uptake of BPNPs into macrophages (Kupffer cells) and hepatocytes. In the meantime, organs devoid of major phagocytic cells showed no or only a slight enhancement. Weak contrast in the kidneys and urinary tract was visible 1–2 days after the administration of BPNPs, presumably due to renal excretion. In preliminary observations, they found no residual bismuth in any organ 2 months after the administration of a single dose. In order to investigate if BPNPs could be used for lymph-node delivery to improve cancer staging, mice were injected subcutaneously with 50 μl BPNPs (~11.4 μmol Bi^{3+}) and repeatedly imaged up to 140 h after injection. Regional lymph nodes were clearly contrasted (Fig. 3).

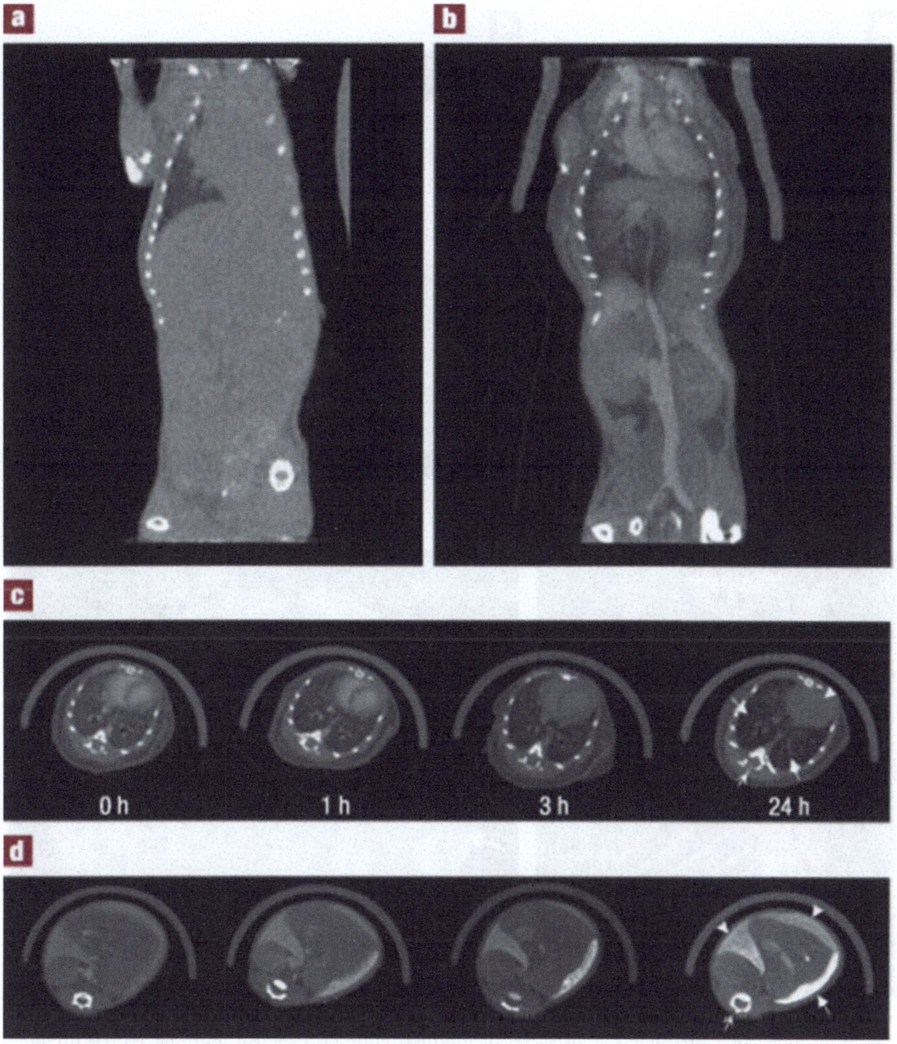

Fig. 3 Serial in vivo imaging of vasculature. (**a**, **b**) Coronal curved multiplanar reformatted reconstructions of a live Balb/c mouse before (**a**) and after (**b**) intravenous BPNP administration, showing the vascular enhancement of large vessels and the heart, and the organ delineation, achieved with the nanoparticle agent. The length of the reconstruction in **a** and **b** is 6.5 cm. (**c**, **d**) Serial CT scans of a live Balb/c mouse following tail-vein injection of 250 μl BPNP suspension (0.228 M bismuth). Series **c** shows transverse slices through the heart (*arrowhead*), the lungs (*solid arrow*), and a vertebra (*open arrow*). Series (**d**) are transverse slices including the liver (*arrowhead*), the spleen (*solid arrow*), and a vertebral body (*open arrow*). The diameter of the mouse in **c** and **d** is 1.6 cm

They also performed comparative dose-response cytotoxicity studies of BPNPs in human macrophage (U937) and hepatocyte (HepG2) cell lines (Fig. 4). For macrophages, the LD_{50} is 8 mM for free bismuth ions, and 100 mM for BPNPs,

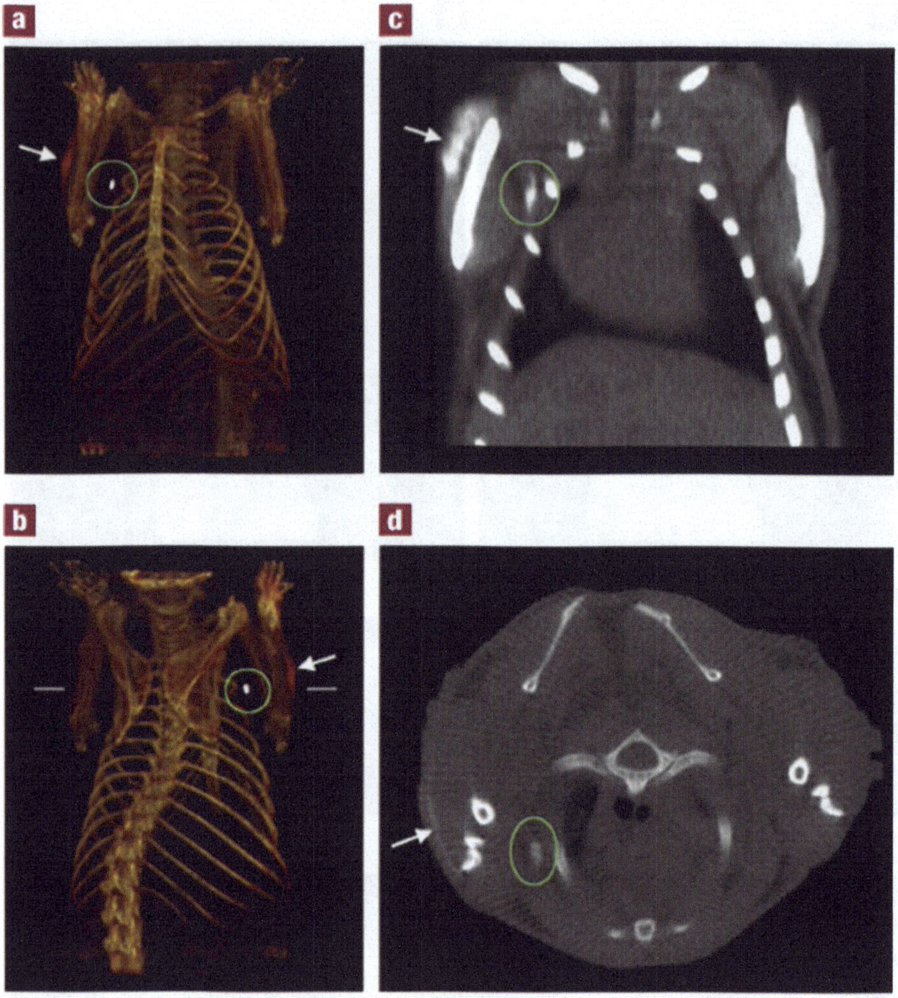

Fig. 4 Lymph-node imaging. CT imaging of a lymph node of a mouse with the BPNP imaging agent. (**a**, **b**) Three-dimensional volume renderings of the CT data set, the length of the reconstruction is 3.8 cm. (**c**) Coronal slice (length of the slice 2.3 cm). (**d**) Transverse slice at the height indicated by the horizontal lines in (**b**). The maximal diameter of the mouse 1.8 cm. The position of the lymph node under the right shoulder is indicated by the ovals, and the injection site is shown by the *arrows*. Note the lack of contrast in the corresponding contralateral (left shoulder) lymph node

suggesting that nanoparticles are at least one order of magnitude less toxic to cells than free bismuth ions. When LD_{50} was compared with the effective dose (the ED/LD_{50} ratio), BPNPs show a similar profile to that of an iodinated agent in clinical use. For hepatocytes, the LD_{50} is 5 mM for free bismuth ions, and 114 mM for BPNPs, showing a superior profile compared with the iodinated agent.

Recently, a modified method for a large-scale synthesis of Bi_2S_3 nanoparticles with remarkable size uniformity (2–3 nm) and monodispersity was reported as an attempt to address several problems associated with the previously reported synthesis of this material [27, 28]. According to these authors, the use of oleic acid (OA) in the synthesis offers a big advantage in that the carboxylic group of OA can strongly bind to Bi^{3+} ions, thus effectively preventing Bi_2S_3 nanoparticles from aggregating, as well as avoiding the reduction of Bi^{3+} ions. Additionally, the surface of such NPs can be readily modified with polymers, such as PVP, to offer improved biocompatibility, increased colloidal stability, and reduced cytotoxicity. They found that the X-ray contrast efficiency of such NPs is 1.9 times higher than that of Iobitridol. Cell viability studies on HeLa cells show a quite high viability of >80 % even at contrast media concentrations of 3 mg Bi/ml. Their in vivo imaging studies in rats show an uptake of the NPs by the liver and spleen, suggesting macrophage activation, as opposed to renal clearance, as expected for the NPs with such ultrasmall size. The nanoparticles exhibit a long blood-circulation time of several hours and are cleared from the rats within 1 month after the intravenous injection.

3.2 *Sugarcoated Elemental Bismuth Nanoparticles for CT Imaging*

Unlike the noble metals (Au or Pt), elemental bismuth nanoparticles (Bi NPs) may undergo oxidative decomposition to release Bi^{3+} ions under the physiological conditions at controlled rates for renal excretion. This feature makes Bi NPs as X-ray imaging agents highly attractive. Goforth and coworkers conducted intensive investigations into the synthesis of ultrahigh payload Bi NPs in 1,2-propanediol (PPD) using a borane reducing agent and various sugar molecules as biocompatible surface stabilizers [29]. They found that the use of PPD and glucose as biocompatible small molecule surfactants resulted in the formation of highly uniform, aqueous nanoparticles with a mean hydrodynamic diameter of 86 nm (74 ± 14 nm Bi NP cores) consisting of a thin protective PPD-glucose shell (Fig. 5).

The high ratio of the inorganic core size to the mean hydrodynamic diameter led the authors to conclude that such Bi NPs are ~64 % bismuth by volume and carry ~6 million bismuth atoms per nanoparticle. Furthermore, the Bi NPs synthesized by this method at 80 °C exhibited a high X-ray opacity in the simulated body phantom and in vitro cellular CT imaging studies. Specifically, the CT attenuation of bismuth in the form of aqueous Bi NPs, relative to that of solvated bismuth(III) nitrate ions and an aqueous iodine standard (iopamidol) indicated that Bi NPs show significantly higher X-ray attenuation relative to iodine. The authors also noted that the iodine attenuation rate decreases with increasing X-ray tube voltage, while the attenuation of bismuth(III) ions and Bi NPs was relatively insensitive to tube voltage in the range used in clinical CT scanners (Fig. 6). This observation is consistent with the fact that bismuth has a higher K-edge energy, and thus better energy match with the incident X-ray beam of higher voltage than iodine.

On the other hand, the CT imaging studies on Bi NPs incubated with HeLa cells or murine macrophage J774A.1 cell line showed a clear increase in X-ray attenuation for both cell lines. There is a more marked increase for the macrophages than

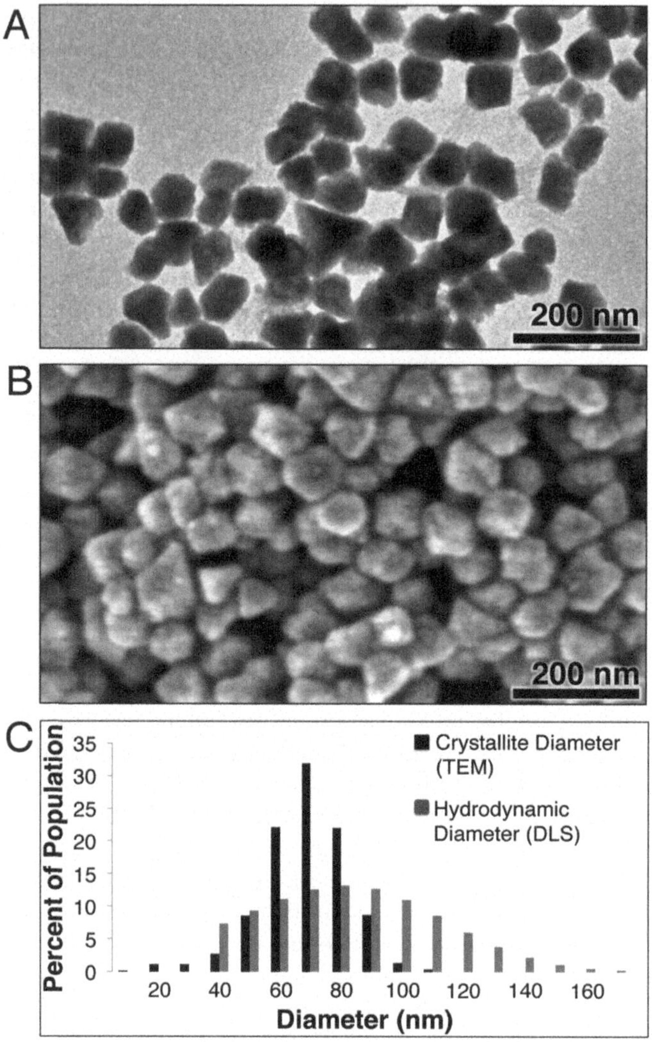

Fig. 5 Size and morphology of Bi NPs synthesized in glucose saturated PPD at 80 °C by (**a**) TEM and (**b**) SEM. (**c**) The core diameter (from TEM) and hydrodynamic diameter (measured by DLS)

for the HeLa cells, indicating the high phagocytotic activity of macrophages. In addition, the cellular internalization by both cell lines was linear with concentration, indicating a nonspecific uptake process (Fig. 7b). The electron microscopic imaging on sections of the cells confirmed the presence of NPs within the HeLa cells and fixed macrophage cells with formation of endosomal/lysosomal structures around Bi NPs (Fig. 7c, d).

The cell viability assays showed that Bi NPs are nontoxic to HeLa cells, but had an LD_{50} of 50 μg/ml (0.2 mM) for the macrophage cells. The authors stated that the observed cytotoxicity of Bi NPs in the macrophages was probably caused by the extremely high concentration of nanoparticles in the macrophages at 24 h, as supported by their CT imaging data.

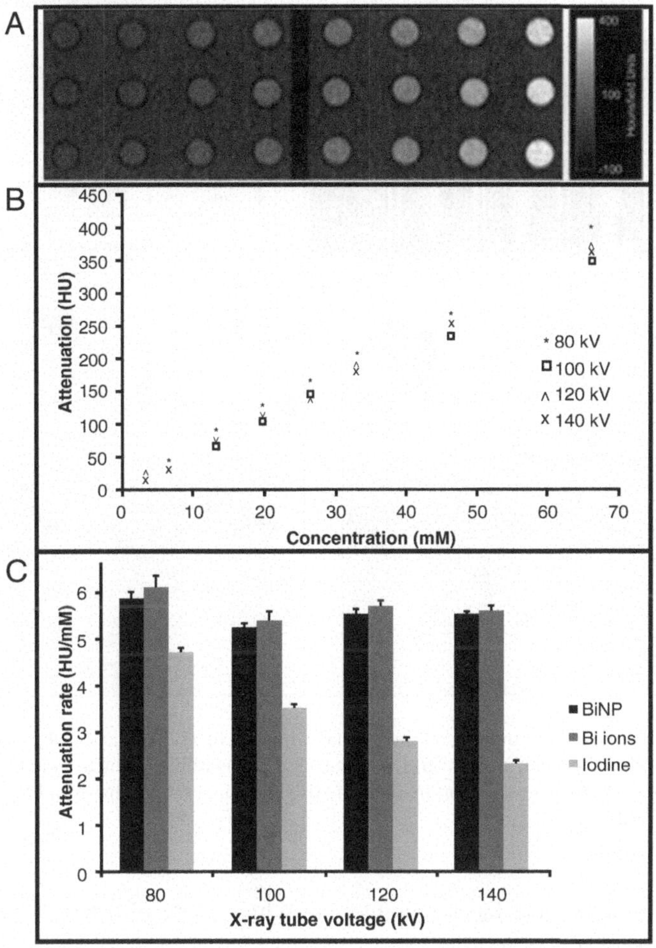

Fig. 6 CT attenuation evaluation of BiNPs. (**a**) CT image of solutions of Bi NP varying in concentration between 1 and 65 mM bismuth. (**b**) The attenuation of bismuth nanoparticles as a function of concentration and

3.3 Bismuth-Containing Coordination Polymer Nanoparticles for CT Imaging

Nanoparticles of a novel potassium bismuth ferrocyanide coordination polymer $KBi(H_2O)_2[Fe(CN)_6]\cdot H_2O$ were prepared and investigated for their potential application as an X-ray imaging agent [30]. This coordination polymer falls into the large class of the so-called Prussian blue (PB) analogues. This class of compounds have emerged as a promising structural platform for developing a variety of theranostic agents including cellular MRI probes, image-guided drug delivery vehicles for cancer treatment, to name but a few [30–33]. In this report, the authors discovered that these coordination-polymer-based nanoparticles containing Bi^{3+} ions (BiFeCN NPs)

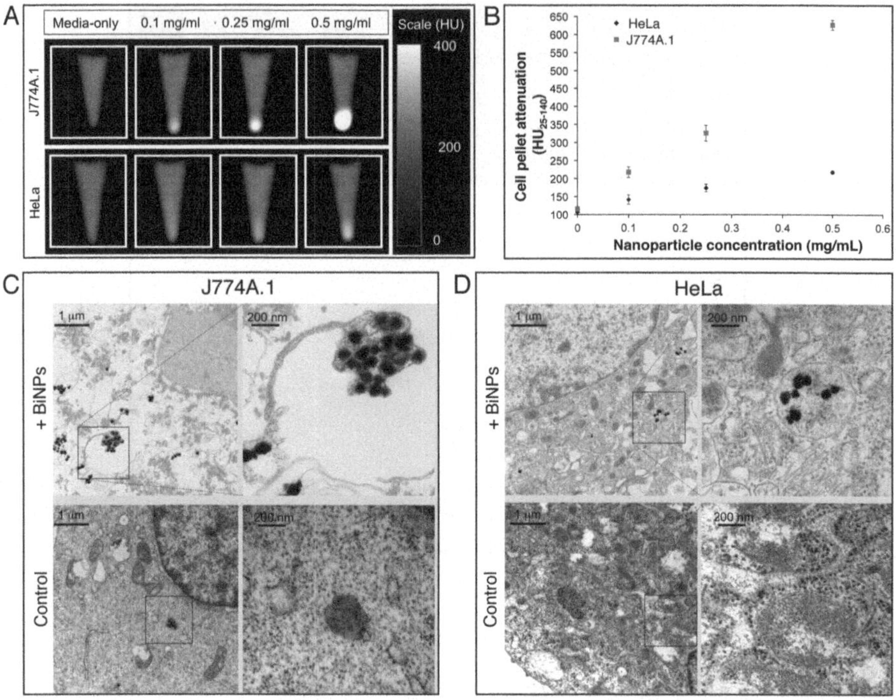

Fig. 7 Uptake of Bi NPs in model cell lines. (**a**) CT images of J774A.1 and HeLa cells incubated with Bi NPs and formed into pellets. (**b**) Quantitative analysis of the attenuation in the cell pellets. TEM shows that Bi NPs are taken up in endosomal/lysosomal compartments in both J774A.1 (**c**) and HeLa cells (**d**)

are extremely stable and biocompatible, but with a wide distribution of size from circa 10 nm to over 30 nm.

The X-ray attenuation of the BiFeCN NPs was measured to be ~2.7 times of the typical iodine-based CT contrast agent, but ~62 % of the Bi_2S_3 nanoparticles at the same molar concentration. They also showed that BiFeCN NPs can be internalized by cells via endocytosis. Cytotoxicity studies in HeLa cells showed cell viability to be over 90 % with concentrations of up to 0.83 mg Bi/ml. Additional studies on leaching of the CN^{-1} and Bi^{3+} ions from the NPs showed vary low levels of CN^{-1} and Bi^{3+} ions released to the outside solution under various conditions.

3.4 PVP-Coated Ultrasmall Nanoparticles for CT Imaging

As the US Food and Drug Administration (FDA) currently requires that all injectable contrast agents to be completely excreted from the body in a reasonable time period, renal clearance provides the most viable pathway to achieve this goal [31, 32]. For

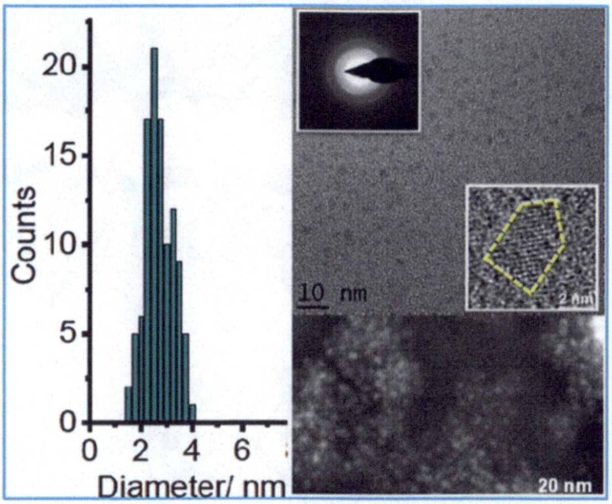

Fig. 8 Histogram of particle size (*left*), TEM image (*top right*) with HRTEM and SAED insets, and Z-contrast STEM image (*bottom right*) of BiOI NPs

this reason, ultrasmall bismuth oxyiodide BiOI nanoparticles (BiOI NPs) were investigated as a potentially renal clearable particulate X-ray imaging agent [33]. Owing to a unique decelerated hydrolytic procedure in aqueous solution, ultrasmall PVP-coated BiOI NPs with a narrow size distribution, i.e., 2.8 ± 0.5 nm, were synthesis by a one-pot procedure (Fig. 8). The essence of this new method is the controlled hydrolysis of the hydrated Bi^{3+} ion under the acidic conditions (i.e., pH = 2.75) to produce the BiO^+ ion that in turn reacts with the I^- ion to form BiOI. It is known that the following reaction is the rate-limiting step in the formation of BiO^+

$$\left[\mathrm{Bi}\left(\mathrm{H_2O}\right)_n\right]^{3+}(\mathrm{aq}) + \mathrm{H_2O}(\mathrm{l}) \rightarrow \left[\mathrm{Bi}\left(\mathrm{H_2O}\right)_{n-1}\left(\mathrm{OH}\right)\right]^{2+}(\mathrm{aq}) + \mathrm{H}^+(\mathrm{aq}) \qquad (3)$$

In the presence of a polymer coating agent, such a hydrolysis reaction carried out in an acidic solution is so slow that it becomes the rate-limiting step for the nucleation of BiOI NPs as well, thus leading to the formation of ultrasmall and monodispersed BiOINPs.

Conceptually, BiOI can be viewed as a compound consisting of the BiO^+ and I^- ions. This notion has led to the development of a controlled hydrolytic procedure for the preparation of biocompatible ultrasmall BiOI NPs, but BiOI is a covalent solid-state compound with a layered structure and a very low solubility-product constant ($3.34 \pm 9 \times 10^{-11}$ mol^2 dm^{-6}), which is a desirable feature to ensure low osmolality if such NPs are to be developed as an intravenously injectable CT contrast agent (Fig. 9).

The X-ray attenuation of PVP-coated BiOI NPs was measured using six different concentrations of NPs dispersed in aqueous media with a microCT scanner operating at 75 kVp and 110 μA. It was found that the X-ray attenuation efficiency of this CT agent has the value of *ca.* 20 HU/mM. In comparison, the PEGylated gold NPs gave the value of 5.3 HU/mM (i.e., 1.27 M of PEG-coated Au NPs has the CT value of 6690 HU), while this value for PVP-coated Bi_2S_3 was reported to be 9.3 HU/mM. Similarly, the X-ray attenuation efficiency was measured and reported for a

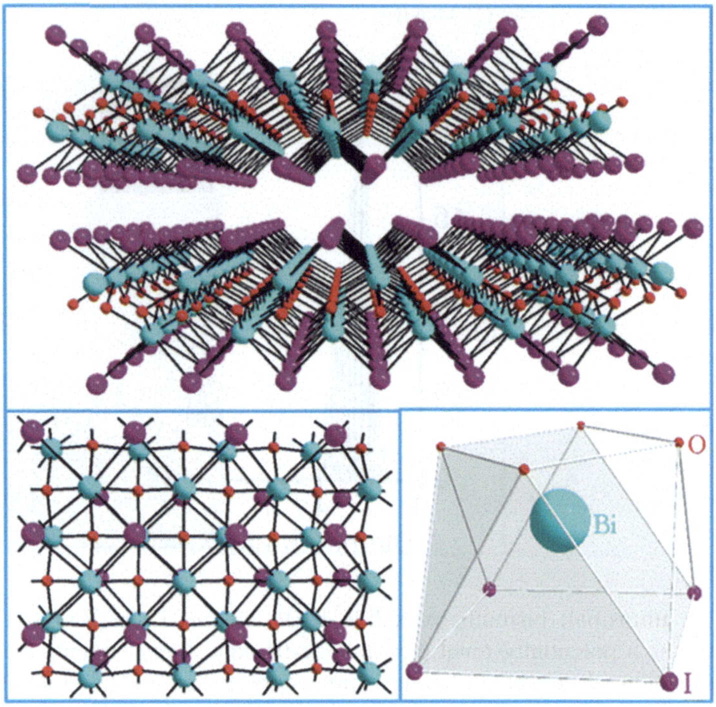

Fig. 9 X-ray structure of BiOI viewed along the *b* axis showing the layers stacked in the direction perpendicular to the *c* axis (*top*), the view of a single layer (*bottom left*), and the coordination geometry of the Bi atom (*bottom right*)

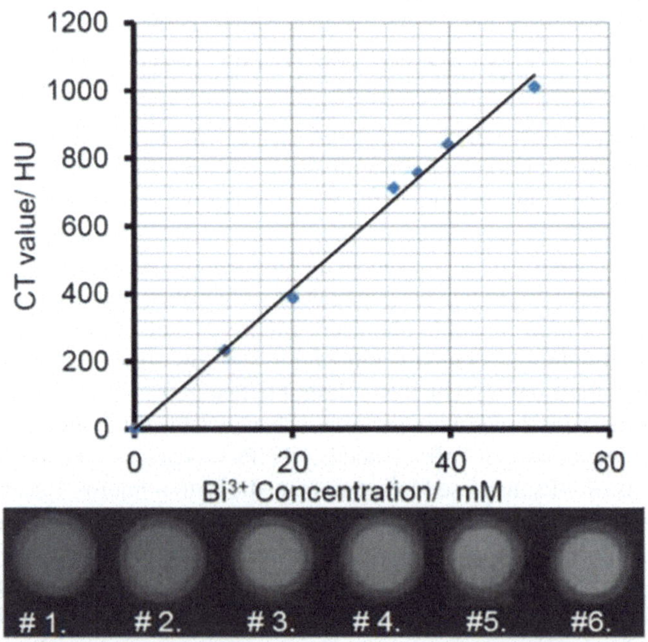

Fig. 10 The CT value (HH) as a function of Bi^{3+} concentration in aqueous solution (*top*), and phantom images of the PVP-coated BiOI NPs for (1) 5.07, (2) 12.7, (3) 25.4, (4) 30.4, (5) 38.1, and (6) 50.7 mM (*top*)

polymer-coated TaO_x NP system to be 6.0 HU/mM. Therefore, the PVP-coated BiOI NPs appear to have the highest value on the molarity basis owing to the combined X-ray attenuation effect of bismuth and iodine. However, the molar mass of BiOI of 351.88 g/mol is also the highest among the above-mentioned inorganic substances, suggesting that when a given molar concentration of each contrast agent is to be injected into the body, the amount of mass required for BiOI NPs would turn out be the largest as well.

To develop BiOI NPs as an intravenous X-ray imaging agent, the authors examined in vitro cytotoxicity in HeLa cells. They found that after 24-h incubation with the highest concentration that can be prepared, 4.15 mg/ml BiOI NPs, cell viability was *ca.* $92 \pm 0.8\,\%$, indicating that the PVP-coated BiOI NPs exhibit no significant cytotoxicity.

4 Concluding Remarks and Outlook

Thus far, all the bismuth-based nanoparticle systems investigated for X-ray imaging applications exhibit superb X-ray attenuation efficiencies. None of them show discernable cytotoxicity at the concentrations required to obtain satisfactory contrast enhancement. Although not an abundant element, bulk bismuth is approximately $19/kg compared with $35,000/kg for gold. Currently, Au nanoparticles are the frontrunner candidate for clinical applications in X-ray CT imaging. Given its scarcity and price, gold may not defeat bismuth in finding its way to CT scanning rooms. In this regard, bismuth-based nanoparticles have a bright future for X-ray CT imaging applications.

References

1. Kalender WA. X-ray computed tomography. Phys Med Biol. 2006;51(13):R29–43.
2. Sandborg M, Christoffersson JO, Carlsson GA, Almén T, Dance DR. The physical performance of different x-ray contrast agents: calculations using a Monte Carlo model of the imaging chain. Phys Med Biol. 1995;40(7):1209–24.
3. Lumbroso P, Dick CE. X-ray attenuation properties of radiographic contrast media. Med Phys. 1987;14(5):752–8.
4. Frenzel T, Lawaczeck R, Taupitz M, Jost G, Lohrke J, Sieber MA, Pietsch H. Contrast media for x-ray and magnetic resonance imaging development, current status and future perspectives. Invest Radiol. 2015;50(9):6710–8.
5. Yu SB, Watson AD. Metal-based x-ray contrast media. Chem Rev. 1999;99(9):2353–78.
6. Havron A, Davis MA, Selter SE, Paskins-Hurlburt AJ, Hessel SJ. Heavy metal particulate contrast materials for computed tomography of the liver. J Comput Assist Tomogr. 1980;4(5):642–8.
7. Seltzer SE, Adams DF, Davis MA, Hessel SJ, Havron A, Judy PF, Paskins-Hurlburt AJ, Hollenberg NK. Hepatic contrast agents for computed tomography: high atomic number particulate material. J Comput Assist Tomogr. 1981;5(3):370–4.
8. Abbatt JD. History of the use and toxicity of thorotrast. Environ Res. 1979;18(1):6–12.
9. Janower ML, Miettinen OS, Flynn MJ. Effects of long-term thorotrast exposure. Radiology. 1972;103(1):13–20.

10. Becker N, Liebermann D, Wesch H, Van Kaick G. Mortality among thorotrast-exposed patients and an unexposed comparison group in the German Thorotrast study. Eur J Cancer. 2008;44(9):1259–68.
11. Grainger RG. Intravascular contrast media-the past, the present and the future. Br J Radiol. 1982;55(649):1–18.
12. McClennan BL. Ionic and nonionic iodinated contrast media: evolution and strategies for use. Am J Roentgenol. 1990;155(2):225–33.
13. Lusic H, Grinstaff MW. X-ray-computed tomography contrast agents. Chem Rev. 2013;113(3):1641–66.
14. Krause W. Delivery of diagnostic agents in computed tomography. Adv Drug Deliv Rev. 1999;37(1-3):159–73.
15. Wiberg E, Holleman AF, Wiberg N (2001) Inorganic chemistry. Academic Press.
16. de Marcillac P, Coron N, Dambier G, Leblanc J, Moalic JP. Experimental detection of alpha-particles from the radioactive decay of natural bismuth. Nature. 2003;422(6934):876–8.
17. Miersch L, Rueffer T, Lang H, Schulze S, Hietschold M, Zahn D, Mehring M. A novel water-soluble hexanuclear bismuth oxido cluster-synthesis, structure and complexation with polyacrylate. Eur J Inorg Chem. 2010;30:4763–9.
18. Briand GG, Burford N. Bismuth compounds and preparations with biological or medicinal relevance. Chem Rev. 1999;99(9):2601–58.
19. Sadler PJ, Li HY, Sun HZ. Coordination chemistry of metals in medicine: target sites for bismuth. Coord Chem Rev. 1999;185-6:689-09.
20. Maile FJ, Pfaff G, Reynders P. Effect pigments: past, present and future. Prog Org Coat. 2005;54(3):150–63.
21. Hainfeld JF, Slatkin DN, Focella TM, Smilowitz HM. Gold nanoparticles: a new x-ray contrast agent. Br J Radiol. 2006;79(939):248–53.
22. Della Rocca J, Liu D, Lin W. Nanoscale metal-organic frameworks for biomedical imaging and drug delivery. Acc Chem Res. 2011;44(10):957–68.
23. Lee N, Choi SH, Hyeon T. Nano-sized CT contrast agents. Adv Mater. 2013;25(19):2641–60.
24. Nune SK, Gunda P, Thallapally PK, Lin YY, Forrest ML, Berkland CJ. Nanoparticles for biomedical imaging. Expert Opin Drug Deliv. 2009;6(11):1175–94.
25. Rosi NL, Mirkin CA. Nanostructures in biodiagnostics. Chem Rev. 2005;105(4):1547–62.
26. Rabin O, Manuel Perez J, Grimm J, Wojtkiewicz G, Weissleder R. An X-ray computed tomography imaging agent based on long-circulating bismuth sulphide nanoparticles. Nat Mater. 2006;5(2):118–22.
27. Ai K, Liu Y, Liu J, Yuan Q, He Y, Lu L. Large-scale synthesis of Bi_2S_3 nanodots as a contrast agent for in vivo X-ray computed tomography imaging. Adv Mater. 2011;23(42):4886–91.
28. Liu Y, Ai K, Lu L. Nanoparticulate X-ray computed tomography contrast agents: from design validation to in vivo applications. Acc Chem Res. 2012;45(10):1817–27.
29. Brown AL, Naha PC, Benavides-Montes V, Litt HI, Goforth AM, Cormode DP. Synthesis, x-ray opacity, and biological compatibility of ultra-high payload elemental bismuth nanoparticle x-ray contrast agents. Chem Mater. 2014;26(7):2266–74.
30. Perera VS, Hao J, Gao M, Gough M, Zavalij PY, Flask C, Basilion JP, Huang SD. Nanoparticles of the novel coordination polymer $KBi(H_2O)_2[Fe(CN)_6]\cdot H_2O$ as a potential contrast agent for computed tomography. Inorg Chem. 2011;50(17):7910–2.
31. Christiansen C. X-ray contrast media-an overview. Toxicology. 2005;209(2):185–7.
32. Liu J, Yu M, Zhou C, Zheng J. Renal clearable inorganic nanoparticles: a new frontier of bionanotechnology. Mater Today. 2013;16(12):477–86.
33. Kandanapitiye MS, Gao M, Molter J, Flask CA, Huang SD. Synthesis, characterization, and X-ray attenuation properties of ultrasmall BiOI nanoparticles: toward renal clearable particulate CT contrast agents. Inorg Chem. 2014;53(19):10189–94.

Hyperspectral Microscopy and Cellular Array Imaging Using Colloidal Quantum Dots

John X.J. Zhang and Kazunori Hoshino

1 Introduction

Quantum dots have unique characteristics suitable for miniaturized imaging systems, especially for integrated hyperspectral microscopy in total analytical systems. The notable advantageous characteristics are: (1) capability of multicolor emission from visible to IR wavelengths, (2) various photo/electrical excitation schemes available for different applications, and (3) compatibility with the microfabrication technology and widely available low cost prototyping techniques.

In this chapter, we first explain basic characteristics of quantum dots and discuss how they are different from conventional organic fluorescent markers. Notable applications of quantum dots in fluorescence imaging and sensing, including the fluorescence resonance energy transfer (FRET)-based measurement, are described. Following sections introduce our recent studies on using colloidal QDs as light sources for absorption hyperspectral microscopy and fluorescence microscopy. We use gold nanorods and nanospheres as biomarkers that work complementarily to QD-based excitation light sources. Hyperspectral analysis of optical signals is presented both theoretically and experimentally. Finally, we describe integration of electrically excited colloidal QDs on silicon substrate. Electrically pumped QD light sources are integrated on silicon microdevices such as microchannels and scanning probes which open the possibilities of a wide variety of nano-optomechanical microscopic systems for imaging applications.

J.X.J. Zhang (✉)
Thayer School of Engineering, Dartmouth College, 14 Engineering Drive, Hanover, NH 03755, USA
e-mail: john.zhang@dartmouth.edu

K. Hoshino
Department of Biomedical Engineering, University of Connecticut, 260 Glenbrook Rd Unit 3247, Storrs, CT 06269, USA

J.W.M. Bulte, M.M.J. Modo (eds.), *Design and Applications of Nanoparticles in Biomedical Imaging*, DOI 10.1007/978-3-319-42169-8_21

2 Fundamentals of Quantum Dots

2.1 Colloidal Quantum Dots

Colloidal quantum dots (QDs) are nanometer scale semiconductor crystals. Due to the quantum confinement, the energy levels of a single QD are discrete and the band gap is highly related to the size and the shape [1]. The confinement becomes significant when the size of the crystal is smaller than the exciton Bohr radius. Colloidal QDs are typically between 5 and 50 nm, and are made of several materials including ZnS, CdS, ZnSe, CdTe, and PbSe [2]. QDs may also be fabricated on bulk semiconductor devices by lithographic approaches [3, 4]. In this article, we discuss chemically synthesized colloidal QDs, and their applications in biomedical imaging.

The most important feature of colloidal QDs is their size-dependent fluorescence emission, which makes colloidal QDs suitable biomarkers and light sources for hyperspectral microscopy. Desirable emission peak wavelength can be obtained by tailoring the diameter of the synthesized crystals. Emission wavelengths ranging from the UV to the infrared have been reported [1]. The mechanism of fluorescence emission is explained with an energy diagram (Fig. 1a). A QD has a conduction band and a valence band with the band gap just as a typical bulk semiconductor. The significant difference from a bulk semiconductor is that the energy levels in the conduction and valence bands are quantized because of the discrete molecular orbitals.

The molecular orbitals in the valence band are bonding orbitals, which are full. The orbitals in the conduction band are antibonding orbitals and are vacant. The energy level difference between the highest occupied molecular orbital (HOMO, the top of valence band) and the lowest unoccupied molecular orbital (LUMO, bottom of conduction band) defines the band gap of the colloidal QD. When an electron in HOMO is excited, it moves to an orbital in the conduction band, leaving a hole in HOMO (Fig. 1a left). The electron eventually goes back to HOMO and fluorescence is observed as a result of the transition, which can be also considered as the recombination of the electron and the hole (Fig. 1a right). The emission wavelength is defined by the band gap, or the difference between the LUMO and the HOMO.

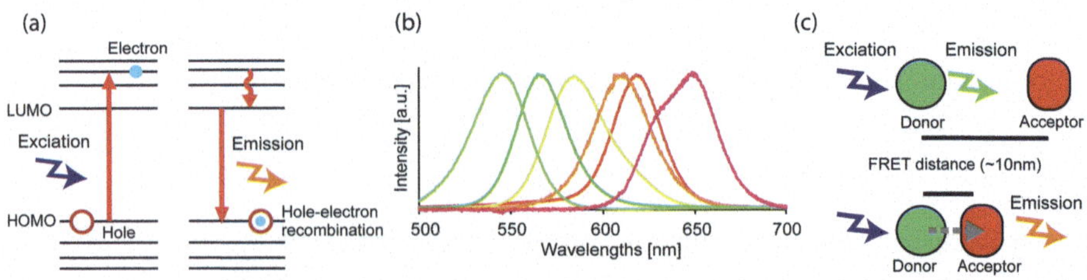

Fig. 1 Fluorescence excitation and emission from quantum dots. (**a**) Energy diagrams of a bulk semiconductor and a quantum dot. (**b**) Fluorescence spectra measured from different types of quantum dots. Reprinted from Ref. [12], copyright (2014), with permission from Elsevier. (**c**) Fluorescence resonance energy transfer (FRET). When the distance between the donor and the acceptor is close enough, FRET occurs and emission from the acceptor is observed

The band gap is dependent on the size of QDs. Smaller quantum dots have larger band gaps because the quantum confinement effect is stronger. Because of this size effect, the emission wavelengths of QDs can be tailored by controlling the diameters of the particles. The emission wavelength becomes shorter for smaller particles because the emission energy is larger. Emission spectra measured from different types of colloidal QDs (Fig. 1b).

For colloidal quantum dots, the mechanism of fluorescence emission is defined by this simple structure of the energy levels, and less affected by the excitation wavelengths. Colloidal QDs can be excited by any wavelength shorter than the emission wavelengths. Typically, a UV light will excite all types of colloidal QDs with visible emission wavelengths. On the other hand, most organic fluorescent materials follow the mirror image rule, where and absorption spectrum is similar to a mirror image of the emission spectrum. An excitation light for a certain fluorescent dye may not excite other fluorescent materials. This difference is important in designing fluorescent optical systems using either colloidal QDs or organic fluorescent materials.

QDs with a core–shell structure is often used to obtain stability. One example is QDs with a CdSe core capped with a 1–2 monolayers of ZnS. The shell is used to passivate the core surface and protect it from surrounding solutions. The structure prevents crystal defects which lead to unwanted electron or hole traps, and substantially improves the photoluminescence intensity and stability [5]. Colloidal QDs are typically prepared in organic solvent. They are usually capped with hydrophilic surface ligands when used in an aqueous solution or further chemical attachment of functional biomolecules is needed.

2.2 Quantum Dots as a Fluorescence Marker

QDs are alternatives to organic fluorescent dyes for biological imaging and sensing. They can be functionalized with biomolecules to label specific biomaterials. A notable early study for immunofluorescence imaging with quantum dots used breast cancer marker Her2 to selectively stain fixed and live cancer cells [6].

Characteristics of organic dyes and QDs have been compared in [7], with important differences for hyperspectral imaging being summarized in Table 1. Along with the tunable emission wavelengths, the photo stability is also a significant advantage of QDs. For example in [6], the nuclei and microtubules of mouse 3 T3 fibroblast cells are stained with either colloidal QDs or the organic dye Alexa 488 with fluorescence intensities being compared by taking time lapse images. After 180 s, the fluorescence intensity of QDs were almost unchanged, while the normalized intensity of the 488 dye became about 50 % because of the photobleaching induced by continuous fluorescence excitation. One of the notable difficulties associated with the use of QDs as a fluorescent marker is the size. QDs behave as nano-colloids rather than as molecules, and the intracellular delivery of QDs is still challenging [7]. Methodologies to deliver QDs into cells for intracellular imaging are behind those developed for organic dyes.

Table 1 Comparison of organic dyes and QDs (adapted from Ref. [7])

Property	Organic dye	QD
Absorption spectra	Discrete. FWHM 35–100 nm	Steady increase toward UV with a single peak
Emission spectra	Asymmetric, often with a tail to long wavelength side	Symmetric Gaussian distribution. FWHM 30–90 nm
Quantum yield	0.5–1.0 (visible), 0.05–0.25 (NIR)	0.1–0.8 (visible), 0.2–0.7 (NIR)
Size	~0.5 nm molecule	6–60 nm in hydrodynamic diameter

2.3 Fluorescence Resonance Energy Transfer (FRET)

QDs can be used as a light source for fluorescence excitation. One of the applications where QDs are used as a excitation source is sensors based on fluorescence resonance energy transfer (FRET). FRET is a nonradiative energy transfer between the excitation source (donor) and the fluorescence molecule (acceptor). FRET occurs when donor-to-acceptor separation is at distances typically smaller than 10 nm (Fig. 1c). Because of this nanoscale sensitivity to distance, FRET can be used to detect changes in distances at the nanometer scale. Medintz et al. developed a chemical sensor that utilized FRET to detect changes in the distances between molecules [8]. The sensor molecules containing a QD, fluorescent molecules and maltose-binding molecules that bind the QD and the fluorescent molecules. Because the distance between QD and the fluorescent molecules are closer than the FRET distances, the emission from the QD is quenched by the fluorescent molecules and not observed in the initial status. When maltose is added, a maltose-binding protein takes maltose molecules and releases fluorescent molecules from the QD sensor molecules, resulting in the increase of QD emission. Zhang et al. developed a single QD-based DNA sensor [9]. A QD molecule is linked to DNA probes to capture DNA targets. When a target strand is sandwiched by a dye-labeled reporter strand (acceptor) and the DNA probes on the QD (donor), molecules form a pair of FRET donor and acceptor, and the excited reporter fluorescent dye emits fluorescence to be observed. In [10], Hohng et al. developed a QD-dye molecule (cy5) pair using a molecular junction composed of four DNA double helices. They measured FRET signals induced by conformational changes of individual molecules.

3 Quantum Dots as a Light Source for Hyperspectral Microscopy

In this section, we describe our recent studies that utilized colloidal QDs as light sources for microscopic spectroscopy. In [11] and [12], we used photo-excited QD light sources for absorption spectroscopy and fluorescence spectroscopy, respectively. In the following Sects. 4.1 and 4.2, we describe our unique efforts to use electrically excited colloidal QDs in microelectromechanical systems (MEMS)-based microscopes.

3.1 Colloidal QDs for Absorption Microscopy

The tunable emission wavelengths described in Sect. 2 make colloidal QDs an excellent light source for hyperspectral microscopy. We have used an array of QD light sources to construct a uniform, solid state swept light source that covers the visible–IR wavelength range. In the systems we reported in [11], QDs are excited by UV lights from LEDs. Figure 2a shows a photograph of QDs packed in a polydimethylsiloxane (PDMS) enclosure that is replicated from a plastic mold fabricated by rapid 3D printing. There is an array of miniature wells, each of which contains a single color of QDs. Figure 2b illustrates the design of the light source. The central concave-shaped circular part with multiple dimples transmits and scatters emission from the QDs. Digital on–off control of arrayed LEDs selectively

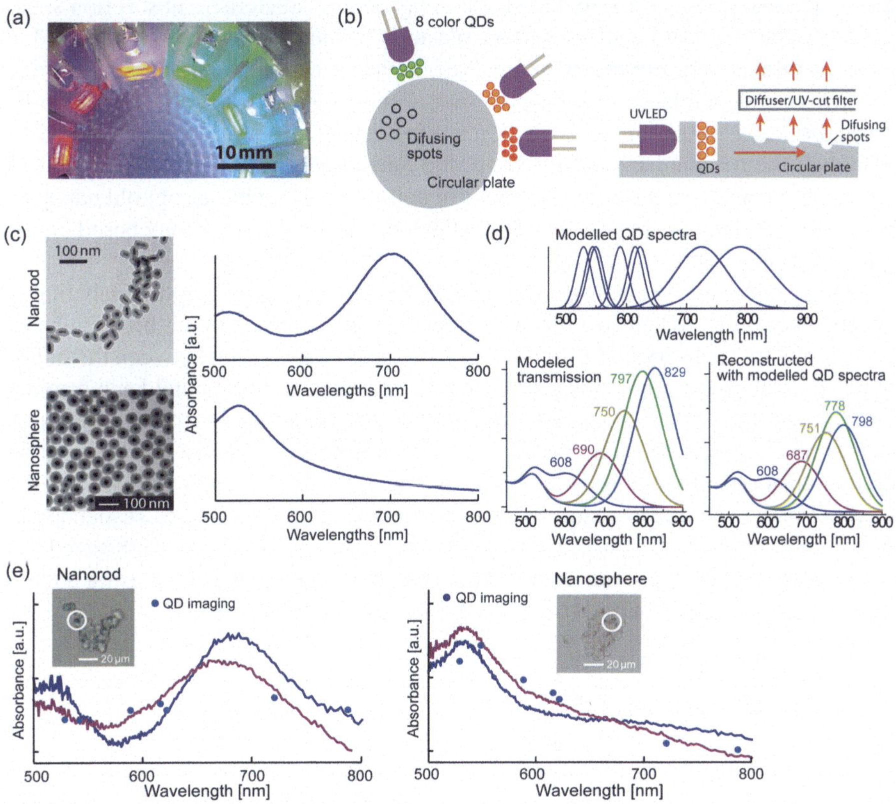

Fig. 2 Absorption spectroscopy with the QD swept light source. (**a**) Photograph of QD-based swept light source (**b**) Schematic of light source. Emission of 520–800 nm can be chosen by digital on–off control of the excitation UVLEDs. (**c**) TEM micrographs along with optical transmission spectra of silica coated (*top*) nanorods and (*bottom*) nanospheres. (**d**) Using the modeled nonuniformly distributed QD spectra (*top*), transmission spectra of different types of gold nanorods (*bottom left*) were reconstructed (*bottom right*). (**e**) Absorption spectra measured from the reference hyperspectral images and the QD swept light source for macrophage cells stained (*left*) with nanorods and (*right*) with nanospheres

excites QDs with desired color. UV-LEDs with the emission peak at 380 nm were used. Emission wavelengths ranging from 520 to 800 nm were obtained by 8 colors of CdSe/ZnS and CdTe/ZnS colloidal quantum dots. A top plate consisting of a diffuser and a 400 nm long-pass filter is placed on top of the concave surface to cut off the excitation UV lights. Typical hyperspectral microscopes employ mechanically controlled color wheels or mechanical stages that make the entire system more expensive. The QD-based swept light source does not contain any moving part, and is simply controlled by LabVIEW™ based digital switches to turn on/off the UVLEDs. It can also create any linear combination of the eight colors by changing duties ratios of excitation UVLEDs.

3.1.1 Gold Nanorods and Nanospheres as Contrast Agents

Metal nanoparticles have size and shape-dependent characteristic absorption spectra due to the effect of localized surface plasmon resonance, and they can be used in hyperspectral absorption spectroscopy [13]. In the case of gold nanorods, one of the two peaks is dependent on the aspect ratio (length-to-diameter ratio), and can be tuned between ~600 nm and ~800 nm by controlled chemical synthesis [14]. Although the mechanisms are different, the tunable absorption peaks of nanorods are complementary to the emission peaks of colloidal QDs. The use of gold nanoparticles as optical markers gives an edge to the hyperspectral microscopy based on the QD swept light source.

One problem associated with the gold nanoparticles used for optical labeling is plasmon coupling. When two metal particles are close to each other, broadening of optical spectra is observed due to the electromagnetic interaction between them. For hyperspectral microscopy, it is crucial to reduce optical coupling and keep original optical properties of nanorods. We used gold nanoparticles that are coated by silica-shells, which successfully reduced plasmon coupling [15]. Figure 2c shows TEM photographs of gold nanorods and nanospheres (20 nm diameter) we used for the measurement. The transmission spectra of nanorods and nanospheres suspended in buffer solutions are also shown in Fig. 2c. The nanospheres were synthesized via citrate reduction of chloroauric acid. The narnorods were synthesized using a reaction described by Jana et al. [16] and Nikoobakht et al. [14].

3.1.2 Theoretical Analysis

We first studied the spectral resolution of the proposed system based on a theoretical analysis. In practice, the emission peak of colloidal QDs may be shifted from the nominal or designed value by about ±5 nm. In the model spectra of QD emission shown in Fig. 2d top, we incorporated errors we found in commercially available colloidal quantum dots (Life Technologies QDot®). Figure 2d bottom left shows modeled absorption curves of nanorods built by fitting a linear combination of Gaussian functions to the curves shown in Nikoobakht et al. [8]. Each curve is built from a fixed peak at 520 nm and an additional peak ranging from 608 to 829 nm.

We simulated the hyperspectral measurement in the following way: Transmitted intensities of each color of QDs were calculated as a product of the emission spectrum and the absorption spectrum. For each type of nanorod suspension, transmitted intensities of eight colors of QDs were calculated. Using the 8 values and the QD spectra, the transmission spectrum of the nanorod can be reconstructed by curve fitting. The curves reconstructed based on the modeled QD spectra is shown in Fig. 2d bottom right. The results suggest that even with errors of about 5°, the swept light source can successfully resolve five different types of nanorods.

3.1.3 Experimental Results

Hyperspectral transmission imaging was conducted by using macrophage cells labeled with nanorods and nanospheres. Macrophage cells were incubated in media containing nanorods or nanospheres for 18 h and allowed to uptake nanoparticles. Cell suspensions are centrifuged and labeled cells are collected as pellets to be washed and fixed on a glass slide using Fluoromount. RGB color images of macrophage cells labeled with nanorods and nanospheres were taken with a 12-bit color mosaic CCD camera (SPOT Pursuit XS, Diagnostic Instruments) and shown in panels of Fig. 2e left and right, respectively. Nanorods stain cells blue, while nanospheres show pink colors. As a reference hyperspectral analysis, the same regions of the cells are imaged using a PARISS spectral imager (Lightform Inc.). The spectra measured from circled areas of the RGB images are shown in Fig. 2e left and right for the nanorod and nanosphere, respectively. The reference spectra were measured with a 20×0.5 NA objective and 100 W halogen light source. The lamp spectrum was obtained from the part of the microscope slide that did not have cells, and was used to normalize the hyperspectral images. For each image, spectra measured for two neighboring areas of 2×3 pixels are shown to indicate spatial deviation typically found in hyperspectral imaging. Curves show the same characteristic as the measurements from the particle suspensions in Fig. 2c.

The QD swept light source was then used to measured absorption characteristics of the same cell samples. Eight images illuminated by eight colors of the QDs were taken with a monochromatic cooled CCD (Pixis 400, Princeton instruments) with the wavelength sensitivity range of 450–1000 nm. The light intensity obtained from the part of the image that did not have cells was used as the denominator, and the absorbance for each LED was calculated for the spot indicated by the circle. The results were plotted as dots in Fig. 2e left and right for nanorods and nanospheres, respectively. Measurements with the QD light source fit within the range of deviation with the curves from the commercial hyperspectral microscope. QD imaging method clearly showed characteristic peaks of both nanorods and nanospheres.

We have shown that the use of QDs allows for low-cost hyperspectral imaging of biosamples with a simple standard microscope configuration. Proper choice of different gold nanoparticles, such as nanorods and nanospheres, will suitably fit absorption-based analysis using QD light multicolor sources. The use of QDs and gold nanoparticles demonstrated hyperspectral imaging for multi-biomarkers recognition at cellular level.

3.2 Colloidal QDs for Fluorescence Excitation

Figure 3a shows another type of QD light source we developed for fluorescence excitation. We adapted the microcontact printing technique reported in Ref. [17] to pattern CdSe/ZnS core–shell QDs onto a glass slide. An UV light (emission peak 365 nm) from a high-power LED (200 mW) is focused and introduced inside a glass slide which works as a wave guide. The patterned QDs are excited in an evanescent field on the surface of the glass slide induced by the total internal reflection of the transmitted UV light. When excited, the QDs work as an illumination source for microscopic imaging. The light source can be used in a standard microscope to add the functions of absorption and fluorescence imaging. The intensity of excitation observed from the microscope is negligible compared to the QD fluorescence, because the UV evanescent field on the glass slide decays in a sub-wavelength distance from the surface and does not transmit energy in the far-field. Emission wavelengths ranging from 520 to 650 nm were tested.

3.2.1 Experimental Setup

The size and the design of patterned QDs can be chosen from millimeter scale to single molecular order [18], and the pattern can be easily tailored to fit lab-on-a-chip cell systems. We utilized low cost printing-based techniques to construct the array of QD light sources, microchambers, and optical filters. Important potential applications for microdevice based systems include real-time cell culture and analysis under controlled microenvironments [19] and immunofluorescence-based detection and analysis of biomarkers. Hyperspectral imaging based on multiple-biomarker Immunofluorescence enables precise identification and quantification of cells [20]. The QD light source fits well with arrayed micro cell culture chambers, providing a suitable design for excitation sources used in such multicolor fluorescence microscopy. Here we used the light source for arrayed microanalytical systems for cell culture, analysis, and imaging. The efficacy of our light source was tested through fluorescence imaging of cancer. Two types of cancer cells (MDA-MB 435 and SKBR3) were cultured in a microwell array and immunofluorescence imaging was conducted to quantify the disease specific protein expression.

Fig. 3 Fluorescence imaging with the QD light source. Reprinted from Ref. [12], copyright (2014), with permission from Elsevier. (**a**) Fluorescence excitation setup. (**b**) Schematic of the experimental setup. The emission from the light source QD (QD540) excites the sample QD (QD620). (**c**) (*Left*) Transmission spectra of the cyan filter and the red (magenta + yellow) filter, which were used as the excitation filter and the emission filter, respectively. (*Middle*) Emission spectra of the QD light source (QD540) before and after the excitation filter. (*Right*) Emission from QD620 before and after the emission filter. (**d**) Bright field (*left*), reference fluorescence (*middle*), and QD-excited fluorescence (*right*) images of SKBR3 cells (*top*) and MDA MB 435 cells (*bottom*)

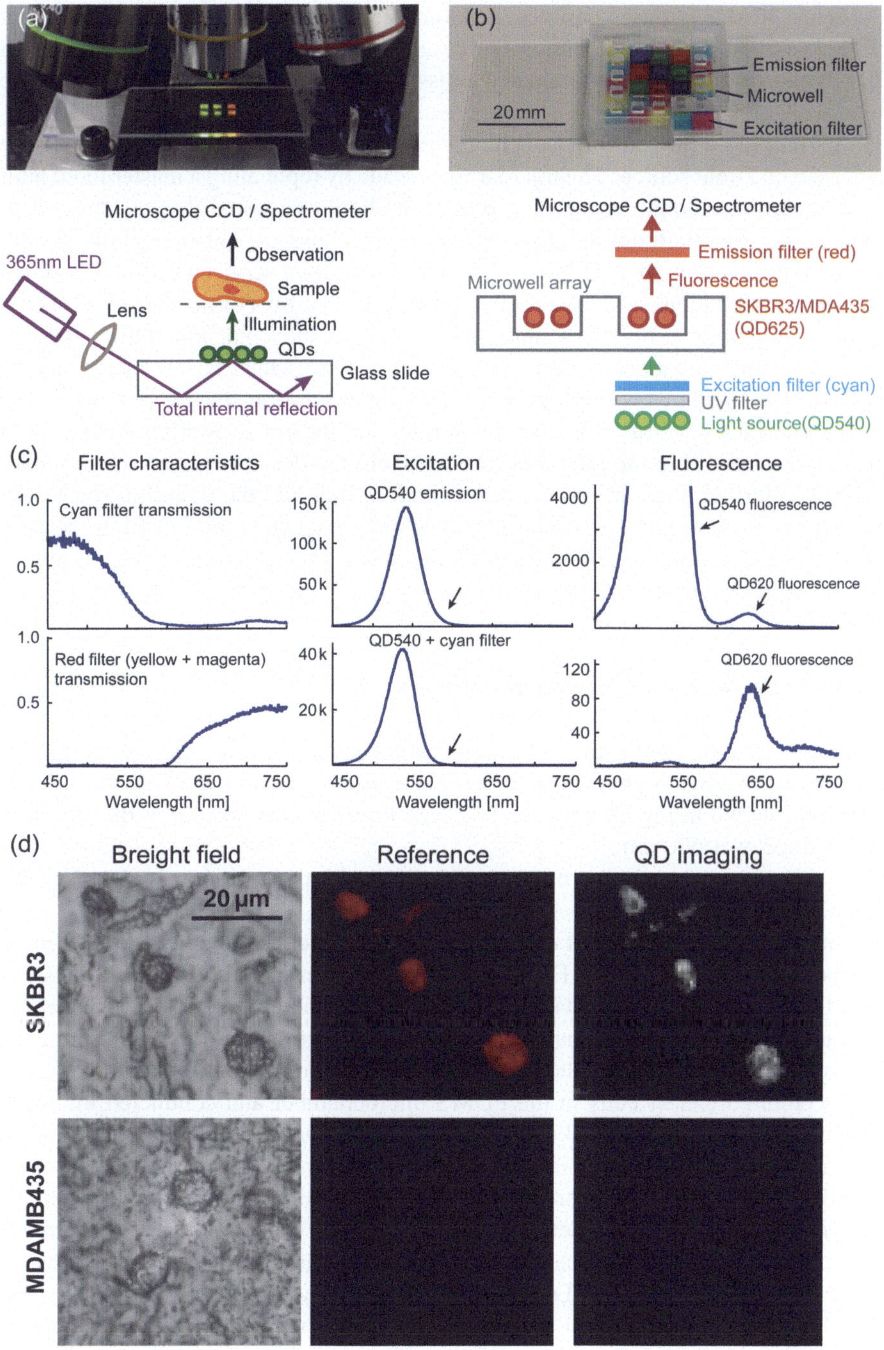
(a)
(b)
Emission filter
Microwell
20 mm
Excitation filter
Microscope CCD / Spectrometer
Observation
365nm LED
Sample
Lens
Illumination
QDs
Glass slide
Total internal reflection
Microscope CCD / Spectrometer
Emission filter (red)
Fluorescence
Microwell array
SKBR3/MDA435
(QD625)
Excitation filter (cyan)
UV filter
Light source(QD540)
(c)
Filter characteristics
Excitation
Fluorescence
Cyan filter transmission
QD540 emission
QD540 fluorescence
QD620 fluorescence
Red filter (yellow + magenta)
transmission
QD540 + cyan filter
QD620 fluorescence
1.0
0.5
150k
100k
50k
40k
20k
4000
2000
120
80
40
450
550
650
750
Wavelength [nm]
(d)
Breight field
Reference
QD imaging
20 µm
SKBR3
MDAMB435

Figure 3b shows the experimental fluorescence imaging setup, where colloidal QDs are used as both the excitation source and the fluorescence marker. It included a PDMS microwell array, excitation and emission filters, and the QD light source. The microwell array contains five by five micro cell culture chambers at 2 mm (w) × 2 mm (l) × 1.5 mm (d), arranged with a pitch of 5 mm. The well array was designed to fit the patterned QD light source. The microarray is made by replicating a master mold built by a three dimensional prototyping printer. It should be noted that all three main components were built through low-cost printing techniques. The optical filters were printed out on a transparent plastic film by a commercial service (PsPrint.com). An example of excitation and mission filters placed on an arrays of microwells is shown in the photographs in Fig. 3b. Combinations of cyan, magenta, and yellow dyes are easily available with most commercial printing systems. Other types of dyes may be used and patterned by inkjet printing for different absorption characteristics.

Transmission spectra of the excitation filter and the emission filter were used in the measurement (Fig. 3c left). In order to obtain the desired transmission characteristics, multiple films are used. The excitation filter and the emission filter were two layers of cyan films, and five layers of red (magenta + yellow) films, respectively. The excitation filter includes a UV-cut plastic thin film (thickness 100 μm) to prevent cross excitation of sample QDs by stray lights from the UV LED.

3.2.2 Fluorescence Excitation and Imaging

Emission from the light source QDs (emission peak at 540 nm, QD540) is first filtered by the excitation filter (cyan filter). The emission from QD540 and the filtered signal are shown in Fig. 3c middle. The cyan filter cuts the portion of the emission in the range over 600 nm.

The fluorescent marker to be imaged is QDs with the emission peak at 620 nm (QD620).

The filtered excitation light excites the sample, which are another type of (emission peak at 620 nm, QD620). Figure 3c right shows measured fluorescence emission from QD620 (top) without and (bottom) with the emission filter (red filter). Note that excitation light from QD540 is found in without the emission filter measurements. The emission filter separates the QD620 fluorescence from the excitation light.

We cultured cancer cells in the PDMS microchamber and conducted immunofluorescence imaging. SKBR3 (breast cancer) and MDA MB 435 (melanoma) were used as model cancer cell lines. We used a breast cancer cell biomarker HER2 as the model marker. HER2 overexpression is found in certain aggressive types of breast cancer and is used to characterize types of breast cancer. There are therapies designed for HER2+ breast cancer. SKBR3 is a breast cancer cell line known to overexpress HER2, while MDA MB 435 shows only marginal or no expression [21]. Grown cells are fixed on the bottom of the microchamber using ice cold acetone and labeled with QD markers. QDs used are with the emission peak at 625 nm (Qdot® 625 Antibody Conjugation Kit, Life Technologies™) functionalized with HER2 (1 mg/mL, provided by the University of Texas Southwestern Medical Center) antibodies as the biomarker.

Reference bright field images and fluorescence images SKBR3 cells and MDA MB 435 cells, are shown in Fig. 3d left and middle panels. For the reference images, a fluorescence filter cube (excitation: peak 535 nm/bandwidth 50 nm, dichroic mirror: 565 nm long-pass and emission: peak 610 nm/bandwidth 75 nm), was used for fluorescence excitation and imaging with a color CCD camera (Olympus DP71). The exposure time was 50 ms.

Figure 3d right panels are the fluorescence images excited with the QD540 light source and recorded with a cooled monochromatic CCD. Instead of the printed emission filter, we used a commercially available bandpass filter (570–650 nm) for imaging purpose. The other components including the excitation QD, the excitation cyan filter, the PDMS microwell array are the same as illustrated in Fig. 3b. Both fluorescence images were taken with the exposure time of 5 s. Fluorescence is observed with SKBR3 cells, while no visible signal was observed for MDA MB 435 cells, indicating the different expression levels of HER2 for the two cell lines (SKBR3: HER2+, MDA MB 435: HER2-).

In conclusion, we have developed a micropatterned QD-based excitation light source for compact cellular screening. QD fluorescence imaging of cancer cells was conducted to demonstrate the capability of immunoassay-based cancer cell analysis. Low cost printing techniques are used to fabricate microwells and optical filters used to construct the experimental setup. Arrayed fluorescence excitation is suitable for patterned immunoassays [22, 23] and cellular microarray imaging [19], where multiple markers are used in a microarrayed configuration.

4 Electrically Excited Quantum Dots for Biomedical Imaging

QDs can be electrically excited to emit fluorescence as in QD light emitting diodes (QDLEDs) [24–26]. QDLEDs have shown great promise for large-area, bright displays due to their excellent electro-photo luminescence properties. Electrical excitation of QDs will further extend the possibility of microscopic illumination for compact imaging and sensing applications. The well-controlled emission wavelengths and narrow bandwidth of QDs gives QDLEDs a considerable potential as excitation sources for fluorescence microscopy. The electrically pumped QDs enable multicolor, self-illuminating systems with no conventional optics needed for light coupling.

QD-based light emitting devices are typically fabricated with organic carriers transporting layers deposited on indium tin oxide (ITO) substrates [27–29]. These organic structures are susceptible to oxygen, moisture, thermal, and electrochemical degradation [30, 31], which makes the use of QDLEDs in integrated microanalytical systems difficult or inefficient. It is desirable to have robust inorganic transport layers for LEDs that can be used for imaging applications. Early examples of inorganic QDLEDs used ITO [32] or NiO and ZnO:SnO_2 [33].

4.1 QDLED on Silicon

We demonstrated electrical excitation of QDs micropatterned on silicon carrier transporting layers [24, 25]. The silicon substrate is used as the hole transporting layers and the inorganic ($ZnO:SnO_2$) electron transporting layer was deposited on top of QDs. Deposition of uniform quantum dots on substrates is critical for creating QDLEDs. It is difficult to use standard lithographic processes to pattern QDs, because they are inhomogeneous in nature and contain a semiconductor material and organic capping. We developed a modified microcontact printing method to create micropatterns of QDs on silicon substrates. In short, the colloidal QDs resuspended in hexane are used in a modified Langmuir Schaffer technique to create monolayers of QD films, which are picked up by PDMS stamps and transfered on silicon substrates [24, 25]. An alternate reported method includes inkjet printing [34], where QD suspensions can be printed onto substrates.

Figure 4a shows examples of electroluminescence observed from QDLEDs. Because our device consists of metals, silicon and metal oxides, it is compatible with a standard silicon microfabrication process for photolithography-based integrated circuits and microelectromechanical systems. The bottom left two panels in Fig. 4a show QDLED light sources, which are based on photolithographically patterned electrodes. In addition we used e-beam lithography to fabricate PDMS microstamps with submicrometer patterns. The size of the patterned nanoscale light source was found to be submicrometer [25]. The compatibility with existing microfabrication techniques will open up several new applications of QDs in biomedical imaging. Fixed MDA435 cancer cells stained with HEMA-3 were placed over the multicolor QDLED source and imaged with a microscope under a different excitation wavelength (Fig. 4b). The experimental setup illustrated in the right panel of Fig. 4b was used to observe cells. We were able to observe the absorption characteristics of different parts in the cells.

4.2 QDLED for Scanning Probe Microscopy

Fraunhofer diffraction of light limit the resolution of a conventional optical systems. Scanning probe microscopy (SPM) is a technique that overcomes the resolution limit by mechanically scanning probe tips. The Atomic Force Microscope (AFM) is the most commonly used SPM tool to image biosamples at nanometer scale resolution. Near-field scanning optical microscopy (NSOM), also known as scanning near-field optical microscopy (SNOM), is a type of scanning microscopy that utilizes a nanometer scale light source created at the tip of an AFM probe to measure maps of sub-wavelength optical properties [35–37]. Since it is based on the AFM technology, it can simultaneously measure nanoscale topographic profiles.

Implementation of typical types of NSOM is shown in Fig. 4c left. The nanoscale light source scans the sample surface to induce a nanoscale optical interaction between the light source and the sample, which is recorded by a photon detector. The optical resolution of NSOM is mainly defined by the size of the light source and the

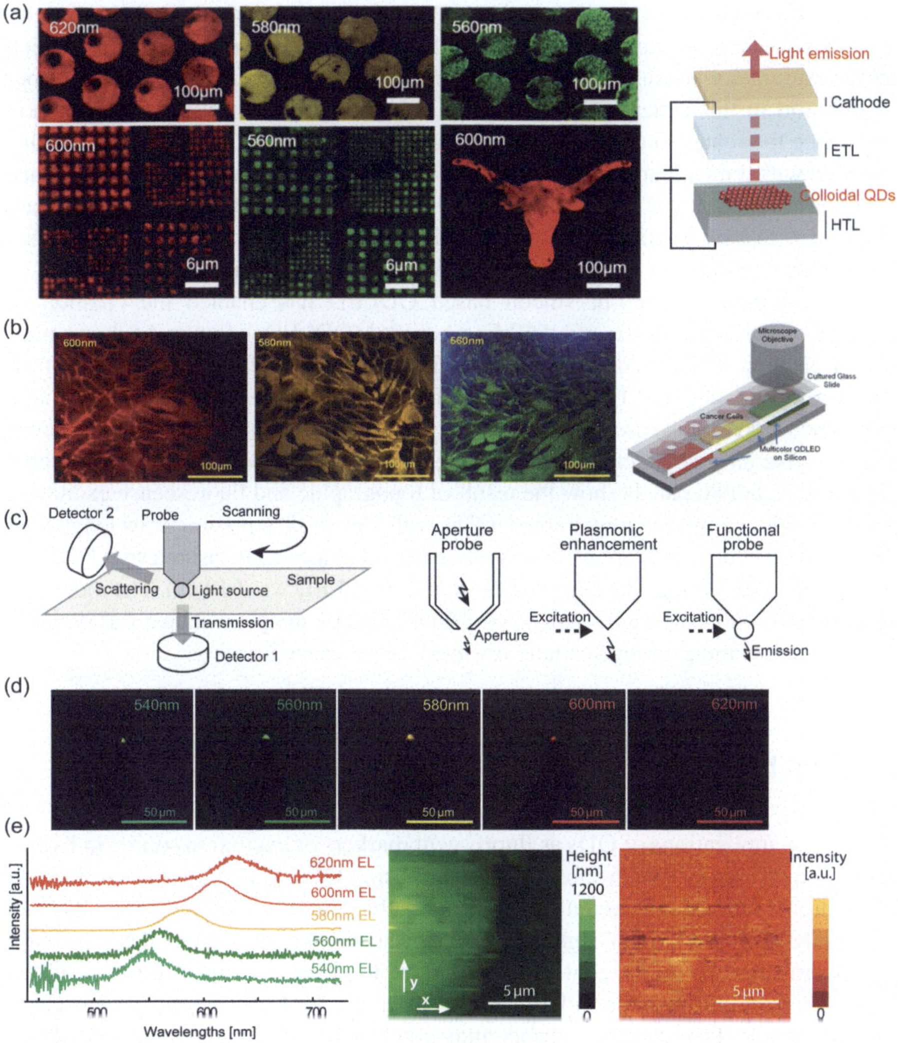

Fig. 4 Electrically excited QD light source for integrated microscopy. (**a**) QD light emitting diodes patterned on silicon hole transporting layers. (**b**) Fixed MDA-MB435 cells imaged by the QDLEDs on silicon. Experimental setup is illustrated on the right panel. (**c**) Overview of the near-field scanning optical microscopy. (**d**) QDLEDs created at the tip of a silicon microprobe. (**e**) (*Left*) Electroluminescence spectra of the QDLEDs. (*Middle*) Topographic image of patterned QDs with the peak emission wavelengths of 780 nm. (*Right*) Simultaneously recorded fluorescence image excited by the tip-embedded QDLED. Figures (**d**) and (**e**) are reprinted from Ref. [26], copyright (2012), with permission from American Institute of Physics

tip–sample distance. The probe tip is usually fixed in the focal point of the detector optics and the sample stage is displaced to keep the tip and the sample surface in a controlled distance. The photo detector is either located under the sample in an inverted microscope configuration, or attached to a side-viewing microscope to detect scattered light. Different types of NSOM tips are summarized in Fig. 4c right. The most common type of NSOM tip utilizes a nanometer-sized aperture, which allows nanometer sized laser emission at the tip. Metal-coated tapered optical fiber is common in this scheme [36]. Plasmonic tips utilize localized surface plasmon resonance to create enhanced electromagnetic field at the metal tip [38]. Probe tips attached with a metal nanoparticle [39] also create similar effect of localized surface plasmon resonance and can be categorized as type this type. Functional probes are attached with a small-sized light emitting element at the tip. Such functional materials include organic dyes [40], and nanodiamonds [41], and quantum dots (QDs) [42].

The compatibility of our silicon based QDLED has enabled the creation of nanoscale QDLED at the tip of NSOM probes [26] (QDLEDs). Figure 4d shows multicolor electrically induced emission from the QDLEDs. The emission spectra measured from the probe QDLEDs are shown in Fig. 4e. The narrow bandwidth of colloidal QDs made it possible to use QDLED at the tip as an excitation sources for fluorescence imaging. We have conducted NSOM fluorescence imaging with the QDLED probe. Figure 4e right two panels show the result of topographic and fluorescence excitation imaging. The sample was patterned QDs with the peak emission wavelengths of 780 nm. The sample QDs are patterned on a glass substrate and measured with QDLED create at the NSOM tip. The emission peak of the QDLED was 610 nm. The measured sensitivity of fluorescence intensity to the QDLED–QD sample distance was down to 50 nm order, demonstrating spatially resolved fluorescence imaging.

5 Conclusion

We review applications of QDs as fluorescent markers or excitation source in hyperspectral microscopy. We propose methods of using QD light sources for solid state swept light sources and excitation sources for fluorescence imaging. The QDs can be simply excited by a UV light or in an electrical excitation setup. The emission wavelength can be easily controlled thanks to the emission tunability of collidal QDs. We used gold nanoparticles as optical markers for hyperspectral microscopy. Due to the effect of localized surface plasmon resonance, gold nanoparticles demonstrate size and shape-dependent absorption spectra. We employed QD light sources in a custom-built microanalytical system developed on a standard microscope, and successfully performed hyperspectral absorption imaging and fluorescence imaging. We further explored the use of electrically excited colloidal QDs in microelectromechanical systems (MEMS)-based microscopes.

Our QD light source can be used for other types of spectroscopy techniques, including pulse oximetry [43] or tissue spectroscopy [44], where the absorption characteristics are correlated with the physical status of the examinee. Further applications beyond biomedical imaging may be the use as an illumination source alternative to conventional RGB-LEDs. Conventional RGB color spaces, such as sRGB or Adobe RGB, cover only a portion of visible colors [45], while the QD-based light sources can simulate any spectrum expressed as linear combinations of multiple QD emission peaks in the visible and IR range. We have shown the strong potential of the QD light source for biomedical and industrial imaging.

References

1. Reed M, Randall J, Aggarwal R, Matyi R, Moore T, Wetsel A. Observation of discrete electronic states in a zero-dimensional semiconductor nanostructure. Phys Rev Lett. 1988;60(6):535–7.
2. Medintz IL, Uyeda HT, Goldman ER, Mattoussi H. Quantum dot bioconjugates for imaging, labelling and sensing. Nat Mater. 2005;4(6):435–46.
3. Leobandung E, Guo L, Wang Y, Chou SY. Observation of quantum effects and Coulomb blockade in silicon quantum-dot transistors at temperatures over 100 K. Appl Phys Lett. 1995;67(7):938–40.
4. Moreau E, Robert I, Gérard J, Abram I, Manin L, Thierry-Mieg V. Single-mode solid-state single photon source based on isolated quantum dots in pillar microcavities. Appl Phys Lett. 2001;79(18):2865–7.
5. Dabbousi B, Rodriguez-Viejo J, Mikulec FV, Heine J, Mattoussi H, Ober R, et al. (CdSe) ZnS core-shell quantum dots: synthesis and characterization of a size series of highly luminescent nanocrystallites. J Phys Chem B. 1997;101(46):9463–75.
6. Wu X, Liu H, Liu J, Haley KN, Treadway JA, Larson JP, et al. Immunofluorescent labeling of cancer marker Her2 and other cellular targets with semiconductor quantum dots. Nat Biotechnol. 2002;21(1):41–6.
7. Resch-Genger U, Grabolle M, Cavaliere-Jaricot S, Nitschke R, Nann T. Quantum dots versus organic dyes as fluorescent labels. Nat Methods. 2008;5(9):763–75.
8. Medintz IL, Clapp AR, Mattoussi H, Goldman ER, Fisher B, Mauro JM. Self-assembled nanoscale biosensors based on quantum dot FRET donors. Nat Mater. 2003;2(9):630–8.
9. Zhang C, Yeh H, Kuroki MT, Wang T. Single-quantum-dot-based DNA nanosensor. Nat Mater. 2005;4(11):826–31.
10. Hohng S, Ha T. Single-molecule quantum-dot fluorescence resonance energy transfer. Chem Phys Chem. 2005;6(5):956–60.
11. Hoshino K, Joshi PP, Bhave G, Sokolov KV, Zhang X. Use of colloidal quantum dots as a digitally switched swept light source for gold nanoparticle based hyperspectral microscopy. Biomed Optics Express. 2014;5(5):1610–5.
12. Hoshino K, Bhave G, Ng E, Zhang X. Micro patterned quantum dots excitation and imaging for cellular microarray screening. Sensor Actuator A Phys. 2014;216:301–7.
13. Huang X, El-Sayed IH, Qian W, El-Sayed MA. Cancer cell imaging and photothermal therapy in the near-infrared region by using gold nanorods. J Am Chem Soc. 2006;128(6):2115–20.
14. Nikoobakht B, El-Sayed MA. Preparation and growth mechanism of gold nanorods (NRs) using seed-mediated growth method. Chem Mater. 2003;15(10):1957–62.
15. Joshi PP, Yoon SJ, Chen Y, Emelianov S, Sokolov KV. Development and optimization of near-IR contrast agents for immune cell tracking. Biomed Optic Express. 2013;4(11):2609–18.
16. Jana NR, Gearheart L, Murphy C. Seed mediated growth approach for shape-controlled synthesis of spheroidal and rod-like gold nanoparticles using a surfactant template. Adv Mater. 2001;13(18):1389–93.
17. Santhanam V, Andres RP. Microcontact printing of uniform nanoparticle arrays. Nano Lett. 2004;4(1):41–4.
18. Hoshino K, Turner TC, Kim S, Gopal A, Zhang X. Single molecular stamping of a sub-10-nm colloidal quantum dot array. Langmuir. 2008;24(23):13804–8.
19. Moeller H, Mian MK, Shrivastava S, Chung BG, Khademhosseini A. A microwell array system for stem cell culture. Biomaterials. 2008;29(6):752–63.
20. Uhr JW, Huebschman ML, Frenkel EP, Lane NL, Ashfaq R, Liu H, et al. Molecular profiling of individual tumor cells by hyperspectral microscopic imaging. Transl Res. 2012;159(5):366–75.
21. Wilson KS, Roberts H, Leek R, Harris AL, Geradts J. Differential gene expression patterns in HER2/ neu-positive and-negative breast cancer cell lines and tissues. Am J Pathol. 2002;161(4):1171–85.
22. Blinka E, Loeffler K, Hu Y, Gopal A, Hoshino K, Lin K, et al. Enhanced microcontact printing of proteins on nanoporous silica surface. Nanotechnology. 2010;21(41):415302.
23. Ng E, Gopal A, Hoshino K, Zhang X. Multicolor microcontact printing of proteins on nanoporous surface for patterned immunoassay. Appl Nanosci. 2011;1(2):79–85.

24. Gopal A, Hoshino K, Kim S, Zhang X. Multi-color colloidal quantum dot based light emitting diodes micropatterned on silicon hole transporting layers. Nanotechnology. 2009;20:235201.
25. Gopal A, Hoshino K, Zhang X. Photolithographic patterning of subwavelength top emitting colloidal quantum dot based inorganic light emitting diodes on silicon. Appl Phys Lett. 2010;96:131109.
26. Hoshino K, Gopal A, Glaz MS, Vanden Bout DA, Zhang X. Nanoscale fluorescence imaging with quantum dot near-field electroluminescence. Appl Phys Lett. 2012;101(4):043118. -043118-4.
27. Anikeeva PO, Halpert JE, Bawendi MG, Bulovic V. Electroluminescence from a mixed red-green-blue colloidal quantum dot monolayer. Nano Lett. 2007;7(8):2196–200.
28. Colvin V, Schlamp M, Alivisatos A. Light-emitting diodes made from cadmium selenide nanocrystals and a semiconducting polymer. Nature. 1994;370(6488):354–7.
29. Ray K, Badugu R, Lakowicz JR. Metal-enhanced fluorescence from CdTe nanocrystals: a single-molecule fluorescence study. J Am Chem Soc. 2006;128(28):8998–9.
30. Price SP, Henzie J, Odom TW. Addressable, large-area nanoscale organic light-emitting diodes. Small. 2007;3(3):372–4.
31. Fenter P, Schreiber F, Bulović V, Forrest S. Thermally induced failure mechanisms of organic light emitting device structures probed by X-ray specular reflectivity. Chem Phys Lett. 1997;277(5):521–6.
32. Hikmet R, Talapin D, Weller H. Study of conduction mechanism and electroluminescence in CdSe/ZnS quantum dot composites. J Appl Phys. 2003;93(6):3509–14.
33. Caruge J, Halpert J, Wood V, Bulović V, Bawendi M. Colloidal quantum-dot light-emitting diodes with metal-oxide charge transport layers. Nat Photonics. 2008;2(4):247–50.
34. Wood V, Panzer MJ, Chen J, Bradley MS, Halpert JE, Bawendi MG, et al. Inkjet-printed quantum dot-polymer composites for full-color AC-driven displays. Adv Mater. 2009;21(21):2151–5.
35. Betzig E, Trautman J, Harris T, Weiner J, Kostelak R. Breaking the diffraction barrier: optical microscopy on a nanometric scale. Science. 1991;251(5000):1468–70.
36. Hecht B, Sick B, Wild UP, Deckert V, Zenobi R, Martin OJ, et al. Scanning near-field optical microscopy with aperture probes: fundamentals and applications. J Chem Phys. 2000;112:7761.
37. Dunn RC. Near-field scanning optical microscopy. Chem Rev. 1999;99(10):2891–928.
38. Inouye Y, Kawata S. Near-field scanning optical microscope with a metallic probe tip. Opt Lett. 1994;19(3):159–61.
39. Kalkbrenner T, Ramstein M, Mlynek J. V S. A single gold particle as a probe for apertureless scanning near-field optical microscopy. J Microsc. 2001;202(1):72–6.
40. Michaelis J, Hettich C, Mlynek J, Sandoghdar V. Optical microscopy using a single-molecule light source. Nature. 2000;405(6784):325–8.
41. Cuche A, Drezet A, Sonnefraud Y, Faklaris O, Treussart F, Roch J, et al. Near-field optical microscopy with a nanodiamond-based single-photon tip. Opt Express. 2009;17(22): 19969–80.
42. Shubeita G, Sekatskii S, Dietler G, Potapova I, Mews A, Basché T. Scanning near-field optical microscopy using semiconductor nanocrystals as a local fluorescence and fluorescence resonance energy transfer source. J Microsc. 2003;210(3):274–8.
43. Fantini S, Barbieri BB, Gratton E, Franceschini M, Maier JS, Walker SA. Frequency-domain multichannel optical detector for noninvasive tissue spectroscopy and oximetry. Opt Eng. 1995;34(1):32–42.
44. Bevilacqua F, Berger AJ, Cerussi AE, Jakubowski D, Tromberg BJ. Broadband absorption spectroscopy in turbid media by combined frequency-domain and steady-state methods. Appl Opt. 2000;39(34):6498–507.
45. Standard RGB color spaces. Color and imaging conference: Society for Imaging Science and Technology; 1999.

Index

J.W.M. Bulte, M.M.J. Modo (eds.), *Design and Applications of Nanoparticles in Biomedical Imaging*, DOI 10.1007/978-3-319-42169-8

Q

R

S